Multivariate Calculus

Multivariate Calculus

Dr. Samiran Karmakar
Assistant Professor of Mathematics
Bankura Sammilani College, Bankura
Former Assistant Professor of Mathematics and Statistics
St. Xavier's College, Kolkata

Prof. Sibdas Karmakar
Retd. Associate Professor of Mathematics
Ramananda College, Bishnupur, Bankura
Former Lecturer in Mathematics
Achruram Memorial College, Jhalda, Purulia

Levant Books
India

CRC Press is an imprint of the
Taylor & Francis Group, an **informa** business

First published 2023
by CRC Press
4 Park Square, Milton Park, Abingdon, Oxon, OX14 4RN

and by CRC Press
6000 Broken Sound Parkway NW, Suite 300, Boca Raton, FL 33487-2742

British Library Cataloguing-in-Publication Data
A catalogue record for this book is available from the British Library

Library of Congress Cataloging-in-Publication Data
A catalog record has been requested

ISBN: 9781032526898 (hbk)
ISBN: 9781032526904 (pbk)
ISBN: 9781003407874 (ebk)

DOI: 10.4324/9781003407874

Typeset in Knuth Computer Modern 10.5 pt
by levant Books

LEVANT

This book is dedicated to PIYASA, *the eldest daughter of Samiran Karmakar.*

Preface

This book entitled "**Multivariate Calculus**" is a compilation of all basic topics of Calculus and provides us with the tools to do so by extending the concepts that we find in Calculus, such as the computation of the rate of change to multiple variables, determination of the gradient of a multivariate function by finding its derivatives in different directions, extensive use in Neural Networks to update the model parameters etc. It is intended to serve as an introductory text aimed towards undergraduate and postgraduate students in Science and Technology covering the newly introduced B.A./B.Sc. Honours syllabus for Mathematics of Indian universities, recommended by the University Grants Commission (UGC) in Choice Based Credit System (CBCS).

The treatments of theories are a special feature of this book, which are presented in a systematic and interesting manner. Care has been taken to explain the subject matter in a clear and perspicuous style, so that even an average student can understand it independently. The purpose in writing this book has been to provide a development of the subject matter which is well-motivated, rigorous and up-to-date as best as possible.

This book contains nine chapters including the chapter of **Preliminaries (Chapter 0)**, in which we have tried to make a thorough systematic discussion on this subject. The topics of this chapter have frequently been used in the subsequent chapters of this book.

Sincere attempts have been made to present the topics in a simple and lucid manner to create interest into the subject along with various types of worked out examples. A large number of problems have been suitably framed, properly graded and supplied with answers. Exercises contain motivated problems and are given in right places. Majority of them are straight forward; hints are occasionally given for harder once.

Authors hope that the book will prove useful not only to the Mathematics Honours students for whom it is intended but also to the Engineering students and professionals and to the candidates of different competitive examinations as well.

Criticisms and constructive suggestions for improvement, modification and correction of this book, if any, from the teachers, students and readers will be gratefully acknowledged. Correspondence may be made through our e-mail address skmath.rnc@gmail.com in this regard.

Finally, the authors are thankful to Mr. Milinda De, Director, Levant Books to encourage us to bring out this book.

Kolkata   Authors
Mahalaya,   Karmakar & Karmakar
25th September, 2022.

Contents

Chapter 0

Preliminaries

0.1 Introduction

One of the most powerful methods in modern Mathematics is that of the Calculus. This branch of Mathematics focused on functions and its limits, continuity, derivatives, integrals and infinite series etc. It has widespread applications in modern science and technologies, economics and business sectors and presently in medical sciences also. In the last two centuries, calculus has been developed to such an extent that it is now used to deal with problems in every branch of technical sciences.

In this present chapter we will discuss certain fundamental concepts and properties of differential and integral calculus of functions of one independent variable and give a list of important formulae as ready-reckoner which will be very helpful in discussing the calculus of functions of several variables.

0.2 Differential Calculus

0.2.1 Function

Let A and B be two non-empty sets. A mapping or function from A to B, written as $f : A \to B$ is a rule that assigns to every element $a \in A$, a unique element $b \in B$. In this case, we call the set A as the domain set and that B as the co-domain set of the function f.

In this section we mainly consider real valued functions whose domain and co-domain sets are all subsets of the set of real numbers $\mathbb{R}$, i.e., the functions $f : \mathbb{R} \to \mathbb{R}$.

0.2.2 Hyperbolic functions

$$(i) \quad \cosh x = \frac{e^x + e^{-x}}{2}; \quad (ii) \ \sinh x = \frac{e^x - e^{-x}}{2}$$

$$(iii) \quad \tanh x = \frac{\sinh x}{\cosh x} = \frac{e^x - e^{-x}}{e^x + e^{-x}}; \quad (iv) \ \coth x = \frac{\cosh x}{\sinh x} = \frac{1}{\tanh x}$$

$$(v) \quad \text{sech } x = \frac{1}{\cosh x} = \frac{2}{e^x + e^{-x}}; \quad (vi) \ \text{cosech } x = \frac{1}{\sinh x} = \frac{2}{e^x - e^{-x}}$$

(vii) $e^x = \cosh x + \sinh x$; $(viii)$ $e^{-x} = \cosh x - \sinh x$

(ix) $\sin(ix) = i\sinh x$; $\cos(ix) = \cosh x$; $\tan(ix) = i\tanh x$

(x) $\cosh^2 x - \sinh^2 x = 1$; $\operatorname{sech}^2 x + \tanh^2 x = 1$;

$\coth^2 x - \operatorname{cosech}^2 x = 1$

(xi) $\sinh 2x = 2\sinh x \cosh x$

(xii) $\cosh 2x = \cosh^2 x + \sinh^2 x = 2\cosh^2 x - 1 = 1 + 2\sinh^2 x$

$(xiii)$ $\tanh 2x = \dfrac{2\tanh x}{1 + \tanh^2 x}$

(xiv) $\sinh 0 = 0$; $\cosh 0 = 1$; $\tanh 0 = 0$

(xv) $\sinh(-x) = -\sinh x$; $\cosh(-x) = \cosh x$; $\tanh(-x) = -\tanh x$.

Osborn's rule

The Osborn's rule is used to construct the formulae connecting to the hyperbolic functions from the corresponding trigonometric function formulae. This rule states that,

Any formula related to trigonometric functions, replace each trigonometric function by the corresponding hyperbolic functions and change the sign of every product or implied product of two sine functions.

For instance, let us consider the following formulas:

(i) $\sin(A + B) = \sin A \cos B + \cos A \sin B$

(ii) $\cos(A - B) = \cos A \cos B + \sin A \sin B$

(iii) $\cos 2A = \cos^2 A - \sin^2 A$.

Now by the above rule the corresponding hyperbolic function formulas are

(i) $\sinh(A + B) = \sinh A \cosh B + \cosh A \sinh B$

(ii) $\cosh(A - B) = \cosh A \cosh B - \sinh A \sinh B$

(iii) $\cosh 2A = \cosh^2 A + \sinh^2 A$.

Remark: The hyperbolic functions can be expressed in terms of circular functions (i.e., trigonometric functions) of ix as given below:

(i) $\sinh x = -i\sin ix$ (ii) $\cosh x = \cos ix$

(iii) $\tanh x = -i\tan ix$ (iv) $\operatorname{cosech} x = i\operatorname{cosech} ix$

(v) $\operatorname{sech} x = \operatorname{sech} ix$ (vi) $\coth x = i\cot ix$.

0.2.3   Inverse hyperbolic functions

(i) $\sinh^{-1} x = \log\left(x + \sqrt{x^2 + 1}\right)$, $\forall\ x \in \mathbb{R}$

(ii) $\cosh^{-1} x = \log\left(x + \sqrt{x^2 - 1}\right)$, $x \geq 1$

(iii) $\tanh^{-1} x = \dfrac{1}{2}\log\dfrac{1 + x}{1 - x}$, $x^2 < 1$

(iv) $\coth^{-1} x = \dfrac{1}{2}\log\dfrac{1 + x}{1 - x}$, $x^2 > 1$

$(v) \quad \operatorname{cosech}^{-1} x = \log \dfrac{1 + \sqrt{1 + x^2}}{x}, \quad x \neq 0$

$(vi) \quad \operatorname{sech}^{-1} x = \log \dfrac{1 + \sqrt{1 - x^2}}{x}, \quad 0 < x \leq 1$

$(vii) \quad \sinh^{-1} x = -i \sin^{-1}(ix) \qquad (viii) \ \cosh^{-1} x = -i \cos^{-1}(ix)$

$(ix) \quad \tanh^{-1} x = -i \tan^{-1}(ix).$

0.2.4 Limit of a function

Let $f : A(\subseteq \mathbb{R}) \to \mathbb{R}$ be a function and a be a limit point of A. Then f is said to tend to a finite limit l as x tends to a if corresponding to each $\epsilon > 0$ there exists a $\delta > 0$, such that

$|f(x) - l| < \epsilon$ for all values of x satisfying $0 < |x - a| < \delta$

i.e., whenever $a - \delta < x < a + \delta$, but $x \neq a$.

In this case, we write it symbolically as $\lim\limits_{x \to a} f(x) = l$.

Right and left hand limits

When $x \to a$, from the values of x greater than a, then corresponding limit is called the *right hand limit (RHL)* of $f(x)$ and is written as $\lim\limits_{x \to a+} f(x)$ or $\lim\limits_{x \to a+0} f(x)$ or $f(a + 0)$ or $f(a+)$. In other words, l_r is said to be the right hand limit of $f(x)$ if corresponding to each $\epsilon > 0$ there exists a $\delta_r > 0$, such that $|f(x) - l_r| < \epsilon$ whenever $a < x < a + \delta_r$.

Similarly, when $x \to a$, from the values of x smaller than a, then the corresponding limit is called the *left hand limit (LHL)* of $f(x)$ and is written as $\lim\limits_{x \to a-} f(x)$ or $\lim\limits_{x \to a-0} f(x)$ or $f(a - 0)$ or $f(a-)$. In other words, l_l is said to be the left hand limit of $f(x)$ if corresponding to each $\epsilon > 0$ there exists a $\delta_l > 0$, such that $|f(x) - l_l| < \epsilon$ whenever $a < x < a + \delta_l$.

Remark 1: It is worth mentioning to note that $\lim\limits_{x \to a} f(x)$ exists if and only if both the above two limits $f(a + 0)$ and $f(a - 0)$ *exist and are equal in value*, i.e., l is the limit of $f(x)$ as $x \to a$ if and only if $f(a + 0) = l = f(a - 0)$.

Remark 2: The limit of $f(x)$ as $x \to a$ *does not exist even if both these limits exist but are not equal in value.*

Infinite limits

A function f is said to tend to $+\infty$ as x tends to a if corresponding to each $K > 0$ (large enough) there exists a $\delta > 0$, such that $f(x) > K$ for all values of x satisfying $0 < |x - a| < \delta$. It is symbolised as $\lim\limits_{x \to a} f(x) = +\infty$.

Similarly, a function f is said to tend to $-\infty$ as x tends to a if corresponding to each $K > 0$ (large enough) there exists a $\delta > 0$, such that $f(x) < -K$ for all values of x satisfying $0 < |x - a| < \delta$. It is written as $\lim\limits_{x \to a} f(x) = -\infty$.

Limits at infinity

A function f is said to tend to a finite value l_1 as x tends to $+\infty$ if corresponding to each $\epsilon > 0$ (small enough) there exists a large number G, such that $|f(x) - l_1| < \epsilon$ for all values of x satisfying $x > G$. It is symbolised as $\lim\limits_{x \to +\infty} f(x) = l_1$.

Similarly, a function f is said to tend to a finite value l_2 as x tends to $-\infty$ if corresponding to each $\epsilon > 0$ (small enough) there exists a large number G, such that $|f(x) - l_1| < \epsilon$ for all values of x satisfying $x < -G$. It is symbolised as $\lim\limits_{x \to -\infty} f(x) = l_2$.

Also, the function f is said to tend to ∞ as x tends to ∞ if corresponding to each $K > 0$ (large enough) there exists a large $G > 0$, such that $f(x) > K$ for all values of x satisfying $x > G$. It is written as $\lim\limits_{x \to \infty} f(x) = \infty$.

Cauchy's criterion for limits

Theorem 0.2.1 *The necessary and sufficient condition that the limit of the function $f(x)$ exists finitely as x tends to a is that, corresponding to any $\epsilon > 0$ (where ϵ is small enough), there exist another positive quantity δ and two arbitrary values x_1 and x_2 of the variable x such that*

$$|f(x_1) - f(x_2)| < \epsilon \quad whenever \quad 0 < |x_1 - a| < \delta \text{ and } 0 < |x_2 - a| < \delta.$$

When the limiting value of the function is not known in advance the above criterion is very much helpful to prove the existence of the limit.

Frequently used limits

$(i) \quad \lim\limits_{x \to \infty} \left(1 + \dfrac{1}{x}\right)^x = \lim\limits_{y \to 0} (1 + y)^{\frac{1}{y}} = e$

$(ii) \quad \lim\limits_{x \to \infty} \left(1 + \dfrac{a}{x}\right)^x = e^a = \lim\limits_{y \to 0} (1 + ay)^{\frac{1}{y}}$

$(iii) \quad \lim\limits_{x \to a} \dfrac{x^n - a^n}{x - a} = na^{n-1}, \quad \text{when } n \in \mathbb{Q}$

$(iv) \quad \lim\limits_{\theta \to 0} \dfrac{\sin \theta^c}{\theta^c} = 1, \ \lim\limits_{\theta \to 0} \dfrac{\sin \theta^\circ}{\theta} = \dfrac{\pi}{180^\circ} \text{ and } \lim\limits_{\theta \to 0} \dfrac{\tan \theta^c}{\theta^c} = 1$

$(v) \quad \lim\limits_{x \to 0} \dfrac{e^x - 1}{x} = 1, \ \lim\limits_{x \to 0} \dfrac{a^x - 1}{x} = \log_e a \ (a > 0)$

$(vi) \quad \lim\limits_{x \to 0} \dfrac{(1 + x)^m - 1}{x} = m, \quad (vii) \ \lim\limits_{x \to 0} \dfrac{1}{x} \log_e (1 + x) = 1$

$(viii) \quad \lim\limits_{x \to 0} \dfrac{1}{x} \log_a (1 + x) = \log_a e \ (a > 0, \ a \ne 1)$

$(ix) \quad \lim\limits_{x \to 0} \dfrac{x^n}{n!} = 0 \quad (x) \ \lim\limits_{n \to \infty} x^n = 0 \ (-1 < x < 1)$

(xi) $\quad \lim_{x \to a} \{f(x)\}^{g(x)} = e^{\lim_{x \to a} \left\{ \frac{f(x) - 1}{g(x)} \right\}}$,

$$\text{provided } g(x) \to \infty, \; f(x) \to 1 \text{ as } x \to a$$

(xii) $\quad \lim_{x \to \infty} a^x = \begin{cases} \infty, & \text{i.e., does not exist if } a > 1 \\ 1, & \text{if } a = 1 \\ 0, & \text{if } -1 < a < 1 \\ \text{does not exist if } a < -1. \end{cases}$

Sandwich theorem or Squeeze play theorem

Theorem 0.2.2 *If f, g and h are three functions such that $f(x) \leq g(x) \leq h(x)$, then $\lim_{x \to a} f(x) \leq \lim_{x \to a} g(x) \leq \lim_{x \to a} h(x)$.*

0.2.5 Indeterminate forms

If a function $f(x)$ takes any of the following forms at any particular point $x = a$ given by $\frac{0}{0}$, $\frac{\infty}{\infty}$, $\infty - \infty$, $0 \times \infty$, 0^0, ∞^0 or 1^∞, then $f(x)$ is said to be indeterminate at $x = a$.

De L' Hospital's rule

If $f(x)$, $g(x)$ and also their derivatives $f'(x)$, $g'(x)$ are continuous at $x = a$ such that $f(a) = g(a) = 0$ $\left[\text{i.e., } \lim_{x \to a} f(x) = \lim_{x \to a} g(x) = 0\right]$ then $\lim_{x \to a} \frac{f(x)}{g(x)} = \lim_{x \to a} \frac{f'(x)}{g'(x)}$, provided $g'(a) \neq 0$.

Note 0.2.1 *For all other indeterminate forms, we are to convert to $\frac{0}{0}$ or $\frac{\infty}{\infty}$ and the L' Hospital's rule. In many cases, we are to repeat the process, if it be of the form $\frac{0}{0}$ or $\frac{\infty}{\infty}$ again.*

0.2.6 Continuity of a function

The function $f(x)$ is said to be *continuous at the point $x = a$*, provided $\lim_{x \to a} f(x)$ exists, is finite and is equal to $f(a)$, i.e., $\lim_{x \to a+} f(x) = \lim_{x \to a-} f(x) = f(a)$, i.e., RHL=LHL= functional value at $x = a$.

Analytical definition of continuity

A function f is said to be continuous at $x = a$ if corresponding to each $\epsilon > 0$ (small enough) there exists another number $\delta > 0$, such that $|f(x) - f(a)| < \epsilon$ for all values of x satisfying $|x - a| < \delta$.

Continuity on an interval

- **Continuity on an open interval:** A function f is said to be continuous on an open interval (a, b) if and only if it is continuous at every point on the interval (a, b).

- **Continuity on a closed interval:** A function f is said to be continuous on a closed interval $[a, b]$ if and only if it is continuous on the open interval (a, b), $f(a + 0) = f(a)$ and $f(b - 0) = f(b)$.

Discontinuity

A function which is not continuous at any point of its domain then it is said to be *discontinuous function at that point* and the said point is known as *point of discontinuity.*

Types of discontinuity

(i) **Removable discontinuity:** If the functional limit exists finitely but the limiting value is not equal to the functional value then the discontinuity is known as removable discontinuity. This type of discontinuity can be removed by properly defining the function value at the said discontinuous point.

(ii) **First kind discontinuity:** If $f(a - 0)$ and $f(a + 0)$ both exist but $f(a - 0) \neq f(a + 0)$ then this kind of discontinuity is known as non-removable first kind or simply first kind discontinuity or jump discontinuity. $f(a + 0) - f(a - 0)$ is the height of the jump of f at $x = a$.

(iii) **Second kind discontinuity:** If one or both of $f(a - 0)$ and $f(a + 0)$ does not exist then this kind of discontinuity is known as non-removable second kind or simply second kind discontinuity.

(iv) **Infinite discontinuity:** If one or both of $f(a - 0)$ and $f(a + 0)$ tend to $-\infty$ or $+\infty$ then this kind of discontinuity is known as infinite discontinuity.

(v) **Oscillatory discontinuity:** Any type of discontinuity which is not of the above four categories, is called oscillatory discontinuity. At a point of oscillatory discontinuity, the function may oscillate finitely or infinitely and does not tend to a finite limit or to $-\infty$ or $+\infty$.

Properties of Continuous functions on closed and bounded interval

Theorem 0.2.3 *Boundedness Property*

If a function is continuous in a closed and bounded interval, then it is bounded therein and it attains its bounds at least once in the said interval.

Theorem 0.2.4 *Neighbourhood Property*

If a function f is continuous at $c \in [a, b]$ where $a \neq c \neq b$, then there exists a neighbourhood of the point c where $f(x)$ and $f(c)$ have the same sign.

Theorem 0.2.5 *Bolzano's Property*

If the function $f(x)$ is continuous in a closed and bounded interval $[a, b]$ and $f(a).f(b) < 0$, then there exists at least one point $k \in (a, b)$ such that $f(k) = 0$.

Theorem 0.2.6 *Intermediate Value Theorem*

If the function $f(x)$ is continuous in a closed and bounded interval $[a, b]$ and $f(a) \neq f(b)$, then f assumes every value between $f(a)$ and $f(b)$.

Theorem 0.2.7 *Fixed Point Theorem*

If $f : [a, b] \to [a, b]$ be a continuous function then f has a fixed point, i.e., there exist a point $k \in [a, b]$ such that $f(k) = k$.

0.2.7 Uniform Continuity

A real valued function f defined on an interval I is said to be uniformly continuous on I if for every $\epsilon > 0$ there exists a $\delta > 0$ (depending on ϵ only but not on x) and for any two $x_1, x_2 \in I$ such that $|f(x_1) - f(x_2)| < \epsilon$ whenever $|x_2 - x_1| < \delta$.

Note 0.2.2 *Uniform continuity is actually defined over an interval (or sometimes on a set) whereas simple continuity is defined at a point. So it can be said that simple continuity of a function is a local property but uniform continuity is a global property of a function.*

Theorem 0.2.8 *A function continuous on a closed and bounded interval is uniformly continuous thereat.*

0.2.8 Differentiation

Let $y = f(x)$ be a finite, single valued and continuous function defined in any interval $[a, b]$. Let Δy be the change of the dependent variable y corresponding to the change Δx of the independent variable x in $[a, b]$. Now, if the ratio $\dfrac{\Delta y}{\Delta x}$ of these changes tends to a definite finite limit as Δx tends to zero, i.e., if $\lim\limits_{\Delta x \to 0} \dfrac{\Delta y}{\Delta x}$ exists and finite, then this limit is called the *Differential Coefficient* or *Derivative* of $f(x)$ or y, for the particular values of x. This is denoted by $f'(x)$ or $y'(x)$ or y' or $\dfrac{d}{dx}\{f(x)\}$ or $\dfrac{dy}{dx}$ or $D\{f(x)\}$ or Dy etc.

For the function $y = f(x)$, we have $y + \Delta y = f(x + \Delta x)$.

$\therefore \ \Delta y = f(x + \Delta x) - y = f(x + \Delta x) - f(x)$

or, $\dfrac{\Delta y}{\Delta x} = \dfrac{f(x + \Delta x) - f(x)}{\Delta x}$

$\therefore \ \dfrac{dy}{dx} = \lim\limits_{\Delta x \to 0} \dfrac{\Delta y}{\Delta x} = \lim\limits_{\Delta x \to 0} \dfrac{f(x + \Delta x) - f(x)}{\Delta x}$, if it exists.

Putting $\Delta x = h$, we get $\dfrac{dy}{dx} = \lim\limits_{h \to 0} \dfrac{f(x + h) - f(x)}{h}$, if it exists.

Derivative at a point: The derivative of $f(x)$, for any particular value of x, say $x = a$, written as $f'(a)$ or $\left[f'(x)\right]_{x=a}$ or $\left[\dfrac{dy}{dx}\right]_{x=a}$ and is given from the above formula as $f'(a) = \lim\limits_{h \to 0} \dfrac{f(a + h) - f(a)}{h}$, if it exists.

Alternative Definition

Let $f : A \to \mathbb{R}$ be a function, where $A \subseteq \mathbb{R}$ and let $a \in A$ be a limit point of A. Then we say that f is differentiable at $x = a$, if $\lim\limits_{x \to a} \dfrac{f(x) - f(a)}{x - a}$ exists and we write it as $f'(a) = \lim\limits_{x \to a} \dfrac{f(x) - f(a)}{x - a}$, if it exists.

Right hand and left hand derivatives

The *progressive derivative* or *right hand derivative* (RHD) of $f(x)$ at $x = a$ is given by $\lim\limits_{h \to 0} \dfrac{f(a + h) - f(a)}{h}$, $h > 0$, if it exists finitely. It is denoted by $Rf'(a)$ or $f'(a + 0)$ or $f'(a+)$.

The *regressive derivative* or *left hand derivative* (LHD) of $f(x)$ at $x = a$ is given by $\lim\limits_{h \to 0} \dfrac{f(a + h) - f(a)}{h}$, $h < 0$, if it exists finitely. It is denoted by $Lf'(a)$ or $f'(a - 0)$ or $f'(a-)$.

Remark: The function $f(x)$ is said to be *differentiable* at $x = a$ if $Rf'(a)$ and $Lf'(a)$ both exist finitely and are equal and their *common value* is called the *derivative* or *differential co-efficient* at $x = a$.

Critical points

The points on the curve $y = f(x)$ at which $\dfrac{dy}{dx} = 0$ or $\dfrac{dy}{dx}$ does not exist are known as the *Critical points*.

0.2.9 Monotonicity

A function f defined on an interval $[a, b]$ is said to be

(i) **monotonically increasing** if $x_2 > x_1 \Rightarrow f(x_2) \geq f(x_1)$ for all $x_2, x_1 \in [a, b]$ and **strictly monotonically increasing** if $x_2 > x_1 \Rightarrow f(x_2) > f(x_1)$ for all $x_2, x_1 \in [a, b]$.

(ii) **monotonically decreasing** if $x_2 > x_1 \Rightarrow f(x_2) \leq f(x_1)$ for all $x_2, x_1 \in [a, b]$ and **strictly monotonically decreasing** if $x_2 > x_1 \Rightarrow f(x_2) < f(x_1)$ for all $x_2, x_1 \in [a, b]$.

Test for monotonicity

(i) The function $f(x)$ is monotonically increasing on $[a, b]$ if $f'(x) \geq 0$ on $[a, b]$ and is strictly monotonically increasing thereon if $f'(x) > 0$.

(ii) The function $f(x)$ is monotonically decreasing on $[a, b]$ if $f'(x) \leq 0$ on $[a, b]$ and is strictly monotonically decreasing thereon if $f'(x) < 0$.

0.2.10 Higher order derivatives

Let $y = f(x)$ be a real valued differentiable function. Then its first order derivative is $\frac{dy}{dx} = f'(x)$. If this first order derivative $f'(x)$ is again a differentiable function of x, we can get a second order derivative $\frac{d^2y}{dx^2} = \frac{d}{dx}f'(x) = f''(x)$. In a similar manner, we can obtain third order derivative $\frac{d^3y}{dx^3} = f'''(x)$, if it exists. The n-th order successive derivative $\frac{d^ny}{dx^n}$, if it exists, can be obtained from the $(n-1)$-th order derivative for $n = 1, 2, 3, ...$

The higher order derivatives or successive derivatives are denoted by $y_1, y_2, ..., y_n$ ... or $y', y'', y''', ..., y^{(n)}, ...$ etc.

Theorem 0.2.9 *Leibnitz Rule*

If u and v be two functions differentiable n times then their product function (uv) is also differentiable n times and the n th derivative of the product (uv) is given by

$$(uv)_n = u_nv + {}^nC_1u_{n-1}v_1 + {}^nC_2u_{n-2}v_2 + ... + {}^nC_ru_{n-r}v_r + ... + uv_n$$
$$= \sum_{r=0}^{n} {}^nC_ru_{n-r}v_r$$

where the suffixes of u and v denote the order of derivatives of the functions u and v respectively with respect to x.

0.2.11 The curve

The locus of variable points on a plane or space obeying certain fixed property is termed as a *curve*. The equation of a curve given by the algebraic relation between the coordinates of every points on the locus which satisfy the given fixed property and is not satisfied by the coordinates of any point that is not on the locus.

Curves can be described mathematically by *non-parametric* or *parametric equations*. Non-parametric equations can be *explicit* or *implicit*. For a non-parametric curve in space, the coordinates y and z of a point on the curve are expressed as two separate functions of the third coordinate x as the independent variable. In two dimensional Cartesian coordinate system, locus of the points (x, y) may be represented in the following ways:

(i) **Explicit representation**: Equations of a curve may explicitly be written as $y = f(x)$ or $x = \phi(y)$, e.g., $y = x^2 + 2x + 3$, $y = 5x + 9$, $x = \frac{1}{y}$ etc. are explicit representation of a parabola, a straight line, a rectangular hyperbola etc. respectively.

(ii) **Implicit representation**: We can represent a curve by writing it in the form $f(x, y) = 0$. Thus $x^2 + y^2 = 16$, $x^2 - 3y = 0$, $y^2 - 2y = 0$ and $5x + 6y + 7 = 0$ etc. are respectively the implicit representation of a circle, a parabola, again a parabola and a straight line etc.

(iii) Parametric representation: Sometimes in the equation of a curve x and y are expressed in terms of a third variable t, say, called a parameter. Thus $x = a\cos t, y = a\sin t;\ x = at^2, y = 2at;\ x = a\cos t, y = b\sin t$ etc. are respectively the parametric equations of a circle, a parabola, an ellipse etc. The form of parametric equations are $x = \phi(t), y = \psi(t),\ t$ being the parameter.

Remark 1: In two dimensional polar coordinate system, the equation of a curve can have the following three representation (i) $r = f(\theta)$ (Explicit representation), (ii) $f(r,\theta) = 0$ (Implicit representation) and (iii) $r = \phi(s), \theta = \psi(s)$, (Parametric representation), s being the parameter.

Remark 2: In three dimensional system of coordinates, the equation of a curve may also have the following three representation (i) $z = f(x,y)$ (Explicit representation), (ii) $f(x,y,z) = 0$ (Implicit representation) and (iii) $x = \phi(t), y = \psi(t), z = \chi(t)$ (Parametric representation), t being the parameter and (x,y,z) being a point on the locus representing the curve in space.

Rules on derivative

(i) $\dfrac{d}{dx}\{k.f(x)\} = k.\dfrac{d}{dx}f(x)$, where k is any constant

(ii) $\dfrac{d}{dx}\{c_1 f_1(x) \pm c_2 f_2(x) \pm ...\} = c_1\dfrac{d}{dx}f_1(x) + c_2\dfrac{d}{dx}f_2(x) \pm ...$

where $c_1, c_2, ...$ are constants

(iii) $\dfrac{d}{dx}\{f_1(x).f_2(x)....f_n(x)\} = \dfrac{d}{dx}\{f_1(x)\}.\{f_2(x).f_3(x)...f_n(x)\}$

$+\dfrac{d}{dx}\{f_2(x)\}.\{f_1(x).f_3(x)...f_n(x)\} + ...$

$+\dfrac{d}{dx}\{f_n(x)\}.\{f_1(x).f_2(x)...f_{n-1}(x)\}$[Product rule]

(iv) $\dfrac{d}{dx}\left\{\dfrac{f(x)}{g(x)}\right\} = \dfrac{g(x)\frac{d}{dx}f(x) - f(x)\frac{d}{dx}g(x)}{\{g(x)\}^2}$, $g(x) \neq 0$ [Quotient rule]

(v) If $y = f_1(u_1), u_1 = f_2(u_2), u_2 = f_3(u_3), ..., u_n = f_{n+1}(x)$, then

$\dfrac{dy}{dx} = \dfrac{dy}{du_1}.\dfrac{du_1}{du_2}.\dfrac{du_2}{du_3}...\dfrac{du_n}{dx}$ [Chain rule]

(vi) If $x = \phi(t), y = \psi(t)$ be the parametric equation, then $\dfrac{dy}{dx} = \dfrac{\frac{dy}{dt}}{\frac{dx}{dt}} = \dfrac{\psi'(t)}{\phi'(t)}$, $\phi'(t) \neq 0$.

(vii) If $f(x,y) = 0$ be the implicit form of equation, then differentiate it with respect to x, collect the terms of $\dfrac{dy}{dx}$ and then solve it. Alternatively,

$$\dfrac{dy}{dx} = -\dfrac{\frac{\partial f}{\partial x}}{\frac{\partial f}{\partial y}} = -\dfrac{f_x}{f_y}\ [f_y \neq 0],$$

where f_x, f_y are the partial derivatives of f with respect to x and y respectively.

In generalised implicit form, if $f(x_1, x_2, ..., x_n) = 0$, where $x_2, ..., x_n$ are functions of x_1, then

$$\dfrac{df}{dx_1} = \dfrac{\partial f}{\partial x_1} + \dfrac{\partial f}{\partial x_2}\dfrac{dx_2}{dx_1} + \dfrac{\partial f}{\partial x_3}\dfrac{dx_3}{dx_1} + ... + \dfrac{\partial f}{\partial x_n}\dfrac{dx_n}{dx_1}.$$

($viii$) If $y = \{f(x)\}^{\phi(x)}$, then taking logarithm on both sides, we get

$$\log y = \phi(x)\log\{f(x)\}.$$

Now differentiating both sides with respect to x, we get

$$\frac{1}{y}\frac{dy}{dx} = \phi(x).\frac{1}{f(x)}f'(x) + \phi'(x)\log f(x)$$

$$\therefore \frac{dy}{dx} = \{f(x)\}^{\phi(x)}\left[\phi(x).\frac{f'(x)}{f(x)} + \phi'(x).\log f(x)\right].$$

Note 0.2.3 *If* $y = f_1(x).f_2(x).f_3(x)...$ *or* $y = \dfrac{f_1(x).f_2(x).f_3(x)...}{g_1(x).g_2(x).g_3(x)...}$, *i.e., the functions are given in product or quotient forms respectively, then it is convenient to take the logarithms of the respective functions first and then differentiate.*

Note 0.2.4 *We can also write* $\{f(x)\}^{\phi(x)} = e^{\phi(x)\log_e f(x)}$ *and differentiate easily.*

0.2.12 Properties of Derivative

Theorem 0.2.10 *Darboux's Theorem*

Let $f : [a, b] \rightarrow \mathbb{R}$ *be a differentiable function on* $[a, b]$ *and* $f'(a).f'(b) < 0$ *then there exists at least one point* $k \in [a, b]$ *such that* $f'(k) = 0$.

Theorem 0.2.11 *Intermediate Value Theorem for Derivative*

Let $f : [a, b] \rightarrow \mathbb{R}$ *be a differentiable function on* $[a, b]$ *and* $f'(a) \neq f'(b)$ *then* $f'(x)$ *assumes every value between* $f'(a)$ *and* $f'(b)$ *at least once in the open interval* (a, b).

In particular, monotone derived function in any open interval is continuous therein.

0.2.13 Mean Value Theorems and Expansion of Functions

Theorem 0.2.12 *Roll's Theorem*

If a function f *be defined and continuous in a closed interval* $[a, b]$, *differentiable in the open interval* (a, b) *and* $f(a) = f(b)$ *then there exists at least one real number* $\xi \in (a, b)$ *such that* $f'(\xi) = 0$.

Theorem 0.2.13 *Lagrange's Mean Value Theorem*

If a function f *be defined and continuous in a closed interval* $[a, b]$ *and differentiable in the open interval* (a, b), *then there exists at least one point* $\xi \in (a, b)$ *such that* $f(b) - f(a) = (b - a)f'(\xi)$.

Note 0.2.5 *Put* $b - a = h$ *and* $\xi = a + \theta(b - a) = a + \theta h$ *in the above Mean Value theorem then for* $0 < \theta < 1$ *we get* $f(a + h) = f(a) + hf'(a + \theta h)$.

Note 0.2.6 *Again put* $a = x$ *then* $f(x + h) = f(x) + hf'(x + \theta h)$, $0 < \theta < 1$.

Now putting $x = 0, h = x$ *in the above form we get*

$$f(x) = f(0) + hf'(\theta h), \ 0 < \theta < 1.$$

Theorem 0.2.14 *Cauchy's Mean Value Theorem*

If the functions f and g be defined and continuous in a closed interval $[a, b]$, differentiable in the open interval (a, b) and $g'(x) \neq 0$ in (a, b) then there exists at least one point $\xi \in (a, b)$ such that

$$\frac{f(b) - f(a)}{g(b) - g(a)} = \frac{f'(\xi)}{g'(\xi)}.$$

Note 0.2.7 *Put $b - a = h$ and $\xi = a + \theta(b - a) = a + \theta h$ in the Cauchy's Mean Value theorem above then for $0 < \theta < 1$ we get*

$$\frac{f(a + h) - f(a)}{g(a + h) - g(a)} = \frac{f'(a + \theta h)}{g'(a + \theta h)}.$$

Theorem 0.2.15 *Generalized Mean Value Theorem or Taylor's Theorem with Remainder (Finite form)*

A function $f : [a, b] \to \mathbb{R}$ be such that its $n - 1$ th derivative f^{n-1} is continuous in the closed interval $[a, b]$ and the n th derivative exists in the open interval (a, b) then there exists a $\theta \in (0, 1)$ such that

$$f(b) = f(a) + (b - a)f'(a) + \frac{(b - a)^2}{\underline{2}} f''(a) + ... + \frac{(b - a)^{n-1}}{\underline{n-1}} f^{n-1}(a) + R_n.$$

where R_n is the remainder term which will take the following forms:

Schlomilch-Roche's form of remainder:

$$\text{Here} \quad R_n = \frac{(b - a)^n (1 - \theta)^{n-p}}{\underline{n-1} \; p} f^n(a + \theta(b - a)), \; p \in \mathbb{R} - \{0\}.$$

Lagrange's form of remainder:

$$\text{Here} \quad p = n, \; R_n = \frac{(b - a)^n}{\underline{n}} f^n(a + \theta(b - a)).$$

Cauchy's form of remainder:

$$\text{Here} \quad p = 1, \; R_n = \frac{(b - a)^n (1 - \theta)^{n-1}}{\underline{n-1}} f^n(a + \theta(b - a)).$$

Note 0.2.8 *Put $b = a + h$ then for $0 < \theta < 1$ Taylor's theorem with Lagrange's form of remainder becomes*

$$f(a + h) = f(a) + hf'(a) + \frac{h^2}{\underline{2}} f''(a) + ... + \frac{h^{n-1}}{\underline{n-1}} f^{n-1}(a) + \frac{h^n}{\underline{n}} f''(a + \theta h).$$

Replacing a by x we get

$$f(x + h) = f(x) + hf'(x) + \frac{h^2}{\underline{2}} f''(x) + ... + \frac{h^{n-1}}{\underline{n-1}} f^{n-1}(x) + \frac{h^n}{\underline{n}} f''(x + \theta h).$$

Note 0.2.9 *Taylor's Series Expansion in Infinite form*

If f possesses continuous derivatives of any order in $[a, a + h]$ and $R_n \to 0$ as $n \to \infty$ then we get

$$f(a + h) = f(a) + hf'(a) + \frac{h^2}{\underline{2}} f''(a) + ... + \frac{h^{n-1}}{\underline{n-1}} f^{n-1}(a) + \frac{h^n}{\underline{n}} f^n(0) + ...$$

Theorem 0.2.16 *Maclaurin's Theorem in Finite form*

A function $f : [a, b] \to \mathbb{R}$ be such that its $n-1$ th derivative f^{n-1} is continuous in the closed interval $[0, h]$ and the n th derivative exists in the open interval $(0, h)$ then there exists a $\theta \in (0, 1)$ such that

$$f(x) = f(0) + xf'(0) + \frac{x^2}{\lfloor 2} f''(0) + ... + \frac{x^{n-1}}{\lfloor n-1} f^{n-1}(0) + \frac{x^n}{\lfloor n} f''(\theta x), \ x \in [0, h].$$

Note 0.2.10 *Maclaurin's Series in Infinite form*

If f possesses continuous derivatives of any order in $[0, h]$ then for any $x \in [0, h]$ we get

$$f(x) = f(0) + xf'(0) + \frac{x^2}{\lfloor 2} f''(0) + ... + \frac{x^{n-1}}{\lfloor n-1} f^{n-1}(0) + \frac{x^n}{\lfloor n} f^n(0) + ...$$

Infinite Series Expansion of some well-known Functions

$(a) \quad e^x = 1 + x + \frac{x^2}{\lfloor 2} + ... + \frac{x^n}{\lfloor n} + ... \ \forall \ x \in \mathbb{R}.$

$(b) \quad \sin x = x - \frac{x^3}{\lfloor 3} + \frac{x^5}{\lfloor 5} - ... + (-1)^n \frac{x^{2n+1}}{\lfloor 2n+1} + ... \ \forall \ x \in \mathbb{R}.$

$(c) \quad \cos x = 1 - \frac{x^2}{\lfloor 2} + \frac{x^4}{\lfloor 4} - ... + (-1)^n \frac{x^{2n}}{\lfloor 2n} + ... \ \forall \ x \in \mathbb{R}.$

$(d) \quad \sinh x = x + \frac{x^3}{\lfloor 3} + \frac{x^5}{\lfloor 5} + ... + \frac{x^{2n+1}}{\lfloor 2n+1} + ... \ \forall \ x \in \mathbb{R}.$

$(e) \quad \cosh x = 1 + \frac{x^2}{\lfloor 2} + \frac{x^4}{\lfloor 4} + ... + \frac{x^{2n}}{\lfloor 2n} + ... \ \forall \ x \in \mathbb{R}.$

$(f) \quad \log(1 + x) = x - \frac{x^2}{2} + \frac{x^3}{3} - ... + (-1)^{n-1} \frac{x^n}{n} + ... \ \forall \ x \in (-1, 1].$

0.2.14 Maxima and Minima of a Function of One Variable

Definition 0.2.1 *A function $f(x)$ is said to have a **maximum** at $x = c$, if in any δ-neighbourhood of c*

$$f(x) < f(c) \ \forall \ x \ in \ 0 < |x - c| < \delta, \ i.e., \ in \ c - \delta < x < c + \delta.$$

Definition 0.2.2 *A function $f(x)$ is said to have a **minimum** at $x = c$, if in any δ-neighbourhood of c*

$$f(x) > f(c) \ \forall \ x \ in \ 0 < |x - c| < \delta, \ i.e., \ in \ c - \delta < x < c + \delta.$$

Note 0.2.11 *The single term **extreme value** or **turning value** is used both for a maximum as well as for a minimum value. The tangent to the curve at these points are parallel to the x-axis, i.e., $\dfrac{dy}{dx} = 0$ at these points.*

Local Maxima and Local Minima

Definition 0.2.3 *If c is a number in the domain of f, then $f(c)$ is a **local maximum value** of f if $f(c) > f(x)$ when x is 'near' c. Likewise $f(c)$ is a **local minimum value** of f, if $f(c) < f(x)$ when x is 'near' c.*

In any interval $[a, b]$, a function $f(x)$ may have more than one maximum or one minimum. Even in any interval, one maximum value at one point may be less than its minimum value at other point in that interval.

Global Maxima and Global Minima

The maximum or minimum over the entire function is called an '**Absolute**' or '**Global**' maximum or minimum. In an interval $[a, b]$, there is only one global maximum and one global minimum of $f(x)$, but there can have more than one local maximum or minimum.

$$\text{Global Maximum} = \max\{f(a), f(b), \text{all maxima}\}$$
$$\text{Global Minimum} = \min\{f(a), f(b), \text{all minima}\}.$$

Note 0.2.12 *A necessary condition for the existence of a extreme value of the function $f(x)$ at a point $x = c$ is that $f'(c) = 0$, if it exists. Remember that there may exist such function whose derivative vanishes at $x = c$, but it has no extreme value at $x = c$.*

Working rule for finding maxima and minima

(a) First derivative test

To check the maxima or minima at $x = a$

(i) If $f'(x) > 0$ for $x < a$ and $f'(x) < 0$ for $x > a$, i.e., the sign of $f'(x)$ changes from positive to negative, then $f(x)$ has a local maximum at $x = a$.

(ii) If $f'(x) < 0$ for $x < a$ and $f'(x) > 0$ for $x > a$, i.e., the sign of $f'(x)$ changes from negative to positive, then $f(x)$ has a local minimum at $x = a$.

(iii) If the sign of $f'(x)$ does not change, then $f(x)$ has neither a local maximum nor a local minimum at $x = a$ and the point 'a' is called a *Point of inflexion*.

(b) Second derivative test

(i) If $f'(a) = 0$, $f''(a) < 0$, then $x = a$ is a point of local maximum.

(ii) If $f'(a) = 0$, $f''(a) > 0$, then $x = a$ is a point of local minimum.

(iii) If $f'(a) = 0$, $f''(a) = 0$, then we need further differentiation and obtain $f'''(a)$.

(iv) If $f'(a) = f''(a) = f'''(a) = \ldots = f^{n-1}(a) = 0$, $f^n(a) \neq 0$, then for $n =$ odd, $f(x)$ has neither a local maximum nor a local minimum at $x = a$, i.e., $x = a$ is a point of inflexion. On the other hand, if n is even, then $f^n(a) < 0$ implies $f(x)$ has a local maximum at $x = a$ and $f^n(a) > 0$ implies $f(x)$ has a local minimum at $x = a$.

0.3 Integral Calculus

0.3.1 Fundamental results

(i) If $\dfrac{d}{dx}F(x) = f(x)$ then $\displaystyle\int f(x)dx = F(x) + C,$

where C is known as integration constant.

(ii) $\displaystyle\int \left(f_1(x) \pm f_2(x) \pm f_3(x) \pm ...\text{to } n \text{ terms}\right) dx$

$$= \int f_1(x)dx \pm \int f_2(x)dx \pm \int f_3(x)dx \pm ...\text{to } n \text{ terms}$$

(iii) $\displaystyle\int kf(x)dx = k\int f(x)dx$ where k is a constant.

(iv) $\displaystyle\int x^n dx = \dfrac{x^{n+1}}{n+1}, \; n \neq -1$ (v) $\displaystyle\int \dfrac{dx}{x} = \log|x|$ (vi) $\displaystyle\int e^{mx}dx = \dfrac{e^{mx}}{m}$

(vii) $\displaystyle\int a^{mx} dx = \dfrac{a^{mx}}{m \log_e a}, \; a > 0$

$(viii)$ $\displaystyle\int \cos mx\,dx = \dfrac{\sin mx}{m}$ (ix) $\displaystyle\int \sin mx\,dx = -\dfrac{\cos mx}{m}$

(x) $\displaystyle\int \sec^2 mx\,dx = \dfrac{\tan mx}{m}$ (xi) $\displaystyle\int \operatorname{cosec}^2 mx\,dx = -\dfrac{\cot mx}{m}$

(xii) $\displaystyle\int \sec mx \tan mx\,dx = \dfrac{\sec mx}{m}$

$(xiii)$ $\displaystyle\int \operatorname{cosec} mx \cot mx\,dx = -\dfrac{\operatorname{cosec} mx}{m}$

(xiv) $\displaystyle\int \sinh mx\,dx = \dfrac{\cosh mx}{m}$ (xv) $\displaystyle\int \cosh mx = \dfrac{\sinh mx}{m}$

(xvi) $\displaystyle\int \tanh mx\,dx = \dfrac{\log|\cosh mx|}{m}$ $(xvii)$ $\displaystyle\int \coth mx\,dx = \dfrac{\log|\sinh mx|}{m}$

$(xviii)$ $\displaystyle\int \operatorname{cosech} mx\,dx = \dfrac{\log\left|\tanh\left(\frac{mx}{2}\right)\right|}{m}$ (xix) $\displaystyle\int \operatorname{sech} mx\,dx = \dfrac{2\tan^{-1}(e^{mx})}{m}$

(xx) $\displaystyle\int \operatorname{sech}^2 mx\,dx = \dfrac{\tanh mx}{m}$ (xxi) $\displaystyle\int \operatorname{cosech}^2 mx\,dx = -\dfrac{\coth mx}{m}.$

0.3.2 Standard integrals

(i) $\displaystyle\int \dfrac{f'(x)}{f(x)}dx = \log|f(x)|$ (ii) $\displaystyle\int \tan mx\,dx = \dfrac{\log|\sec mx|}{m}$

(iii) $\displaystyle\int \cot mx\,dx = \dfrac{\log|\sin mx|}{m}$

(iv) $\displaystyle\int \dfrac{dx}{x^2 + a^2} = \dfrac{1}{a}\tan^{-1}\dfrac{x}{a}, \; a \neq 0$

(v) $\displaystyle\int \dfrac{dx}{x^2 - a^2} = \dfrac{1}{2a}\log\left|\dfrac{x-a}{x+a}\right|, \; |x| > |a|$

(vi) $\displaystyle\int \dfrac{dx}{a^2 - x^2} = \dfrac{1}{2a}\log\left|\dfrac{a+x}{a-x}\right|, \; |x| < |a|$

(vii) $\displaystyle\int \frac{dx}{\sqrt{x^2+a^2}} = \log\left|x+\sqrt{x^2+a^2}\right| = \sinh^{-1}\frac{x}{a}$

$(viii)$ $\displaystyle\int \frac{dx}{\sqrt{x^2-a^2}} = \log\left|x+\sqrt{x^2-a^2}\right| = \cosh^{-1}\frac{x}{a}$

(ix) $\displaystyle\int \frac{dx}{\sqrt{a^2-x^2}} = \sin^{-1}\frac{x}{a}, \quad |x|<|a| \quad (x) \int \frac{dx}{x\sqrt{x^2-a^2}} = \frac{1}{a}\sec^{-1}\frac{x}{a}.$

0.3.3 Integration by parts

(i) $\displaystyle\int u(x)v(x)dx = u(x)\int v(x)dx - \int\left(\frac{du}{dx}\cdot\int v(x)dx\right)dx$

(ii) $\displaystyle\int e^{ax}\cos bx\, dx = \frac{e^{ax}(a\cos bx + b\sin bx)}{a^2+b^2}$

$\displaystyle \qquad = \frac{1}{\sqrt{a^2+b^2}}e^{ax}\cos\left(bx-\tan^{-1}\frac{b}{a}\right)$

(iii) $\displaystyle\int e^{ax}\sin bx\, dx = \frac{e^{ax}(a\sin bx - b\cos bx)}{a^2+b^2}$

$\displaystyle \qquad = \frac{1}{\sqrt{a^2+b^2}}e^{ax}\sin\left(bx-\tan^{-1}\frac{b}{a}\right)$

(iv) $\displaystyle\int \sqrt{x^2+a^2}\, dx = \frac{x\sqrt{x^2+a^2}}{2} + \frac{a^2}{2}\log\left|x+\sqrt{x^2+a^2}\right|$

$\displaystyle \qquad = \frac{x\sqrt{x^2+a^2}}{2} + \frac{a^2}{2}\sinh^{-1}\frac{x}{a}$

(v) $\displaystyle\int \sqrt{x^2-a^2}\, dx = \frac{x\sqrt{x^2-a^2}}{2} - \frac{a^2}{2}\log\left|x+\sqrt{x^2-a^2}\right|$

$\displaystyle \qquad = \frac{x\sqrt{x^2-a^2}}{2} - \frac{a^2}{2}\cosh^{-1}\frac{x}{a}$

(vi) $\displaystyle\int \sqrt{a^2-x^2}\, dx = \frac{a\sqrt{a^2-x^2}}{2} + \frac{a^2}{2}\sin^{-1}\frac{x}{a}$

(vii) $\displaystyle\int \operatorname{cosec} x\, dx = \log\left|\tan\frac{x}{2}\right| = \log|\operatorname{cosec} x - \cot x|$

$(viii)$ $\displaystyle\int \sec x\, dx = \log\left|\tan\left(\frac{\pi}{4}-\frac{x}{2}\right)\right| = \log|\sec x + \tan x|.$

0.3.4 Definite integral

1. Definite integration as the limit of a sum:

$$\int_a^b f(x)dx = \lim_{n\to\infty} \frac{b-a}{n}\sum_{r=0}^{n-1} f\left[a+(b-a)\frac{r}{n}\right]$$

$$= \lim_{n\to\infty} \frac{b-a}{n}\sum_{r=1}^{n} f\left[a+(b-a)\frac{r}{n}\right]$$

$$= \lim_{h\to 0} h\sum_{r=0}^{n-1} f(a+rh) = \lim_{h\to 0} h\sum_{r=1}^{n} f(a+rh),$$

where $nh = b - a$.

2. Generalised definition of definite integral as the limit of a sum:

Let $f : [a,b] \to \mathbb{R}$ be a bounded function on $[a,b]$. Let this interval $[a,b]$ be divided in any manner into n subintervals, equal or unequal lengths $\delta x_1, \delta x_2, ..., \delta x_n$ given by $[a = x_0, x_1], [x_1, x_2], [x_2, x_3], ..., [x_{n-1}, x_n = b]$.

In each subinterval, a perfectly arbitrary number is chosen. These arbitrary points may be within or at the end points of the intervals. Let these be considered to be $\xi_1, \xi_2, ..., \xi_n$.

Let us consider the sum $S_n = \sum_{i=1}^{n} f(\xi_i)\delta x_i$. Let n be increased indefinitely in such a way that the largest of the lengths δx_i, $i = 1, 2, ..., n$ tends to zero. If, in this case S_n tends to a definite finite limit, which does not depend on the mode of subdivision, we say that the function $f(x)$ is integrable on the interval $[a,b]$. The value of this limit is called definite integral or the ordinary single integral of $f(x)$ on $[a,b]$. This is denoted by $\int_a^b f(x)dx$.

Thus $\int_a^b f(x)dx = \lim_{\substack{n\to\infty \\ \delta x_i \to 0}} \sum_{i=1}^{n} f(\xi_i)\delta x_i = \lim_{n\to\infty} \sum_{i=1}^{n} f(\xi_i)(x_i - x_{i-1})$.

Definition 0.3.1 *Let $f(x)$ be a bounded single-valued function defined in the interval $[a,b]$, a and b both finite. Let the interval $[a,b]$ be divided into n subintervals, then $\int_a^b f(x)dx = \lim_{\delta\to 0} \sum_{r=1}^{n} f(\xi_r)\delta_r$, where δ is norm of the partition $P = [a, \xi_1, \xi_2, ..., \xi_n = b]$ i.e., $||P||$.*

Note 0.3.1 *Since $\xi_1, \xi_2, ..., \xi_n$ may coincide with the end points $x_1, x_2, ..., x_n$ of the subintervals, so the above may also be given by*

$$\int_a^b f(x)dx = \lim_{\substack{n\to\infty \\ \delta x_i \to 0}} \sum_{i=1}^{n} f(x_i)\delta x_i.$$

Theorem 0.3.1 *Fundamental Theorem of Calculus*

If $f(x)$ be a bounded, continuous and single-valued function defined in the interval $[a,b]$, a and b both finite, $f(x)$ is integrable in $[a,b]$ and if there exists a function $\phi(x)$, such that $\phi'(x) = f(x)$ in $[a,b]$ then

$$\int_a^b f(x)dx = \phi(b) - \phi(a).$$

Theorem 0.3.2 *If m and M be the least and greatest values of $f(x)$ respectively in $[a,b]$, then*

$$m(b-a) \le \int_a^b f(x)dx \le M(b-a).$$

Theorem 0.3.3 *If $g(x)$ and $h(x)$ be two functions defined in $[a,b]$ and if $g'(x)$ and $h'(x)$ both exists in (a,b) and if $f(t)$ be continuous in $t \in [g(x), h(x)]$, then*

$$\frac{d}{dx}\int_{g(x)}^{h(x)} f(t)dt = f(h(x))\frac{dh(x)}{dx} - f(g(x))\frac{dg(x)}{dx}.$$

0.3.5 Properties of definite integrals

(i) $\displaystyle\int_a^b f(x)dx = \int_a^b f(z)dz$ (ii) $\displaystyle\int_a^b f(x)dx = -\int_b^a f(x)dx$

(iii) $\displaystyle\int_a^b f(x)dx = \int_a^c f(x)dx + \int_c^b f(x)dx,\;\; a < c < b$

(iv) $\displaystyle\int_0^a f(x)dx = \int_0^a f(a-x)dx$

(v) $\displaystyle\int_0^{na} f(x)dx = n\int_0^a f(x)dx,\;\; \text{if}\;\; f(a+x) = f(x)$

(vi) $\displaystyle\int_0^{2a} f(x)dx = 2\int_0^a f(x)dx,\;\; \text{if}\;\; f(2a-x) = f(x)$

$\qquad\qquad = 0, \qquad\qquad\qquad \text{if}\;\; f(2a-x) = -f(x)$

(vii) $\displaystyle\int_{-a}^a f(x)dx = 2\int_0^a f(x)dx,\;\; \text{if}\;\; f(-x) = f(x)$

$\qquad\qquad = 0, \qquad\qquad\qquad \text{if}\;\; f(-x) = -f(x).$

0.3.6 Useful reduction formulae

(i) $\displaystyle I_n = \int x^n e^{ax}dx\;\; \Rightarrow\;\; I_n = \frac{x^n e^{ax}}{a} - \frac{n}{a}I_{n-1}$

(ii) $\displaystyle I_n = \int \sin^n x\,dx\;\; \Rightarrow\;\; I_n = -\frac{\sin^{n-1}x\cos x}{n} + \frac{n-1}{n}I_{n-2}$

$\qquad$ $\displaystyle J_n = \int_0^{\frac{\pi}{2}} \sin^n x\,dx = \frac{n-1}{n}J_{n-2},\;\; n > 1$

(iii) $\displaystyle I_n = \int \cos^n x\,dx\;\; \Rightarrow\;\; I_n = -\frac{\cos^{n-1}x\sin x}{n} + \frac{n-1}{n}I_{n-2}$

$\qquad$ $\displaystyle J_n = \int_0^{\frac{\pi}{2}} \cos^n x\,dx = \frac{n-1}{n}J_{n-2},\;\; n > 1$

(iv) $\displaystyle\int_0^{\frac{\pi}{2}} \sin^n x\,dx = \int_0^{\frac{\pi}{2}} \cos^n x\,dx$

$\qquad$ $\displaystyle = \frac{n-1}{n}\cdot\frac{n-3}{n-2}\cdot\frac{n-5}{n-4}\cdots\frac{3}{4}\cdot\frac{1}{2}\cdot\frac{\pi}{2},\;\;$ if n is an even integer, or

$\qquad$ $\displaystyle = \frac{n-1}{n}\cdot\frac{n-3}{n-2}\cdot\frac{n-5}{n-4}\cdots\frac{4}{5}\cdot\frac{2}{3}\cdot 1,\;\;$ if n is an odd integer.

(v) $\displaystyle I_n = \int \tan^n x\,dx = \frac{\tan^{n-1}x}{n-1} - I_{n-2}$

$\qquad$ $\displaystyle J_n = \int_0^{\frac{\pi}{4}} \tan^n x\,dx = \frac{1}{n-1} - J_{n-2}$

(vi) $\displaystyle I_n = \int \sec^n x\,dx = \frac{\sec^{n-2}x\tan x}{n-1} + \frac{n-2}{n-1}I_{n-2}$

(vii) $\displaystyle I_n = \int \cot^n x\,dx = \frac{\cot^{n-1}x}{1-n} - I_{n-2}$

(vii) $\quad I_{m,n} = \displaystyle\int \sin^m x \cos^n x dx, \quad m, n \in \mathbb{N}$

$$= \frac{\sin^{m+1} x \cos^{n-1} x}{m+n} + \frac{n-1}{m+n} I_{m,n-2}$$

$$= -\frac{\sin^{m-1} x \cos^{n+1} x}{m+n} + \frac{m-1}{m+n} I_{m-2,n}$$

$(viii)$ $\quad J_{m,n} = \displaystyle\int_0^{\frac{\pi}{2}} \sin^m x \cos^n x dx = \frac{n-1}{m+n} J_{m,n-2} = \frac{m-1}{m+n} J_{m-2,n}$

(ix) $\quad \displaystyle\int_0^{\frac{\pi}{2}} \sin^m x \cos^n x dx = \int_0^{\frac{\pi}{2}} \cos^m x \sin^n x dx, \quad m, n \in \mathbb{N}$

$$= \frac{1.3.5...(m-1).1.3.5...(n-1)}{2.4.6...(m+n)} \frac{\pi}{2}, \text{ if } m, \ n \text{ both even}$$

$$= \frac{2.4.6...(m-1)}{(n+1)(n+3)...(n+m)}, \text{ if any one say } m \text{ is odd.}$$

0.3.7   Beta and Gamma functions

(i) $\quad B(m,n) = B(n,m) = \displaystyle\int_0^1 x^{m-1}(1-x)^{n-1} dx = \int_0^1 x^{n-1}(1-x)^{m-1} dx$

$$= \int_0^{\infty} \frac{x^{m-1}}{(1+x)^{m+n}} dx = \int_0^{\infty} \frac{x^{n-1}}{(1+x)^{m+n}} dx, \quad (m > 0, \ n > 0)$$

(ii) $\quad \Gamma(n) = \displaystyle\int_0^{\infty} e^{-x} x^{n-1} dx = a^n \int_0^{\infty} e^{-ax} x^{n-1} dx$

$$= 2\int_0^{\infty} e^{-x^2} x^{2n-1} dx, \quad (n > 0, a > 0)$$

(iii) $\quad B(m,n) = \dfrac{\Gamma(m)\Gamma(n)}{\Gamma(m+n)}$, [Relation between Beta and Gamma functions]

(iv) $\quad B(m,n) = \dfrac{\underline{|m-1}}{n(n+1)(n+2)...(n+m-1)},$

$$\text{when } m \text{ is a positive integer}$$

$$= \frac{\underline{|n-1}}{m(m+1)(m+2)...(m+n-1)},$$

$$\text{when } n \text{ is a positive integer}$$

$$= \frac{\underline{|m-1}\ \underline{|n-1}}{\underline{|m+n-1}}, \text{ when both } m \text{ and } n \text{ are positive integers}$$

(v) $\quad B(m,n) = 2\displaystyle\int_0^{\frac{\pi}{2}} \sin^{2m-1}\theta \cos^{2n-1}\theta d\theta$

$$= 2\int_0^{\frac{\pi}{2}} \cos^{2m-1}\theta \sin^{2n-1}\theta d\theta$$

$$= \frac{1}{(a-b)^{m+n-1}} \int_b^a (x-b)^{m-1}(a-x)^{n-1} dx$$

(vi) $\quad \Gamma(n+1) = n\Gamma(n)$

$$= \underline{|n}, \quad \text{if } n \text{ is a positive integer}$$

(vii) $\Gamma(1) = 1,$ $(viii)$ $\Gamma\left(\dfrac{1}{2}\right) = \sqrt{\pi}$

(ix) $\Gamma(m)\Gamma(1 - m) = \dfrac{\pi}{\sin m\pi},$ $(0 < m < 1)$

[Legendre's Duplication Formula]

(x) $\displaystyle\int_0^{\frac{\pi}{2}} \sin^p \theta \cos^q \theta\, d\theta = \dfrac{1}{2}\dfrac{\Gamma\left(\frac{p+1}{2}\right)\Gamma\left(\frac{q+1}{2}\right)}{\Gamma\left(\frac{p+q+2}{2}\right)},$ $(p > -1,\ q > -1)$

$= \dfrac{1}{2}B\left(\dfrac{p+1}{2}, \dfrac{q+1}{2}\right)$

(xi) $\displaystyle\int_0^{\frac{\pi}{2}} \sin^p \theta\, d\theta = \int_0^{\frac{\pi}{2}} \cos^p \theta\, d\theta = \dfrac{\Gamma\left(\frac{1}{2}\right)\Gamma\left(\frac{p+1}{2}\right)}{2\Gamma\left(\frac{p+1}{2}\right)}$

$= \dfrac{\sqrt{\pi}}{2}\dfrac{\Gamma\left(\frac{p+1}{2}\right)}{\Gamma\left(\frac{p+2}{2}\right)},$ $(p > -1)$

(xii) $\displaystyle\int_0^{\infty} e^{-x^2}\, dx = \dfrac{\sqrt{\pi}}{2}.$

Note 0.3.2 *From previous result* (xi) *we can get*

$$\int_0^{\frac{\pi}{2}} \sin^{n-1} \theta\, d\theta = \int_0^{\frac{\pi}{2}} \cos^{n-1} \theta\, d\theta = \dfrac{\Gamma\left(\frac{1}{2}\right)\Gamma\left(\frac{n}{2}\right)}{2\Gamma\left(\frac{n+1}{2}\right)}$$

[with the assumption $n - 1 = p$*].*

Chapter 1

Functions of Several Variables: Limits & Continuity

1.1 Introduction

Multivariate Calculus is the *extension* of calculus in *one variable* (i.e., calculus in real line $\mathbb{R}$ which is also known as *one dimensional Euclidean space* $\mathbb{R}^1$) to calculus with functions of *several variables* which is in the real Euclidean space of n dimensions $\mathbb{R}^n$. Therefore, to study multivariate calculus we shall be mainly concerned with the applications of calculus to functions of *more than one variable* and for this, the knowledge of functions of single independent variable is essential, which, at this stage, readers must have already obtained. This important branch of Mathematics deals with the *limits, continuities*, the *differentiations* and *integrations* of functions *involving multiple variables, rather than just one.*

In our current text, for the sake of brevity, we shall be mainly confined ourselves to discuss all these in the Euclidean spaces $\mathbb{R}^2$ or $\mathbb{R}^3$. For this we shall have to study the followings:

- Brief discussion of Point sets in $\mathbb{R}^2$ and $\mathbb{R}^3$,

- Discussion on functions of several variables and

- Limits and Continuities of functions of two or more variables.

1.2 Some Concepts on Point Sets in $\mathbb{R}^2$ and $\mathbb{R}^3$

1. Neighbourhood of a point

Let $(\alpha, \beta) \in \mathbb{R}^2$ be any point and $\delta > 0$ be chosen arbitrarily. The δ-*neighbourhood* (or simply neighbourhood) of the point (α, β) is a subset of $\mathbb{R}^2$ which is of various shape and size in two dimensional plane. It is generally denoted by δ-*nbd* of (α, β) or $N_\delta(\alpha, \beta)$ or simply $N(\alpha, \beta)$.

The square δ-*nbd* of (α, β) (See **Figure** 1.2) is defined by the set $\{(x, y) \in \mathbb{R}^2 : |x - \alpha| < \delta, |y - \beta| < \delta\}$. The set $\{(x, y) \in \mathbb{R}^2 : (x - \alpha)^2 + (y - \beta)^2 < \delta^2\}$ is giving the circular neighbourhood of (α, β) which is shown in the **Figure** 1.1. In a same way, the set $\{(x, y) \in \mathbb{R}^2 : |x - \alpha| < \delta_1, |y - \beta| < \delta_2\}$ defines the rectangular neighbourhood of the point (α, β) for arbitrarily chosen $\delta_1 > 0, \delta_2 > 0$. In this way, we can define other arbitrary shaped neighbourhoods of a point. However, we use most commonly the square or circular neighbourhoods to serve our purpose.

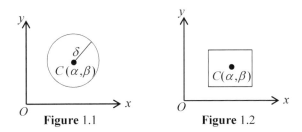

Figure 1.1   **Figure** 1.2

If $(\alpha, \beta, \gamma) \in \mathbb{R}^3$ be any point and $\delta > 0$ then δ-*neighbourhood* of (α, β, γ) is similarly a subset of $\mathbb{R}^3$. Here the set $\{(x, y, z) \in \mathbb{R}^3 : |x - \alpha| < \delta, |y - \beta| < \delta, |z - \gamma| < \delta\}$ gives a cubical neighbourhood $N_\delta(\alpha, \beta, \gamma)$ and the set $\{(x, y, z) \in \mathbb{R}^3 : (x - \alpha)^2 + (y - \beta)^2 + (z - \gamma)^2 < \delta^2\}$ will express the spherical δ-*nbd* of (α, β, γ).

Deleted Neighbourhood

Let $N(\alpha, \beta)$ be the neighbourhood of the point (α, β). The set $N(\alpha, \beta) \smallsetminus \{(\alpha, \beta)\}$ is called the deleted neighbourhood of (α, β) which is denoted by $N'(\alpha, \beta)$. In $\mathbb{R}^3$, the deleted neighbourhood of the point (α, β, γ), denoted by $N'(\alpha, \beta, \gamma)$, is defined by the set $N(\alpha, \beta, \gamma) \smallsetminus \{(\alpha, \beta, \gamma)\}$. For example, if $\{(x, y) \in \mathbb{R}^2 : |x - \alpha| < \delta, |y - \beta| < \delta\}$ be a δ-*nbd* of (α, β) then its corresponding deleted δ-*nbd* is the set $\{(x, y) \in \mathbb{R}^2 : 0 < |x - \alpha| < \delta, 0 < |y - \beta| < \delta\}$.

2. Interior point

Let $D \subset \mathbb{R}^2$. The point $(\alpha, \beta) \in D$ is called an *interior point* of the set D if there exists a δ-*nbd* of the point (α, β) which is contained in D, i.e., if $N_\delta(\alpha, \beta) \in D$ for any $\delta > 0$.

3. Open sets

A subset D of $\mathbb{R}^2$ is said to be open if every point of D is an interior point of D. In a similar manner we can define interior points and open sets in $\mathbb{R}^3$ and can be extended for n-dimensional Euclidean Space $\mathbb{R}^n$.

For example, the set $S_1 = \{(x, y) : (x - a)^2 + (y - b)^2 < r^2\}$ is an open set in $\mathbb{R}^2$. Clearly all points of S_1 are interior points of S_1 for no boundary points belong to the set S_1. S_1 is known as *Open disc* centered at (a, b) with radius r. Similarly, the set $S_2 = \{(x, y, z) : (x - a)^2 + (y - b)^2 + (z - c)^2 < r^2\}$ is an open set in $\mathbb{R}^3$ which is known as the *Open ball* centered at (a, b, c) with radius r.

3. Limit point or Cluster point or Point of accumulation

A point $(\xi, \eta) \in \mathbb{R}^2$ is a limit point or cluster point or accumulation point or point of accumulation of a set $D \subseteq \mathbb{R}^2$ if every neighbourhood of the point (ξ, η) contains an infinite number of members of D. Thus, if for any neighbourhood $N(\xi, \eta)$ of the point (ξ, η), $N(\xi, \eta) \cap D$ is an infinite set then (ξ, η) is a limit point.

Alternative definition

A point $(\xi, \eta) \in \mathbb{R}^2$ is a limit point or cluster point or accumulation point or point of accumulation of a set $D \subseteq \mathbb{R}^2$ if for every arbitrary $\delta > 0$, there exists an δ-*nbd* of (ξ, η) which contains at least one member of D other than the point (ξ, η).

4. Closed sets

A set $D \subseteq \mathbb{R}^2$ is said to be closed if each limit point of D is a member of the set D. The concepts of limit points and closed sets can very easily be extended for n-dimensional Euclidean Space $\mathbb{R}^n$.

For example, the sets $E_1 = \{(x, y) : (x-a)^2 + (y-b)^2 \le r^2\}$ and $E_2 = \{(x, y, z) : (x-a)^2 + (y-b)^2 + (z-c)^2 \le r^2\}$ are closed sets in $\mathbb{R}^2$ and $\mathbb{R}^3$ respectively. These closed sets E_1 and E_2 are respectively known as closed disc and closed ball in $\mathbb{R}^2$ and $\mathbb{R}^3$.

5. Bounded sets

A set $D \subset \mathbb{R}^2$ is said to be bounded if there exist a circle $R = \{(x, y) : (x-a)^2 + (y-b)^2 \le r^2\}$ with centre (a, b) and radius r such that $D \subseteq R$.

Now we state an important theorem without proof.

Theorem 1.2.1 *Bolzano-Weierstrass Theorem*

Every infinite bounded set $D \subset \mathbb{R}^2$ has at least one limit point in $\mathbb{R}^2$.

Example 1.2.1 *Show that the set $S = \{(x, y) \in \mathbb{R}^2 : 2x^2 + 4y^2 < 1\}$ is an open set and not a closed one. Find the boundary of the set S.*

Solution: The given set of points in $\mathbb{R}^2$ actually represents the set of all interior points but not the boundary points of the ellipse $2x^2 + 4y^2 = 1$, i.e., the set S itself is a neighbourhood of any of its points. Hence S is an open set.

Clearly, the set S is not closed. Let us take any point on $2x^2 + 4y^2 = 1$, say, $\left(0, \dfrac{1}{2}\right)$. Any δ-nbd of $\left(0, \dfrac{1}{2}\right)$ contains an infinite number of points of S which implies $\left(0, \dfrac{1}{2}\right)$ is a limit point of S but $\left(0, \dfrac{1}{2}\right) \notin S$, S is not closed.

The boundary of S is the set $S_b = \{(x, y) \in \mathbb{R}^2 : 2x^2 + 4y^2 = 1\}$. $\qquad\square$

Example 1.2.2 *If $B = \{(a, 0) : a \in \mathbb{R}\}$, show that B is a closed subset but not an open subset of $\mathbb{R} \times \mathbb{R}$.*

Solution: The set B actually gives the set of points on the x-axis. Obviously, each point of B is a limit point of it and no other point of $\mathbb{R} \times \mathbb{R}$ is a limit point of B. Hence B is a closed set.

Again the set B has no interior points, because if we consider any δ-nbd, say $N_\delta(a, 0)$ of an arbitrary point $(a, 0) \in B$, then that neighbourhood is not wholly contained in B. Hence B is not an open set. $\qquad\square$

1.3   Functions of Two Independent Variables

Let there be a set of points $R \subset \mathbb{R}^2$. If there exists a law or rule by which a single definite value of u can be assigned for each pair of values of $(x, y) \in R$, we say that u is a function of the two independent variables x and y over the domain of definition R. It is a function $f : \mathbb{R}^2 \to \mathbb{R}$ and is denoted by $u = f(x, y)$.

Note 1.3.1 *The definition of function may be extended to n independent variables $x_1, x_2, ..., x_n$ in a similar manner which is a function $f : \mathbb{R}^n \to \mathbb{R}$ and is written as $u = f(x_1, x_2, ..., x_n)$.*

Example 1.3.1 *Area of a triangle is a function of two independent variables, viz., its base (b) and altitude (h), i.e., Area $A = A(b, h)$.*

Example 1.3.2 *Volume V of a gas is a function of its pressure (p) and temperature (T), i.e., $V = V(p, T)$.*

Example 1.3.3 *The volume of the parallelepiped is a function of three independent variables, viz., its length, breath and height.*

1.4   Types of Functions

1.4.1   Single and Multiple valued functions

If to each pair of values of x and y, u has a single definite value, u is called a *single-valued function* and if to each pair of values of x and y, u has more than one definite value, u is called a *multiple-valued function.*

Note 1.4.1 *We are mainly concerned in all mathematical investigations with single-valued functions. Also it is to be noticed that a multiple-valued function with proper limitations imposed on its value can, in general, be treated as defining two or more single-valued functions.*

1.4.2   Explicit and Implicit functions

Let us consider a set of n independent variables $x_1, x_2, ..., x_n$ and one dependent variables u. The equation

$$u = f(x_1, x_2, ..., x_n) \tag{1.1}$$

denotes a functional relation, which is called an *explicit* function. In this case if $x_1, x_2, ..., x_n$ are the n arbitrary assigned values of the independent variables, the corresponding values of the dependent variable u are determined by the functional relation (1.1). But in many cases, it may be inconvenient or even impossible to obtain an equation of type (1.1), expressing one of the variables explicitly in terms of the other. However, we can get a functional relation of the form

$$\phi(x_1, x_2, ..., x_n, u) = 0 \tag{1.2}$$

connecting the $n + 1$ variables $x_1, x_2, ..., x_n, u$, but it is not in general to solve it to find an explicit function, which expresses one of these variables, say, u, in terms of the other n variables. This type of functional relation (1.2) will be called an *implicit* function.

Remark: Examples of some other types of functions of two and three variables are given below without giving their formal definitions.

(i) **Linear function:** $u = f(x, y) = ax + by + c$ is defined for all points $(x, y) \in \mathbb{R}^2$, i.e., in the xy-plane.

Similarly $u = f(x, y, z) = ax + by + cz + d$ is a linear function of three independent variables x, y, z. It is defined for all points $(x, y, z) \in \mathbb{R}^3$, i.e., in the three dimensional space.

(ii) **Rational function:** The functional relations of the following form of quotient of two polynomials given by

$$u = \frac{f(x, y)}{g(x, y)} = \frac{ax + by + c}{\alpha x + \beta y + \gamma}$$

is defined for all points (x, y) in the xy plane excepting the points which lie on the straight line $\alpha x + \beta y + \gamma = 0$.

(iii) **Algebraic function:** The function

$$u = f(x, y) = \sqrt{(x - 1)(2 - x)(y - 1)(2 - y)}$$

is defined at all points (x, y) except where the radicals are negative. Here the independent variables x, y can assume any value within the closed square $\{(x, y) \in \mathbb{R}^2 : 1 \leq x \leq 2, 1 \leq y \leq 2\}$, so as to make u a real valued function of x and y.

The function $u = x^2 y^2 + y^2 z^2 + z^2 x^2$ is a function of three independent variables x, y, z and is defined for all points (x, y, z) in three dimensional Euclidean Space $\mathbb{R}^3$.

1.5 Geometrical Representation of a Function of the form $u = f(x, y)$

Let us given a single-valued function $z = f(x, y)$. For each point of values of x and y, there corresponds a point Q in the plane XOY shown in **Figure** 1.3. Let a perpendicular QP of length equal to the value of z is erected from Q. Then the ordered pair $(u, z) \equiv ((x, y), z)$ is expressed as the *ordered triplet* (x, y, z) which represents a point P, obtained from the given relation. The points like P describe what is called a *surface* in three dimensional Euclidean space $\mathbb{R}^3$. Thus to a functional relation between three variables x, y, z, therefore corresponds a surface referred to coordinate axes OX, OY, OZ in space. Hence all the points of

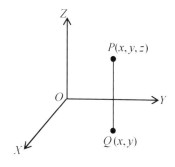

Figure 1.3

f may be supposed to lie on a *surface* which is linear in z.

In particular, if $z = ax + by + c$, where $a, b, c \in \mathbb{R}$, then z is a function of two real variables x and y, representing a plane.

Similarly $z = \sqrt{1 - (x^2 + y^2)}$, where the radicals are not negative, geometrically represents a hemisphere.

Note 1.5.1 *For functions of three or more variables, there is no suitable geometrical representation. However the concept of domain of definition and neighbourhood can easily be extended to such functions. Thus $u = ax + by + cz + d$, $a, b, c, d \in \mathbb{R}$ represents a function of three variables x, y and z and is called a hyperplane but this cannot be represented geometrically by means of a diagram.*

1.6   Examples of Functions

1. $f(x, y) = x + y$ is defined over the entire xy-plane.

2. $f(x, y) = \sqrt{(x-1)(3-x)(y-1)(3-y)}$ is defined, where the radicals are not negative. In this case, two independent variables x and y can assume arbitrary chosen values independent of each other within the set of values $\{(x, y) \in \mathbb{R}^2 : 1 \le x \le 3 \text{ and } 1 \le y \le 3\}$.

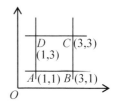

Figure 1.4

The aggregate of the pair of values (x, y) on or within the rectangle with vertices $A(1, 1)$, $B(3, 1)$, $C(3, 3)$ and $D(1, 3)$ is the domain of definition of the function $f(x, y)$ (ref. **Figure** 1.4).

3. $f(x, y, z) = \sqrt{(x-1)(3-x)(y-1)(3-y)(z-1)(3-z)}$ is defined, where the radicals are not negative. Here the three independent variables x, y and z can assume any values independent of each other within the set of values $\{(x, y, z) \in \mathbb{R}^3 : 1 \le x \le 3, 1 \le y \le 3 \text{ and } 1 \le z \le 3\}$.

Geometrically, the ranges of values of x, y, z can be represented by a cube in three dimensional space. The aggregate of ordered triplet (x, y, z) on or within the rectangular parallelepiped obtained is the domain of definition.

4. $f(x_1, x_2, ..., x_n) = \sqrt{(x_1 - 1)(3 - x_1)(x_2 - 1)(3 - x_2)...(x_n - 1)(3 - x_n)}$ where the radicals are not negative. The domain of definition of f is the closed set $\{(x_1, x_2, ..., x_n) \in \mathbb{R}^n : 1 \le x_1 \le 3, 1 \le x_2 \le 3, ..., 1 \le x_n \le 3\}$. The system of n ordered variables $(x_1, x_2, ..., x_n)$ is assumed to correspond to a point in the n-dimensional space. The aggregate of all such points in space constitute the domain or region of definition of f. Obviously this region is closed.

5. $f(x, y) = \dfrac{xy}{x^2 + y^2}$ is defined over the entire xy-plane excluding only the point $(0, 0)$.

6. $f(x, y) = \sqrt{1 - (x^2 + y^2)}$ is defined over the closed circular domain $x^2 + y^2 \le 1$.

7. $f(x, y, z) = \sqrt{a^2 - (x^2 + y^2 + z^2)}$ is defined over the closed spherical domain $x^2 + y^2 + z^2 \le a^2$.

8. $f(x, y) = \sqrt{x + y - 1}$ is defined over the closed domain $x + y \ge 1$.

9. $f(x, y) = \dfrac{1}{\sqrt{x + y - 1}}$ is defined over the open domain $x + y > 1$.

10. $f(x, y) = \log(1 - x^2 - y^2)$ is defined over the open domain $x^2 + y^2 < 1$.

1.7 Limit of a Function

Let $f(x, y)$ be a function of two independent variables x and y. We can define several kinds of limits. If (a, b) be the limiting point of a set of values of x and y in $\mathbb{R}^2$, then we have the following limits $\lim\limits_{\substack{x \to a \\ y \to b}} f(x, y)$, $\lim\limits_{x \to a} \lim\limits_{y \to b} f(x, y)$ and $\lim\limits_{y \to b} \lim\limits_{x \to a} f(x, y)$.

We speak the first type of limit as a simultaneous limit or double limit and the last two types as repeated or iterated limits.

1.7.1 Simultaneous or Double limits

The function $f(x, y)$ is said to tend to the limit l as $(x, y) \to (a, b)$ [but $(x, y) \neq (a, b)$] if and only if the limit l is independent of the path followed by the point $(x, y) \to (a, b)$ and we write it as $\lim\limits_{\substack{x \to a \\ y \to b}} f(x, y) = l$ or $\lim\limits_{(x,y) \to (a,b)} f(x, y) = l$.

Analytically, the simultaneous limit $\lim\limits_{\substack{x \to a \\ y \to b}} f(x, y)$ or $\lim\limits_{(x,y) \to (a,b)} f(x, y)$ is said to exist and is equal to l, provided given any pre-assigned positive quantity ϵ, however small, we can determine another positive quantity δ, such that $|f(x, y) - l| < \epsilon$, for all points (x, y) belongs to a deleted δ-nbd $N'_\delta(a, b)$.

Conveniently, we choose the δ-nbd $N'_\delta(a, b)$ either as a square neighbourhood $\{(x, y) \in \mathbb{R}^2 : 0 < |x - a| < \delta \text{ and } 0 < |y - \beta| < \delta\}$ or $\{(x, y) \in \mathbb{R}^2 : 0 < |x - a| + |y - \beta| < \delta\}$ or a circular neighbourhood $\{(x, y) \in \mathbb{R}^2 : 0 < (x - a)^2 + (y - b)^2 < \delta^2\}$.

Remark: In our definition, we have allowed the point (x, y) to vary in the whole xy-plane. If we so desire, however, we can impose restrictions on the point (x, y). For example, we may require it to *lie along a specific curve*. But the essential condition for the existence of a simultaneous limit is that the limiting value must be *unique* in whatever way an approach to (a, b) is made.

Therefore, to establish *non-existence* of a limit, we must find two methods of approach to the limiting point, which will give two different limiting values. To illustrate this, let us take the following example.

Example 1.7.1 *Test the existence of the limit* $\lim\limits_{\substack{x \to 0 \\ y \to 0}} \dfrac{xy^3}{x^2 + y^6}$.

Solution: Here $f(x, y) = \dfrac{xy^3}{x^2 + y^6}$.

First, let (x, y) approaches $(0, 0)$ along the line $y = x$. To find the limit, we first put $y = x$ in the function and then allow x to approach 0 for the single variable function and we get

$$\lim_{\substack{x \to 0 \\ y \to 0}} \frac{xy^3}{x^2 + y^6} = \lim_{x \to 0} \frac{x x^3}{x^2 + x^6} = \lim_{x \to 0} \frac{x^4}{x^2 + x^6} = \lim_{x \to 0} \frac{x^2}{1 + x^4} = 0.$$

Next, let (x, y) approaches $(0, 0)$ along the curve $x = y^3$. We then get

$$\lim_{\substack{x \to 0 \\ y \to 0}} \frac{xy^3}{x^2 + y^6} = \lim_{y \to 0} \frac{y^3 y^3}{y^6 + y^6} = \lim_{y \to 0} \frac{1}{2} = \frac{1}{2}.$$

Thus the limits obtained by two different methods of approach are different. Therefore the simultaneous limit does not exist. □

Note 1.7.1 *We can easily verify that, if we approach the origin along any line $y = mx$, then the limit comes to be zero. So, although for different values of m gives same value of the limits, equal to zero, yet we should not conclude that the simultaneous limit* $\lim\limits_{\substack{x \to 0 \\ y \to 0}} \dfrac{xy^3}{x^2 + y^6}$ *exists and is equal to zero, as we have seen that the limits along $x = y^3$ comes out to be* $\dfrac{1}{2}$.

Example 1.7.2 *Show that the double limit* $\lim\limits_{\substack{x \to 0 \\ y \to 0}} \tan^{-1}\left(\dfrac{y}{x}\right)$ *does not exist.*

Solution: Here, if (x, y) approaches to $(0,0)$ along the path $y = mx$ then

$$\lim_{\substack{x \to 0 \\ y \to 0}} \tan^{-1}\left(\frac{y}{x}\right) = \lim_{x \to 0} \tan^{-1}\left(\frac{mx}{x}\right) = \lim_{x \to 0} \tan^{-1} m = \tan^{-1} m.$$

Hence the limiting value depends on m. For different values of m, the path of approach $y = mx$ will be different and then the limiting value will be different. So, the limit does not exist. □

Example 1.7.3 *Show that* $\lim\limits_{\substack{x \to 0 \\ y \to 0}} \dfrac{x^2 y^2}{x^2 + y^2} = 0$.

Solution: Here we have to show that the limit exists. For existence of a double limit we must use ϵ-δ definition of limit.

Let $\epsilon > 0$ be chosen arbitrarily and $f(x, y) = \dfrac{x^2 y^2}{x^2 + y^2}$.

$$\therefore \quad |f(x, y) - 0| = \left|\frac{x^2 y^2}{x^2 + y^2} - 0\right| = \frac{x^2 y^2}{x^2 + y^2}.$$

Using $A.M \geq G.M.$ for the terms x^2 and y^2 we get

$$\frac{x^2 + y^2}{2} \geq |x||y| \quad \text{or,} \quad \frac{(x^2 + y^2)^2}{4} \geq x^2 y^2.$$

$$\therefore \quad \frac{x^2 y^2}{x^2 + y^2} \leq \frac{(x^2 + y^2)^2}{4} \cdot \frac{1}{x^2 + y^2} = \frac{(x^2 + y^2)}{4} < \epsilon \text{ (say)}.$$

Now $\dfrac{(x^2 + y^2)}{4} < \epsilon$ whenever $0 < x^2 + y^2 < (2\sqrt{\epsilon})^2$ $(\because (x, y) \neq (0,0))$ is true. If we take $\delta = 2\sqrt{\epsilon}$ we say that,

$$|f(x, y) - 0| < \epsilon \quad \text{whenever} \quad 0 < (x - 0)^2 + (y - 0)^2 < \delta^2 \text{ i.e., } (x, y) \in N'_\delta(0, 0).$$

Hence the limit exist and is equal to 0.

Note 1.7.2 *It is already mentioned that the proof of existence of simultaneous limit is more difficult than the proof of non-existence one. For proving existence we can sometimes use the transformation $x = r\cos\theta$ and $y = r\sin\theta$ to establish the ϵ-δ definition. Let us try to apply the method for this example as alternative method.*

Let $\epsilon > 0$ and $f(x,y) = \dfrac{x^2 y^2}{x^2 + y^2}$. Put $x = r\cos\theta$ and $y = r\sin\theta$, then

$$|f(x,y) - 0| = \left|\frac{x^2 y^2}{x^2 + y^2}\right| = \left|\frac{r^2\cos^2\theta . r^2\sin^2\theta}{(r\cos\theta)^2 + (r\sin\theta)^2}\right| = r^2|\cos^2\theta\sin^2\theta|$$

$$= \frac{r^2}{4}|\sin^2 2\theta| \le \frac{r^2}{4} = \frac{(x^2 + y^2)}{4} < \epsilon \text{ (say)}.$$

Now proceeding as above we can consider $\delta = 2\sqrt{\epsilon}$ we say that,

$$|f(x,y) - 0| < \epsilon \quad \text{whenever} \quad 0 < (x-0)^2 + (y-0)^2 < \delta^2 \text{ i.e., } (x,y) \in N'_\delta(0,0).$$

Hence the limit exist and $\lim\limits_{\substack{x\to 0 \\ y\to 0}} \dfrac{x^2 y^2}{x^2 + y^2} = 0.$ $\square$

Note 1.7.3 *To use ϵ-δ definition for proving existence of simultaneous limit we need to know the value of the limit in advance.*

Example 1.7.4 *Show that* $\lim\limits_{(x,y)\to(0,0)} \dfrac{x^2 - y^2}{1 + x^2 + y^2} = 0.$

Solution: Let $\epsilon > 0$ be chosen arbitrarily and $f(x,y) = \dfrac{x^2 - y^2}{1 + x^2 + y^2}$. Put $x = r\cos\theta$ and $y = r\sin\theta$, then

$$|f(x,y) - 0| = \left|\frac{x^2 - y^2}{1 + x^2 + y^2}\right| = \left|\frac{r^2(\cos^2\theta - \sin^2\theta)}{1 + r^2}\right| = \frac{r^2}{1 + r^2}|\cos 2\theta|$$

$$\le r^2 = x^2 + y^2 < \epsilon \text{ (say)} \left(\because |\cos 2\theta| \le 1 \text{ and } \frac{1}{1 + r^2} \le 1\right).$$

Let us take $\delta = \sqrt{\epsilon}$, and then

$$|f(x,y) - 0| < \epsilon \quad \text{whenever} \quad 0 < (x-0)^2 + (y-0)^2 < \delta^2 \text{ i.e., } (x,y) \in N'_\delta(0,0).$$

Hence the limit exist and $\lim\limits_{(x,y)\to(0,0)} \dfrac{x^2 - y^2}{1 + x^2 + y^2} = 0.$ $\square$

1.8 Simultaneous Limit of a Function of Three Variables

Let $f : \mathbb{R}^3 \to \mathbb{R}$ be a function defined in a domain $R \subseteq \mathbb{R}^3$. Then f is said to tend to a limit l, as the point (x,y,z) tends to a given point (a,b,c), which lie, entirely within the domain R if to each $\epsilon > 0 \; \exists \; \delta > 0$ such that

$$|f(x,y,z) - l| < \epsilon \text{ for all } (x,y,z) \in N'_\delta(a,b,c).$$

Like two variable functions here also in general we use either a cube neighbourhood $\{(x,y,z) \in \mathbb{R}^3 : 0 < |x - a| < \delta, 0 < |y - b| < \delta \text{ and } 0 < |z - c| < \delta\}$ or $\{(x,y,z) \in \mathbb{R}^3 : 0 < |x - a| + |y - b| + |z - c| < \delta\}$ or a spherical neighbourhood $\{(x,y,z) \in \mathbb{R}^3 : 0 < (x - a)^2 + (y - b)^2 + (z - c)^2 < \delta^2\}$.

Example 1.8.1 *Discuss the following function for the existence of simultaneous limit* $\lim\limits_{(x,y,z)\to(0,0,0)} f(x,y,z)$ *where*

$$f(x,y,z) = \begin{cases} x\sin\dfrac{1}{x} + y\sin\dfrac{1}{y} + z\sin\dfrac{1}{z} & \text{when} \quad xyz \ne 0 \\ 0 & \text{when} \quad xyz = 0. \end{cases}$$

Solution: Let $\epsilon > 0$ and $(x, y, z) \neq (0, 0, 0)$ then

$$|f(x, y, z) - 0| = \left| x \sin \frac{1}{x} + y \sin \frac{1}{y} + z \sin \frac{1}{z} \right|$$

$$\leq |x| \left| \sin \frac{1}{x} \right| + |y| \left| \sin \frac{1}{y} \right| + |z| \left| \sin \frac{1}{z} \right| \leq |x| + |y| + |z| < \epsilon \text{ (say)}.$$

Therefore $|x| + |y| + |z| < \epsilon$ is true whenever $0 < |x| < \frac{\epsilon}{3}, 0 < |y| < \frac{\epsilon}{3}, 0 < |y| < \frac{\epsilon}{3}$.
If we consider $\delta = \frac{\epsilon}{3}$ then we can write

$$|f(x, y, z) - 0| < \epsilon \text{ whenever } 0 < |x| < \delta, \ 0 < |y| < \delta, \ 0 < |z| < \delta.$$

Hence the limit exist and it is equal to 0. $\square$

1.9   Repeated or Iterated Limits

Let $f(x, y)$ be defined in a certain neighbourhood of a point (a, b). Then $\lim_{x \to a} f(x, y)$, if it exists, is a function of y, say $\phi(y)$. If now $\lim_{y \to b} \phi(y)$ exists and is equal to l, then we say that the iterated or repeated limit $\lim_{x \to a} \lim_{y \to b} f(x, y)$ exists and its value is equal to l, i.e., $\lim_{x \to a} \lim_{y \to b} f(x, y) = l$.

A change in the order of obtaining limits may produce a change in the final result, i.e., if we consider first $y \to b$ and then $x \to a$, we may have a different result, provided both the limits exist. In other words, $\lim_{y \to b} \lim_{x \to a} f(x, y)$ will give different limit and may not be equal to $\lim_{x \to a} \lim_{y \to b} f(x, y)$.

For example, let $f(x, y) = \dfrac{x - y}{x + y}, \ (x, y) \neq (0, 0)$ then

$$\lim_{y \to 0} \lim_{x \to 0} \frac{x - y}{x + y} = \lim_{y \to 0} \frac{-y}{y} = -1 \text{ but } \lim_{x \to 0} \lim_{y \to 0} \frac{x - y}{x + y} = \lim_{x \to 0} \frac{x}{x} = 1.$$

Example 1.9.1 *Show that for the following function the double limit exists but none of the repeated limits exist*

$$f(x, y) = \begin{cases} x \sin \dfrac{2}{y} + y \sin \dfrac{3}{x^3} & \text{when} \quad xy \neq 0 \\ 0 & \text{when} \quad xy = 0. \end{cases}$$

Solution: First we check that the existence of the double limit.

Let $\epsilon > 0$. When $(x, y) \neq (0, 0)$ we have

$$|f(x, y) - 0| = \left| x \sin \frac{2}{y} + y \sin \frac{3}{x^3} \right| \leq |x| \left| \sin \frac{2}{y} \right| + |y| \left| \sin \frac{3}{x^3} \right|$$

$$\leq |x| + |y| < \epsilon \text{ (say) which is true if } |x| < \frac{\epsilon}{2}, \ |y| < \frac{\epsilon}{2}.$$

Let us take $\delta = \frac{\epsilon}{2}$. Then we have

$$\therefore \ |f(x, y) - 0| < \epsilon \text{ whenever } 0 < |x| < \delta, \ 0 < |y| < \delta \text{ or } (x, y) \in N'_\delta(0, 0),$$

showing that the double limit exists and equal to 0.

Now we check the existence of the repeated limit $\lim\limits_{x \to 0} \lim\limits_{y \to 0} f(x, y)$.

$$\lim_{x \to 0} \left\{ \lim_{y \to 0} \left(x \sin \frac{2}{y} + y \sin \frac{3}{x^3} \right) \right\} = \lim_{x \to 0} x \left(\lim_{y \to 0} \sin \frac{2}{y} + 0 \right).$$

By Cauchy's Criterion of limit we have $\lim\limits_{y \to 0} \sin \dfrac{2}{y}$ doesn't exist and consequently the repeated limit doesn't exist.

By a similar consideration we get the limit $\lim\limits_{x \to 0} \sin \dfrac{3}{x^3}$ doesn't exist and hence the other repeated limit $\lim\limits_{y \to 0} \lim\limits_{x \to 0} f(x, y)$ also doesn't exist. $\quad\square$

Important Observation: In the previous example we have noticed that the double limit exists for a function without being any of the repeated limits exist. It can easily be proved that if the simultaneous limit exists, the two repeated limits, *if exist*, must *necessarily be equal* and the *three limits have the same value but the converse is not true.* If, however, the repeated limits exist and are *unequal*, then *the double limit cannot exist.*

Example 1.9.2 *Show that for the function* $f(x, y) = \dfrac{xy}{x^2 + y^2}$ *the two repeated limits exists and are equal but the double limit doesn't.*

Solution: We have $\lim\limits_{x \to 0} \lim\limits_{y \to 0} f(x, y) = \lim\limits_{x \to 0} \left\{ \lim\limits_{y \to 0} \dfrac{xy}{x^2 + y^2} \right\} = \lim\limits_{x \to 0} 0 = 0$. Similarly, we can show that $\lim\limits_{y \to 0} \lim\limits_{x \to 0} f(x, y) = 0$.

To prove that the double limit $\lim\limits_{(x,y) \to (0,0)} f(x, y) = \lim\limits_{(x,y) \to (0,0)} \dfrac{xy}{x^2 + y^2}$ does not exist, let us assume that $(x, y) \to (0, 0)$ along the line $y = mx$ which is a line through the origin.

$$\therefore \quad \lim_{(x,y) \to (0,0)} \frac{xy}{x^2 + y^2} = \lim_{x \to 0} \frac{x.mx}{x^2 + m^2 x^2} = \lim_{x \to 0} \frac{1}{1 + m^2} = \frac{1}{1 + m^2}.$$

which is different for different finite values of m. Therefore, the double limit $\lim\limits_{(x,y) \to (0,0)} \dfrac{xy}{x^2 + y^2}$ doesn't exist.

Alternative Approach

Let us take $x = r \cos \theta$ and $y = r \sin \theta$, $\therefore$ $r = \sqrt{x^2 + y^2}$. So when $(x, y) \to (0, 0)$ then $r \to 0$.

Now $f(x, y) = \dfrac{xy}{x^2 + y^2} = \dfrac{r^2 \sin \theta \cos \theta}{r^2} = \sin \theta \cos \theta = \dfrac{1}{2} \sin 2\theta$.

$$\therefore \quad |f(x, y) - 0| = \left| \frac{1}{2} \sin 2\theta \right| \leq \frac{1}{2} \nless \epsilon \quad \text{when } \epsilon < \frac{1}{2}.$$

Hence the double limit does not exists, although both the repeated limits exist and are equal. $\quad\square$

Note 1.9.1 *The above example shows that the converse of first part of '**Important observation**' is not true.*

Example 1.9.3 *Given the function* $f(x, y) = \dfrac{x^2 - y^2}{(x - y)^2 + x^2 y^2}$. *Discuss the existence of the repeated and double limits at* $(0, 0)$. *What conclusion can you draw from it?*

Solution: Here we observe that both the repeated limits exists and we get

$$\lim_{x \to 0} \lim_{y \to 0} \frac{x^2 - y^2}{(x - y)^2 + x^2 y^2} = \lim_{x \to 0} \frac{x^2}{(x)^2 + 0} = 1$$

and $\quad \displaystyle \lim_{y \to 0} \lim_{x \to 0} \frac{x^2 - y^2}{(x - y)^2 + x^2 y^2} = \lim_{y \to 0} \frac{0 - y^2}{(0 - y)^2 + 0} = -1.$

i.e., the two repeated limits are unequal.

Now for double limit $\displaystyle \lim_{\substack{x \to 0 \\ y \to 0}} f(x, y)$ let us first set $(x, y) \to (0, 0)$ along the line

$y = 0$. Then $\displaystyle \lim_{\substack{x \to 0 \\ y \to 0}} \frac{x^2 - y^2}{(x - y)^2 + x^2 y^2} = \lim_{x \to 0} \frac{x^2}{(x)^2 + 0} = 1$. In another approach we

set $(x, y) \to (0, 0)$ along the line $x = y$, then $\displaystyle \lim_{\substack{x \to 0 \\ y \to 0}} \frac{x^2 - y^2}{(x - y)^2 + x^2 y^2} = 0$. Hence the

limiting values are different for different approaches and the double limit doesn't exist.

This example shows that if the two repeated limits exist but not equal then simultaneous limit cannot exist. ☐

Example 1.9.4 *Discuss the existence of double limit and repeated limits of the function* $f(x, y) = x + y + \dfrac{xy}{\sqrt{x^2 + y^2}}$, $x > 0, y > 0$ *at* $(0, 0)$.

Solution: For repeated limits we have

$$\lim_{x \to 0} \lim_{y \to 0} f(x, y) = \lim_{x \to 0} \lim_{y \to 0} \left[x + y + \frac{xy}{\sqrt{x^2 + y^2}} \right] = \lim_{x \to 0} x = 0$$

and $\quad \displaystyle \lim_{y \to 0} \lim_{x \to 0} f(x, y) = \lim_{y \to 0} \lim_{x \to 0} \left[x + y + \frac{xy}{\sqrt{x^2 + y^2}} \right] = \lim_{y \to 0} y = 0.$

So both the repeated limits exist at $(0, 0)$ and the limiting values are 0.

Now we check for the simultaneous limit. Let $\epsilon > 0$ be chosen arbitrarily. We

have $\displaystyle \lim_{\substack{x \to 0 \\ y \to 0}} \left[x + y + \frac{xy}{\sqrt{x^2 + y^2}} \right] = \lim_{\substack{x \to 0 \\ y \to 0}} (x + y) + \lim_{\substack{x \to 0 \\ y \to 0}} \frac{xy}{\sqrt{x^2 + y^2}}$. Clearly the first

part of the limit exists and is equal to 0. We are going to check only the second part. Let $x = r \cos \theta$, $y = r \sin \theta$ then

$$\left| \frac{xy}{\sqrt{x^2 + y^2}} - 0 \right| = \left| \frac{r^2 \sin \theta \cos \theta}{r} \right| = \frac{r}{2} |\sin 2\theta| \leq \frac{r}{2} = \frac{\sqrt{x^2 + y^2}}{2} < \epsilon \text{ (say)}.$$

Now $\dfrac{\sqrt{x^2 + y^2}}{2} < \epsilon$ is true whenever $x^2 + y^2 < 4\epsilon^2$. So if $\delta = 2\epsilon$ then

$$\left| \frac{xy}{\sqrt{x^2+y^2}} - 0 \right| < \epsilon \quad \text{whenever} \quad 0 < (x-0)^2 + (y-0)^2 < \delta^2.$$

Hence the double limit exists and its value is equal to 0. □

Example 1.9.5 *Show that the simultaneous limit* $\lim\limits_{\substack{x\to 0 \\ y\to 0}} y\sin\dfrac{1}{x}$ *exists and is equal to 0 but the single limit* $\lim\limits_{x\to 0} y_1 \sin\dfrac{1}{x}$, *where* $y_1 \neq 0$ *is a constant, does not exist.*

Solution: Let $\epsilon > 0$ be given. We have $\left| \sin\dfrac{1}{x} \right| \leq 1$. Then for $(x,y) \neq (0,0)$,

$$\left| y\sin\frac{1}{x} - 0 \right| = \left| y\sin\frac{1}{x} \right| = |y| \left| \sin\frac{1}{x} \right| \leq |y| < \epsilon \text{ (say)}.$$

So if we take $\delta = \epsilon$ then we say $\left| y\sin\dfrac{1}{x} - 0 \right| < \epsilon$ whenever $0 < |y-0| < \delta$ and for any $x > 0$. Hence the simultaneous limit $\lim\limits_{\substack{x\to 0 \\ y\to 0}} y\sin\dfrac{1}{x}$ exists and is equal to 0.

For any constant value $y = y_1 \neq 0$ we have $\lim\limits_{x\to 0} y_1 \sin\dfrac{1}{x} = y_1 . \lim\limits_{x\to 0}\sin\dfrac{1}{x}$ and we know, by Cauchy's criterion of limit that $\lim\limits_{x\to 0}\sin\dfrac{1}{x}$ doesn't exist. Hence the result. □

1.10 Algebra of Limits

If $\lim\limits_{(x,y)\to(a,b)} f(x,y) = l$ and $\lim\limits_{(x,y)\to(a,b)} g(x,y) = m$, then

1. $\lim\limits_{(x,y)\to(a,b)} [f(x,y) \pm g(x,y)] = l \pm m$

2. $\lim\limits_{(x,y)\to(a,b)} [f(x,y).g(x,y)] = lm$

3. $\lim\limits_{(x,y)\to(a,b)} \left[\dfrac{f(x,y)}{g(x,y)} \right] = \dfrac{l}{m}, \quad$ provided $m \neq 0$.

Proof: The proofs of the above results are exactly same as in the case of functions of single real variable. □

Example 1.10.1 *Prove that* $\lim\limits_{(x,y)\to(1,2)} (x^2 + 2y) = 5$.

Solution: Using the above results given in § 1.10 we get

$$\lim\limits_{(x,y)\to(1,2)} (x^2 + 2y) = \lim\limits_{(x,y)\to(1,2)} x^2 + \lim\limits_{(x,y)\to(1,2)} 2y = 1^2 + 2.2 = 5. \quad □$$

Example 1.10.2 *Evaluate the following double limits (assuming the existence)*

(i) $\lim\limits_{(x,y)\to(1,1)} (xy + x + y)$ (ii) $\lim\limits_{(x,y)\to(2,2)} \dfrac{x(y-1)}{y(x-1)}$

(iii) $\lim\limits_{(x,y)\to(\infty,2)} \dfrac{xy+1}{x^2+2y^2}$ (iv) $\lim\limits_{(x,y)\to(1,2)} \dfrac{2x^2 y}{x^2+y^2+1}$.

Solution: (i) $\lim\limits_{\substack{x\to 1 \\ y\to 1}} (xy + x + y) = \lim\limits_{\substack{x\to 1 \\ y\to 1}} xy + \lim\limits_{\substack{x\to 1 \\ y\to 1}} x + \lim\limits_{\substack{x\to 1 \\ y\to 1}} y$

$$= 1 \times 1 + 1 + 1 = 3.$$

(ii) $\lim\limits_{(x,y)\to(2,2)} \dfrac{x(y-1)}{y(x-1)} = \dfrac{\lim\limits_{(x,y)\to(2,2)} x(y-1)}{\lim\limits_{(x,y)\to(2,2)} y(x-1)} = \dfrac{2(2-1)}{2(2-1)} = 1.$

(iii) $\lim\limits_{(x,y)\to(\infty,2)} \dfrac{xy+1}{x^2 + 2y^2} = \lim\limits_{(x,y)\to(\infty,2)} \dfrac{x\left(y + \frac{1}{x}\right)}{x^2 \left(1 + 2\frac{y^2}{x^2}\right)}$

$$= \lim\limits_{(x,y)\to(\infty,2)} \frac{1}{x} \frac{y + \frac{1}{x}}{1 + 2\frac{y^2}{x^2}} = \lim\limits_{(x,y)\to(\infty,2)} \frac{1}{x} \times \frac{2+0}{1+0} = 0.$$

(iv) **Do Yourself.** ☐

1.11 Continuity of Functions of Several Variables

Definition 1

A function $f(x,y)$ is said to be continuous at the point (a,b) in the domain of f, provided $\lim\limits_{(x,y)\to(a,b)} f(x,y)$ *exists, is finite and is equal to $f(a,b)$.*

If a function is *continuous at all points of its domain of definition*, then it is said to be *continuous in that domain*.

A function which is *not continuous* at a point in the domain is said to be *discontinuous* thereat.

Definition 2 (Analytical Definition)

Corresponding to the analytical definition of limit, we have the following analytical definition of continuity of a function at a point as given below:

A function $f(x,y)$ is said to be continuous at the point (a,b) in the domain of f, if corresponding to a pre-assigned positive number ϵ, however small, there exists another positive number δ such that

$$|f(x,y) - f(a,b)| < \epsilon \;\; whenever \;\; (x,y) \in N_\delta(a,b).$$

The δ-neighbourhood of the point (a,b), i.e., $N_\delta(a,b)$ can be chosen conveniently.

Definition 3

Let R be the domain of the function $f(x,y)$ and $(a,b) \in R$ then $f(x,y)$ be continuous at the point (a,b) if $\lim\limits_{(h,k)\to(0,0)} f(a+h, b+k) = f(a,b)$ *where $(a+h, b+k) \in R$.*

Note 1.11.1 *The definition of continuity of a function f at a point (a,b) requires that f is not only defined at (a,b) but also defined in a certain neighbourhood of the point (a,b).*

Example 1.11.1 *Examine the continuity of the following function at* $(0,0)$

$$f(x,y) = \frac{3xy^2}{x^3 + y^3} \quad for \quad (x,y) \neq (0,0)$$

$$= 0 \quad for \quad (x,y) = (0,0).$$

Solution: For continuity, we are to show that $\lim\limits_{(x,y)\to(0,0)} f(x,y) = f(0,0)$. Now

for the simultaneous limit $\lim\limits_{(x,y)\to(0,0)} f(x,y) = \lim\limits_{(x,y)\to(0,0)} \frac{3xy^2}{x^3 + y^3}$, let us use the

approach to the point $(0,0)$ through the straight line $x = my$, then

$$\lim\limits_{(x,y)\to(0,0)} \frac{3xy^2}{x^3 + y^3} = \lim\limits_{y\to 0} \frac{3my.y^2}{(my)^3 + y^3} = \frac{3m}{m^3 + 1}$$

which gives different values for different values of m. Hence the simultaneous limit does not exist and consequently the function is not continuous at $(0,0)$. □

Example 1.11.2 *Show that the following function is continuous at* $(0,0)$

$$f(x,y) = \frac{x^3 y^3}{x^2 + y^2} \quad for \quad (x,y) \neq (0,0)$$

$$= 0 \quad for \quad (x,y) = (0,0).$$

Solution: Let $\epsilon > 0$ be chosen in advance. Also let $x = r\cos\theta, y = r\sin\theta$.

$$\text{Now} \quad |f(x,y) - f(0,0)| = \left| \frac{x^3 y^3}{x^2 + y^2} - 0 \right| = \left| \frac{(r\cos\theta)^3 (r\sin\theta)^3}{(r\cos\theta)^2 + (r\sin\theta)^2} \right|$$

$$= |r^4 \sin^3\theta \cos^3\theta| = \left| \frac{r^4}{2^3}(2\sin\theta\cos\theta)^3 \right| = \left| \frac{r^4}{8}\sin^3 2\theta \right| \leq \frac{r^4}{8}$$

$$= \frac{(x^2 + y^2)^2}{8} < \epsilon \text{ (say), whenever } x^2 + y^2 < \sqrt{8\epsilon}.$$

If we consider $\sqrt{8\epsilon} = \delta^2$ then we get

$$|f(x,y) - f(0,0)| < \epsilon \quad \text{whenever } x^2 + y^2 < \delta^2.$$

Hence $f(x,y)$ is continuous at $(0,0)$. □

Remark: If a function of more than one variable is continuous at a point then it is continuous at that point when considered as a function of a single variable. This may be stated for functions of two variables in the following theorem:

Theorem 1.11.1 *If $f(x,y)$ be continuous at the point (a,b) then it will be continuous at (a,k) and (k,b) where k is any arbitrary real value.*

Proof: Let $\epsilon > 0$ be given. Since $f(x,y)$ is continuous at (a,b), there exists $\delta > 0$ such that

$$|f(x,y) - f(a,b)| < \epsilon \quad \text{whenever } (x,y) \in N_\delta(a,b) \tag{1.3}$$

To show that $f(x,y)$ is continuous at any arbitrary point (a,k) we consider the single variable function $g(y) = f(a,y)$ by keeping $x = a$ fixed in $f(x,y)$.

Now from the relation (1.3) we can write

$$|f(a, y) - f(a, b)| < \epsilon \quad \text{whenever} \quad (a, y) \in N_\delta(a, b)$$

$$\text{or,} \quad |g(y) - g(b)| < \epsilon \quad \text{whenever} \quad |y - b| < \delta$$

which means that the function $g(y) = f(a, y)$ is continuous at $y = b$.

In a similar way, keeping $y = b$ fixed and considering the single variable function $h(x) = f(x, b)$ we can show that $f(x, y)$ will be continuous at any arbitrary point (k, b) where k is any arbitrary real value.   □

Note 1.11.2 *The converse of the above theorem is not necessarily true, i.e., there may exist a function $f(x, y)$ which is not continuous at (a, b) but is continuous at the point (a, k) or (k, b), where k is a constant, specified to either of the variable x or y. As an illustration, we have the following example:*

Example 1.11.3 *Test the continuity at $(0, 0)$ of*

$$f(x, y) \quad = \quad \frac{xy^3}{x^2 + y^6} \quad for \quad (x, y) \neq (0, 0)$$

$$= \quad 0 \quad for \quad (x, y) = (0, 0).$$

The above function is not continuous at $(0, 0)$, for $\lim\limits_{(x,y) \to (0,0)} f(x, y)$ does not exist. [*vide.* illustrated **Example 1.7.1**]

But $\lim\limits_{x \to 0} f(x, 0) = 0 = f(0, 0)$ and $\lim\limits_{y \to 0} f(0, y) = 0 = f(0, 0)$, showing that f is continuous at $(0, 0)$ when considered as a function of a single variable x or that of y.

Example 1.11.4 *Examine whether the function:*

$$f(x, y) = \begin{cases} x^2 + 4y & when & (x, y) \neq (1, 2) \\ 0 & when & (x, y) = (1, 2). \end{cases}$$

(i) has a double limit as $x \to 1$ and $y \to 2$, (ii) is continuous at $(1, 2)$.

Solution: (i) $\lim\limits_{\substack{x \to 1 \\ y \to 2}} (x^2 + 4y) = 1^2 + 4.2 = 1 + 8 = 9$. Hence the double limit exists

and equal to 9.

(ii) Since $f(1, 2) = 0$ (given) and $\lim\limits_{\substack{x \to 1 \\ y \to 2}} (x^2 + 4y) = 9$.

So, $\lim\limits_{\substack{x \to 1 \\ y \to 2}} f(x, y) \neq f(1, 2)$. Hence the function is not continuous at $(1, 2)$.   □

Observation: The above function has a *point of discontinuity* at $(1, 2)$ which is the *only* point of discontinuity. Let us redefine the function as follows:

$$f(x, y) = \begin{cases} x^2 + 4y & when & (x, y) \neq (1, 2) \\ 9 & when & (x, y) = (1, 2). \end{cases}$$

Now the discontinuity of the function *has been removed* because in such situation

$$\lim\limits_{(x,y) \to (1,2)} (x^2 + 4y) = 9 = [f(x, y)]_{(1,2)}$$

i.e., $\lim\limits_{(x,y) \to (1,2)} f(x, y) = f(1, 2).$

So, it is possible to redefine the value of the function the point of discontinuity so that the new function is continuous. We say, in this case that the point of discontinuity is a **removable discontinuity** of the original function.

1.12 Some Properties of Continuous Functions of Several Variables

Property 1: *The sum, difference and product of two functions, each continuous at a given point or in a given region is continuous at that point or in that region.*

Note 1.12.1 *The result may be extended to any finite number of functions.*

Property 2: *The quotient of two continuous functions, both of which are continuous at a point or in a given region is continuous at that point or in that region, except at those points where the denominators vanish.*

Proof: Proof of **Property 1** and **Property 2** are exactly similar as in the case of single variable functions. □

Note 1.12.2 *If one of the two functions is continuous at a point and other is not, then their sum, difference and product will be discontinuous thereat and behaves like the later function.*

Property 3: *If $f(x, y)$ is continuous at (a, b) and $f(a, b) \neq 0$, then in the neighbourhood of the point (a, b), $f(x, y)$ has the same sign as that of $f(a, b)$ i.e., we get a positive number δ such that $f(x, y)$ preserves the same sign as that of $f(a, b)$ whenever $0 < |x - a| \leq \delta$ and $0 < |y - b| \leq \delta$.*

Proof: Since $f(x, y)$ is continuous in (a, b), corresponding to any pre-assigned positive quantity ϵ, however small, there exists a positive number δ, such that

$$|f(x, y) - f(a, b)| < \epsilon \text{ whenever } 0 < |x - a| \leq \delta \text{ and } 0 < |y - b| \leq \delta$$

i.e., $\quad f(a, b) - \epsilon < f(x, y) < f(a, b) + \epsilon$

$$\text{whenever } 0 < |x - a| \leq \delta \text{ and } 0 < |y - b| \leq \delta \quad (1.4)$$

Now two cases may arise.

Case 1: Let $f(a, b) > 0$.
If we now choose ϵ such that $\epsilon < f(a, b)$ then $f(a, b) - \epsilon = +ve$.
Also $f(a, b) + \epsilon$ is always positive. So form (1.4) we find that for every point (x, y) in the square domain $R'(a - \delta \leq x \leq a + \delta, b - \delta \leq y \leq b + \delta)$, the function $f(x, y)$ lies between two positive quantities and so itself is positive. Therefore $f(x, y)$ has the same sign as $f(a, b)$.

Case 2: Let $f(a, b) < 0$ then $-f(a, b) > 0$.
Let us choose $\epsilon < -f(a, b)$, then $f(a, b) + \epsilon < 0$.
Also $f(a, b) - \epsilon < 0$ [∵ $f(a, b) < 0$]. So form (1.4) we find that for every pair (x, y) in the square domain R', the function $f(x, y)$ lies between two negative quantities and so itself is negative. Therefore $f(x, y)$ has the same sign as $f(a, b)$.
Hence the proposition. □

Property 4: *A continuous function of one or more continuous function is again a continuous function.*

To be specific, let $u = g(x, y)$, $v = h(x, y)$ be two continuous functions at (x_0, y_0) and let $u_0 = g(x_0, y_0)$, $v_0 = h(x_0, y_0)$. If now $z = f(u, v)$ be continuous in (u, v) and (u_0, v_0), then $z = f(g(x_0, y_0), h(x_0, y_0))$ is also continuous in x, y at (x_0, y_0).

1.13   Uniform Continuity

A function $f : R \to \mathbb{R}^2$ is continuous at any point $(a,b) \in R$ if for any $\epsilon > 0$, $\exists$ a $\delta > 0$ such that $|f(x,y) - f(a,b)| < \epsilon$ when $(x,y) \in N_\delta(a,b)$. This δ, in general, depends not only on the pre-defined ϵ but also on the choice of the point (a,b). However, if it is possible to find a more general δ which works for arbitrary points of the domain of the function, i.e., δ depends only on the pre-defined ϵ but not on the choice of the point, then f is said to be uniformly continuous on R.

Thus, continuity of a function at any particular point is a local property, whereas uniform continuity in a domain is a global property of the function.

Definition 1.13.1 *A function $f(x,y)$ defined in the domain R, is said to be uniformly continuous in that domain, if corresponding to any pre-assigned $\epsilon > 0$, however small, there exists another number $\delta > 0$ such that $|f(x_1,y_1) - f(x_2,y_2)| < \epsilon$, where $(x_1,y_1), (x_2,y_2) \in R$ for which $\sqrt{(x_2 - x_1)^2 + (y_2 - y_1)^2} \leq \delta$.*

1.13.1   Some important properties of continuous functions defined over a closed domain

Below we state some important properties of continuous functions defined over a closed domain without their rigorous proof. These are mainly extensions of the properties on continuity for functions of single variable to functions of several variables.

Property 1: *A function continuous in a closed domain is bounded and attains its bounds at least once in the domain.*

Property 2: *If $f(x,y)$ is continuous in a closed domain then it must assume at least once every value between its upper and lower bounds.*

Property 3: *If $f(x,y)$ is a uniformly continuous function in a domain then it must be a simply continuous function in that domain.*

A function $f(x,y)$ is continuous in x and y simultaneously if it must have the same limiting value by all possible approaches to the point in consideration. Therefore, we can conclude that a necessary and sufficient condition for continuity is that the function is not only continuous in every direction but also it will be uniformly continuous in all directions.

For if we put $x = x_1 + r\cos\theta$, $y = y_1 + r\sin\theta$, we have from the definition of continuity

$$|f(x_1 + r\cos\theta, y_1 + r\sin\theta) - f(x_1,y_1)| < \epsilon \text{ for any } \epsilon > 0$$

which must hold for all values of r less than some number r_0 which is θ independent. This is equivalent to saying that the transformed function must be uniformly continuous in r, for all values of $|\theta| \leq 2\pi$.

However, the converse of the above property is not always true, i.e., there exists continuous functions in a domain which is not uniformly continuous in that domain. The following property is important in this context.

Property 4: *If a function of two variables is continuous at every point of a closed domain, then it is uniformly continuous in that domain.*

1.14 Miscellaneous Illustrative Examples

Example 1.14.1 *Show that* $S = \{(x, y) | x \text{ is rational and } y \text{ is irrational}\}$ *is neither open nor closed set.*

Solution: Let $(\alpha, \beta) \in S$ be any arbitrary point such that α is rational and β is irrational and $N_\delta(\alpha, \beta)$ be any neighbourhood of (α, β). Now, $N_\delta(\alpha, \beta)$ contains points whose both coordinates are rationals or both irrationals and obviously such points do not belong to S. Therefore $N_\delta(\alpha, \beta)$ is not a subset of S and consequently (α, β) is not an interior point of S.

Hence S is not an open set.

Let $(a, b) \in \mathbb{R}^2$ be such that either both a and b are rationals or both are irrationals. Then any neighbourhood of (a, b) contains infinitely many points of S. So, (a, b) is a limit point of the set S, but $(a, b) \notin S$. Therefore S is not a closed set. □

Example 1.14.2 *Let* $S = \{(x, y) | x^2 + y^2 < 1\}$

and $N = \left\{(x, y) | -\dfrac{1}{2} < x \text{ and } y < \dfrac{1}{2}\right\}$. *Is* N

an open neighbourhood of $(0, 0)$ *lying in* S? *Justify your answer.*

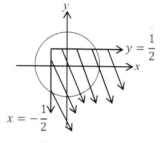

Figure 1.5

Solution: Here S is a set of all interior points of a circle with unit radius and N is a set of points bounded by the straight lines $x = -\dfrac{1}{2}$ and $y = \dfrac{1}{2}$. Clearly N is an unbounded open set. So, it can be considered as an open neighbourhood of the point $(0, 0)$, but it is not lying in S for $(0, -3) \in N$ but $(0, -3) \notin S$. □

Example 1.14.3 *Find the domain of definition of the following functions:*

(i) $f(x, y) = \dfrac{2x + 5y}{\sqrt{15 - x^2 - y^2}}$

(ii) $f(x, y) = \log_e(16 - x^2 - y^2) + \log_e(x^2 + y^2 - 1)$

(iii) $f(x, y) = \sin^{-1}(x^2 + y^2)$.

Solution: (i) $f(x, y)$ is defined if $15 - x^2 - y^2 > 0$ or $x^2 + y^2 < 15$. Hence the domain of definition is the set $\{(x, y) \in \mathbb{R}^2 : x^2 + y^2 < 15\}$ which is an open circular disc centered at origin with radius $\sqrt{15}$.

(ii) Here $f(x, y)$ is defined if $16 - x^2 - y^2 > 0$ and $x^2 + y^2 - 1 > 0$, i.e., the domain of definition is the set $\{(x, y) \in \mathbb{R}^2 : 1 < x^2 + y^2 < 16\}$ which is an annular domain between two open circular discs centered at origin and radii 1 and 4.

(iii) For the function $f(x, y) = \sin^{-1}(x^2 + y^2)$ the inverse circular function is defined if $-1 \leq (x^2 + y^2) \leq 1$. Hence the domain in this case is $\{(x, y) \in \mathbb{R}^2 : 0 < x^2 + y^2 \leq 1\}$ because $-1 \leq x^2 + y^2 < 0$ is inadmissible here. □

Example 1.14.4 *Show that*

(i) $\lim\limits_{\substack{x \to 0 \\ y \to 0}} \left(\dfrac{1}{|x|} + \dfrac{1}{|y|}\right) = \infty$

(ii) $\lim\limits_{\substack{x \to 0 \\ y \to 0}} \dfrac{\sin(x^2 y + xy^2)}{xy} = 0$

$(iii) \lim\limits_{\substack{x\to 2 \\ y\to 1}} \dfrac{\sin^{-1}(xy-2)}{\tan^{-1}(3xy-6)} = \dfrac{1}{3}$ $(iv) \lim\limits_{\substack{x\to 4 \\ y\to \pi}} x^2 \sin\dfrac{y}{x} = 8\sqrt{2}$

$(v) \lim\limits_{\substack{x\to 0 \\ y\to 1}} \exp\left\{-\dfrac{1}{x^2(y-1)^2}\right\} = 0$ $(vi) \lim\limits_{\substack{x\to 0 \\ y\to 1}} \dfrac{x+y-1}{\sqrt{x}-\sqrt{1-y}} = 0$

$$(x>0,\ y<1)$$

Solution: (i) Let $G > 0$ be a sufficiently large number. For $(x,y) \neq (0,0)$

$$\left|\frac{1}{|x|}+\frac{1}{|y|}\right| > G \text{ whenever } \frac{1}{|x|} > \frac{G}{2} \text{ and } \frac{1}{|y|} > \frac{G}{2}$$

$$\text{i.e., whenever } |x| < \frac{2}{G} \text{ and } |y| < \frac{2}{G}.$$

Let $\delta = \dfrac{2}{G}$. Then for sufficiently large $G > 0$, $\exists$ sufficiently small $\delta > 0$ such that

$$\left|\frac{1}{|x|}+\frac{1}{|y|}\right| > G \text{ whenever } 0 < |x-0| < \delta \text{ and } 0 < |y-0| < \delta.$$

Hence $\lim\limits_{\substack{x\to 0 \\ y\to 0}} \left(\dfrac{1}{|x|}+\dfrac{1}{|y|}\right) = \infty$.

(ii) When $(x,y) \to (0,0)$ then $(x^2y+xy^2) \to 0$

$$\therefore \quad \lim\limits_{\substack{x\to 0 \\ y\to 0}} \frac{\sin(x^2y+xy^2)}{xy} = \lim\limits_{\substack{x\to 0 \\ y\to 0}} \left[\frac{\sin(x^2y+xy^2)}{xy(x+y)} \times (x+y)\right]$$

$$= \lim\limits_{(x^2y+xy^2)\to 0} \frac{\sin(x^2y+xy^2)}{x^2y+xy^2} \times \lim\limits_{\substack{x\to 0 \\ y\to 0}} (x+y) = 1 \times 0 = 0.$$

(iii) Put $t = xy - 2$. Then $(x,y) \to (2,1) \ \Rightarrow \ t \to 0$

$$\therefore \quad \lim\limits_{\substack{x\to 2 \\ y\to 1}} \frac{\sin^{-1}(xy-2)}{\tan^{-1}3(xy-2)} = \lim\limits_{t\to 0} \frac{\sin^{-1}t}{\tan^{-1}3t} \left[\frac{0}{0} - \text{form}\right] = \lim\limits_{t\to 0} \frac{\frac{1}{\sqrt{1-t^2}}}{\frac{3}{1+9t^2}} = \frac{1}{3}.$$

(iv) $\lim\limits_{\substack{x\to 4 \\ y\to \pi}} x^2 \sin\dfrac{y}{x} = 4^2 \sin\dfrac{\pi}{4} = 16 \times \dfrac{1}{\sqrt{2}} = 8\sqrt{2}.$

(v) Let $\epsilon > 0$. For $(x,y) \neq (0,1)$,

$$\left|\exp\left\{-\frac{1}{x^2(y-1)^2}\right\} - 0\right| < \epsilon \ \Rightarrow \ \exp\left\{\frac{1}{x^2(y-1)^2}\right\} > \frac{1}{\epsilon}$$

$$\Rightarrow \ \frac{1}{x^2(y-1)^2} > \log_e\left(\frac{1}{\epsilon}\right) \ \Rightarrow \ x^2(y-1)^2 < \frac{1}{\log_e\left(\frac{1}{\epsilon}\right)}$$

$$\Rightarrow \ |x||y-1| < \sqrt{1/\left[\log_e\left(\frac{1}{\epsilon}\right)\right]} - \text{ this is true whenever}$$

$$|x| < \sqrt[4]{1/\left[\log_e\left(\frac{1}{\epsilon}\right)\right]} \text{ and } |y-1| < \sqrt[4]{1/\left[\log_e\left(\frac{1}{\epsilon}\right)\right]} \text{ are true.}$$

Let $\quad \delta = \sqrt[4]{1 / \left[\log_e \left(\dfrac{1}{\epsilon} \right) \right]}\quad$ then we can write

$$\left| \exp\left\{ -\frac{1}{x^2(y-1)^2} \right\} - 0 \right| < \epsilon \text{ whenever } 0 < |x|, |y-1| < \delta.$$

Hence $\quad \lim\limits_{\substack{x \to 0 \\ y \to 1}} \exp\left\{ -\dfrac{1}{x^2(y-1)^2} \right\} = 0.$

(vi) $\lim\limits_{\substack{x \to 0 \\ y \to 1}} \dfrac{x+y-1}{\sqrt{x} - \sqrt{1-y}} = \lim\limits_{\substack{x \to 0 \\ y \to 1}} \dfrac{(x+y-1)(\sqrt{x} + \sqrt{1-y})}{x - (1-y)}$

$= \lim\limits_{\substack{x \to 0 \\ y \to 1}} (\sqrt{x} + \sqrt{1-y}) = 0 \text{ [As } x > 0, \ y < 1].$ $\qquad\square$

Example 1.14.5 *Show that the function*

$$f(x,y) = \begin{cases} xy\dfrac{x^2 - y^2}{x^2 + y^2} & when \quad x^2 + y^2 \neq 0 \\ 0 & when \quad x^2 + y^2 = 0 \end{cases}$$

is continuous at $(0,0)$.

Solution: Let $\epsilon > 0$ be given arbitrarily. Also take $x = r\cos\theta$ and $y = r\sin\theta$, then

$$|f(x,y) - f(0,0)| = \left| xy\frac{x^2 - y^2}{x^2 + y^2} - 0 \right| = r^2|\sin\theta\cos\theta(\cos^2\theta - \sin^2\theta)|$$

$$= \frac{r^2}{2}|\sin 2\theta \cos 2\theta| = \frac{r^2}{4}|\sin 4\theta| \leq \frac{r^2}{4} = \frac{x^2 + y^2}{4} < \epsilon \ \text{(say)}.$$

It is true whenever $x^2 + y^2 < 4\epsilon$, i.e., $|x| < \sqrt{2\epsilon}$, $|y| < \sqrt{2\epsilon}$.

Let $\delta = \sqrt{2\epsilon}$. Then $|f(x,y) - f(0,0)| < \epsilon$ whenever $|x| < \delta, |y| < \delta$. Hence the function is continuous at the origin. $\qquad\square$

Example 1.14.6 *Examine the continuity of the function* $f(x,y) = \sqrt{|xy|}$ *at* $(0,0)$.

Solution: Let $\epsilon > 0$ be chosen arbitrarily in advance. Then $|f(x,y) - f(0,0)| = |\sqrt{|xy|} - 0| < \epsilon \Rightarrow |xy| < \epsilon^2$ which is true whenever $|x| < \epsilon$ and $|y| < \epsilon$. Considering $\delta = \epsilon$ we can write $|\sqrt{|xy|} - 0| < \epsilon$ whenver $|x| < \delta$ and $|y| < \delta$. Hence $f(x,y) = \sqrt{|xy|}$ is continuous at $(0,0)$. $\qquad\square$

Example 1.14.7 *Find* $\lim\limits_{\substack{x \to 0 \\ y \to 0}} \dfrac{x^3 + y^3}{x^2 + y}$, *if it exists.*

Solution: Let us consider the approach to $(0,0)$ along the line $y = x$, then

$$\lim\limits_{\substack{x \to 0 \\ y \to 0}} \frac{x^3 + y^3}{x^2 + y} = \lim\limits_{x \to 0} \frac{x^3 + x^3}{x^2 + x} = \lim\limits_{x \to 0} \frac{2x^2}{x+1} = 0.$$

Now we take another approach to the point $(0,0)$ along the curve $y = -x^2 e^x$, then

$$\lim_{\substack{x \to 0 \\ y \to 0}} \frac{x^3 + y^3}{x^2 + y} = \lim_{x \to 0} \frac{x^3 + (-x^2 e^x)^3}{x^2 - x^2 e^x} = \lim_{x \to 0} \frac{x^3 - x^6 e^{3x}}{x^2 - x^2 e^x}$$

$$= \lim_{x \to 0} \frac{x}{1 - e^x} \cdot \frac{x^4 e^{3x}}{} \quad \left[\frac{0}{0} - form\right] = \lim_{x \to 0} \frac{1 - (4x^3 e^{3x} + x^4 \cdot 3 e^{3x})}{-e^x}$$

$$= -1 \qquad\qquad\qquad\qquad\qquad \text{[Using L' Hospital Rule]}$$

Since two approaches give two different limits, so the given double limit does not exist. □

Example 1.14.8 *Show that the function*

$$f(x,y) = \begin{cases} \dfrac{xy}{\sqrt{x^2 + y^2}} & for \quad x \neq 0, y \neq 0 \\ 0 & for \quad (x,y) = (0,0). \end{cases}$$

is continuous at the origin in (x,y) together.

Solution: Let $\epsilon > 0$ be chosen arbitrarily. Put $x = r\cos\theta$, $y = r\sin\theta$, then

$$|f(r\cos\theta, r\sin\theta) - f(0,0)| = |f(r\cos\theta, r\sin\theta)| = \left| \frac{r^2 \sin\theta \cos\theta}{r\sqrt{\sin^2\theta + \cos^2\theta}} \right|$$

$$= r|\sin\theta \cos\theta| = \frac{1}{2}r|\sin 2\theta| \leq \frac{1}{2}r \ \forall \ \theta \ (\because \ |\sin 2\theta| \leq 1)$$

It follows that if we put $r_0 = 2\epsilon$, then any ϵ be chosen (however small it may be), $|f(r\cos\theta, r\sin\theta)|$ be made always less than $\epsilon \ \forall \ \theta$ and for $r < r_0$, i.e., for all paths within a circle with centre at the origin and radius $r_0 = 2\epsilon$. The transformed expression is therefore uniformly continuous in r for all values of θ. Hence the criterion set forth in the explanation of **Property 3** of § **1.13.1** is satisfied and $f(x,y)$ is continuous in (x,y) together at the origin.

An Alternative Approach:

Let $\epsilon > 0$. Put $x = r\cos\theta$, $y = r\sin\theta$, then

$$|f(x,y) - f(0,0)| = \left| \frac{xy}{\sqrt{x^2 + y^2}} \right| = \left| \frac{r^2 \sin\theta \cos\theta}{r\sqrt{\sin^2\theta + \cos^2\theta}} \right|$$

$$= r|\sin\theta \cos\theta| = \frac{1}{2}r|\sin 2\theta| \leq \frac{1}{2}r = \frac{\sqrt{x^2 + y^2}}{2} < \epsilon \text{ if } \frac{x^2 + y^2}{4} < \epsilon^2$$

$$\text{i.e., if} \quad x^2 < \frac{\epsilon^2}{2}, \ y^2 < \frac{\epsilon^2}{2} \quad \text{i.e., if } |x| < \frac{\epsilon}{\sqrt{2}}, \ |y| < \frac{\epsilon}{\sqrt{2}}$$

$$\text{Thus} \quad \left| \frac{xy}{\sqrt{x^2 + y^2}} - 0 \right| < \epsilon \text{ whenever } |x| < \frac{\epsilon}{\sqrt{2}}, \ |y| < \frac{\epsilon}{\sqrt{2}}$$

$$\Rightarrow \quad \lim_{(x,y) \to (0,0)} f(x,y) = 0 = f(0,0)$$

asserting that $f(x,y)$ is continuous at $(0,0)$. □

Example 1.14.9 *Discuss the continuity of the following function*

$$f(x,y) = \begin{cases} 0 & for \quad (x,y) = (2y,y) \\ \exp\left\{\dfrac{|x-2y|}{x^2-4xy+4y^2}\right\} & for \quad (x,y) \neq (2y,y). \end{cases}$$

Solution: Let $\epsilon > 0$ be an arbitrary sufficiently small number given in advance. For $(x,y) \to (2y,y)$ and for arbitrary y we get

$$|f(x,y) - f(2y,y)| = \left|\exp\left\{\frac{|x-2y|}{x^2-4xy+4y^2}\right\} - 0\right| = \exp\left\{\frac{|x-2y|}{|x-2y|^2}\right\}$$

$$= \exp\left\{\frac{1}{|x-2y|}\right\} < \epsilon \text{ if } \frac{1}{|x-2y|} < \log_e \epsilon \text{ or if } |x-2y| > \frac{1}{\log_e \epsilon}.$$

Now if we consider $G = \dfrac{1}{\log_e \epsilon}$ which will be a large positive for given ϵ then

$$|f(x,y) - f(2y,y)| < \epsilon \text{ whenever } |x-2y| > G$$

i.e., the function is discontinuous at such points. $\square$

1.15   Exercises

1. Define (i) neighbourhood of a point in plane, (ii) interior point and (iii) limit point of a set in $\mathbb{R}^2$ with proper examples.

2. Define an *open set* and a *closed set* in $\mathbb{R}^2$. Give examples in each case.

3. If $S = \{(x, y) : 0 < x < 1, 0 < y < 1\}$, show that S is an open set in $\mathbb{R} \times \mathbb{R}$. Mention one accumulation point of S which does not belong to S.

4. Show that a subset in $\mathbb{R}^2$ containing a finite number of points is a closed set.

[*Hints:* A set containing a finite number of points cannot have any limit point. Again, a set having no limit points must be closed. Hence proved.]

5. Determine whether each of the following sets is *open, closed* or *bounded*:

(i) $\{(x, y, z) \in \mathbb{R}^3 : x^2 + y^2 + z^2 = 1\}$ (ii) $\{(x, y) \in \mathbb{R}^2 : \dfrac{x^2}{2^2} + \dfrac{y^2}{3^2} < 5\}$

(iii) $\{(x, y) \in \mathbb{Q}^2\}$ (iv) $\{(x, y) : |x| \leq 1 \text{ and } |y| \leq 1\}$.

6. Find the domain of definition of the following functions:

(i) $f(x, y) = \sqrt{|x| + |y| - 5}$ (ii) $f(x, y) = \dfrac{1}{\sqrt{x + y - k}\sqrt{x - y + k'}}$

(iii) $f(x, y) = xy \sin \dfrac{1}{x^2} + y \cos \dfrac{1}{x}$.

7. (a) Let $\lim\limits_{(x,y) \to (a,b)} f(x, y) = l$ and $y = \phi(x)$ be any function such that $\lim\limits_{x \to a} \phi(x) = b$, show that $\lim\limits_{x \to a} f(x, \phi(x)) = l$.

(b) If $f(x, y)$ is a real valued function continuous at (a, b) and functions g and h are defined by $g(x) = f(x, b)$ and $h(x) = f(a, x)$. Show that g and h are continuous at $x = a$ and $x = b$ respectively.

[*Hints:* (a) $\lim\limits_{x \to a} \phi(x) = b$ $\therefore$ $\lim\limits_{x \to a} y = b$, i.e., $x \to a \Rightarrow y \to b$

$\therefore$ $\lim\limits_{x \to a} f(x, \phi(x)) = \lim\limits_{\substack{x \to a \\ y \to b}} f(x, y) = \lim\limits_{(x,y) \to (a,b)} f(x, y) = l$.

(b) Since $f(x, y)$ is a real valued continuous functions at (a, b), so $f(x, b)$ is continuous at $x = a$ $\therefore$ $g(x) = f(x, b)$ is continuous at $x = a$ and similarly $h(x) = f(a, x)$ is continuous at $x = b$.]

8. Let $f(x, y)$ be continuous at an interior point (a, b) of domain of definition of f and $f(a, b) \neq 0$. Show that there exists a neighbourhood of the point (a, b) in which $f(x, y)$ retains the same sign as that of $f(a, b)$.

[*Hints:* See **Property 3** of § **1.12**]

9. When is a point (a, b) said to be an accumulation point of a set S of points in $\mathbb{R}^2$? If f is a real valued function defined on S and if (a, b) is an accumulation point of S, when is it said that $f(x, y)$ tends to l as $(x, y) \to (a, b)$?

Find l if $(a, b) = (0, 0)$ and $f(x, y) = xy \dfrac{x^2 - y^2}{x^2 + y^2}$ for $(x, y) \neq (0, 0)$.

10. A function f of two independent variables x and y is continuous at (a, b). Prove that $f(x, b)$ is continuous at $x = a$ and $f(a, y)$ is continuous at $y = b$. Is the converse true? Justify your answer.

11. Use (δ, ϵ)-definition to justify the following statements or limits:

(i) $\lim_{(x,y)\to(0,0)} (3x + 2y) = 0$

(ii) $\lim_{(x,y)\to(0,0)} xy\dfrac{x^2 - y^2}{x^2 + y^2} = 0$

(iii) $\lim_{\substack{x\to 0 \\ y\to 0}} \dfrac{1 - \cos(x^2 + y^2)}{(x^2 + y^2)^2} = \dfrac{1}{2}$

(iv) $\lim_{\substack{x\to 0 \\ y\to 0}} \left(y\sin\dfrac{1}{x} + x\sin\dfrac{1}{y} \right) = 0$

(v) $\lim_{(x,y)\to(0,0)} \dfrac{x^3 - y^3}{x^2 + y^2} = 0$

(vi) $\lim_{(x,y)\to(0,0)} \dfrac{xy}{\sqrt{x^2 + y^2}} = 0$

(vii) $\lim_{(x,y)\to(0,0)} \dfrac{x^4 + y^4}{x^2 + y^2} = 0$

(viii) $\lim_{(x,y)\to(-1,-1)} (xy - 2x) = 1$

(ix) $\lim_{\substack{x\to 0 \\ y\to 0}} (1 + y^2)\dfrac{\sin x}{x} = 1$

(x) $\lim_{\substack{x\to 0 \\ y\to 0}} e^{-\frac{1}{x^2+y^2}} = 0$

(xi) $\lim_{\substack{x\to 0 \\ y\to 0}} e^{-(x^2+y^2)} = 0.$

12. Show that the following limits do not exist:

(i) $\lim_{(x,y)\to(0,0)} \dfrac{2xy^2}{x^2 + y^4}$

(ii) $\lim_{(x,y)\to(0,0)} \dfrac{x^2 - y^2}{x^2 + y^2}$

(iii) $\lim_{\substack{x\to 0 \\ y\to 0}} \dfrac{xy}{x^2 + y^2}$

(iv) $\lim_{\substack{x\to 0 \\ y\to 0}} \dfrac{xy^3}{x^2 + y^6}$

(v) $\lim_{(x,y)\to(0,0)} \dfrac{y + (x + y)^2}{y - (x + y)^2}$

(vi) $\lim_{\substack{x\to 0 \\ y\to 0}} \dfrac{x^2}{x^2 + y^2 - x}$

(vii) $\lim_{\substack{x\to 0 \\ y\to 0}} \dfrac{x^3 + y^3}{x - y}$

(viii) $\lim_{\substack{x\to 0 \\ y\to 0}} \dfrac{x^2 y}{x^4 + y^2}$

(ix) $\lim_{\substack{x\to 0 \\ y\to 0}} \dfrac{x^2 y^2}{x^2 y^2 + (x^2 - y^2)^2}$

(x) $\lim_{(x,y)\to(0,0)} (x + y)\dfrac{y + (x + y)^2}{y - (x + y)^2}.$

[*Hints:* (i) Put $x = my^2$, (iv) Put $x = my^3$, (v) Put $y = 0$ and $y = x$ separately, (vii) Put $y = x - mx^3$, (ix) Take the approach $(x, y) \to (0, 0)$ along $y = 0$ and $y = x$ separately, (x) Put $y = 0$ and $y = x^2$ separately.]

13. Verify that $\lim_{\substack{x\to 0 \\ y\to 0}} f(x, y) = 0$, where

(i) $f(x, y) = x\sin\dfrac{1}{x} + y\sin\dfrac{1}{y}, \quad x > 0, \ y > 0$

(ii) $f(x, y) = (x + y)\sin\dfrac{1}{x}\sin\dfrac{1}{y}, \quad x > 0, \ y > 0.$

14. Given $f(x, y) = \dfrac{x^2 y^2}{x^2 y^2 + (x^2 - y^2)^2}$. Show that the two repeated limits $\lim_{x\to 0}\lim_{y\to 0} f(x, y)$ and $\lim_{y\to 0}\lim_{x\to 0} f(x, y)$ both exist and are equal, but the simultaneous limit $\lim_{(x,y)\to(0,0)} f(x, y)$ does not exists.

15. Discuss the existence of the repeated limits of the forgoing examples **13.** (i) and (ii).

16. If $f(x, y) = \begin{cases} 1 & \text{for} \quad xy \neq 0 \\ 0 & \text{for} \quad xy = 0. \end{cases}$ Show that $f(x, y)$ is not continuous at $(0, 0)$ but $f(x, 0)$ is continuous for all x.

17. If $f(x, y)$ is continuous at the point (x, y), then show that $f(x, y_0)$ is continuous at $x = x_0$.

18. If $f(x, y) = \dfrac{y - x^2}{y + x^2}$ for $(x, y) \neq (0, 0)$ examine whether

$$\lim_{x \to 0} \lim_{y \to 0} f(x, y) = \lim_{y \to 0} \lim_{x \to 0} f(x, y).$$

19. Is $f(x, y) = \begin{cases} (px + qy) \sin\left(\dfrac{x}{y}\right) & \text{for} \quad y \neq 0 \\ 0 & \text{for} \quad y - 0 \end{cases}$ continuous at $(0, 0)$?

20. If $f(x, y)$ is continuous at a point (a, b) in its domain of definition and if $f(a, b) > 0$ then show that $f(x, y) > 0$ in some neighbourhood of (a, b) and also show that $f(x, y)$ is bounded in some neighbourhood of (a, b).

21. Let $\lim_{(x,y) \to (0,0)} f(x, y) = A \ (\in \mathbb{R})$. Prove that

$$\lim_{x \to 0} \lim_{y \to 0} f(x, y) = A = \lim_{y \to 0} \lim_{x \to 0} f(x, y),$$

if $\lim_{x \to a} f(x, y)$ exists for each y in a neighbourhood of b and $\lim_{y \to b} f(x, y)$ exists for each x in a neighbourhood of a.

22. Verify that $f(x, y) = \begin{cases} \dfrac{x^2 y^2}{x^2 + y^2 + (x - y)^2} & \text{for} \quad (x, y) \neq (0, 0) \\ 0 & \text{for} \quad (x, y) \neq (0, 0), \end{cases}$ is not

continuous at $(0, 0)$

23. If $f(x, y) = \begin{cases} 1 & \text{for} \quad xy \neq 0 \\ 0 & \text{for} \quad xy = 0 \end{cases}$ then show that $f(x, 0)$ and $f(0, y)$ are

continuous functions of x and y respectively and examine the continuity of f at $(0, 0)$ and at $(1, 0)$.

24. Check the continuity of the function $f(x, y)$ at $(0, 0)$

$$f(x, y) = \begin{cases} xy \log(x^2 + y^2 + 1) & \text{for} \quad x^2 + y^2 \neq 0 \\ 0 & \text{for} \quad x^2 + y^2 = 0. \end{cases}$$

25. Let $f(x, y) = \begin{cases} 3xy & \text{when} \quad (x, y) \neq (2, 3) \\ 6 & \text{when} \quad (x, y) = (0, 0). \end{cases}$ Is f continuous at $(2, 3)$?

If discontinuous, can we remove the discontinuity? How?

26. (a) If $f(x, y) = \dfrac{x^2(1 - y^2)}{y^2 + x^2}$, $(x, y) \neq (0, 0)$ then show that

$$\lim_{x \to 0} \lim_{y \to 0} f(x, y) \neq \lim_{y \to 0} \lim_{x \to 0} f(x, y).$$

(b) If $f(x, y) = \begin{cases} \dfrac{2xy^2}{\sqrt{x^3 + y^3}} & \text{for} \quad (x, y) \neq (0, 0) \\ 0 & \text{for} \quad (x, y) = (0, 0) \end{cases}$ then examine that f is

continuous or not at the point $(0, 0)$.

(c) If $f(x, y) = \begin{cases} \dfrac{x^3 - y^3}{x^2 + y^2} & \text{for} \quad x^2 + y^2 \neq 0 \\ 0 & \text{for} \quad x^2 + y^2 = 0 \end{cases}$ then show that f is continuous

at $(0, 0)$.

27. (a) If $f(x, y) = \begin{cases} x & \text{if } xy \text{ is rational} \\ y & \text{if } xy \text{ is irrational}, \end{cases}$ show that $\lim_{x \to 0} f(x, y)$ does not

exist.

(b) Show that the function f defined on $\mathbb{R}$ by $f(x, y) = \begin{cases} x & \text{if } xy \text{ is rational} \\ -x & \text{if } xy \text{ is irrational} \end{cases}$

is continuous at $x = 0$.

(c) Examine if $\lim\limits_{x\to 0}\lim\limits_{y\to 0} f(x,y) = \lim\limits_{y\to 0}\lim\limits_{x\to 0} f(x,y)$, where $f(x,y) = \dfrac{y+x^2}{y-x^2}$ for $(x,y) \neq (0,0)$.

28. (a) Let $f(x,y) = \begin{cases} y\sin\dfrac{1}{x} + \dfrac{xy}{x^2+y^2} & \text{for} \quad x \neq 0 \\ 0 & \text{for} \quad x = 0. \end{cases}$ Examine the existence of the double limit and the repeated limits of this function at $(0,0)$.

(b) Let $f(x,y) = \begin{cases} x\sin y^{-1} + \dfrac{x^2-y^2}{x^2+y^2} & \text{for} \quad y \neq 0 \\ 0 & \text{for} \quad y = 0 \text{ but } x \neq 0. \end{cases}$ Show that $\lim\limits_{y\to 0}\lim\limits_{x\to 0} f(x,y)$ exists but neither $\lim\limits_{(x,y)\to(0,0)} f(x,y)$ nor $\lim\limits_{x\to 0}\lim\limits_{y\to 0} f(x,y)$ exists.

29. (a) Let $f(x,y) = \begin{cases} x + y\sin\dfrac{1}{x} & \text{for} \quad x \neq 0 \\ 0 & \text{for} \quad x = 0. \end{cases}$ Verify that the simultaneous limit $\lim\limits_{(x,y)\to(0,0)} f(x,y)$ and the repeated limit $\lim\limits_{x\to 0}\lim\limits_{y\to 0} f(x,y)$ exist but the repeated limit $\lim\limits_{y\to 0}\lim\limits_{x\to 0} f(x,y)$ does not exist.

(b) Let $f(x,y) = \begin{cases} x\sin\dfrac{1}{y} + y\sin\dfrac{1}{x} & \text{for} \quad xy \neq 0 \\ 0 & \text{for} \quad xy = 0. \end{cases}$ Show that at $(0,0)$, the double limit exists but the two repeated limits do not exist.

30. (a) Let $f(x,y) = \begin{cases} \dfrac{2xy}{x^2+y^2} & \text{for} \quad x^2+y^2 \neq 0 \\ 0 & \text{for} \quad x^2+y^2 = 0. \end{cases}$ Prove that f is a continuous function of either variable when the other variable is given a fixed value. Is the function continuous at $(0,0)$?

(b) Let $f(x,y) = \begin{cases} \dfrac{2xy}{(x^2+y^2)^n} & \text{for} \quad x^2+y^2 \neq 0 \\ 0 & \text{for} \quad x^2+y^2 = 0. \end{cases}$ Prove that f is a continuous at $(0,0)$ if $n = \dfrac{1}{2}$.

(c) Find a function $f(x,y)$ which is a function of x^2+y^2 and is also a product of the form $\psi(x)\psi(y)$.

[*Hints:* Let $\psi(x) = a^{x^2}$, then $\psi(y) = a^{y^2}$,

$$\therefore \quad \psi(x).\psi(y) = a^{x^2}.a^{y^2} = a^{x^2+y^2} = f(x^2+y^2).]$$

ANSWERS

5. (i) Not open, closed, bounded; (ii) Open, not closed, bounded; (iii) Neither open nor closed and unbounded; (iv) Not open, closed, bounded; **6.** (i) $\{(x, y) \in \mathbb{R} : |x| + |y| \geq 5\}$; (ii) $\{(x, y) \in \mathbb{R} : (x + y - k < 0$ and $x - y + k' < 0)||(x + y - k > 0$ and $x - y + k' > 0)\}$; (iii) $\{(x, y) \in \mathbb{R} : x \neq 0\}$; **9.** $l = 0$; **15.** (i) No repeated limits exist; (ii) No repeated limits exists; **19.** Yes; **23.** Discontinuous at $(0, 0)$ and continuous at $(1, 0)$; **24.** Continuous; **25.** Discontinuous. This can be removed by redefining $f(2, 3) = 18$; **26.** **(b)** Continuous; **27.** **(c)** Not equal; **28.** **(a)** $\lim_{x \to 0} \lim_{y \to 0} f(x, y)$ exists but $\lim_{(x,y) \to (0,0)} f(x, y)$ and $\lim_{y \to 0} \lim_{x \to 0} f(x, y)$ do not exist; **30.** **(a)** Not continuous, **(c)** $f(x, y) = a^{(x^2 + y^2)}$.

Chapter 2

Functions of Several Variables: Differentiation - I

2.1 Introduction

The details of multivariate functions, its different kinds of limits, continuities etc. have been discussed in **Chapter 1** along with few topological aspects of subsets of $\mathbb{R}^2$ and $\mathbb{R}^3$. In this chapter, we are going to discuss the analytical definitions of first and higher order partial derivatives, the total differentiations and directional derivatives etc. The inter relations among the limits, continuities and different kinds of derivatives of functions of several variables have also been discussed with numerous appropriate examples in analytical point of view.

2.2 Partial Derivatives

The partial derivatives of a function of several variables with respect to one of the independent variables is an ordinary derivative with respect to that particular variable and considering all other independent variables as constant. Thus

(i) If in a function $u = f(x, y, z, ...)$, the variables $y, z, ...$ are kept fixed, then

$$\lim_{\Delta x \to 0} \frac{f(x + \Delta x, y, z, ...) - f(x, y, z, ...)}{\Delta x}, \text{ if it exists, is called the } \textit{partial derivative}$$

of u with respect to x at the point $(x, y, z, ...)$. It is denoted by $\dfrac{\partial u}{\partial x}$, u_x, f_x etc.

Similarly, the partial derivatives with respect to $y, z, ...$ are defined which are denoted by $\dfrac{\partial u}{\partial y}$, $\dfrac{\partial u}{\partial z}$ ... etc respectively.

(ii) For the function $u = f(x, y)$ of the two independent variables x and y, the partial derivatives at (a, b) are given by

$$\left(\frac{\partial u}{\partial x}\right)_{(a,b)} = u_x(a, b) = f_x(a, b) = \lim_{h \to 0} \frac{f(a + h, b) - f(a, b)}{h}$$

$$\text{and} \quad \left(\frac{\partial u}{\partial y}\right)_{(a,b)} = u_y(a, b) = f_y(a, b) = \lim_{k \to 0} \frac{f(a, b + k) - f(a, b)}{k},$$

provided the limits exist.

2.2.1   Successive partial derivatives of higher order

Like ordinary derivatives, the arguments of partially derived functions are same and consequently those can further be differentiable partially with respect to the same arguments as above. Let $u = f(x, y)$. Then the following are the definitions of second order partial derivatives.

1.
$$\left[\frac{\partial^2 u}{\partial x^2}\right]_{(a,b)} = \left[\frac{\partial}{\partial x}\left(\frac{\partial u}{\partial x}\right)\right]_{(a,b)} = \frac{\partial}{\partial x}(f_x(a,b)) = f_{xx}(a,b)$$

$$= \lim_{h \to 0}\left[\frac{f_x(a+h,b) - f_x(a,b)}{h}\right]$$

$$= \lim_{h \to 0}\frac{f(a+2h,b) - 2f(a+h,b) + f(a,b)}{h^2}.$$

2.
$$\left[\frac{\partial^2 u}{\partial y^2}\right]_{(a,b)} = \left[\frac{\partial}{\partial y}\left(\frac{\partial u}{\partial y}\right)\right]_{(a,b)} = \frac{\partial}{\partial y}(f_y(a,b)) = f_{yy}(a,b)$$

$$= \lim_{k \to 0}\left[\frac{f_y(a,b+k) - f_y(a,b)}{k}\right]$$

$$= \lim_{k \to 0}\frac{f(a,b+2k) - 2f(a,b+k) + f(a,b)}{k^2}.$$

3.
$$\left[\frac{\partial^2 u}{\partial y \partial x}\right]_{(a,b)} = \left[\frac{\partial}{\partial y}\left(\frac{\partial u}{\partial x}\right)\right]_{(a,b)} = \frac{\partial}{\partial y}(f_x(a,b)) = f_{yx}(a,b)$$

$$= \lim_{k \to 0}\left[\frac{f_x(a,b+k) - f_x(a,b)}{k}\right]$$

$$= \lim_{k \to 0}\lim_{h \to 0}\frac{f(a+h,b+k) - f(a,b+k) - f(a+h,b) + f(a,b)}{hk}.$$

4.
$$\left[\frac{\partial^2 u}{\partial x \partial y}\right]_{(a,b)} = \left[\frac{\partial}{\partial x}\left(\frac{\partial u}{\partial y}\right)\right]_{(a,b)} = \frac{\partial}{\partial x}(f_y(a,b)) = f_{xy}(a,b)$$

$$= \lim_{h \to 0}\left[\frac{f_y(a+h,b) - f_y(a,b)}{h}\right]$$

$$= \lim_{h \to 0}\lim_{k \to 0}\frac{f(a+h,b+k) - f(a+h,b) - f(a,b+k) + f(a,b)}{hk}.$$

Note 2.2.1 *The values of $f_{xy}(a,b)$ and $f_{yx}(a,b)$ may not necessarily always be equal. The details of the existence and equality of these mixed partial derivatives will be discussed in § 2.8.*

2.3   Continuity and Partial Derivatives

In forming the partial derivatives of $f(x,y)$ at (a,b), we vary x and y along the lines $y = b$ and $x = a$ respectively and these depend upon the existence and values of (x,y) at the points in the neighbourhood of (a,b) along the lines $x = a$ and $y = b$. For continuity of $f(x,y)$ at (a,b), we have to consider the existence and values of the function and every point in the neighbourhood of (a,b). It is, therefore, just possible, that the partial derivatives at (a,b) exists, although the function is not continuous thereat. Thus *a function may posses partial derivatives without being continuous at that point.* [*vide* **Example 2.3.3**]

If, however, the partial derivatives are bounded in the domain of definition of the function, then the function must be continuous in that domain. [*vide* **Theorem 2.3.2**]

On the contrary, there exist functions $f(x, y)$ whose partial derivatives do not exist at some point in the domain but the function is continuous at that points. [*vide* **Example 2.3.5**]

Note 2.3.1 *The existence of higher derivatives implies the existence of the corresponding derivatives of lower order. Thus, in order that f_{yx} to exist at a point, it is necessary that the partial derivative f_x should exist in the neighbourhood of the point. However, it is possible for the limit defining f_{yx} to exist, without the partial derivative f_x being exist. In such cases, higher derivatives cannot be said to exist.* [*vide* **Example 2.3.2**]

Example 2.3.1 *If* $f(x, y) = \begin{cases} \dfrac{x^2 - xy}{x + y} & \text{for} \quad (x, y) \neq (0,0) \\ 0 & \text{for} \quad (x, y) = (0,0), \end{cases}$ *find $f_x(0,0)$ and* $f_y(0,0)$.

Solution: $f_x(0,0) = \lim\limits_{h \to 0} \dfrac{f(0 + h, 0) - f(0,0)}{h} = \lim\limits_{h \to 0} \dfrac{f(h, 0) - f(0,0)}{h}$

$$= \lim_{h \to 0} \frac{1}{h} \left(\frac{h^2}{h} \right) = 1,$$

and $f_y(0,0) = \lim\limits_{k \to 0} \dfrac{f(0, 0 + k) - f(0,0)}{k} = \lim\limits_{k \to 0} \dfrac{f(0, k) - f(0,0)}{k}$

$$= \lim_{k \to 0} \frac{1}{k} \left(\frac{0.k}{k} \right) = 0. \qquad \square$$

Example 2.3.2 *Let* $f(x, y) = \begin{cases} x \sin \dfrac{1}{x} + y \sin \dfrac{1}{y} & \text{for} \quad x \neq 0, y \neq 0, \\ y \sin \dfrac{1}{y} & \text{for} \quad x = 0, y \neq 0, \\ x \sin \dfrac{1}{x} & \text{for} \quad x \neq 0, y = 0, \\ 0 & \text{for} \quad x = 0, y = 0. \end{cases}$

Examine the existence of f_x and f_{yx} at $x = 0, y = 0$.

Solution: We have $f_x(0,0) = \lim\limits_{h \to 0} \dfrac{f(h,0) - f(0,0)}{h} = \lim\limits_{h \to 0} \dfrac{h \sin \frac{1}{h}}{h} = \lim\limits_{h \to 0} \left(\sin \dfrac{1}{h} \right)$

which clearly does not exist due to Cauchy's limiting criterion.

Also, we have

$$f_{yx}(a, b) = \lim_{k \to 0} \lim_{h \to 0} \frac{f(a+h, b+k) - f(a, b+k) - f(a+h, b) + f(a, b)}{hk}$$

$$\therefore \ f_{yx}(0,0) = \lim_{k \to 0} \lim_{h \to 0} \frac{f(h, k) - f(0, k) - f(h, 0) + f(0, 0)}{hk}$$

$$= \lim_{k \to 0} \lim_{h \to 0} \frac{h \sin \frac{1}{h} + k \sin \frac{1}{k} - k \sin \frac{1}{k} - h \sin \frac{1}{h} + 0}{hk} = \lim_{k \to 0} \lim_{h \to 0} \frac{0}{hk} = 0.$$

In this case, in spite of the fact that the limit is zero, the derivative $f_{yx}(0,0)$ cannot be said to exist, since the lower derivative $f_x(0,0)$ does not exist. □

Example 2.3.3 *Let* $f(x,y) = \dfrac{x^2 y}{x^4 + y^2}$ *for* $x \neq 0$, $y \neq 0$ *and* $f(0,0) = 0$. *Show that the partial derivatives* f_x, f_y *exist everywhere in the domain* $\{(x,y) \in \mathbb{R} : -1 \le x \le 1, -1 \le y \le 1\}$ *although* $f(x,y)$ *is discontinuous at the origin.*

Solution: For $x \neq 0, y \neq 0$, we have $f_x(x,y) = \lim\limits_{h \to 0} \dfrac{f(x+h,y) - f(x,y)}{h}$.

Here $f(x,y) = \dfrac{x^2 y}{x^4 + y^2}$.

$$\therefore \quad f_x(x,y) = \lim_{h \to 0} \frac{1}{h} \left[\frac{(x+h)^2 y}{(x+h)^4 + y^2} - \frac{x^2 y}{x^4 + y^2} \right]$$

$$= \lim_{h \to 0} \frac{1}{h} \left[\frac{(x^4 + y^2)(x+h)^2 y - x^2 y\{(x+h)^4 + y^2\}}{(x^4 + y^2)\{(x+h)^4 + y^2\}} \right]$$

$$= \lim_{h \to 0} \frac{1}{h(x^4 + y^2)\{(x+h)^4 + y^2\}} [(x^4 + y^2)(x^2 y + 2xhy + h^2 y)$$
$$- x^2 y (x^4 + 4x^3 h + 6x^2 h^2 + 4xh^3 + h^4 + y^2)]$$

Now $(x^4 + y^2)(x^2 y + 2xhy + h^2 y) = x^6 y + 2x^5 hy + x^4 h^2 y + x^2 y^3$
$$+ 2xhy^3 + h^2 y^3$$

and $x^2 y (x^4 + 4x^3 h + 6x^2 h^2 + 4xh^3 + h^4 + y^2) = x^6 y + 4x^5 hy$
$$+ 6x^4 h^2 y + 4x^3 h^3 y + x^2 h^4 y + x^2 y^3$$

So we get

$$
\begin{aligned}
f_x(x,y) &= \lim_{h \to 0} \frac{1}{h} \left[\frac{-2x^5 hy - 5x^4 h^2 y + 2xhy^3 + h^2 y^3 - 4x^3 h^3 y - x^2 h^4 y}{(x^4 + y^2)\{(x+h)^4 + y^2\}} \right] \\
&= \lim_{h \to 0} \frac{1}{h} \left[\frac{hy(-2x^5 - 5x^4 h + 2xy^2 + hy^2 - 4x^3 h^2 - x^2 h^3)}{(x^4 + y^2)\{(x+h)^4 + y^2\}} \right] \\
&= \lim_{h \to 0} \frac{y(-2x^5 - 5x^4 h + 2xy^2 + hy^2 - 4x^3 h^2 - x^2 h^3)}{(x^4 + y^2)\{(x+h)^4 + y^2\}} \\
&= \frac{y(-2x^5 + 2xy^2)}{(x^4 + y^2)(x^4 + y^2)} = 2xy \frac{y^2 - x^4}{(x^4 + y^2)^2}
\end{aligned}
$$

i.e., $f_x(x,y) = 2xy\dfrac{y^2 - x^4}{(x^4 + y^2)^2}$, $x \neq 0$, $y \neq 0$. and similarly, we shall get

$$f_y(x,y) = x^2 \frac{x^4 - y^2}{(x^4 + y^2)^2}, \quad x \neq 0, \ y \neq 0.$$

For $x = 0, y = 0$, we obtain

$$f_x(0,0) = \lim_{h \to 0} \frac{f(h,0) - f(0,0)}{h} = \lim_{h \to 0} \frac{1}{h} \frac{h.0}{h^4 + 0} = \lim_{h \to 0} \frac{0}{h^4} = 0$$

$$f_y(0,0) = \lim_{k \to 0} \frac{f(0,k) - f(0,0)}{k} = \lim_{k \to 0} \frac{1}{k} \frac{0}{k^2} = \lim_{h \to 0} \frac{0}{k^3} = 0.$$

Similarly, we can show that

$$f_x(x,y) = 0 \text{ for } x = 0, y \neq 0; \ f_x(x,y) = 0 \text{ for } x \neq 0, y = 0;$$

$$f_y(x,y) = 0 \text{ for } x = 0, y \neq 0; \ f_y(x,y) = \frac{1}{x^2} \text{ for } x \neq 0, y = 0$$

and consequently the partial derivatives f_x, f_y exist at all points of the given region. The function $f(x,y)$ is, however, not continuous in (x,y) at the origin, as can be seen easily by comparing the limiting values along the curve $y = mx^2$ towards the origin for different values of m. □

Example 2.3.4 *Cite an example to show that mere existence of the partial derivatives of a function at all points within a domain for which it is defined, is not enough to ensure the continuity of the function for all points within the domain.*

Solution: Let us consider the function $f(x,y) = \begin{cases} \dfrac{2xy}{x^2 + y^2}, & (x,y) \neq (0,0) \\ 0, & (x,y) = (0,0). \end{cases}$

Proceeding as in the previous **Example 2.3.3** for partial derivatives of the function $f(x,y)$ we can easily see that the first partial derivatives exist at all points within the domain of the function. [See **Example 2.3.6**]

That the function is discontinuous at the origin will be seen by approaching (x,y) to $(0,0)$ along the line $y = mx$ and then obtaining

$$\lim_{(x,y) \to (0,0)} f(x,y) = \lim_{(x,y) \to (0,0)} \frac{2x.mx}{x^2 + m^2x^2} = \lim_{(x,y) \to (0,0)} \frac{2m}{1 + m^2} = \frac{2m}{1 + m^2}$$

giving different limiting values for different values of m. □

Example 2.3.5 *Give an example of a function which is continuous but all the first partial derivatives does not exist for any particular point.*

Solution: Let us consider the function $f(x,y) = x|x| + |y|, \ (x,y) \in \mathbb{R}^2$.

Let $\epsilon > 0$. Then

$$|f(x,y) - f(0,0)| = |(x|x| + |y|) - 0| \leq |x|x|| + |y| = x^2 + |y| < \epsilon$$

which is true whenever $x^2 < \dfrac{\epsilon}{2}$ or $|x| < \sqrt{\dfrac{\epsilon}{2}}$ and $|y| < \dfrac{\epsilon}{2}$.

If we consider $\delta_1 = \sqrt{\dfrac{\epsilon}{2}}$ and $\delta_2 = \dfrac{\epsilon}{2}$ then we can write

$$|f(x,y) - f(0,0)| < \epsilon \text{ whenever } |x| < \delta_1, \ |y| < \delta_2.$$

Hence $f(x,y)$ is continuous at $(0,0)$.

Now,

$$f_x(0,0) = \lim_{h \to 0} \frac{f(h,0) - f(0,0)}{h} = \lim_{h \to 0} \frac{h|h|}{h} = \lim_{h \to 0} |h| = 0$$

$$f_y(0,0) = \lim_{k \to 0} \frac{f(0,k) - f(0,0)}{k} = \lim_{k \to 0} \frac{|k|}{k} \text{ -which does not exsit.}$$

So the function is continuous at the origin, but the first partial derivative $f_y(0,0)$ does not exists. However the other partial derivative $f_x(0,0)$ exists here. □

Theorem 2.3.1 *Mean Value Theorem (M.V.T) for a Function of Two Independent Variables*

Let a function f of two independent variables x and y possesses f_x in a certain neighbourhood N of (a, b) and let f_y exists at (a, b), then for any point $(a+h, b+k) \in N$

$$f(a + h, b + k) - f(a, b) = h f_x(a + \theta h, b + k) + k\left[f_y(a, b) + \eta\right]$$

where $0 < \theta < 1$ and η is a function of k such that $\eta \to 0$ as $k \to 0$.

Proof: We have

$$f(a+h, b+k) - f(a, b) = \{f(a+h, b+k) - f(a, b+k)\} + \{f(a, b+k) - f(a, b)\} \quad (2.1)$$

We assume that h, k are small enough, so that all the points $(a+h, b+k), (a, b+k)$ and $(a + h, b + k) \in N$.

Since f_x exists at all points of N, applying Mean Value Theorem of Lagrange, we get,

$$f(a + h, b + k) - f(a, b + k) = h f_x(a + \theta h), \quad 0 < \theta < 1 \quad (2.2)$$

Again, $f_y(a, b)$ exists, so

$$\lim_{k \to 0} \frac{f(a, b + k) - f(a, b)}{k} = f_y(a, b)$$

$$\therefore \quad \frac{f(a, b + k) - f(a, b)}{k} = f_y(a, b) + \eta$$

where η is a function of k and as $k \to 0$, $\eta \to 0$

$$\therefore \quad f(a, b + k) - f(a, b) = k[f_y(a, b) + \eta] \quad (2.3)$$

(2.1), (2.2) and (2.3) together give the required result. $\qquad \square$

Note 2.3.2 *Another useful form of* **Mean Value Theorem** *is given later in §* **2.10.1**.

Theorem 2.3.2 *Sufficient condition of Continuity of functions of two variables*

If a function $f(x, y)$ of two independent variables x and y defined in a domain R have partial derivatives f_x and f_y everywhere in R and further if these derivatives are bounded in R, then $f(x, y)$ is continuous in R.

Proof: Let (x, y) is an interior point of R.

We assume that h, k are small enough, so that all the points $(x+h, y+k), (x+h, y), (x, y+k) \in R$. We now write

$$f(x + h, y + k) - f(x, y) = \{f(x + h, y + k) - f(x + h, y)\}$$
$$+ \{f(x + h, y) - f(x, y)\} = \lambda + \mu \quad \text{(say)} \quad (2.4)$$

where in $\lambda = f(x + h, y + k) - f(x + h, y)$, the two terms differ only in y and since f_y exists we can apply M.V.T on the function $f(x + h, y)$ of one variable y ($x + h$ remaining constant) in the interval $(y, y + k)$ and thus we get

$$f(x + h, y + k) - f(x + h, y) = k f_y(x + h, y + \theta_1 k), \quad 0 < \theta_1 < 1 \quad (2.5)$$

Similarly, the two terms in $\mu = f(x + h, y) - f(x, y)$ differ in x (y remaining constant here) and since f_x exists we can apply M.V.T. on the function $f(x, y)$ of one variable x ($\because y$ remains constant) in the interval $[x, x + h]$ and thereby we obtain

$$f(x + h, y) - f(x, y) = h f_x(x + \theta_2 h, y), \quad 0 < \theta_2 < 1 \tag{2.6}$$

So, form (2.4), using (2.5) and (2.6) we get

$$f(x + h, y + k) - f(x, y) = k f_y(x + h, y + \theta_1 k) + h f_x(x + \theta_2 h, y),$$
$$0 < \theta_1 < 1, \ 0 < \theta_2 < 1$$

Now it is given that the partial derivatives are bounded in R, i.e., the partial derivatives $|f_x|$ and $|f_y|$ are everywhere less than M, where M is the greatest of the upper bounds of f_x and f_y, i.e., we have $|f_x| < M$ and $|f_y| < M$. So we get

$$|f(x + h, y + k) - f(x, y)| \leq M(|k| + |h|).$$

The right hand side can be made less than any pre-assigned positive number ϵ, if we take $(|k| + |h|) < \dfrac{\epsilon}{M}$, i.e., if we take $|h| < \delta, |k| < \delta$ where $\delta = \dfrac{\epsilon}{2M}$. Thus

$$|f(x + h, y + k) - f(x, y)| < \epsilon \text{ if } |h| < \delta, |k| < \delta, \text{ where } \delta = \frac{\epsilon}{2M}$$

which shows that $f(x, y)$ is continuous at any point $(x, y) \in R$. $\square$

Corollary 2.3.1 *If f_x, f_y are continuous at (x, y) in closed domain R, they must be bounded in R and so $f(x, y)$ is continuous in R by the **Theorem** 2.3.2.*

Example 2.3.6 *Prove that the function $f(x, y)$ defined by*

$$f(x, y) = \begin{cases} \dfrac{2xy}{x^2 + y^2} & \text{for} \quad x^2 + y^2 \neq 0 \\ 0 & \text{for} \quad x^2 + y^2 = 0 \end{cases}$$

has first partial derivatives everywhere. Is it continuous everywhere?

Solution: When $x \neq 0$, $y \neq 0$, we have

$$\begin{aligned}
f_x &= \lim_{h \to 0} \frac{f(x + h, y) - f(x, y)}{h} = \lim_{h \to 0} \frac{1}{h}\left[\frac{2(x + h)y}{(x + h)^2 + y^2} - \frac{2xy}{x^2 + y^2}\right] \\
&= \lim_{h \to 0} 2 \times \frac{1}{h}\left[\frac{(xy + hy)(x^2 + y^2) - xy(x^2 + y^2 + h^2 + 2xh)}{(x + h)^2(x^2 + y^2 + h^2 + 2xh)}\right] \\
&- \lim_{h \to 0} \frac{2}{h} \times \frac{h(y^3 - x^2 y - xhy)}{(x^2 + y^2)(x^2 + y^2 + h^2 + 2xh)} \\
&= \lim_{h \to 0} \frac{2(y^3 - x^2 y - xhy)}{(x^2 + y^2)(x^2 + y^2 + h^2 + 2xh)} = \frac{2y(y^2 - x^2)}{(x^2 + y^2)^2}
\end{aligned}$$

which exists for all values of x other than $x = 0, y = 0$.

Similarly, we can show that f_y exists for all values of x other than $x = 0, y = 0$ and whose value is $f_y = \dfrac{2x(x^2 - y^2)}{(x^2 + y^2)^2}$.

The function is continuous everywhere except at the origin and this can be proved by using the discontinuity criterion discussed in the previous **Example 2.3.4**. $\square$

2.4   Differentiability or Total Differentiability

Let $u = f(x, y)$ be a single valued function. Consider (x, y) be any point in the domain of u and $(x + \Delta x, y + \Delta y)$ be any neighbouring point of (x, y) belongs to the same domain. For the argumentative change from the point (x, y) to $(x + \Delta x, y + \Delta y)$, the functional change Δf (or Δu) is given by

$$\Delta f = f(x + \Delta x, y + \Delta y) - f(x, y)$$

Now, the function f is called differentiable (or total differentiable) at a point (x, y) if the functional change Δf be expressed in the form

$$\Delta f = A.\Delta x + B.\Delta y + \Delta x.\phi(\Delta x, \Delta y) + \Delta y.\psi(\Delta x, \Delta y) \tag{2.7}$$

where A and B are constants independent of Δx and Δy and ϕ, ψ are two functions such that

$$\lim_{\substack{x \to 0 \\ y \to 0}} \phi(\Delta x, \Delta y) = 0 \quad \text{and} \quad \lim_{\substack{x \to 0 \\ y \to 0}} \psi(\Delta x, \Delta y) = 0.$$

The portion $(A.\Delta x + B.\Delta y)$ in the expression (2.7) is called the total differential df of f at (x, y) and $(\Delta x.\phi(\Delta x, \Delta y) + \Delta y.\psi(\Delta x, \Delta y))$ is the error part when df is taken in place of Δf.

Let us replace Δx and Δy by h and k respectively and denote $\rho = \sqrt{\Delta x^2 + \Delta y^2}$ $= \sqrt{h^2 + k^2}$, then using these notation alternatively we say that $u = f(x, y)$ is differentiable if we express Δf at (x, y) as

$$\Delta f = Ah + Bk + \epsilon \rho$$

where ϵ tends to 0 when ρ tends to 0, i.e., $(h, k) \to (0, 0)$.

For a differentiable function from the error part we can write

$$\epsilon \rho = \Delta f - (Ah + bk) = \Delta f - df$$

$$\text{or} \quad \epsilon = \frac{\Delta f - df}{\rho} \to 0 \quad \text{as} \quad \rho \to 0 \text{ i.e., } (h, k) \to (0, 0).$$

This criterion is very useful to verify a function is differentiable or not.

2.4.1   Differential

The differential of the function $u = f(x, y)$ at (x, y) as seen above is given by

$$du = A\Delta x + B\Delta y = f_x \Delta x + f_y \Delta y = \frac{\partial f}{\partial x} \Delta x + \frac{\partial f}{\partial y} \Delta y$$

Taking $u = x$, we get $dx = \Delta x$ and similarly taking $u = y$, we get $dy = \Delta y$. Thus the differentials dx, dy of x and y are respectively Δx and Δy and we get

$$du = \frac{\partial f}{\partial x} \Delta x + \frac{\partial f}{\partial y} \Delta y = \frac{\partial f}{\partial x} dx + \frac{\partial f}{\partial y} dy = f_x dx + f_y dy \tag{2.8}$$

This is the differential of u at (x, y).

Next, we give two important theorems regarding differentiability of a two variable function:

Theorem 2.4.1 *If a function be differentiable at a point in its domain then the first partial derivatives must exist thereat for it.*

Proof: Let $f(x, y)$ be a differentiable function at the point (x, y). Also let $(x + h, y + k)$ be a neighbouring point in the domain of f such that the change of f can be expressed as

$$\Delta f = f(x + h, y + k) - f(x, y) = Ah + Bk + \epsilon\rho \tag{2.9}$$

where $\rho = \sqrt{h^2 + k^2}$ and ϵ tends to 0 when ρ tends to 0, i.e., $(h, k) \to (0, 0)$.

Let $k = 0$, then form (2.9) we get

$$Ah + \epsilon h = f(x + h, y) - f(x, y) \text{ or, } A + \epsilon = \frac{f(x + h, y) - f(x, y)}{h}$$

Now letting $h \to 0$,

$$\lim_{h \to 0}(A + \epsilon) = A = \lim_{h \to 0}\frac{f(x + h, y) - f(x, y)}{h} = f_x(x, y) \; [\because \lim_{h \to 0}\epsilon = 0]$$

Similarly if we put $h = 0$ and letting $k \to 0$ we get

$$B = \lim_{k \to 0}\frac{f(x, y + k) - f(x, y)}{k} = f_y(x, y).$$

Hence both the first partial derivatives exists and $f_x(x, y) = A$ and $f_y(x, y) = B$. $\square$

Note 2.4.1 *Converse of the above theorem is obviously not true, i.e., **there exists functions whose first partial derivatives exist but the function is not differentiable** (vide. **Example 2.4.1**). However, in **contra-positive state-ments, if any of the first partial derivatives does not exist then the func-tion cannot be differentiable** (vide **Example 2.4.3**).*

Theorem 2.4.2 *A function which is differentiable at a point in its domain is always continuous thereat.*

Proof: Let $f(x, y)$ be a differentiable at (x, y), then the instantaneous change in its value can be written as

$$\Delta f = f(x + h, y + k) - f(x, y) = Ah + Bk + \epsilon\rho \tag{2.10}$$

where $\rho = \sqrt{h^2 + k^2}$ and ϵ tends to 0 when ρ tends to 0, i.e., $(h, k) \to (0, 0)$.

Now taking simultaneous limit $(h, k) \to (0, 0)$ we get form (2.10),

$$\lim_{(h,k)\to(0,0)}\{f(x + h, y + k) - f(x, y)\} = \lim_{(h,k)\to(0,0)}(Ah + Bk + \epsilon\rho)$$

$$\text{or,} \quad \lim_{(h,k)\to(0,0)}\{f(x + h, y + k) - f(x, y)\} = 0$$

$$\text{or,} \quad \lim_{(h,k)\to(0,0)}f(x + h, y + k) = f(x, y)$$

Hence $f(x, y)$ is continuous at the said point. $\square$

Note 2.4.2 *Converse of the above theorem is not always true, i.e., **continuous functions may not be differentiable** [vide **Example 2.4.1**].*

Note 2.4.3 *It is clear that* **differentiability of a function implies continuity** *and in* **contra-positive statement** *it is also true that* **a discontinuous function at a point in its domain cannot be differentiable thereat** *(vide.* **Example** *2.4.2).*

Note 2.4.4 *The above two* **Theorems 2.4.1** *and* **2.4.2** *give the two* **necessary** *conditions for total differentiability of $f(x, y)$, i.e.,* **Continuity** *of a function at a point and the* **existence of both the first partial derivatives** *at a point in the domain of definition of the function can be considered as the necessary condition for the existence of total derivative.*

Example 2.4.1 *Let $f(x, y) = \dfrac{xy}{\sqrt{x^2 + y^2}}, (x, y) \neq (0, 0)$ and $f(0, 0) = 0$. Show that $f(x, y)$ is continuous and possesses first order partial derivatives at $(0, 0)$ but is not differentiable at that point.*

Solution: The given function is continuous at the point $(0, 0)$. [See **Example 1.14.8**.]

Now, $f_x(0, 0) = \lim\limits_{h \to 0} \dfrac{f(h, 0) - f(0, 0)}{h} = 0$ and similarly $f_y(0, 0) = 0$, i.e., the first partial derivatives do also exist.

For checking total differentiability we have to consider the expression of instantaneous change in functional value Δf, i.e.,

$$\Delta f = Ah + Bk + \epsilon\rho = df + \epsilon\rho, \text{ where } \rho = \sqrt{h^2 + k^2}$$

$$\text{or, } \quad \epsilon = \frac{\Delta f - df}{\sqrt{h^2 + k^2}}$$

where $\Delta f = f(h, k) - f(0, 0) = \dfrac{hk}{\sqrt{h^2 + k^2}} - 0 = \dfrac{hk}{\sqrt{h^2 + k^2}}$ and

$df = hf_x(0,0) + kf_y(0,0) = 0$. If the function is differentiable then the simultaneous

limit $\lim\limits_{(h,k) \to (0,0)} \epsilon = \lim\limits_{(h,k) \to (0,0)} \dfrac{\Delta f - df}{\sqrt{h^2 + k^2}}$ should be 0.

Now, $\lim\limits_{(h,k) \to (0,0)} \dfrac{\Delta f - df}{\sqrt{h^2 + k^2}} = \lim\limits_{(h,k) \to (0,0)} \dfrac{\frac{hk}{\sqrt{h^2+k^2}} - 0}{\sqrt{h^2 + k^2}} = \lim\limits_{(h,k) \to (0,0)} \dfrac{hk}{h^2 + k^2}.$

Clearly, this limit does not exist which can be shown bu considering the approach $k = hm$ towards $(h, k) \to (0, 0)$. Hence the given function is not differentiable at the origin. □

Example 2.4.2 *Show that $f(x, y) = \begin{cases} \dfrac{\sin(xy)}{x^2 + y^2} & \text{for} \quad x^2 + y^2 \neq 0 \\ 0 & \text{for} \quad x^2 + y^2 = 0 \end{cases}$ is not differentiable at the origin.*

Solution: Since we have to check the differentiability at the origin so we have to consider the functional behaviour at the neighbourhood of the origin. When $(x, y) \to (0, 0)$, $xy \to 0$ then $\sin(xy) \simeq (xy)$. Then the function can be redefined as

$$f(x, y) = \begin{cases} \dfrac{\sin(xy)}{x^2 + y^2} \simeq \dfrac{xy}{x^2 + y^2} & \text{for} \quad x^2 + y^2 \neq 0 \\ 0 & \text{for} \quad x^2 + y^2 = 0. \end{cases}$$

Now very easily we can show that $\lim\limits_{(x,y)\to(0,0)} \dfrac{xy}{x^2+y^2}$ does not exist by straight line approach $y = mx$. Hence the function is not continuous at the origin and consequently, by the contra-positive statement in **Note 2.4.3** we say that the function is not differentiable at $(0,0)$. $\qquad\square$

Example 2.4.3 *Show that $f(x,y) = |x| + |y|$ is not differentiable at $(0,0)$.*

Solution: Clearly, for the given function $f(x,y) = |x|+|y|$ the first partial derivatives $f_x(0,0)$ and $f_y(0,0)$ do not exist (cf. **Example 2.3.5**) and consequently, by the contra-positive statement in **Note 2.4.1** we say that the function is not differentiable at $(0,0)$. $\qquad\square$

2.5 Conditions for Total Differentiability

Let $u = f(x,y)$ be the function and the differential form du can be written as

$$du = \frac{\partial f}{\partial x}dx + \frac{\partial f}{\partial y}dy = f_x dx + f_y dy \qquad (2.11)$$

Let x and y be connected by a functional relation

$$y = \phi(x) \qquad (2.12)$$

or x and y are expressed in terms of a parameters t by the function relation

$$x = \psi_1(t),\ y = \psi_2(t) \qquad (2.13)$$

We may find the total derivatives of u with respect to x and with respect to t by the formulae

$$\frac{du}{dx} = \frac{\partial u}{\partial x} + \frac{\partial u}{\partial y}\cdot\frac{dy}{dx} \qquad (2.14)$$

$$\frac{du}{dt} = \frac{\partial u}{\partial x}\frac{dx}{dt} + \frac{\partial u}{\partial y}\cdot\frac{dy}{dt} \qquad (2.15)$$

We may also find these derivatives by differentiating directly the transformed equation

$$u = f\{x,\phi(x)\} \qquad (2.16)$$
$$u = f\{\psi_1(x),\psi_2(x)\} \qquad (2.17)$$

In order that relation (2.11) shall constitute a satisfactory definition of total differential, the numerical values of $\dfrac{du}{dx}$ and $\dfrac{du}{dt}$ as given by (2.14) and (2.15) must be the same as obtained by differentiating directly the relations (2.16) and (2.17) respectively.

That, this is not always true, if we assume (2.11) merely the existence of the partial derivatives $\dfrac{\partial u}{\partial x}$ and $\dfrac{\partial u}{\partial y}$, is illustrated by the following examples:

Example 2.5.1 *Let $u = f(x,y) = \dfrac{x^2 y}{3x^2 + y^3}$ for $x \neq 0, y \neq 0$ and $f(0,0) = 0$.*

Here it is easy to verify that $f_x(0,0) = 0$ and $f_y(0,0) = 0$.

If we now assume $y = \phi(x) = x$ as the curve along which we propose to restrict the change in x and y, we have from relation (2.14)

$$\frac{du}{dx} = \frac{\partial u}{\partial x} + \frac{\partial u}{\partial y} \cdot \frac{dy}{dx} = 0 + 0.1 = 0.$$

On the other hand, putting $y = x$ in the given equation, we get $u = \dfrac{x}{3+x}$ for $x \neq 0, u = 0$ for $x = 0$.

In this case, for $x > 0$, we have

$$\frac{du}{dx} = \lim_{h \to 0} \frac{f(0+h) - f(0)}{h} = \lim_{h \to 0} \frac{h}{(3+h)h} = \lim_{h \to 0} \frac{1}{3+h} = \frac{1}{3}.$$

Thus the value of $\dfrac{du}{dx}$ obtained by the two methods is not the same.

Example 2.5.2 *Let $u = \sqrt{|xy|}$.*

The partial derivatives $f_x(0,0)$ and $f_y(0,0)$ both exist, each being equal to zero. [easily verifiable]

Hence (2.15) gives

$$\frac{du}{dt} = \frac{\partial u}{\partial x}\frac{dx}{dt} + \frac{\partial u}{\partial y} \cdot \frac{dy}{dt} = 0. \frac{dx}{dt} + 0. \frac{dy}{dt} = 0.$$

On the other hand, if we transform the given function by the relation $x = t, y = t$, we have

$$u = \sqrt{\psi_1(t)\psi_2(t)} = \sqrt{|t^2|} = t \quad and \quad \frac{du}{dt} = 1$$

which is different from 0 as obtained before.

Hence, in these cases, relations (2.11) and (2.14) cannot be said to give the total differential.

Remarks: The above examples show that any statement of the conditions which involves merely the existence of the partial derivatives f_x and f_y is not sufficient. If, however, we assume the continuity of these partial derivatives with respect to two variables (x, y) together, then we can show by the aid of the law of the mean the existence of the total differential as given by (2.11) and consequently the total derivatives in a particular direction as given by (2.14) and (2.15).

Theorem 2.5.1 *Sufficient Condition for Differentiability*

If (x_0, y_0) be a point in the domain of definition of a function $f(x, y)$ of two independent variables x and y such that

i) one of the first partial derivatives f_x or f_y exists finitely at (x_0, y_0) and

ii) the other is continuous at (x_0, y_0),

then $f(x, y)$ is differentiable at (x_0, y_0).

Proof: Let $f_x(x_0, y_0)$ exists finitely and $f_y(x, y)$ is continuous at (x_0, y_0).

Let $\epsilon > 0$ be given. Since $f_x(x_0, y_0)$ exists finitely, $\exists$ a $\delta_1 > 0$, such that for $|h| < \delta_1$, we have

$$\left| \frac{f(x_0 + h, y_0) - f(x_0, y_0)}{h} - f_x(x_0, y_0) \right| < \frac{\epsilon}{2}.$$

Let $\dfrac{f(x_0 + h, y_0) - f(x_0, y_0)}{h} - f_x(x_0, y_0) = \eta$ (2.18) whence $|\eta| < \dfrac{\epsilon}{2}$.

Since $f_y(x, y)$ is continuous at (x_0, y_0), so $f_y(x, y)$ exists $\forall$ (x, y) belongs to a neighbourhood of the point (x_0, y_0). Applying M.V.T we get

$$f(x_0 + h, y_0 + k) - f(x_0 + h, y_0) = kf_y(x_0 + h, y_0 + \theta k),\ 0 < \theta < 1 \quad (2.19)$$

$$\therefore \quad \left| \frac{1}{\rho}[f(x_0 + h, y_0 + k) - f(x_0, y_0) - hf_x(x_0, y_0) - kf_y(x_0, y_0)] \right|,$$

$$\text{where } \rho = \sqrt{h^2 + k^2}.$$

$$= \left| \frac{1}{\rho}[f(x_0 + h, y_0 + k) - f(x_0 + h, y_0) + f(x_0 + h, y_0) - f(x_0, y_0) \right.$$

$$\left. -hf_x(x_0, y_0) - kf_y(x_0, y_0)] \right|$$

$$= \left| \frac{1}{\rho}[kf_y(x_0 + h, y_0 + k\theta) + \eta h + hf_x(x_0, y_0) - hf_x(x_0, y_0) - kf_y(x_0, y_0)] \right|$$

$$[\text{Using (2.18) and (2.19)}]$$

$$= \left| \frac{k}{\rho}[f_y(x_0 + h, y_0 + k\theta) - f_y(x_0, y_0)] + \eta\frac{h}{\rho} \right|$$

$$\leq \left| \frac{k}{\rho} \right| |[f_y(x_0 + h, y_0 + k\theta) - f_y(x_0, y_0)]| + \left| \eta\frac{h}{\rho} \right| \quad (2.20)$$

Again since $f_y(x, y)$ is continuous in (x, y) at (x_0, y_0), $\exists$ a $\delta_2 > 0$, such that

$$|f_y(x_0 + h, y_0 + k\theta) - f_y(x_0, y_0)| < \frac{\epsilon}{2} \text{ for } |h| < \delta_2 \text{ and } |k| < \delta_2 \quad (2.21)$$

Moreover, since $h \neq 0, k \neq 0$, we have

$$\left| \frac{h}{\rho} \right| = \left| \frac{h}{\sqrt{h^2 + k^2}} \right| < 1, \text{ similarly, } \left| \frac{k}{\rho} \right| < 1 \quad (2.22)$$

Let $\delta = \min(\delta_1, \delta_2)$. Then for $|h| < \delta$ and $|k| < \delta$, the inequality (2.20), with the help of (2.21) and (2.22), becomes

$$\left| \frac{1}{\rho}[f(x_0 + h, y_0 + k) - f(x_0, y_0) - hf_x(x_0, y_0) - kf_y(x_0, y_0)] \right| < \frac{\epsilon}{2} + \frac{\epsilon}{2} = \epsilon$$

$$\left(\because |\eta| < \frac{\epsilon}{2} \right)$$

i.e., $\quad \lim\limits_{\substack{h \to 0 \\ k \to 0}} \dfrac{f(x_0 + h, y_0 + k) - f(x_0, y_0) - [hf_x(x_0, y_0) + kf_y(x_0, y_0)]}{\rho} = 0$

i.e., $\quad \lim\limits_{\substack{h \to 0 \\ k \to 0}} \dfrac{\Delta f - df}{\rho} = 0$

which follows that $f(x, y)$ is differentiable at (x_0, y_0). $\qquad\qquad\qquad$ $\square$

Note 2.5.1 *The second condition of the above theorem, i.e., **the continuity of any one of the first partial derivatives** is only a sufficient condition but not necessary. Let us consider the following example:*

Example 2.5.3 *Let*

$$f(x,y) = \begin{cases} x^2 \sin \dfrac{1}{x} + y^2 \sin \dfrac{1}{y} & for \quad x \neq 0, y \neq 0, \\[2mm] y^2 \sin \dfrac{1}{y} & for \quad x = 0, y \neq 0, \\[2mm] x^2 \sin \dfrac{1}{x} & for \quad x \neq 0, y = 0, \\[2mm] 0 & for \quad x = 0, y = 0. \end{cases}$$

Show that both the first partial derivatives are discontinuous at the origin but the function is totally differentiable at that point.

Solution: From the definition of the function it is clear that if it has any discontinuity or non existence of any types of derivatives or any kind of analytical disturbances then it will occur at the points on x or y axis and for the other points the function will be smooth.

To check the continuity of the of the function $f_x(x,y)$ and $f_y(x,y)$ at $(0,0)$ we have to define those functions in the neighbourhood of $(0,0)$.

$$f_x(0,0) = \lim_{h \to 0} \frac{f(h,0) - f(0,0)}{h} = \lim_{h \to 0} \frac{h^2 \sin \frac{1}{h} - 0}{h} = \lim_{h \to 0} h \sin \frac{1}{h} = 0$$

and similarly $f_y(0,0) = 0$.

Now $f_x(x,y) = 2x \sin \dfrac{1}{x} - \cos \dfrac{1}{x}, x \neq 0$ and y remains constant. Hence

$$f_x(x,y) = \begin{cases} 2x \sin \dfrac{1}{x} - \cos \dfrac{1}{x} & for \quad x \neq 0, \\[2mm] 0 & for \quad x = 0. \end{cases}$$

Clearly $f_x(x,y)$ is not continuous at the origin due to the existence of the term $\cos \dfrac{1}{x}$. Similarly,

$$f_y(x,y) = \begin{cases} 2y \sin \dfrac{1}{y} - \cos \dfrac{1}{y} & for \quad y \neq 0, \\[2mm] 0 & for \quad y = 0. \end{cases}$$

is also not continuous at $(0,0)$.

Now for total differentiation we have to show that $\lim\limits_{\substack{h \to 0 \\ k \to 0}} \dfrac{\Delta f - df}{\rho} = 0$.

$$\Delta f = f(h,k) - f(0,0) = h^2 \sin \frac{1}{h} + k^2 \sin \frac{1}{k}$$

$$df = h f_x(0,0) + k f_y(0,0) = 0.$$

$$\therefore \quad \frac{\Delta f - df}{\rho} = \frac{h^2 \sin \frac{1}{h} + k^2 \sin \frac{1}{k} - 0}{\sqrt{h^2 + k^2}} = \frac{1}{\sqrt{h^2 + k^2}} \left(h^2 \sin \frac{1}{h} + k^2 \sin \frac{1}{k} \right)$$

For arbitrary $\epsilon > 0$ and for $(h,k) \neq (0,0)$,

$$\left| \frac{1}{\sqrt{h^2 + k^2}} \left(h^2 \sin \frac{1}{h} + k^2 \sin \frac{1}{k} \right) - 0 \right|$$

$$\leq \frac{1}{\sqrt{h^2 + k^2}} \left[|h^2| \left| \sin \frac{1}{h} \right| + |k^2| \left| \sin \frac{1}{k} \right| \right] \leq \frac{h^2 + k^2}{\sqrt{h^2 + k^2}}$$

$$= \sqrt{h^2 + k^2} < \epsilon \text{ (say) whenever } 0 < h^2 + k^2 < \epsilon^2$$

from which we can say that $\lim\limits_{\substack{h\to 0 \\ k\to 0}} \dfrac{1}{\sqrt{h^2+k^2}}\left(h^2\sin\dfrac{1}{h}+k^2\sin\dfrac{1}{k}\right)=0$.

This shows that the function is totally differentiable at $(0,0)$ in spite of its both the first partial derivatives are discontinuous thereat. □

Remark: A necessary and sufficient condition for total differentiability can be expressed analytically by means of polar coordinates. From the definition of total differentiability we get

$$\lim_{\substack{h\to 0 \\ k\to 0}} \frac{\Delta f - df}{\rho}=0$$

i.e., $\quad\lim\limits_{\substack{h\to 0 \\ k\to 0}} \dfrac{f(x_0+h,y_0+k)-f(x_0,y_0)-[hf_x(x_0,y_0)+kf_y(x_0,y_0)]}{\rho}=0$

If we put $h=r\cos\theta, k=r\sin\theta$ in the above limiting expression

$$\frac{f(x_0+h,y_0+k)-f(x_0,y_0)-[hf_x(x_0,y_0)+kf_y(x_0,y_0)]}{\sqrt{h^2+k^2}}$$

$$=\quad \frac{1}{r}[f(x_0+r\cos\theta,y_0+r\sin\theta)-f(x_0,y_0)$$

$$-\{r\cos\theta f_x(x_0,y_0)+r\sin\theta f_y(x_0,y_0)\}]$$

$$=\quad \frac{f(x_0+r\cos\theta,y_0+r\sin\theta)-f(x_0,y_0)}{r}$$

$$-\{\cos\theta f_x(x_0,y_0)+\sin\theta f_y(x_0,y_0)\}$$

Now $h\to 0, k\to 0$ means $r\to 0$, i.e., the necessary and sufficient condition for total differentiability at the point (x_0,y_0) is written as

$$\lim_{r\to 0}\frac{f(x_0+r\cos\theta,y_0+r\sin\theta)-f(x_0,y_0)}{r}$$

$$=\{\cos\theta f_x(x_0,y_0)+\sin\theta f_y(x_0,y_0)\}.$$

2.6 Directional Derivatives

Definition 1: The directional derivative of a function at a point P is the rate of variation of the function in a particular direction at P. If P is any point in the region of definition of the scalar point function $f(P)$ and PQ is the line segment drawn from P on one side of P only, then the directional derivative of $f(P)$ at P in the direction of PQ is defined by the scalar quantity $\lim\limits_{Q\to P}\dfrac{f(Q)-f(P)}{PQ}$, provided the limit exists.

The directional derivative of a vector point function, which is also a vector quantity, may be similarly defined.

Definition 2: The directional derivative of a real valued function f of two independent variables x, y in the direction of a unit vector $\hat{\beta}=(l,m)$, where $l^2+m^2=1$ at the point (x_0,y_0), denoted by $D_{\hat\beta}\,f(x_0,y_0)$, is given by

$$D_{\hat\beta}\,f(x_0,y_0)=\lim_{t\to 0}\frac{f(x_0+lt,y_0+mt)-f(x_0,y_0)}{t},$$

provided the limit exists.

Definition 3: In a three dimensional space, the directional derivative of a real valued function f of three independent variables x, y, z in the direction of a unit vector $\hat{a}$ having the direction cosines l, m, n (i.e., $l^2 + m^2 + n^2 = 1$) at the point (x_0, y_0, z_0), denoted by $D_{\hat{a}}\, f(x_0, y_0, z_0)$, is given by

$$D_{\hat{a}}\, f(x_0, y_0, z_0) = \lim_{t \to 0} \frac{f(x_0 + lt, y_0 + mt, z_0 + nt) - f(x_0, y_0, z_0)}{t},$$

provided the limit exists.

Example 2.6.1 *Show that for the function*

$$f(x, y) = \begin{cases} \dfrac{x^3 y}{x^6 + y^3} & for \quad x^2 + y^2 \neq 0, \\ 0 & for \quad x^2 + y^2 = 0 \end{cases}$$

directional derivative at $(0,0)$ exists in every direction but f is not continuous at $(0,0)$.

Solution: Let us find the directional derivative any arbitrary direction of a unit vector $\hat{\beta} = (l, m)$. Then, by definition,

$$
\begin{aligned}
D_{\hat{\beta}} f(0,0) &= \lim_{t \to 0} \frac{f(0 + lt, 0 + mt) - f(0,0)}{t} = \lim_{t \to 0} \frac{\dfrac{l^3 t^3 mt}{l^6 t^6 + m^3 t^3} - 0}{t} \\
&= \lim_{t \to 0} \frac{l^3 mt^4}{l^6 t^7 + m^3 t^4} = \lim_{t \to 0} \frac{l^3 m}{l^6 t^3 + m^3} = \frac{l^3}{m^2}, \text{ if } m \neq 0.
\end{aligned}
$$

If $m = 0$ i.e., if $\hat{\beta} = (l, 0)$ then clearly $D_{\hat{\beta}} f(0,0) = 0$. Hence directional derivative along any arbitrary direction at the origin of $f(x, y)$ exists.

We can prove very easily that the function is discontinuous at the origin. (Proof is left for the readers.) □

Example 2.6.2 *Show that for the function*

$$f(x, y) = \begin{cases} \dfrac{\sqrt{|xy|}}{|x| + |y|} & for \quad (x, y) \neq (0,0), \\ 0 & for \quad (x, y) = (0,0) \end{cases}$$

directional derivative at $(0,0)$ does not exist for any direction but the first order partial derivatives exist at the origin.

Solution: We have the first order partial derivatives as

$$f_x(0,0) = \lim_{h \to 0} \frac{f(h, 0) - f(0,0)}{h} = 0 \text{ and similarly } f_y(0,0) = 0.$$

Now we try to find the directional derivative along any arbitrary direction of a unit vector $\hat{\beta} = (l, m)$. Then

$$D_{\hat{\beta}} f(0,0) = \lim_{t \to 0} \frac{f(lt, mt) - f(0,0)}{t} = \lim_{t \to 0} \frac{\dfrac{\sqrt{|lt.mt|}}{|lt| + |mt|} - 0}{t} = \lim_{t \to 0} \frac{\sqrt{|lm|}}{|l| + |m|} \times \frac{1}{t}.$$

If $l \neq 0, m \neq 0$, then clearly the directional derivative does not exist for any arbitrary direction. □

Note 2.6.1 *If $l = 0, m \neq 0$ or $l \neq 0, m = 0$ then from the above example we shall see that the directional derivatives exist for those directions. We shall discuss later that the directional derivatives along these directions will give the partial derivatives with respect to the variables x and y respectively.*

Note 2.6.2 *From the above examples it is clear that a discontinuous function at a point can have directional derivatives in every directions at that point (vide **Example 2.6.1**). We can also show that a function continuous at a point cannot have directional derivatives in every direction thereat (vide **Example 2.6.3**). This means that continuity and the existence of directional derivative are two independent analytical properties of functions of several variables.*

*On the other hand non existence of any of the first partial derivatives at any point implies the non existence of the directional derivatives in every direction thereat, but the converse is obviously not true (see **Example 2.6.3**).*

Example 2.6.3 *Let* $f(x, y) = \begin{cases} y \sin \dfrac{1}{x} & for \quad x \neq 0, \\ 0 & for \quad x = 0 \end{cases}$. *Discuss the continuity, existence of directional derivatives in arbitrary direction and first order partial derivatives at the origin.*

Solution: Let $\epsilon > 0$. For $x \neq 0$ and for arbitrary y,

$$|f(x, y) - f(0, 0)| = \left| y \sin \frac{1}{x} - 0 \right| = |y| \left| \sin \frac{1}{x} \right| \leq |y| < \epsilon \text{ (say)}$$

If we take $\delta = \epsilon$ then it can be written as

$$|f(x, y) - f(0, 0)| < \epsilon \quad \text{whenever} \quad (x, y) \in N_\delta(0, 0)$$

i.e., the given function is continuous at the origin.

Now we consider the directional derivatives at $(0, 0)$ in any arbitrary direction of a unit vector $\widehat{\beta} = (l, m)$ where $l \neq 0$. Then

$$D_{\widehat{\beta}} f(0, 0) = \lim_{t \to 0} \frac{f(lt, mt) - f(0, 0)}{t} = \lim_{t \to 0} \frac{1}{t} \left(mt \sin \frac{1}{lt} \right) = m \lim_{t \to 0} \sin \frac{1}{lt}.$$

By Cauchy criterion of limit it can be proved that the last limit does not exist and consequently the said directional derivative does not exist at $(0, 0)$.

Now we check the first order partial derivatives,

$$f_x(0, 0) = \lim_{h \to 0} \frac{f(h, 0) - f(0, 0)}{h} = \lim_{h \to 0} \frac{0 - 0}{h} = 0 \text{ and } f_y(0, 0) = 0.$$

Hence first partial derivatives exist though the directional derivative does not exist in any arbitrary direction. □

Example 2.6.4 *Find the directional derivative of $f(x, y) = 2y^2 - xy + 5$ at $(1, 1)$ in the direction of the unit vector $\widehat{\beta} = \dfrac{1}{5}(-4, 3)$.*

Solution: By definition,

$$D_{\hat{\beta}}f(1,1) = \lim_{t\to0} \frac{f\left(1 - \frac{4}{5}t, 1 + \frac{3}{5}t\right) - f(1,1)}{t}$$

$$= \lim_{t\to0} \frac{\left\{2\left(1 + \frac{3}{5}t\right)^2 - \left(1 - \frac{4}{5}t\right)\left(1 + \frac{3}{5}t\right) + 5\right\} - 6}{t}$$

$$= \lim_{t\to0} \frac{2\left(1 + \frac{6}{5}t + \frac{9}{25}t^2\right) - \left(1 - \frac{1}{5}t - \frac{12}{25}t^2\right) + 5 - 6}{t}$$

$$= \lim_{t\to0} \frac{\frac{6}{5}t^2 + \frac{13}{5}t}{t} = \lim_{t\to0}\left(\frac{6}{5}t + \frac{13}{5}\right) = \frac{13}{5}. \qquad \square$$

Example 2.6.5 *Find the directional derivative of $f(x,y,z) = x^2 + y^2 + z^2$ at the point $(1,1,1)$ in the direction of the vector $(1,2,-2)$.*

Solution: The unit vector of the vector $(1,2,-2)$ is given by

$$\hat{\alpha} = \frac{(1,2,-2)}{\sqrt{1^2 + 2^2 + (-2)^2}} = \frac{1}{3}(1,2,-2) \text{ i.e., } l = \frac{1}{3}, m = \frac{2}{3}, n = -\frac{2}{3}.$$

So by definition the directional derivative is

$$D_{\hat{\alpha}}f(1,1,1) = \lim_{t\to0} \frac{f\left(1 + \frac{1}{3}t, 1 + \frac{2}{3}t, 1 - \frac{2}{3}t\right) - f(1,1,1)}{t}$$

$$= \lim_{t\to0} \frac{\left(1 + \frac{1}{3}t\right)^2 + \left(1 + \frac{2}{3}t\right)^2 + \left(1 - \frac{2}{3}t\right)^2 - 3}{t}$$

$$= \lim_{t\to0} \frac{1 + \frac{2}{3}t + \frac{1}{9}t^2 + 1 + \frac{4}{3}t + \frac{4}{9}t^2 + 1 - \frac{4}{3}t + \frac{4}{9}t^2 - 3}{t}$$

$$= \lim_{t\to0} \frac{\frac{2}{3}t + t^2}{t} = \frac{2}{3}. \qquad \square$$

2.7 Directional Derivatives: An Alternative Approach

In a three dimensional space, let us consider a rectangular Cartesian system of axes OX, OY and OZ through the origin O. Let (x,y,z) and $(x + \Delta x, y + \Delta y, z + \Delta z)$ be the coordinates of two adjacent points P and Q respectively, where distance $PQ = \Delta s$ and let this line segment PQ makes angles α, β and γ with the axes OX, OY and OZ

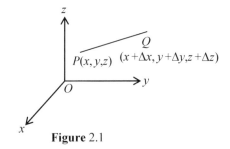

Figure 2.1

respectively whose direction cosines are l, m, n,

i.e., $\dfrac{\Delta x}{\Delta s} = \cos\alpha = l,\ \dfrac{\Delta y}{\Delta s} = \cos\beta = m$ and $\dfrac{\Delta z}{\Delta s} = \cos\gamma = n$

where $\Delta s = \sqrt{(\Delta x)^2 + (\Delta y)^2 + (\Delta z)^2}$.

Therefore, the directional derivative of a scalar point function $f(P)$ at a given point $P(x, y, z)$ in the direction of PQ is

$$\frac{df}{ds} = \lim_{Q \to P} \frac{f(Q) - f(P)}{PQ} = \lim_{\Delta s \to 0} \frac{\Delta f}{\Delta s}$$

$$= \lim_{\Delta s \to 0} \frac{f(x + \Delta x, y + \Delta y, z + \Delta z) - f(x, y, z)}{\Delta s}$$

$$= \lim_{\Delta s \to 0} \frac{1}{\Delta s} \left[\frac{\partial f}{\partial x} \Delta x + \frac{\partial f}{\partial y} \Delta y + \frac{\partial f}{\partial z} \Delta z + \epsilon_1 \Delta(x, y) \right.$$

$$\left. + \epsilon_2 \Delta(y, z) + \epsilon_3 \Delta(z, x) \right]$$

where $\epsilon_1, \epsilon_2, \epsilon_3$ are the errors in assuming df for Δf, i.e., when $f(x, y, z)$ is differentiable at (x, y, z), the terms $\epsilon_1 \Delta(x, y), \epsilon_2 \Delta(y, z)$ and $\epsilon_3 \Delta(z, x)$ vanish as $\Delta s \to 0$

$$\therefore \frac{df}{ds} = \frac{\partial f}{\partial x} \lim_{\Delta s \to 0} \frac{\Delta x}{\Delta s} + \frac{\partial f}{\partial y} \lim_{\Delta s \to 0} \frac{\Delta y}{\Delta s} + \frac{\partial f}{\partial z} \lim_{\Delta s \to 0} \frac{\Delta z}{\Delta s}$$

$$= \frac{\partial f}{\partial x} \frac{dx}{ds} + \frac{\partial f}{\partial y} \frac{dy}{ds} + \frac{\partial f}{\partial z} \frac{dz}{ds}$$

$$= l f_x(x, y, z) + m f_y(x, y, z) + n f_z(x, y, z)$$

Thus we get an alternative formula for finding the directional derivative of a scalar point function f at a given point $P(x, y, z)$ in the direction to the unit vector $\hat{\alpha} = (l, m, n)$ as given below

$$\frac{df}{ds} = l f_x(x, y, z) + m f_y(x, y, z) + n f_z(x, y, z)$$

Note 2.7.1 *The directional derivative of a vector point function $\vec{f}$ at the point P in the direction of the vector $\vec{\alpha}$ may be obtained similarly.*

Note 2.7.2 *In vector notation the above expression can be written as*

$$\frac{d\vec{f}}{ds} = (l, m, n) \cdot (f_x, f_y, f_z)$$

where '.' denotes the dot product and

$$(l, m, n) \equiv l\hat{i} + m\hat{j} + n\hat{k} \ \text{and} \ (f_x, f_y, f_z) \equiv f_x \hat{i} + f_y \hat{j} + f_z \hat{k}.$$

Special Cases:

I) In the direction $\hat{\beta} = (1, 0)$ in two dimensional domain, the directional derivative of f with respect to x at the point (x_0, y_0) which we have denoted by $f_x(x_0, y_0)$ or $\left(\frac{\partial f}{\partial x} \right)_{(x_0, y_0)}$. Thus $f_x(x_0, y_0) = \lim_{t \to 0} \frac{f(x_0 + t, y_0) - f(x_0, y_0)}{t}$, provided the limit exists.

Similarly, $f_y(x_0, y_0) = \lim_{t \to 0} \frac{f(x_0, y_0 + t) - f(x_0, y_0)}{t}$, provided the limit exists is the partial derivative of f with respect to y at the point (x_0, y_0), where the unit vector $\hat{\beta} = (0, 1)$ in this case.

II) In three dimensional space, if the line PQ is parallel to the x axis, then $\hat{\alpha} = (1, 0, 0)$; the directional derivative of f in the direction of $\hat{\alpha}$ with respect to x at the point (x_0, y_0, z_0) is $f_x(x_0, y_0, z_0)$ and those parallel to y and z axes are given by $f_y(x_0, y_0, z_0)$ and $f_z(x_0, y_0, z_0)$ respectively.

Note 2.7.3 *That the two results of § 2.6 and 2.7 are equivalent can be verified by considering any of the above example of § 2.6. Let us consider the* **Example** *2.6.3 where we have taken $f(x, y, z) = x^2 + y^2 + z^2$, the point is $(1, 1, 1)$ and the direction is $(1, 2, -2)$.*

We observe, here that the unit vector along $(1, 2, -2)$ is $\frac{1}{3}(1, 2, -2)$, i.e., (l, m, n)

$$= \left(\frac{1}{3}, \frac{2}{3}, -\frac{2}{3} \right) \text{ and } (f_x, f_y, f_z) = (2x, 2y, 2z) \text{ which is } (2, 2, 2) \text{ at the point } (1, 1, 1).$$

So the directional derivative at the point $(1, 1, 1)$ is obtained as

$$\frac{df}{ds} = \left(\frac{1}{3}, \frac{2}{3}, -\frac{2}{3} \right) . (2, 2, 2) = \frac{2}{3} + \frac{4}{3} - \frac{4}{3} = \frac{2}{3}$$

which is same as before.

2.8   Higher Order Partial Derivatives

If a function $f(x, y)$ has partial derivatives of the first order at each point (x, y) of a certain region, then each of the partial derivatives f_x, f_y, is, in general, a function of x and y. Therefore, these may be differentiated further partially with respect to x and y. These are called *second order partial derivatives*.

The partial derivatives of $f_x(x, y)$ with respect to x is denoted by $f_{xx}(x, y)$ or $\dfrac{\partial^2 f}{\partial x^2}$ and is defined as

$$
\begin{aligned}
f_{xx}(x, y) &= \lim_{h \to 0} \frac{f_x(x + h, y) - f_x(x, y)}{h} \\
&= \lim_{h \to 0} \frac{1}{h} \left[\lim_{h \to 0} \frac{f(x + h + h, y) - f(x + h, y)}{h} \right. \\
&\qquad\qquad \left. - \lim_{h \to 0} \frac{f(x + h, y) - f(x, y)}{h} \right] \\
&= \lim_{h \to 0} \frac{f(x + 2h, y) - 2f(x + h, y) + f(x, y)}{h^2}.
\end{aligned}
$$

In a similar manner, we shall get

$$
\begin{aligned}
f_{yy}(x, y) &= \lim_{k \to 0} \frac{f_y(x, y + k) - f_y(x, y)}{k} \\
&= \lim_{k \to 0} \frac{f(x, y + 2k) - 2f(x, y + k) + f(x, y)}{k} \\
f_{xy}(x, y) &= \lim_{h \to 0} \frac{f_y(x + h, y) - f_y(x, y)}{h} \quad \text{[vide § 2.2.1]} \\
&= \lim_{h \to 0} \frac{1}{h} \left[\lim_{k \to 0} \frac{1}{k} \{ f(x + h, y + k) - f(x + h, y) \right. \\
&\qquad\qquad \left. - f(x, y + k) + f(x, y) \} \right] \\
&= \lim_{h \to 0} \lim_{k \to 0} \frac{f(x + h, y + k) - f(x + h, y) - f(x, y + k) + f(x, y)}{hk} \\
&= \lim_{h \to 0} \lim_{k \to 0} \frac{F(h, k)}{hk} \quad \text{(say)}
\end{aligned}
$$

and

$$\begin{aligned}
f_{yx} &= \lim_{k \to 0} \frac{f_x(x, y+k) - f_x(x, y)}{k} \quad \text{[vide § \textbf{2.2.1}]} \\
&= \lim_{k \to 0} \frac{1}{k} \left[\lim_{h \to 0} \left\{ \frac{f(x+h, y+k) - f(x, y+k)}{h} \right. \right. \\
&\qquad\qquad\qquad\qquad \left. \left. - \frac{f(x+h, y) - f(x, y)}{h} \right\} \right] \\
&= \lim_{k \to 0} \lim_{h \to 0} \frac{f(x+h, y+k) - f(x, y+k) - f(x+h, y) + f(x, y)}{hk} \\
&= \lim_{k \to 0} \lim_{h \to 0} \frac{F(h, k)}{hk} \quad \text{(say).}
\end{aligned}$$

Since repeated limits of a function may or may not be equal, $f_{xy}(x, y)$ and $f_{yx}(x, y)$ may or may not be equal, $f_{xy}(x, y)$ and $f_{yx}(x, y)$ may or may not be equal.

Although *for most of the functions, that occur in applications, we have*, $f_{yx}(x, y) = f_{xy}(x, y)$, i.e., the partial derivatives have the same value whether we differentiate partially first with respect to y and then with respect to x or the reverse, *it must not be supposed that the above relation of equality holds good for all functions*, because the equality implies that the two limiting operations involved therein *should be commutative* which may *not be true always*. The conditions under which these are equal, are given below in the form of the following theorems:

Theorem 2.8.1 *Schwarz's Theorem (Form - 1)*

If (i) the partial derivatives f_x, f_y, f_{xy} and f_{yx} of the function $f(x, y)$ defined in some neighbourhood of (x, y), exist and
(ii) f_{xy} (or f_{yx}) is continuous at that point, then $f_{xy} = f_{yx}$ at that point.

Proof: Choose $h \neq 0, k \neq 0$ be small enough. Let $(x+h, y), (x, y+k)$ and $(x+h, y+k)$ be three neighbouring points belong to any neighbourhood N of the point (x, y).

Clearly, $\{f(x+h, y+k) - f(x, y+k)\} - \{f(x+h, y) + f(x, y)\}$ is a function of h and k, say $F(h, k)$ defined in N, i.e., we can write it as

$$\begin{aligned}
F(h, k) &= \{f(x+h, y+k) - f(x, y+k)\} - \{f(x+h, y) - f(x, y)\} \\
&= g(y+k) - g(y) \tag{2.23}
\end{aligned}$$

by putting $g(y) = f(x+h, y) - f(x, y)$.

By condition (i), f_y exists in some neighbourhood of (x, y). Therefore, $g(y)$ is differentiable in the neighbourhood of (x, y) and we get

$$g'(y) = f_y(x+h, y) - f_y(x, y).$$

Now applying Mean Value Theorem in (2.23) we get

$$\begin{aligned}
F(h, k) &= g(y+k) - g(y) = kg'(y+\theta k), \ 0 < \theta < 1 \\
&= k\{f_y(x+h, y+\theta k) - f_y(x, y+\theta k)\} \tag{2.24}
\end{aligned}$$

By condition (ii) of the theorem, if we consider that f_{xy} is continuous at (x, y), then f_{xy} exists in some neighbourhood of (x, y) and so $f_y(x, y)$ is derivable with respect to x in some neighbourhood of (x, y), i.e., $f_y(x, y+\theta k)$ is derivable with respect to x in $(x, x+h)$. Therefore using M.V. Theorem, we get

$$f_y(x+h, y+\theta k) - f_y(x, y+\theta k) = hf_{xy}(x + \theta'h, y + \theta k), \ 0 < \theta' < 1.$$

So (2.24) gives $F(h, k) = hk f_{xy}(x + \theta' h, y + \theta k)$.

Also since f_{xy} is continuous at the point (x, y), we get

$$f_{xy}(x, y) = \lim_{k \to 0} \lim_{h \to 0} f_{xy}(x + \theta' h, y + \theta k) = \lim_{k \to 0} \lim_{h \to 0} \frac{F(x, y)}{hk}.$$

Again $\qquad \lim_{k \to 0} \lim_{h \to 0} \dfrac{F(x, y)}{hk}$

$$= \lim_{k \to 0} \lim_{h \to 0} \frac{f(x + h, y + k) - f(x, y + k) - f(x + h, y) - f(x, y)}{hk}$$

$$= \lim_{k \to 0} \frac{1}{k} \left[\lim_{h \to 0} \frac{f(x + h, y + k) - f(x, y + k)}{h} - \lim_{h \to 0} \frac{f(x + h, y) - f(x, y)}{h} \right]$$

$$= \lim_{k \to 0} \frac{f_x(x, y + k) - f_x(x, y)}{k} = f_{yx}(x, y).$$

Hence $f_{xy}(x, y) = f_{yx}(x, y)$. $\qquad\square$

Theorem 2.8.2 *Schwarz's Theorem (Form - 2)*

If the derivatives f_y, f_{yx} of the function $f(x, y)$ exist in the neighbourhood of (x, y) and f_{yx} is continuous at that point then $f_{xy}(x, y) = f_{yx}(x, y)$ thereat.

Proof: If f_{yx} is continuous at (x, y), then f_x, f_{yx} exist in the neighbourhood of the point (x, y). Hence this form is same as **Form 1**. $\qquad\square$

Theorem 2.8.3 *Schwarz's Theorem (Form - 3)*

If the mixed partial derivatives f_{yx} and f_{xy} of the function $f(x, y)$ both are continuous in the neighbourhood of (x, y), then $f_{xy} = f_{yx}$ at the point (x, y).

Proof: If f_{yx} and f_{xy} are continuous at (x, y), then f_x, f_y, f_{xy} and f_{yx} exist in the neighbourhood of the point (x, y) which is same as **Form 1**. $\qquad\square$

Theorem 2.8.4 *Young's Theorem*

If (i) $f_x(x, y)$ and $f_y(x, y)$ exist in a certain neighbourhood of the point (x, y) and (ii) $f_x(x, y)$ and $f_y(x, y)$ both are differentiable thereat, then $f_{xy}(x, y) = f_{yx}(x, y)$.

Proof: By condition (ii) of the theorem the first partial derivatives f_x and f_y both are differentiable at (x, y), so all the second partial derivatives f_{xx}, f_{xy}, f_{yx} and f_{yy} exist at the said point.

For any arbitrary $h(\neq 0)$, small enough, let us consider the neighbouring points $(x + h, y)$, $(x + h, y + h)$ and $(x, y + h)$ of (x, y). Clearly, the expression

$$f(x + h, y + h) - f(x, y + h) - f(x + h, y) + f(x, y)$$

is a function of h only in the said neighbourhood.

Let $F(h, h) = f(x + h, y + h) - f(x, y + h) - f(x + h, y) + f(x, y)$

and $g(x) = f(x + h, y) - f(x, y)$,

i.e., $F(h, h) = g(x + h) - g(x)$.

Since f_x exists in the neighbourhood of (x, y), $g(x)$ is differentiable in the closed interval $[x, x + h]$. So applying M.V. Theorem for the function $g(x)$ we get

$$
\begin{aligned}
F(h, h) &= g(x + h) - g(x) = h g'(x + \theta h),\ 0 < \theta < 1 \\
&= h[f_x(x + \theta h, y + h) - f_x(x + \theta h, y)] \qquad (2.25)
\end{aligned}
$$

Now, we know that if a function $f(x, y)$ is differentiable in a certain neighbour-hood of a point (x, y) then we can write

$$f(x + h, y + k) - f(x, y) = h f_x(x, y) + k f_y(x, y) + h\phi(h, k) + k\psi(h, k) \quad (2.26)$$

where $\phi(h, k)$ and $\psi(h, k)$ tend to 0 separately as $(h, k) \to 0$.

Here the function f_x is differentiable at the neighbourhood of (x, y), so by relation (2.26) we get

$$f_x(x + \theta h, y + h) - f_x(x, y) = \theta h f_{xx}(x, y) + h f_{yx}(x, y)$$
$$+ \theta h \phi(h, h) + h\psi(h, h)$$

[replacing f by f_x, h by θh, and k by h]

and $\quad f_x(x + \theta h, y) - f_x(x, y) = \theta h f_{xx}(x, y) + \theta h\, \xi(h, h)$

[replacing f by f_x, h by θh, and k by 0 in (2.26)]

where $\phi(h, h)$, $\psi(h, h)$ and $\xi(h, h)$ tend to zero as h tends to zero. So, from (2.25) we get

$$
\begin{aligned}
\frac{F(h, h)}{h} &= f_x(x + \theta h, y + h) - f_x(x + \theta h, y) \\
&= f_x(x + \theta h, y + h) - f_x(x, y) - \{f_x(x + \theta h, y) - f_x(x, y)\} \\
&= \theta h f_{xx}(x, y) + h f_{yx}(x, y) + \theta h\phi(h, h) + h\psi(h, h) \\
&\qquad\qquad - \theta h f_{xx}(x, y) - \theta h\, \xi(h, h)
\end{aligned}
$$

$$\frac{F(h, h)}{h^2} = f_{yx}(x, y) + \theta\phi(h, h) + \psi(h, h) - \theta\, \xi(h, h)$$

or, $\quad \lim_{h \to 0} \dfrac{F(h, h)}{h^2} = f_{yx}(x, y).$

Similarly, writing $u(y) = f(x + h, y) - f(x, y)$ and proceeding in a similar manner, we get

$$\lim_{h \to 0} \frac{F(h, h)}{h^2} = f_{xy}(x, y).$$

Thus we get $f_{xy}(x, y) = f_{yx}(x, y)$. $\square$

Note 2.8.1 *It is to be noticed that the conditions of Schwartz's and Young's theorem are* **sufficient** *but* **not necessary** *which means that if the conditions of Schwartz's and Young's theorem are satisfied then $f_{yx} = f_{xy}$ at a point (x, y). But if the conditions are* **not satisfied**, *we* **can not draw any conclusion regarding the equality of f_{yx} and f_{xy}**, *i.e., they may or may not be equal.*

The following example will illustrate this.

Example 2.8.1 *If $f(x, y) = \begin{cases} \dfrac{x^2 y^2}{x^2 + y^2} & when \quad x^2 + y^2 \neq 0 \\ 0 & when \quad x^2 + y^2 = 0, \end{cases}$ show that*

$f_{xy}(0, 0) = f_{yx}(0, 0)$, *even though the conditions of Schwartz's and Young's theorem are not satisfied.*

Solution: We have

$$f_x(0,0) = \lim_{h \to 0} \frac{f(h,0) - f(0,0)}{h} = \lim_{h \to 0} \frac{0}{h} = 0$$

and $\quad f_y(0,0) = \lim_{k \to 0} \frac{f(0,k) - f(0,0)}{k} = \lim_{k \to 0} \frac{0}{k} = 0.$

Also $\quad f_x(0,k) = \lim_{h \to 0} \frac{f(h,k) - f(0,k)}{h} = \lim_{h \to 0} \frac{k^2 h}{h^2 + k^2} = 0$

and $\quad f_y(h,0) = \lim_{k \to 0} \frac{f(h,k) - f(h,0)}{k} = \lim_{k \to 0} \frac{kh^2}{h^2 + k^2} = 0.$

Hence $\quad f_{xy}(0,0) = \lim_{h \to 0} \frac{f_y(h,0) - f_y(0,0)}{h} = \lim_{h \to 0} \frac{0}{h} = 0$

and $\quad f_{yx}(0,0) = \lim_{k \to 0} \frac{f_x(0,k) - f_x(0,0)}{k} = \lim_{k \to 0} \frac{0}{k} = 0.$

Hence $\quad f_{yx}(0,0) = f_{xy}(0,0).$

Further, we note that, for $(x,y) \neq (0,0)$, we have

$$f_{yx}(x,y) = \frac{8xy^3(x^2 + y^2) - 2xy^4.4y(x^2 + y^2)}{(x^2 + y^2)^4} = \frac{8x^3 y^3}{(x^2 + y^2)^3}.$$

Now, if (x,y) approaches to $(0,0)$ along the straight line $y = mx$, then

$$\lim_{(x,y) \to (0,0)} f_{yx}(x,y) = \lim_{(x,y) \to (0,0)} \frac{8x^3 y^3}{(x^2 + y^2)^3} = \lim_{x \to 0} \frac{8m^3 x^6}{(1 + m^2)^3 x^6}$$

$$= \lim_{x \to 0} \frac{8m^3}{(1 + m^2)^3} = \frac{8m^3}{(1 + m^2)^3},$$

which is evidently not unique.

Hence $\lim\limits_{(x,y) \to (0,0)} f_{yx}(x,y)$ does not exist. So $f_{yx}(x,y)$ is not continuous at the origin. Similarly, it can be shown that $f_{xy}(x,y)$ is not continuous at the origin, i.e., the conditions of Schwartz's theorem are not satisfied. Hence the first part is proved.

Next, we show that the conditions of Young's theorem are also not satisfied. From the given function, we get

$$f_x(x,y) = \begin{cases} \dfrac{2xy^4}{(x^2 + y^2)^2} & \text{when} \quad (x,y) \neq (0,0), \\ 0 & \text{when} \quad (x,y) = (0,0). \end{cases}$$

$$f_{yx}(0,0) = \lim_{k \to 0} \frac{f_x(0,k) - f_x(0,0)}{k} = \lim_{k \to 0} \frac{0}{h} = 0$$

and $\quad f_{xx}(0,0) = \lim_{h \to 0} \frac{f_x(h,0) - f_x(0,0)}{h} = \lim_{h \to 0} \frac{0}{h} = 0.$

Now, we have to check that $f_x(0,0)$ is differentiable or not, i.e., we consider the limit $\lim\limits_{\substack{h \to 0 \\ k \to 0}} \dfrac{\Delta f_x - df_x}{\sqrt{h^2 + k^2}}$.

$$\Delta f_x = f_x(h,k) - f_x(0,0) = \frac{2hk^4}{(h^2 + k^2)^2}$$

$$df_x = h f_{xx}(0,0) + k f_{yx}(0,0) = 0.$$

$$\therefore \quad \lim\limits_{\substack{h \to 0 \\ k \to 0}} \frac{\Delta f_x - df_x}{\sqrt{h^2 + k^2}} = \lim\limits_{\substack{h \to 0 \\ k \to 0}} \frac{\dfrac{2hk^4}{(h^2 + k^2)^2}}{\sqrt{h^2 + k^2}} = \lim\limits_{\substack{h \to 0 \\ k \to 0}} \frac{2hk^4}{(h^2 + k^2)^{\frac{5}{2}}}$$

$$= \lim\limits_{h \to 0} \frac{2h(mh)^4}{(h^2 + (mh)^2)^{\frac{5}{2}}} = \frac{2m^4}{(1 + m^2)^{\frac{5}{2}}}.$$

[By approaching the limit $(h,k) \to (0,0)$ along the line $k = mh$.]

This shows that the limit doesn't exist and consequently the function f_x is not differentiable at the origin.

Similarly, it can be shown that f_y is not differentiable at the origin.

Thus the conditions of Young's theorem are also not satisfied, but, as shown above $f_{xy}(0,0) = f_{yx}(0,0)$. This proves the second part. $\qquad\square$

2.9 Differential of Higher Order

If $u = f(x,y)$ be a differentiable function of two independent variables x and y in a certain domain then the *first differential* of u, denoted by du and defined by

$$du = \frac{\partial u}{\partial x} dx + \frac{\partial u}{\partial y} dy = \left(dx \frac{\partial}{\partial x} + dy \frac{\partial}{\partial y} \right) u \qquad (2.27)$$

If $\dfrac{\partial u}{\partial x}$ and $\dfrac{\partial u}{\partial y}$ are differentiable at (x,y) and if dx and dy are considered as constants then du will be a differentiable function of x and y. The *second differential* of u or first differential of du which is denoted by $d^2 u$ can be calculated in a similar way,

$$\text{i.e.,} \quad d^2 u = d(du) = d\left(\frac{\partial u}{\partial x} \right) dx + d\left(\frac{\partial u}{\partial y} \right) dy \qquad (2.28)$$

Replacing u by $\dfrac{\partial u}{\partial x}$ and $\dfrac{\partial u}{\partial y}$ respectively in succession in (2.27), we get

$$d\left(\frac{\partial u}{\partial x} \right) = \frac{\partial^2 u}{\partial x^2} dx + \frac{\partial^2 u}{\partial y \partial x} dy \quad \text{and} \quad d\left(\frac{\partial u}{\partial y} \right) = \frac{\partial^2 u}{\partial x \partial y} dx + \frac{\partial^2 u}{\partial y^2} dy.$$

Since $\dfrac{\partial u}{\partial x}$ and $\dfrac{\partial u}{\partial y}$ are differentiable, therefore by Young's theorem we have $\dfrac{\partial^2 u}{\partial y \partial x} = \dfrac{\partial^2 u}{\partial x \partial y}$. So we get

$$d^2 u = \frac{\partial^2 u}{\partial x^2} dx^2 + 2 \frac{\partial^2 u}{\partial x \partial y} dx dy + \frac{\partial^2 u}{\partial y^2} dy^2$$

which can be written in abbreviated notation as

$$d^2u = \left(dx\frac{\partial}{\partial x} + dy\frac{\partial}{\partial y} \right)^2 u.$$

In a similar manner, we can get the third, fourth, ... differentials as follows:

$$
\begin{aligned}
d^3u &= \frac{\partial^3 u}{\partial x^3}dx^3 + 3\frac{\partial^3 u}{\partial x^2 \partial y}dx^2 dy + 3\frac{\partial^3 u}{\partial x \partial y^2}dxdy^2 + \frac{\partial^3 u}{\partial y^3}dy^3 \\
&= \left(dx\frac{\partial}{\partial x} + dy\frac{\partial}{\partial y} \right)^3 u \\
d^4u &= \left(dx\frac{\partial}{\partial x} + dy\frac{\partial}{\partial y} \right)^4 u \\
\cdots \quad & \quad \cdots \\
d^n u &= \left(dx\frac{\partial}{\partial x} + dy\frac{\partial}{\partial y} \right)^n u.
\end{aligned}
$$

Note 2.9.1 *If u is a function of three independent variables x, y, z, i.e., $u = f(x, y, z)$, then the first differential du is given by*

$$
\begin{aligned}
du &= \frac{\partial u}{\partial x}dx + \frac{\partial u}{\partial y}dy + \frac{\partial u}{\partial x}dx = f_x dx + f_y dy + f_z dz \\
&= \left(dx\frac{\partial}{\partial x} + dy\frac{\partial}{\partial y} + dz\frac{\partial}{\partial z} \right) u
\end{aligned}
$$

and the higher order differentials are obtained similarly as follows:

$$
\begin{aligned}
d^2u &= \left(dx\frac{\partial}{\partial x} + dy\frac{\partial}{\partial y} + dz\frac{\partial}{\partial z} \right)^2 u = \frac{\partial^2 u}{\partial x^2}dx^2 + \frac{\partial^2 u}{\partial y^2}dy^2 + \frac{\partial^2 u}{\partial z^2}dz^2 \\
&\quad + 2\frac{\partial^2 u}{\partial x \partial y}dxdy + 2\frac{\partial^2 u}{\partial y \partial z}dydz + 2\frac{\partial^2 u}{\partial z \partial x}dzdx \\
&= u_{xx}(dx)^2 + u_{yy}(dy)^2 + u_{zz}(dz)^2 + 2u_{xy}dxdy \\
&\qquad\qquad\qquad\qquad + 2u_{yz}dydz + 2u_{zx}dzdx.
\end{aligned}
$$

Note 2.9.2 *It should be noted that the relation (2.27) above are true only when x and y are independent variables. If x and y are not independent but functions of a third independent variable t (say), i.e., $x = \phi(t), y = \psi(t)$, then $dx = \phi'(t)dt$ and $dy = \psi'(t)dt$ when $dt = \triangle t$.*

2.10   Expansion of Functions of Several Variables

In calculus of functions of one independent variable, the Taylor's series and the other corresponding series expansions of functions of single variable have been discussed. The extension of the series expansion is considered in this section for the functions of several variables. In **Theorem 2.3.1**, a particular form of Mean Value Theorem for double variable functions has been discussed. Here a more generalized form of M. V. Theorem is taken into consideration using the concept of total differentiability. Taylor's theorem and the corresponding Maclaurin's theorem for the functions of two independent variables are given and finally its generalized form for n independent variables is also discussed.

2.10.1 Mean Value theorem in total differential form

Theorem 2.10.1 *Let $D \subseteq \mathbb{R}^2$ be an open set. Also let $f : D \to \mathbb{R}$ be such that $f(x, y)$ have continuous first order partial derivatives in D. Then for any neighbouring point $(a+h, b+k) \in D$ of the point $(a, b) \in D$, $\exists$ a $\theta \in \mathbb{R}$, $0 < \theta < 1$, depending on (h, k) such that*

$$f(a+h, b+k) - f(a, b) = h f_x(a+\theta h, b+\theta k) + k f_y(a+\theta h, b+\theta k).$$

Proof: Let us put $x = a + ht, y = b + kt$. Clearly, t varies from 0 to 1 as the point (a, b) traverses to $(a+h, b+k)$. Let us define the following function

$$F(t) = f(x, y) = f(a+ht, b+kt).$$

Given, $f(x, y)$ have continuous first order partial derivatives in D, i.e., it is differentiable in some neighbourhood of (a, b). Hence $F(t)$ is differentiable in $[0, 1]$. By applying Mean Value theorem to the function $F(t)$ in $[0, 1]$ we get

$$F(1) - F(0) = (1-0) f'(\theta) \text{ where } 0 < \theta < 1. \tag{2.29}$$

Now, $F(1) = f(a+h, b+k)$, $F(0) = f(a, b)$ and by using chain rule

$$
\begin{aligned}
F'(t) &= \frac{\partial f}{\partial x}\frac{dx}{dt} + \frac{\partial f}{\partial y}\frac{dy}{dt} = h\frac{\partial f}{\partial x} + k\frac{\partial f}{\partial y} \\
&= h f_x(a+ht, b+kt) + k f_y(a+ht, b+kt) \\
\therefore F'(\theta) &= h f_x(a+h\theta, b+k\theta) + k f_y(a+h\theta, b+k\theta).
\end{aligned}
$$

Putting these values in relation (2.29), we get

$$f(a+h, b+k) - f(a, b) = h f_x(a+\theta h, b+\theta k) + k f_y(a+\theta h, b+\theta k).$$

where $0 < \theta < 1$. $\qquad \square$

Note 2.10.1 *Clearly, the arbitrary point $(a+\theta h, b+\theta k)$ is indicating an intermediate point on the line segment joining (a, b) and $(a+h, b+k)$. Hence the above form of the Mean Value theorem signifies that **the differential of $f(x, y)$ at some intermediate point $(a+\theta h, b+\theta k)$, $0 < \theta < 1$ on the line segment joining (a, b) and $(a+h, b+k)$ is equal to the difference between the values of the function at the extremities (a, b) and $(a+h, b+k)$.***

2.10.2 Taylor's theorem for functions of two independent variables

Theorem 2.10.2 *Let $D \subseteq \mathbb{R}^2$ be an open set. Also let $f : D \to \mathbb{R}$ be such that $f(x, y)$ have continuous partial derivative of n th order in D. Then for any neighbouring point $(a+h, b+k) \in D$ of the point $(a, b) \in D$, $\exists$ a $\theta \in \mathbb{R}$, $0 < \theta < 1$, depending on (h, k) such that*

$$f(a+h, b+k) = f(a, b) + \left(h\frac{\partial}{\partial x} + k\frac{\partial}{\partial y} \right) f(a, b)$$

$$+ \frac{1}{\underline{|2}}\left(h\frac{\partial}{\partial x} + k\frac{\partial}{\partial y} \right)^2 f(a, b) + ... + \frac{1}{\underline{|n-1}}\left(h\frac{\partial}{\partial x} + k\frac{\partial}{\partial y} \right)^{n-1} f(a, b)$$

$$+ \frac{1}{\underline{|n}}\left(h\frac{\partial}{\partial x} + k\frac{\partial}{\partial y} \right)^n f(a+\theta h, b+\theta k),$$

where

$$\left(h\frac{\partial}{\partial x}+k\frac{\partial}{\partial y}\right)^r f(a,b) = \left(\sum_{m=0}^{r}{}^rC_m h^m \frac{\partial^m}{\partial x^m}k^{r-m}\frac{\partial^{r-m}}{\partial y^{r-m}}\right)f(a,b)$$

$$= \sum_{m=0}^{r}{}^rC_m h^m k^{r-m}\frac{\partial^m}{\partial x^m}\frac{\partial^{r-m}}{\partial y^{r-m}}f(x,y)\Bigg|_{(x,y)=(a,b)}, \quad r=1,2,...,n.$$

Proof: Let $x=a+ht, y=b+kt$ for any variable $t \in \mathbb{R}$ so that

$$f = f(x,y) = f(a+ht,b+kt) = F(t) \tag{2.30}$$

become a single variable composite function. Therefore, by chain rule we get,

$$f'(t) = \frac{\partial f}{\partial x}\frac{dx}{dt}+\frac{\partial f}{\partial y}\frac{dy}{dt} = h\frac{\partial f}{\partial x}+k\frac{\partial f}{\partial y} = \left(h\frac{\partial}{\partial x}+k\frac{\partial}{\partial y}\right)f$$

$$f''(t) = \frac{d}{dt}\left(\frac{df}{dt}\right) = \left(h\frac{\partial}{\partial x}+k\frac{\partial}{\partial y}\right)\left(h\frac{\partial f}{\partial x}+k\frac{\partial f}{\partial y}\right) = \left(h\frac{\partial}{\partial x}+k\frac{\partial}{\partial y}\right)^2 f$$

$$\cdots \qquad \cdots \quad \cdots$$

$$\cdots \qquad \cdots \quad \cdots$$

In a similar way

$$f^r(t) = \left(h\frac{\partial}{\partial x}+k\frac{\partial}{\partial y}\right)^r f, \quad r=1,2,3,...,n \tag{2.31}$$

Applying Maclaurin's theorem for the function $F(t)$ of the relation (2.30),

$$F(t) = F(0) + tF'(0) + \frac{t^2}{\lfloor 2}F''(0) + ... + \frac{t^{n-1}}{\lfloor n-1}F^{n-1}(0) + \frac{t^n}{\lfloor n}F^n(\theta t) \tag{2.32}$$

where $0 < \theta < 1$.

Putting $t = 1$, we get from (2.32),

$$F(1) = F(0) + F'(0) + \frac{1}{\lfloor 2}F''(0) + ... + \frac{1}{\lfloor n-1}F^{n-1}(0) + \frac{1}{\lfloor n}F^n(\theta) \tag{2.33}$$

Now from the relation (2.30),

$$F(1) = f(a+h,b+k) \quad \text{and} \quad F(0) = f(a,b)$$

and from (2.31), $F^r(t) = \left(h\dfrac{\partial}{\partial x}+k\dfrac{\partial}{\partial y}\right)^r f(x,y)$.

For $t = 0$, $x = a$ and $y = b$,

$$\therefore \ F'(0) = \left(h\frac{\partial}{\partial x}+k\frac{\partial}{\partial y}\right)f(a,b)$$

$$F''(0) = \left(h\frac{\partial}{\partial x}+k\frac{\partial}{\partial y}\right)^2 f(a,b)$$

$$\cdots \qquad \cdots \quad \cdots$$

$$\cdots \qquad \cdots \quad \cdots$$

$$F^{n-1}(0) = \left(h\frac{\partial}{\partial x}+k\frac{\partial}{\partial y}\right)^{n-1}f(a,b)$$

$$\text{and } F^n(0) = \left(h\frac{\partial}{\partial x}+k\frac{\partial}{\partial y}\right)^n f(a+\theta h,b+\theta k).$$

Putting these values in (2.33), we get

$$f(a+h, b+k) = f(a,b) + \left(h\frac{\partial}{\partial x} + k\frac{\partial}{\partial y}\right)f(a,b)$$

$$+ \frac{1}{\lfloor 2}\left(h\frac{\partial}{\partial x} + k\frac{\partial}{\partial y}\right)^2 f(a,b) + ... + \frac{1}{\lfloor n-1}\left(h\frac{\partial}{\partial x} + k\frac{\partial}{\partial y}\right)^{n-1}f(a,b)$$

$$+ \frac{1}{\lfloor n}\left(h\frac{\partial}{\partial x} + k\frac{\partial}{\partial y}\right)^n f(a+\theta h, b+\theta k), \quad \text{where } 0 < \theta < 1.$$

This is **Taylor's theorem with Remainder** or **Finite Taylor's series expansion** about the point (a,b). ☐

Note 2.10.2 *The term* $\frac{1}{\lfloor n}\left(h\frac{\partial}{\partial x} + k\frac{\partial}{\partial y}\right)^n f(a+\theta h, b+\theta k) = R_n$ *(say) is called*

the **Remainder term** *after n terms of the series.*

Note 2.10.3 *The series expansion given in the above Taylor's theorem can be put in the following form:*

$$f(a+h, b+k) = f(a,b) + df(a,b) + \frac{1}{\lfloor 2}d^2 f(a,b) + ...$$

$$+ \frac{1}{\lfloor n-1}d^{n-1}f(a,b) + \frac{1}{\lfloor n}d^n f(a+\theta h, b+\theta k), \quad 0 < \theta < 1.$$

where $d^r \equiv \left(h\frac{\partial}{\partial x} + k\frac{\partial}{\partial y}\right)^r$ *for* $r = 1, 2, ..., n.$

Note 2.10.4 *Putting* $a = x$ *and* $b = y$, *the above Taylor's expansion formula becomes as follows:*

$$f(x+h, y+k) = f(x,y) + \left(h\frac{\partial}{\partial x} + k\frac{\partial}{\partial y}\right)f(x,y)$$

$$+ \frac{1}{\lfloor 2}\left(h\frac{\partial}{\partial x} + k\frac{\partial}{\partial y}\right)^2 f(x,y) + ... + \frac{1}{\lfloor n-1}\left(h\frac{\partial}{\partial x} + k\frac{\partial}{\partial y}\right)^{n-1}f(x,y)$$

$$+ \frac{1}{\lfloor n}\left(h\frac{\partial}{\partial x} + k\frac{\partial}{\partial y}\right)^n f(x+\theta h, y+\theta k), \quad \text{where } 0 < \theta < 1.$$

Note 2.10.5 *Replacing 'a + h' and 'b + k' by 'x' and 'y' respectively, i.e., 'h' and 'k' by 'x − a' and 'y − b', we get:*

$$f(x,y) = f(a,b) + \left[(x-a)\frac{\partial}{\partial x} + (y-b)\frac{\partial}{\partial y}\right]f(a,b)$$

$$+ \frac{1}{\lfloor 2}\left[(x-a)\frac{\partial}{\partial x} + (y-b)\frac{\partial}{\partial y}\right]^2 f(a,b) + ...$$

$$+ \frac{1}{\lfloor n-1}\left[(x-a)\frac{\partial}{\partial x} + (y-b)\frac{\partial}{\partial y}\right]^{n-1}f(a,b)$$

$$+ \frac{1}{\lfloor n}\left[(x-a)\frac{\partial}{\partial x} + (y-b)\frac{\partial}{\partial y}\right]^n f(a+(x-a)\theta, b+(y-b)\theta),$$

where $0 < \theta < 1.$ *This is the Taylor's series expansion of* $f(x,y)$ *about* (a,b) *in power of* $(x-a)$ *and* $(y-b)$.

Note 2.10.6 *Deduction of Mean Value Theorem from Taylor's Theorem:*

Putting $n = 1$ *in Taylor's series expansion, we get*

$$f(a+h, b+k) = f(a,b) + \left(h\frac{\partial}{\partial x} + k\frac{\partial}{\partial y}\right) f(a+\theta h, b+\theta k)$$

or, $\quad f(a+h, b+k) - f(a,b) = h f_x(a+\theta h, b+\theta k) + k f_y(a+\theta h, b+\theta k),$

where $0 < \theta < 1$.

This is the Mean Value Theorem of two independent variables.

2.10.3 Generalised Taylor's theorem for functions of m independent variables

Theorem 2.10.3 *Let* $D \subseteq \mathbb{R}^m$ *be an open set and* $(a_1, a_2, ..., a_m) \in D$ *be an interior point. Also let* $f : D \to \mathbb{R}$ *be such that* f *possesses n-th order continuous partial derivatives in some neighbourhood of the point* $(a_1, a_2, ..., a_m)$. *Then for any neighbouring point* $(a_1 + h_1, a_2 + h_2, ..., a_m + h_m) \in D$, *there is a real* θ, $0 < \theta < 1$, *such that*

$$f(a_1 + h_1, a_2 + h_2, ..., a_m + h_m) = f(a_1, a_2, ..., a_m)$$

$$+ \left(\sum_{i=1}^{m} h_i \frac{\partial}{\partial x_i}\right) f(a_1, a_2, ..., a_m) + \frac{1}{\underline{2}} \left(\sum_{i=1}^{m} h_i \frac{\partial}{\partial x_i}\right)^2 f(a_1, a_2, ..., a_m)$$

$$+ ... + \frac{1}{\underline{n-1}} \left(\sum_{i=1}^{m} h_i \frac{\partial}{\partial x_i}\right)^{n-1} f(a_1, a_2, ..., a_m)$$

$$+ \frac{1}{\underline{n}} \left(\sum_{i=1}^{m} h_i \frac{\partial}{\partial x_i}\right)^n f(a_1 + \theta h_1, a_2 + \theta h_2, ..., a_m + \theta h_m).$$

Proof: The proof is similar to that of the **Theorem 2.10.2**. □

2.10.4 Maclaurin's theorem for the function of two independent variables

Theorem 2.10.4 *Let* $D \subseteq \mathbb{R}^2$ *be an open neighbourhood about the origin* $(0,0)$ *and* $(x, y) \in D$. *Also let* $f : D \to \mathbb{R}$ *be such that* f *possesses n-th order continuous partial derivatives in* D. *Then for any neighbouring point* $(h, k) \in D$, $\exists$ *a* $\theta \in \mathbb{R}$, $0 < \theta < 1$ *depending on* (h, k) *such that*

$$f(x, y) = f(0,0) + \left(x\frac{\partial}{\partial x} + y\frac{\partial}{\partial y}\right) f(0,0) + \frac{1}{\underline{2}} \left(x\frac{\partial}{\partial x} + y\frac{\partial}{\partial y}\right)^2 f(0,0)$$

$$+ ... + \frac{1}{\underline{n-1}} \left(x\frac{\partial}{\partial x} + y\frac{\partial}{\partial y}\right)^{n-1} f(a,b) + \frac{1}{\underline{n}} \left(x\frac{\partial}{\partial x} + y\frac{\partial}{\partial y}\right)^n f(\theta x, \theta y).$$

Proof: In Taylor's expansion formula of **Theorem 2.10.2**, putting $a = 0, b = 0$ and replacing h and k by x and y respectively we get the required Maclaurin's expansion formula about the origin $(0,0)$. □

Note 2.10.7 *Replacing the point* $(0,0)$ *by any arbitrary point* (p,q), *we get the Maclaurin's theorem in the following form:*

$$f(x,y) = f(p,q) + \left(x\frac{\partial}{\partial p} + y\frac{\partial}{\partial q}\right)f(p,q) + \frac{1}{\lfloor 2}\left(x\frac{\partial}{\partial p} + y\frac{\partial}{\partial q}\right)^2 f(p,q)$$

$$+\ldots + \frac{1}{\lfloor n-1}\left(x\frac{\partial}{\partial p} + y\frac{\partial}{\partial q}\right)^{n-1} f(p,q) + \frac{1}{\lfloor n}\left(x\frac{\partial}{\partial p} + y\frac{\partial}{\partial q}\right)^n f(p,q).$$

Example 2.10.1 *Show that for* $0 < \theta < 1$,

$$\cos x \sin y = y + \frac{1}{6}[x(x^2 + 3y^2)\sin\theta x \sin\theta y - y(y^2 + 3x^2)\cos\theta x \cos\theta y].$$

Solution: Let $f(x,y) = \cos x \sin y \;\Rightarrow\; f(0,0) = 0.$

$\therefore\; f_x(x,y) = -\sin x \sin y \;\Rightarrow\; f_x(0,0) = 0.$

$\therefore\; f_y(x,y) = \cos x \cos y \;\Rightarrow\; f_y(0,0) = 1.$

$\therefore\; f_{xx}(x,y) = -\cos x \sin y \;\Rightarrow\; f_{xx}(0,0) = 0.$

$\therefore\; f_{xy}(x,y) = -\sin x \cos y \;\Rightarrow\; f_{xy}(0,0) = 0.$

$\therefore\; f_{yy}(x,y) = -\cos x \sin y \;\Rightarrow\; f_{yy}(0,0) = 0.$

$\therefore\; f_{xxx}(x,y) = \sin x \sin y \;\Rightarrow\; f_{xxx}(\theta x,\theta y) = \sin\theta x \sin\theta y.$

$\therefore\; f_{xxy}(x,y) = -\cos x \cos y \;\Rightarrow\; f_{xxx}(\theta x,\theta y) = -\cos\theta x \cos\theta y.$

$\therefore\; f_{xyy}(x,y) = \sin x \sin y \;\Rightarrow\; f_{xyy}(\theta x,\theta y) = \sin\theta x \sin\theta y.$

$\therefore\; f_{yyy}(x,y) = -\cos x \cos y \;\Rightarrow\; f_{yyy}(\theta x,\theta y) = -\cos\theta x \cos\theta y.$

Applying Maclaurin's theorem up to third order,

$$f(x,y) = f(0,0) + \left(x\frac{\partial}{\partial x} + y\frac{\partial}{\partial y}\right)f(0,0) + \frac{1}{\lfloor 2}\left(x\frac{\partial}{\partial x} + y\frac{\partial}{\partial y}\right)^2 f(0,0)$$

$$+\frac{1}{\lfloor 3}\left(x\frac{\partial}{\partial x} + y\frac{\partial}{\partial y}\right)^3 f(\theta x,\theta y), \quad 0 < \theta < 1.$$

$\therefore\quad \cos x \sin y = f(0,0) + \{xf_x(0,0) + yf_y(0,0)\} + \frac{1}{2}\{x^2 f_{xx}(0,0)$

$$+2xy f_{xx}(0,0) + y^2 f_{xx}(0,0)\} + \frac{1}{6}\{x^3 f_{xxx}(\theta x,\theta y) + 3x^2 y f_{xxy}(\theta x,\theta y)$$

$$+3xy^2 f_{xyy}(\theta x,\theta y) + y^3 f_{yyy}(\theta x,\theta y)\}$$

$$= \; 0 + \{x.0 + y.1\} + \frac{1}{2}\{x^2.0 + 2xy.0 + y^2.0\} + \frac{1}{6}\{x^3 \sin\theta x \sin\theta y$$

$$+3x^2 y(-\cos\theta x \cos\theta y) + 3xy^2 \sin\theta x \sin\theta y + y^3(-\cos\theta x \cos\theta y)\}$$

$$= y + \frac{1}{6}[x(x^2 + 3y^2)\sin\theta x \sin\theta y - y(y^2 + 3x^2)\cos\theta x \cos\theta y]. \qquad \square$$

Example 2.10.2 *Express* $f(x,y) = x^3 e^y$ *in powers of* $(x-2)$ *and* $(y-1)$ *by Taylor's formula.*

Solution: Given $f(x,y) = x^3 e^y \;\Rightarrow\; f(2,1) = 2^3 e = 8e.$

$\therefore f_x = 3x^2 e^y \Rightarrow f_x(2,1) = 12e$ and $f_y = x^3 e^y \Rightarrow f_y(2,1) = 8e.$

Similarly, $f_{xx} = 6xe^y$, $f_{xy} = 3x^2 e^y$, $f_{yy} = x^3 e^y$

$\therefore\ f_{xx}(2,1) = 12e,\ f_{xy}(2,1) = 12e,\ f_{yy}(2,1) = 8e.$

Now the Taylor's series expansion formula about the point $(x,y) = (2,1)$ is

$$f(x,y) = f(2,1) + \left[(x-2)\frac{\partial}{\partial x} + (y-1)\frac{\partial}{\partial y}\right]f(2,1)$$

$$+ \frac{1}{\lfloor 2}\left[(x-2)\frac{\partial}{\partial x} + (y-1)\frac{\partial}{\partial y}\right]^2 f(2,1) + ...$$

$$+ \frac{1}{\lfloor n-1}\left[(x-2)\frac{\partial}{\partial x} + (y-1)\frac{\partial}{\partial y}\right]^{n-1} f(2,1)$$

$$+ \frac{1}{\lfloor n}\left[(x-2)\frac{\partial}{\partial x} + (y-1)\frac{\partial}{\partial y}\right]^n f(2+(x-2)\theta, 1+(y-1)\theta)$$

$\therefore\ f(x,y) = x^3 e^y = f(2,1) + [(x-2)f_x(2,1) + (y-1)f_y(2,1)]$

$$+ \frac{1}{2}[(x-2)^2 f_{xx}(2,1) + 2(x-2)(y-1)f_{xy}(2,1) + (y-1)^2 f_{yy}(2,1)] +$$

$$... + \frac{1}{\lfloor n}\left[(x-2)\frac{\partial}{\partial x} + (y-1)\frac{\partial}{\partial y}\right]^n f(2+(x-2)\theta, 1+(y-1)\theta)$$

$$= 8e + [(x-2)12e + (y-2)8e] + \frac{1}{2}[(x-2)^2 12e + 2(x-2)(y-1)12e$$

$$+ (y-1)^2 8e] + ... + \frac{1}{\lfloor n}\left[(x-2)\frac{\partial}{\partial x} + (y-1)\frac{\partial}{\partial y}\right]^n \times$$

$$f(2+(x-2)\theta, 1+(y-1)\theta)$$

$$= 8e + 12e(x-2) + 8e(y-1) + 6e(x-2)^2 + 12e(x-2)(y-1)$$

$$+ 4e(y-1)^2 + ... + \frac{1}{\lfloor n}\left[(x-2)\frac{\partial}{\partial x} + (y-1)\frac{\partial}{\partial y}\right]^n f(2+(x-2)\theta, 1+(y-1)\theta). \quad \square$$

2.11   Perfect Differential (or Exact Differential)

The expression

$$M(x,y)dx + N(x,y)dy \tag{2.34}$$

is said to be an exact differential (or perfect differential) if there exists a function u of x, y, such that its differential

$$du, \text{ i.e., } \frac{\partial u}{\partial x}dx + \frac{\partial u}{\partial y}dy \tag{2.35}$$

is equal to the expression (2.34) for all values of dx and dy.

Comparing (2.34) and (2.35) we get $\frac{\partial u}{\partial x} = M$ and $\frac{\partial u}{\partial y} = N$.

Differentiating these with respect to y and x receptively, we get

$$\frac{\partial^2 u}{\partial y \partial x} = \frac{\partial M}{\partial y} \text{ and } \frac{\partial^2 u}{\partial x \partial y} = \frac{\partial N}{\partial x}.$$

Since most of the cases $\frac{\partial^2 u}{\partial y \partial x} = \frac{\partial^2 u}{\partial x \partial y}$, hence, in order that (2.34) may be an exact differential, it is necessary that $\frac{\partial M}{\partial y} = \frac{\partial N}{\partial x}$.

Note 2.11.1 *We can show that, the above condition is, in general, also sufficient.*

2.12 Calculation of Small Errors by Differentials

Let $u = f(x, y, z, t, ...)$ be a function of several variables. If the dependent variables $x, y, z, ...$ are obtained by any measurement subject to infinitesimal errors $dx, dy, dz, ...$ respectively, the corresponding error du of the dependent variable u can be computed approximately by the following formula

$$du = \frac{\partial u}{\partial x}dx + \frac{\partial u}{\partial y}dy + \frac{\partial u}{\partial z}dz + ... = f_x dx + f_y dy + f_z dz + ... \qquad (2.36)$$

We can use total increment of a function equivalent to total differential. Also a close approximation to the error in $f(x, y, z, t, ...)$ corresponding to the small errors $\Delta x, \Delta y, \Delta z, ...$ of $x, y, z, ...$ can be found equivalent to $f_x dx + f_y dy + f_z dz + ...$

Relative and **percentage errors** are respectively given by $\dfrac{du}{u}$ and $\dfrac{du}{u} \times 100$ %.

Example 2.12.1 *Calculate the error, relative error and percentage error of the volume V of a cone with height h and radius of the base r.*

If, in particular, initially $r = 4cm$, $h = 6cm$ and error in $r = 0.4cm$, error in $h = 0.6cm$, find those errors.

Solution: The volume V of the cone is given by $V = \dfrac{1}{3}\pi r^2 h.$

$\therefore \quad \dfrac{\partial V}{\partial r} = \dfrac{2}{3}\pi rh, \quad \dfrac{\partial V}{\partial h} = \dfrac{1}{3}\pi r^2.$

$\therefore$ Error in $V = dV = \dfrac{\partial V}{\partial r}dr + \dfrac{\partial V}{\partial h}dh = \dfrac{2}{3}\pi rh\,dr + \dfrac{1}{3}\pi r^2 dh.$

Relative error in $V = \dfrac{dV}{V} = \dfrac{\frac{2}{3}\pi rh\,dr + \frac{1}{3}\pi r^2 dh}{\frac{1}{3}\pi r^2 h} = \dfrac{2}{r}dr + \dfrac{1}{h}dh$

and percentage error in $V = \dfrac{dV}{V} \times 100\% = \left(\dfrac{2}{r}dr + \dfrac{1}{h}dh\right) \times 100\%.$

Second Part: When $r = 4$cm, $h = 6$cm, $dr = 0.4$cm and $dh = 0.6$cm, then

$$\begin{aligned}
\text{error in } V = dV &= \frac{2}{3}\pi \times 4 \times 6 \times 0.4 + \frac{1}{3}\pi \times 4^2 \times 0.6 \\
&= \frac{2}{3} \times \frac{22}{7} \times 4 \times 6 \times 0.4 + \frac{1}{3} \times \frac{22}{7} \times 16 \times 0.6 \\
&= \frac{1}{3} \times \frac{22}{7} \times 4\,(12 \times 0.4 + 0.6) = \frac{1}{3} \times \frac{22}{7} \times 4 \times 7.2 \\
&= \frac{1}{3} \times \frac{22}{7} \times 4 \times \frac{72}{10} = \frac{11 \times 4 \times 72}{105} = 30.17 \ \text{(approx.)}
\end{aligned}$$

Relative error in $V = \dfrac{dV}{V} = \dfrac{30.17}{\frac{1}{3} \times \frac{22}{7} \times 16 \times 6} = \dfrac{30.17 \times 3 \times 7}{22 \times 16 \times 6} = 0.3$

and Percentage error $= \dfrac{dV}{V} \times 100\% = 0.3 \times 100\% = 30\%.$ $\qquad\qquad \square$

2.13   Differentiation of Implicit Functions

Let us consider the equation of the form

$$f(x, y) = 0 \tag{2.37}$$

and let this equation defines y as a function of x. Also let f_x, f_y be continuous in the neighbourhood of the point (x, y), then we can find $\dfrac{dy}{dx}$ in terms of $\dfrac{\partial f}{\partial x}, \dfrac{\partial f}{\partial y}$ as follows:

   We have $f(x + \Delta x, y + \Delta y) = 0$

$$\therefore \qquad f(x + \Delta x, y + \Delta y) - f(x, y) = 0 \tag{2.38}$$

Using Mean Value Theorem of Lagrange, we get

$$f(x + \Delta x, y + \Delta y) - f(x, y + \Delta y) = \Delta x \frac{\partial}{\partial x} f(x + \theta_1 \Delta x, y + \Delta y),$$
$$0 < \theta_1 < 1 \tag{2.39}$$

$$\text{and} \quad f(x, y + \Delta y) - f(x, y) = \Delta y \frac{\partial}{\partial y} f(x, y + \theta_2 \Delta y),$$
$$0 < \theta_2 < 1 \tag{2.40}$$

$[\ \because$ in the former case (2.39), f is treated as a function of single variable x and in the later case (2.40), f is a function of a single variable $y]$.

   Adding (2.39) and (2.40) and using relation (2.38) and dividing by Δx, we get

$$\frac{\partial}{\partial x} f(x + \theta_1 \Delta x, y + \Delta y) + \frac{\Delta y}{\Delta x} \frac{\partial}{\partial y} f(x, y + \theta_2 \Delta y) = 0 \tag{2.41}$$

   Now since y is a differentiable function of x, when $\Delta x \to 0, \Delta y \to 0$ and since $\dfrac{\partial f}{\partial x}$ and $\dfrac{\partial f}{\partial y}$ are continuous, making $\Delta x \to 0$ in (2.41) we get

$$\frac{\partial}{\partial x} f(x, y) + \frac{dy}{dx} \frac{\partial}{\partial y} f(x, y) = 0 \text{ i.e., } \frac{\partial f}{\partial x} + \frac{dy}{dx} \frac{\partial f}{\partial x} = 0$$

$$\text{or,} \quad \frac{dy}{dx} = -\frac{\frac{\partial f}{\partial x}}{\frac{\partial f}{\partial y}} = -\frac{f_x}{f_y}, \quad (f_y \neq 0).$$

2.14   Miscellaneous Illustrative Examples

Example 2.14.1 *Let* $f(x, y) = \begin{cases} 1 & \text{when} & (x, y) \neq (0, 0) \\ 0 & \text{when} & \text{either } x = 0 \text{ or } y = 0. \end{cases}$ *Show that* $f(x, y)$ *is discontinuous at* $(0, 0)$ *but the partial derivatives* f_x *and* f_y *exist.*

Solution: Since $|f(x, y) - f(0, 0)| = |1 - 0| = 1$, which cannot be less than an arbitrary positive small number ϵ, showing that $f(x, y)$ is not continuous at the origin.

   Now $f_x(0, 0) = \lim\limits_{h \to 0} \dfrac{f(0 + h, 0) - f(0, 0)}{h} = \lim\limits_{h \to 0} \dfrac{0 - 0}{h} = 0$

   and $f_y(0, 0) = \lim\limits_{k \to 0} \dfrac{f(0, 0 + k) - f(0, 0)}{k} = \lim\limits_{k \to 0} \dfrac{0 - 0}{k} = 0.$

   Thus partial derivatives of $f(x, y)$ with respect to x and y exist at $(0, 0)$ although $f(x, y)$ is discontinuous at the origin.   □

Example 2.14.2 *Given* $f(x, y) = 2x^3 - x^2 y + 2y^2$. *Find* f_x *and* f_y *at the point* $(1, 2)$.

Solution: $f(x, y) = 2x^3 - x^2 y + 2y^2$

$\therefore f_x = 2.3.x^2 - 2x.y$ ($\because y$ is constant) and $f_y = -x^2 + 4y$ ($\because x$ is constant).

$\therefore (f_x)_{(1,2)} = 2.3.1^2 - 2.1.2 = 2$ and $(f_y)_{(1,2)} = -(1)^2 - 4.2 = 7$.   □

Example 2.14.3 *Find the first partial derivatives* f_x *and* f_y *at the origin where*

$$f(x, y) = \begin{cases} \dfrac{x^2 + y^2}{x - y} & when \quad x \neq y \\ 0 & when \quad x = y. \end{cases}$$

Show also that these partial derivatives are not continuous at the origin.

Solution: We have

$$f_x(x, y) = \frac{\frac{\partial}{\partial x}(x^2 + y^2) \times (x - y) - (x^2 + y^2)\frac{\partial}{\partial x}(x - y)}{(x - y)^2} \quad \text{when } x \neq y \quad (2.42)$$

But since the denominator $(x - y)^2$ is 0 at $(0, 0)$, so $f_x(0, 0)$ cannot be calculated from this result. So we have to apply the definition

$$f_x(0, 0) = \lim_{h \to 0} \frac{f(0 + h, 0) - f(0, 0)}{h} = \lim_{h \to 0} \frac{1}{h} \frac{(0 + h)^2 + 0^2}{(0 + h) + 0} = \lim_{h \to 0} \frac{h^2}{h^2} = 1.$$

Similarly, $f_y(0, 0) = \lim_{k \to 0} \frac{1}{k} \frac{k^2 + 0^2}{0 - (0 + k)} = \lim_{k \to 0} \frac{k^2}{-k^2} = -1.$

From (2.42), $f_x(x, y) = \dfrac{2x(x - y) - (x^2 + y^2).1}{(x - y)^2} = \dfrac{x^2 - 2xy - y^2}{(x - y)^2}$

$$= \frac{(x - y)^2 - 2y^2}{(x - y)^2}.$$

Let us put $x - y = my$,

$$f_x(x, y) = f_x(y + my, y) = \frac{(m^2 - 2)y^2}{m^2 y^2} \to \frac{m^2 - 2}{m^2} \quad \text{as } y \to 0.$$

Hence $\lim_{(x,y) \to (0,0)} f_x(x, y)$ does not exist, therefore f_x is not continuous at the origin.

Similarly, it can be shown that f_y is also not continuous at the origin.   □

Example 2.14.4 *Let* $f(x, y) = \begin{cases} \dfrac{x^3 - y^3}{x^2 + y^2} & when \quad (x, y) \neq (0, 0) \\ 0 & when \quad (x, y) = (0, 0). \end{cases}$ *Show that*

at $(0, 0), f(x, y)$ *is continuous, possesses partial derivatives but not differentiable thereat.*

Solution: Let $\epsilon > 0$ be given arbitrarily. Putting $x = r \cos \theta, y = r \sin \theta$, we get

$$\left| \frac{x^3 - y^3}{x^2 + y^2} - 0 \right| = r|\cos^3 \theta - \sin^3 \theta| \leq |r| = \sqrt{x^2 + y^2} < \epsilon \text{ (say)}.$$

It is true whenever $|x| < \dfrac{\epsilon}{\sqrt{2}}, |y| < \dfrac{\epsilon}{\sqrt{2}}$. So $\lim\limits_{\substack{x \to 0 \\ y \to 0}} f(x,y) = 0 = f(0,0)$. Therefore

$f(x,y)$ is continuous at $(0,0)$.

$$\text{We have} \qquad f_x(0,0) = \lim_{h \to 0} \frac{f(h,0) - f(0,0)}{h} = \lim_{h \to 0} \frac{h - 0}{h} = 1$$

$$f_y(0,0) = \lim_{k \to 0} \frac{f(0,k) - f(0,0)}{k} = \lim_{k \to 0} \frac{-k - 0}{k} = -1.$$

Therefore the function possesses first partial derivatives at the origin.

Next, we are going to test the total differentiability of the function at the origin. By definition we consider the expression of instantaneous change Δf of $f(x,y)$ as

$$\Delta f = Ah + Bk + \epsilon\rho = df + \epsilon\rho \quad \text{or,} \quad \epsilon = \frac{\Delta f - df}{\sqrt{h^2 + k^2}} \quad [\because \rho = \sqrt{h^2 + k^2}]$$

where $\Delta f = f(0+h, 0+k) - f(0,0) = \dfrac{h^3 - k^3}{h^2 + k^2}$

and $df = hf_x(0,0) + kf_y(0,0) = h - k.$

$$\therefore \ \epsilon = \frac{\frac{h^3 - k^3}{h^2 + k^2} - (h - k)}{\sqrt{h^2 + k^2}} = \frac{h^3 - k^3 - (h-k)(h^2 + k^2)}{(h^2 + k^2)\sqrt{h^2 + k^2}} = \frac{hk(h-k)}{(h^2 + k^2)^{\frac{3}{2}}}$$

If the function is differentiable then the double limit $\lim\limits_{\substack{h \to 0 \\ k \to 0}} \epsilon = \lim\limits_{\substack{h \to 0 \\ k \to 0}} \dfrac{hk(h-k)}{(h^2 + k^2)^{\frac{3}{2}}}$ should

be 0. Putting the limiting approach along the direction $k = mh$ we get,

$\lim\limits_{h \to 0} \dfrac{mh^2(h - mh)}{(h^2 + m^2 h^2)^{\frac{3}{2}}} = \lim\limits_{h \to 0} \dfrac{m(1-m)}{(1+m^2)^{\frac{3}{2}}}$ which implies that the limit does not exist.

Hence the given function is not differentiable at the origin. $\qquad\qquad\qquad \square$

Note 2.14.1 *A similar example has been illustrated in* **Example 2.4.1**.

Example 2.14.5 *If* $f(x,y) = \begin{cases} xy & when & |x| \geq |y| \\ -xy & when & |x| < |y| \end{cases}$ *then show that*
$f_{yx}(0,0) \neq f_{xy}(0,0)$.

Which conditions of Schwartz's theorem are not satisfied by f?

Solution: Here $f_y(h,0) = \lim\limits_{k \to 0} \dfrac{f(h, 0+k) - f(h,0)}{k} = \lim\limits_{k \to 0} \dfrac{f(h,k) - f(h,0)}{k}$

$$= \lim_{k \to 0} \frac{hk - 0}{k} = k \quad (\because \ h > k, \ |x| > |y|)$$

and $f_y(0,0) = \lim\limits_{k \to 0} \dfrac{f(0, 0+k) - f(0,0)}{k} = \lim\limits_{k \to 0} \dfrac{0 - 0}{k} = 0.$

So, $f_{xy}(0,0) = \lim\limits_{h \to 0} \dfrac{f_y(0+h, 0) - f_y(0,0)}{h} = \lim\limits_{h \to 0} \dfrac{h - 0}{h} = 1$

Similarly, $f_x(0,k) = \lim\limits_{h \to 0} \dfrac{f(0+h, k) - f(0,k)}{h} = \lim\limits_{h \to 0} \dfrac{1}{h}(-hk - 0) = -k$

$$[\because \ k > h, \ |y| > |x|]$$

and $f_x(0,0) = \lim\limits_{h \to 0} \dfrac{f(0+h, 0) - f(0,0)}{h} = \lim\limits_{h \to 0} \dfrac{0 - 0}{h} = 0.$

$\therefore \quad f_{yx}(0,0) = \lim\limits_{k \to 0} \dfrac{f_x(0, 0+k) - f_x(0,0)}{k} = \lim\limits_{k \to 0} \dfrac{-k-0}{k} = -1$

$\therefore \quad f_{yx}(0,0) \neq f_{xy}(0,0).$

Since $f_{xy}(0,0) = 1$ and $f_{yx}(0,0) = -1$, so neither f_{yx} nor f_{xy} is continuous at the origin and therefore the second condition of Schwartz's theorem is lacking here. □

Example 2.14.6 *Show that the function*

$$f(x,y) = \begin{cases} xy\dfrac{x^2-y^2}{x^2+y^2} & when \quad x^2+y^2 \neq 0 \\ 0 & when \quad x^2+y^2 = 0 \end{cases}$$

is continuous at $(0,0)$ but $\dfrac{\partial^2 f}{\partial x \partial y} \neq \dfrac{\partial^2 f}{\partial y \partial x}$ at the origin.

Solution: For continuity of the given function see **Example 1.14.5**.

Second Part: We have $f_y(0,0) = \lim\limits_{k \to 0} \dfrac{f(0,k) - f(0,0)}{k} = \lim\limits_{k \to 0} \dfrac{0}{k} = 0$

$f_y(h,0) = \lim\limits_{k \to 0} \dfrac{f(h,k) - f(h,0)}{k} = \lim\limits_{k \to 0} \dfrac{hk(h^2 - k^2)}{k(h^2 + k^2)} = h$

$\therefore \quad f_{xy}(0,0) = \lim\limits_{h \to 0} \dfrac{f_y(h,0) - f_y(0,0)}{h} = \lim\limits_{h \to 0} \dfrac{h-0}{h} = 1$

Again $f_x(0,0) = \lim\limits_{h \to 0} \dfrac{f(h,0) - f(0,0)}{h} = \lim\limits_{h \to 0} \dfrac{0}{h} = 0$

$f_x(0,k) = \lim\limits_{h \to 0} \dfrac{f(h,k) - f(0,k)}{h} = \lim\limits_{h \to 0} \dfrac{hk(h^2 - k^2)}{k(h^2 + k^2)} = -k$

$\therefore \quad f_{yx}(0,0) = \lim\limits_{k \to 0} \dfrac{f_x(0,k) - f_x(0,0)}{k} = \lim\limits_{k \to 0} \dfrac{-k-0}{k} = -1$

Hence $f_{xy}(0,0) \neq f_{yx}(0,0)$. □

Note 2.14.2 *For the points other than the origin, we get $f_{yx} = f_{xy}$ for if $(x,y) \neq (0,0)$, from $f(x,y) = \dfrac{x^3 y - xy^3}{x^2 + y^2}$ we have*

$$f_x(x,y) = \dfrac{x^4 y + 4x^2 y^3 - y^5}{(x^2 + y^2)^2}, \quad f_y(x,y) = \dfrac{x^5 - 4x^3 y^2 - xy^4}{(x^2 + y^2)^2}.$$

$$\therefore \quad f_{yx}(x,y) = \dfrac{\partial}{\partial y} f_x(x,y) = \dfrac{x^6 - y^6 + 9x^4 y^2 - 9x^2 y^4}{(x^2 + y^2)^3}$$

$$and \quad f_{xy}(x,y) = \dfrac{\partial}{\partial x} f_y(x,y) = \dfrac{x^6 - y^6 + 9x^4 y^2 - 9x^2 y^4}{(x^2 + y^2)^3}$$

i.e., $f_{yx}(x,y) = f_{xy}(x,y)$ when $(x,y) \neq (0,0)$.

Again it can be easily seen by taking $x = r\cos\theta, y = r\sin\theta$ that

$$\lim\limits_{(x,y) \to (0,0)} f_{xy}(x,y) = \lim\limits_{(x,y) \to (0,0)} f_{yx}(x,y)$$

$$= \lim\limits_{r \to 0} r\sin\theta(\cos 2\theta + 1 - \cos^2 2\theta) = 0$$

*But $f_{xy}(0,0) = 1$ and $f_{yx}(0,0) = -1$. Thus **neither f_{xy} nor f_{yx} is continuous at the origin.***

Note 2.14.3 *For the above function we can also prove that neither f_x nor f_y is differentiable at the origin. For, we have*

$$\left.\begin{aligned} f_x(x,y) &= \frac{x^4 y + 4x^2 y^3 - y^5}{(x^2 + y^2)^2} \\[2mm] \text{and} \quad f_y(x,y) &= \frac{x^5 - 4x^3 y^2 - xy^4}{(x^2 + y^2)^2} \end{aligned}\right\} \text{ when } (x,y) \neq (0,0).$$

Also $f_y(0,0) = f_x(0,0) = 0$ and that both f_x and f_y are continuous at $(0,0)$. Now, regarding the differentiability of f_x at the origin, we find that

$$f_{xx}(0,0) = \lim_{h \to 0} \frac{f_x(h,0) - f_x(0,0)}{h} = \lim_{h \to 0} \frac{0}{h^5} = 0 \text{ and } f_{yx}(0,0) = -1.$$

Let $\Delta f_x(0,0) = f_x(h,k) - f_x(0,0) = \dfrac{h^4 k + 4h^2 k^3 - k^5}{(h^2 + k^2)^2}.$

and $df_x(0,0) = h f_{xx}(0,0) + k f_{yx}(0,0) = 0 - k = -k.$

If f_x is total differentiable at $(0,0)$ then by definition the error term $\epsilon = \dfrac{\Delta f_x - df_x}{\sqrt{h^2 + k^2}} \to 0$ as $(h,k) \to (0,0)$.

Now $\dfrac{\Delta f_x - df_x}{\sqrt{h^2 + k^2}} = \dfrac{1}{\sqrt{h^2 + k^2}} \left(\dfrac{h^4 k + 4h^2 k^3 - k^5}{(h^2 + k^2)^2} \right) + k = \dfrac{2kh^2(h^2 + 3k^2)}{(h^2 + k^2)^{\frac{5}{2}}}$

By putting the limiting approach along $k = mh$ we can prove that the simultaneous limit $\lim\limits_{(h,k) \to (0,0)} \dfrac{2kh^2(h^2 + 3k^2)}{(h^2 + k^2)^{\frac{5}{2}}}$ does not exist.

Hence f_x is not differentiable at the origin.
Similarly, it can be shown that f_y is also not differentiable at the origin.

Note 2.14.4 *If $f(x,y)$ is found to violate the conditions of Schwarz's and Young theorem even then it is not necessary that $f_{yx}(0,0) \neq f_{xy}(0,0)$.*

In fact it is shown in the above example that $f_{yx}(0,0) = -1 \neq 1 = f_{xy}(0,0)$.

But for $(x,y) \neq (0,0)$, $f_{xy} = \dfrac{x^6 + 9x^4 y^2 - 9x^2 y^4 - y^6}{(x^2 + y^2)^2} = f_{yx}$.

Also, taking $x = r\cos\theta, y = r\sin\theta$, it can be seen that neither $f_{xy} \to 1$ nor $f_{yx} \to -1$ as $r \to 0$. So, neither f_{xy} nor f_{yx} is continuous at the origin.

Example 2.14.7 *If a triangle ABC be slightly varied but so as to be remained inscribed in the same circle, show that*

$$\frac{da}{\cos A} + \frac{db}{\cos B} + \frac{dc}{\cos C} = 0,$$

da, db, dc being small variations of the sides a, b, c.

Solution: *Here, R, the radius of the circum-circle is fixed,*

$$\text{i.e.,} \quad \frac{abc}{4\triangle} = \text{constant} \quad \text{or,} \quad \frac{\triangle^2}{a^2 b^2 c^2} = \text{constant}$$

$$\text{or,} \quad \frac{s(s-a)(s-b)(s-c)}{a^2 b^2 c^2} = \text{constant}$$

$$\text{or,} \quad \frac{2b^2 c^2 + 2c^2 a^2 + 2a^2 b^2 - a^4 - b^4 - c^4}{a^2 b^2 c^2} = \text{constant}$$

$$\text{or,} \quad \frac{2}{a^2} + \frac{2}{b^2} + \frac{2}{c^2} - \frac{a^2}{b^2 c^2} - \frac{b^2}{c^2 a^2} - \frac{c^2}{a^2 b^2} = \text{constant} = u \text{ (say)}$$

$$\therefore \quad \frac{\partial u}{\partial a} = -\frac{4}{a^3} - \frac{2a}{b^2 c^2} + \frac{2b^2}{c^2 a^3} + \frac{2c^2}{a^3 b^2} = \frac{-4b^2 c^2 - 2a^4 + 2b^4 + 2c^4}{a^3 b^2 c^2}$$

$$= \frac{2[(b^2 - c^2)^2 - (a^2)^2]}{a^3 b^2 c^2} = \frac{2(b^2 - c^2 + a^2)(b^2 - c^2 - a^2)}{a^3 b^2 c^2}$$

$$= -\frac{(b^2 + c^2 - a^2)(c^2 + a^2 - b^2)(a^2 + b^2 - c^2)}{a^3 b^3 c^3} \times \frac{2bc}{(b^2 + c^2 - a^2)}$$

Put $-\dfrac{(b^2 + c^2 - a^2)(c^2 + a^2 - b^2)(a^2 + b^2 - c^2)}{a^3 b^3 c^3} = \alpha, \quad \therefore \quad \dfrac{\partial u}{\partial a} = \dfrac{\alpha}{\cos A}$

Similarly, $\dfrac{\partial u}{\partial b} = \dfrac{\alpha}{\cos B}$ and $\dfrac{\partial u}{\partial c} = \dfrac{\alpha}{\cos C}$.

Again $\dfrac{\partial u}{\partial a} da + \dfrac{\partial u}{\partial b} db + \dfrac{\partial u}{\partial c} dc = du$.

But $du = 0 \quad \therefore \quad \dfrac{\alpha}{\cos A} da + \dfrac{\alpha}{\cos B} db + \dfrac{\alpha}{\cos C} dc = 0$

$$\Rightarrow \quad \frac{da}{\cos A} + \frac{db}{\cos B} + \frac{dc}{\cos C} = 0. \qquad \square$$

Example 2.14.8 *In estimating the cost of a pile of bricks measured as $2m \times 15m \times 1.2m$, the tape is stretched 1% beyond the standard length. If the count is 540 bricks to $1cu.m$ and bricks cost Rs. 530 per 1000, find the approximate error in the cost.*

Solution: Let x, y and z be the length, breadth and height of pile so that its volume $V = xyz$. Taking logarithm we get

$$\log V = \log x + \log y + \log z \quad \text{or,} \quad \frac{dV}{V} = \frac{dx}{x} + \frac{dy}{y} + \frac{dz}{z}.$$

Since $V = 2m \times 15m \times 1.2m = 36m^3$ and $\dfrac{dx}{x} = \dfrac{dy}{y} = \dfrac{dz}{z} = \dfrac{1}{100}$

$$\therefore \quad dV = 36 \times \left(\frac{3}{100}\right) = 1.08m^3.$$

Number of bricks in $dV = 1.08 \times 450 = 486$.

Thus error in the cost $= 486 \times \dfrac{530}{1000} =$ Rs. 257.58 which is a loss to the brick seller. $\qquad \square$

Example 2.14.9 *If the partial derivatives f_x, f_y of a function $f(x, y)$ exist and vanish at every point in its domain, prove that $f(x, y)$ is a constant.*

Solution: Let (x, y) and $(x + h, y + k)$ be two neighbouring points in the domain of definition of $f(x, y)$. By M.V.T. we get for $0 < \theta < 1$,

$$f(x + h, y + k) - f(x, y) = h f_x(x + \theta h, y + \theta k) + k f_y(x + \theta h, y + \theta k) \quad (2.43)$$

The points $(x + \theta h, y + \theta k)$ for $0 < \theta < 1$ wholly lies on the line joining the two points (x, y) and $(x + h, y + k)$ in the said domain of $f(x, y)$. So, by the given condition $f_x(x + \theta h, y + \theta k) = 0 = f_y(x + \theta h, y + \theta k)$. Hence by relation (2.43), we get

$$f(x + h, y + k) - f(x, y) = 0$$

which implies that $f(x, y)$ is constant. $\qquad \square$

Example 2.14.10 *Apply MVT to* $f(x, y) = \sin \pi x + \cos \pi y$ *and express* $f\left(\frac{1}{2}, 0\right) -$

$f\left(0, -\frac{1}{2}\right)$ *in terms of the first order partial derivatives of* f *and deduce that* $\exists\, \theta$ *in* $(0, 1)$ *such that* $\dfrac{4}{\pi} = \cos \dfrac{\pi}{2}\theta + \sin \dfrac{\pi}{2}(1 - \theta).$

Solution: For any two neighbouring points (x, y) and $(x + h, y + k)$ in the domain of definition of $f(x, y)$, the M.V.T is

$$f(x + h, y + k) - f(x, y) = h f_x(x + \theta h, y + \theta k) + k f_y(x + \theta h, y + \theta k) \quad (2.44)$$

where $0 < \theta < 1$.

According to the problem, $x + h = \dfrac{1}{2}, x = 0 \Rightarrow h = \dfrac{1}{2}$ and $y + k = 0, y = -\dfrac{1}{2} \Rightarrow k = \dfrac{1}{2}$, i.e., $x + \theta h = 0 + \theta\dfrac{1}{2} = \dfrac{\theta}{2}$ and $y + \theta k = -\dfrac{1}{2} + \theta\dfrac{1}{2} = \dfrac{\theta}{2} - \dfrac{1}{2}.$

Now, from (2.44) we get

$$f\left(\frac{1}{2}, 0\right) - f\left(0, -\frac{1}{2}\right) = \frac{1}{2}\left[f_x\left(\frac{\theta}{2}, \frac{\theta}{2} - \frac{1}{2}\right) + f_y\left(\frac{\theta}{2}, \frac{\theta}{2} - \frac{1}{2}\right)\right] \quad (2.45)$$

where $0 < \theta < 1$.

Given $f(x, y) = \sin \pi x + \cos \pi y$ $\therefore$ $f_x = \pi \cos \pi x, f_y = -\pi \sin \pi y$.

$\therefore$ $f\left(\dfrac{1}{2}, 0\right) = \sin \dfrac{\pi}{2} + \cos \pi.0 = 2$ and $f\left(0, -\dfrac{1}{2}\right) = 0$

Again, $f_x\left(\dfrac{\theta}{2}, \dfrac{\theta}{2} - \dfrac{1}{2}\right) = \pi \cos \dfrac{\pi}{2}\theta$ and $f_y\left(\dfrac{\theta}{2}, \dfrac{\theta}{2} - \dfrac{1}{2}\right) = \pi \sin \dfrac{\pi}{2}(1 - \theta).$

Now from (2.45) we get

$$2 - 0 = \frac{1}{2}\left[\pi \cos \frac{\pi}{2}\theta + \pi \sin \frac{\pi}{2}(1 - \theta)\right]$$

or, $\dfrac{4}{\pi} = \cos \dfrac{\pi}{2}\theta + \sin \dfrac{\pi}{2}(1 - \theta).$ $\qquad\qquad\qquad\qquad\qquad\qquad\qquad\square$

Example 2.14.11 *Prove that the expansion up to fourth term of the Maclaurin's expansion of* $e^{ax} \cos by$ *is*

$$1 + ax + \frac{a^2 x^2 - b^2 y^2}{2!} + \frac{a^3 x^3 - 3ab^2 xy^2}{3!}.$$

Solution: Let $f(x, y) = e^{ax} \cos by$ $\Rightarrow$ $f(0, 0) = 1$.

$\therefore$ $f_x = a e^{ax} \cos by, f_y = -b e^{ax} \sin by$ $\therefore$ $f_x(0, 0) = a, f_y(0, 0) = 0$.

$\therefore$ $f_{xx} = a^2 e^{ax} \cos by, f_{xy} = -ab e^{ax} \sin by, f_{yy} = -b^2 e^{ax} \cos by$

$\therefore$ $f_{xx}(0, 0) = a^2, f_{xy}(0, 0) = 0, f_{yy}(0, 0) = -b^2$.

$\therefore$ $f_{xxx} = a^3 e^{ax} \cos by, f_{xxy} = -a^2 b e^{ax} \sin by, f_{xyy} = -ab^2 e^{ax} \cos by$

and $f_{yyy} = b^3 e^{ax} \sin by$

$\therefore$ $f_{xxx}(0, 0) = a^3, f_{xxy}(0, 0) = 0, f_{xyy}(0, 0) = -ab^2, f_{yyy} = 0$.

Now Maclaurin's series expansion (without remainder term) up to fourth term is

$$
\begin{aligned}
f(x,y) &= f(0,0) + \{xf_x(0,0) + yf_y(0,0)\} + \frac{1}{2!}\{x^2 f_{xx}(0,0) + 2xy f_{xy}(0,0) \\
&\quad + y^2 f_{yy}(0,0)\} + \frac{1}{3!}\{x^3 f_{xxx}(0,0) + 3x^2 y f_{xxy}(0,0) \\
&\quad + 3xy^2 f_{xyy}(0,0) + y^3 f_{yyy}(0,0)\} + \dots \\
&= 1 + (x.a + y.0) + \frac{1}{2!}(x^2 a^2 + 2xy.0 + y^2.b^2) \\
&\quad + \frac{1}{3!}\{x^3.a^3 + 3x^2 y.0 + 3xy^2(-ab^2) + y^3.0\} + \dots \\
&= 1 + ax + \frac{a^2 x^2 - b^2 y^2}{2!} + \frac{a^3 x^3 - 3ab^2 xy^2}{3!} + \dots \qquad \square
\end{aligned}
$$

Example 2.14.12 *Show that the expansion of* $\sin(xy)$ *in powers of* $(x-1)$ *and* $\left(y - \frac{\pi}{2}\right)$ *up to and including second degree terms is*

$$
1 - \frac{1}{8}\pi^2(x-1)^2 - \frac{1}{2}\pi(x-1)\left(y - \frac{\pi}{2}\right) - \frac{1}{2}\left(y - \frac{\pi}{2}\right)^2.
$$

Solution: Here we have to consider the Taylor's series expansion about the point $(x,y) = \left(1, \frac{\pi}{2}\right)$.

Let $f(x,y) = \sin(xy) \Rightarrow f\left(1, \frac{\pi}{2}\right) = \sin\frac{\pi}{2} = 1.$

$\therefore f_x = y\cos(xy), f_y = x\cos(xy) \Rightarrow f_x\left(1, \frac{\pi}{2}\right) = f_y\left(1, \frac{\pi}{2}\right) = 0.$

Again, $f_{xx} = -y^2 \sin(xy),\ f_{xy} = -xy\sin(xy),\ f_{yy} = -x^2 \sin(xy)$

$\therefore f_{xx}\left(1, \frac{\pi}{2}\right) = -\frac{\pi^2}{4},\ f_{xy}\left(1, \frac{\pi}{2}\right) = -\frac{\pi}{2}$ and $f_{yy}\left(1, \frac{\pi}{2}\right) = -1.$

Now the Taylor's series expansion of $f(x,y)$ about the point $(x,y) = \left(1, \frac{\pi}{2}\right)$ up to second order term is

$$
\begin{aligned}
f(x,y) &= f\left(1, \frac{\pi}{2}\right) + \left\{(x-1)f_x\left(1, \frac{\pi}{2}\right) + \left(y - \frac{\pi}{2}\right)f_y\left(1, \frac{\pi}{2}\right)\right\} \\
&\quad + \frac{1}{2!}\left\{(x-1)^2 f_{xx}\left(1, \frac{\pi}{2}\right) + 2(x-1)\left(y - \frac{\pi}{2}\right)f_{xy}\left(1, \frac{\pi}{2}\right) \right. \\
&\quad \left. + \left(y - \frac{\pi}{2}\right)^2 f_{yy}\left(1, \frac{\pi}{2}\right)\right\} + \dots
\end{aligned}
$$

$$
\begin{aligned}
\therefore \ \sin(xy) &= 1 + \left\{(x-1).0 + \left(y - \frac{\pi}{2}\right).0\right\} + \frac{1}{2!}\left\{(x-1)^2\left(-\frac{\pi^2}{4}\right) \right. \\
&\quad \left. + 2(x-1)\left(y - \frac{\pi}{2}\right)\left(-\frac{\pi}{2}\right) + \left(y - \frac{\pi}{2}\right)^2(-1)\right\} + \dots \\
&= 1 - \frac{1}{8}\pi^2(x-1)^2 - \frac{1}{2}\pi(x-1)\left(y - \frac{\pi}{2}\right) - \frac{1}{2}\left(y - \frac{\pi}{2}\right)^2. \qquad \square
\end{aligned}
$$

Example 2.14.13 *Let* (x,y) *approach towards* $(0,0)$ *along the line* $y = -x$. *Using Taylor's Theorem for functions of two variables, show that*

$$
\lim_{\substack{(x,y)\to(0,0) \\ (y=-x)}} \frac{\sin xy + xe^x - y}{x\cos y + \sin 2y} = -2.
$$

Solution: Let $f(x,y) = \sin xy + xe^x - y$ and $g(x,y) = x\cos y + \sin 2y$. First we expand both the functions f and g in terms of Taylor's series expansion.

$$f(x,y) = \sin xy + xe^x - y = f(0,0) + \{xf_x(0,0) + yf_y(0,0)\}$$
$$+ \frac{1}{2!}\{x^2 f_{xx}(0,0) + 2xy f_{xy}(0,0) + y^2 f_{yy}(0,0)\} + \dots$$

Here, $f(0,0) = 0$, $f_x(0,0) = 1$, $f_y(0,0) = -1$, $f_{xx}(0,0) = 2$, $f_{xy}(0,0) = 1$, $f_{yy}(0,0) = 0$ etc.

So, $\sin xy + xe^x - y = x - y + \dfrac{1}{2}[2x^2 + 2xy] + \dots$

Similarly, $g(0,0) = 0, g_x(0,0) = 1, g_y(0,0) = 2, g_{xx}(0,0) = 0, g_{xy}(0,0) = 0, g_{yy}(0,0) = 0$ etc.

Therefore, $x\cos y + \sin 2y = x + 2y + \dfrac{1}{2}.(0) + \dots = x + 2y$

$$\therefore \quad \lim_{\substack{(x,y)\to(0,0) \\ (y=-x)}} \frac{f(x,y)}{g(x,y)} = \lim_{\substack{(x,y)\to(0,0) \\ (y=-x)}} \frac{x - y + x^2 + xy + \dots}{x + 2y + \dots}$$

$$= \lim_{x\to 0} \frac{2x + x^2 - x^2 + \text{higher power of } x}{x - 2y + \text{higher power of } x}$$

$$= \lim_{x\to 0} \frac{2x + \dots}{-x + \dots} = -2. \qquad \square$$

Example 2.14.14 *Find the infinite series expansion about the point $(0,0)$ of the function $f(x,y) = \cos ax \cos by$ by using Taylor's theorem.*

Solution: We have $\cos ax = 1 - \dfrac{a^2 x^2}{\underline{|2}} + \dfrac{a^4 x^4}{\underline{|4}} + \dots \ \forall \ x \in \mathbb{R}$

and $\cos by = 1 - \dfrac{b^2 y^2}{\underline{|2}} + \dfrac{b^4 b^4}{\underline{|4}} + \dots \ \forall \ y \in \mathbb{R}.$

$$\cos ax \cos bx = \left(1 - \frac{a^2 x^2}{\underline{|2}} + \frac{a^4 x^4}{\underline{|4}} + \dots\right)\left(1 - \frac{b^2 y^2}{\underline{|2}} + \frac{b^4 b^4}{\underline{|4}} + \dots\right)$$

$$= 1 - \left(\frac{a^2 x^2}{2} + \frac{b^2 y^2}{2}\right) + \left(\frac{a^4 x^4}{24} + \frac{a^2 b^2 x^2 y^2}{4} + \frac{b^4 y^4}{24}\right) - \dots$$

$$= 1 - \frac{1}{2}(a^2 x^2 + b^2 y^2) + \frac{1}{24}(a^4 x^4 + 6a^2 b^2 x^2 y^2 + b^4 y^4) - \dots$$

$$\forall \ (x,y) \in \mathbb{R}.$$
$$\square$$

2.15 Exercises

1. (a) Define partial derivatives of a function $f(x, y)$ with respect to x and y.

(b) If a function $f(x, y)$ has partial derivatives f_x and f_y everywhere in a domain $\mathbb{R}$ and these derivatives everywhere satisfy the inequalities $|f_x(x, y)| < M$, $|f_y(x, y)| < M$ where M is independent of x, y, prove that $f(x, y)$ is continuous everywhere in $\mathbb{R}$.

[*Hints:* See **Theorem 2.3.2**]

(c) Prove that the function defined by $f(x, y) = \dfrac{2xy}{x^2 + y^2}$, $x^2 + y^2 \neq 0$ and $f(0, 0) = 0$ has partial derivatives everywhere. Is it continuous everywhere?

2. (a) What do you understand by the term total differential of a function of several variables?

(b) Examine whether $f(x, y) = \sqrt{|xy|}$ is totally differentiable at the origin.

[*Hints:* Clearly, $f_x(0, 0) = f_y(0, 0) = 0$. Let (h, k) be the neighbouring point of the origin where $(h, k) \to (0, 0)$. If f is total differentiable then the error term
$$\epsilon = \frac{\Delta f - df}{\sqrt{h^2 + k^2}} \to 0 \text{ as } (h, k) \to (0, 0), \text{ where } \Delta f = f(h, k) - f(0, 0) = \sqrt{|hk|} \text{ and}$$
$df = h f_x(0, 0) + k f_y(0, 0) = 0$.

Now taking the limiting approach along the curve $k = mh$ we can show that the simultaneous limit $\displaystyle\lim_{(h,k)\to(0,0)} \frac{\Delta f - df}{\sqrt{h^2 + k^2}} = \lim_{(h,k)\to(0,0)} \frac{\sqrt{|hk|}}{\sqrt{h^2 + k^2}}$ does not exist. Hence the function $f(x, y) = \sqrt{|xy|}$ is not total differentiable.]

(c) Find the total differential of $u = (x + y + z)e^z$.

[*Hints:* $du = (dx + dy + dz)e^z + (x + y + z)e^z dz$]

3. (a) State and prove Schwarz's theorem on the commutative property of the order of partial derivatives.

(b) Show that for the following function f neither f_{xy} nor f_{yx} is continuous at the origin and that neither f_x nor f_y is differentiable at the origin

$$f(x, y) = \begin{cases} xy\dfrac{x^2 - y^2}{x^2 + y^2} & \text{when} \quad x^2 + y^2 \neq 0 \\ 0 & \text{when} \quad x^2 + y^2 = 0. \end{cases}$$

If $f_{xy}(0, 0) = f_{yx}(0, 0)$?

If f is found to violate the conditions of Schwarz's and Young's theorem even then it is not necessary that $f_{xy}(0, 0) \neq f_{yx}(0, 0)$.

[*Hints:* See **Example 2.12.6** and follow **Note 2.13.1, 2.13.2, 2.13.3**]

4. Show that if u be a function of two independent variables x and y, then under certain conditions to be stated by you, $\dfrac{\partial^2 u}{\partial x \partial y} \neq \dfrac{\partial^2 u}{\partial y \partial x}$.

[*Hints:* Schwarz's Theorem.]

5. For the function $f(x, y) = \begin{cases} \dfrac{x^2 y^2}{x^2 + y^2} & \text{when} \quad x^2 + y^2 \neq 0 \\ 0 & \text{when} \quad x^2 + y^2 = 0, \end{cases}$ show that in spite of the fact that neither the conditions of Schwarz's theorem nor the conditions of Young's theorem are satisfied, $f_{xy} = f_{yx}$ at $(0, 0)$.

6. Show that the function f defined in the neighbourhood of $(0, 0)$ by $f(x, y) = |x| + |y|$ is continuous but not differentiable thereat.

[*Hints:* We see that $f_x(0,0) = \lim\limits_{h \to 0} \dfrac{f(h,0) - f(0,0)}{h} = \lim\limits_{h \to 0} \dfrac{|h|}{h} = \pm 1$ according as $h \to 0+$ or $h \to 0-$. Thus the limit does not exist and hence $f_x(0,0)$ does not exis. Similarly, $f_y(0,0)$ does not exist.

Since the partial derivatives of the function $f(x, y)$ do not exist at $(0,0)$, so the function is not differentiable thereat.]

7. If $f(x, y) = x|x| + |y|$, for all reals x, y, is $f(x, y)$ differentiable at the origin?

[*Hints:* Show that $f_y(0,0)$ does not exist and consequently $f(x, y)$ is not differentiable.]

8. If $f(x, y) = x|y| + y|x|$, for all real values of x, y, show that f_x and f_y both exist and continuous at the origin. Verify whether f is differentiable at the origin.

[*Hints:* If $x > 0, f(x, y) = x|y| + xy \in \mathbb{R}$ and if $x < 0, f(x, y) = x|y| - xy \in \mathbb{R}$. $\therefore$ $f_x(x, y) = |y| + y$ or $|y| - y$ according as $x > 0$ or $x < 0$. clearly, $f_x(0,0) = 0$. Also $\lim\limits_{(x,y) \to (0,0)} f_x(0,0) = \lim\limits_{y \to 0}(|y| \pm y) = 0 = f_x(0,0)$.

Hence f_x exists everywhere on $\mathbb{R}^2$ and is also continuous everywhere. Since $f(x, y) = f(y, x)$, f_y also exists and is continuous everywhere on $\mathbb{R}^2$ and $f_y(0,0) = 0$.

Now, $\Delta f(0,0) = f(x, y) - f(0,0) = f(x, y) = x|y| + y|x| = 0.x + 0.y + |y|x + |x|y = Ax + By + \phi x + \psi y$ where $A = 0, B = 0$ which are independent of x and y and $\phi = |y|$ and $\psi = |x|$ both tends to 0 as $(x, y) \to (0,0)$.

Hence f is differentiable at the origin.]

9. Show that $u(x, y) = ax^2 + 2hxy + by^2 + 2gx + 2fy + c$ is differentiable everywhere in $\mathbb{R}^2$ for all real values of x, y, where a, b, c, f, g, h are all constants.

10. If $f(x, y) = \begin{cases} \dfrac{x^3 + y^3}{x - y} & \text{when} \quad x \neq y \\ 0 & \text{when} \quad x = y, \end{cases}$ then show that $f(x, y)$ is not

continuous at $(0,0)$, but f_x, f_y exist at $(0,0)$.

11. Show that $f(x, y) = \begin{cases} \dfrac{xy^2}{x^2 + y^4} & \text{if} \quad x \neq 0 \\ 0 & \text{if} \quad x = 0, \end{cases}$ possesses first order partial

derivatives at $(0,0)$, yet it is not differentiable at $(0,0)$.

12. State Young's theorem on the commutativity of the order of partial derivatives. Use the following function to verify whether the conditions for Young's theorem are necessary: $f(x, y) = \begin{cases} \dfrac{x^2 y^2}{x^2 + y^2} & \text{if} \quad x^2 + y^2 \neq 0 \\ 0 & \text{if} \quad x^2 + y^2 = 0. \end{cases}$

13. Answer the following questions:

   (a) Define directional derivative.

   (b) Write the expression for total differential of a function of two variables.

   (c) Define the differentiability of the function $f(x, y)$ at the point (x, y).

14. For $f(x, y) = \begin{cases} (x^2 + y^2) \log(x^2 + y^2) & \text{for} \quad (x, y) \neq (0,0) \\ 0 & \text{for} \quad (x, y) = (0,0), \end{cases}$ show that

the conditions of Schwarz's theorem are not satisfied but $f_{xy} = f_{yx}$ at $(0,0)$.

15. Let $f(x, y) = \begin{cases} \dfrac{xy}{x^2 + y^2} & \text{when} \quad x^2 + y^2 \neq 0 \\ 0 & \text{when} \quad x^2 + y^2 = 0. \end{cases}$ Find the directional

derivative at $(0,0)$ in any arbitrary direction $\hat{\beta} = (l, m), l^2 + m^2 = 1$. What are the partial derivatives with respect to x or y at $(0,0)$?

[*Hints:* $D_{\hat{\beta}}f(0,0) = \lim\limits_{t \to 0} \dfrac{f(lt, mt) - f(0,0)}{t} = \lim\limits_{t \to 0} \dfrac{lm}{l^2 + m^2} \cdot \dfrac{1}{t}$. This limit does not exist always. But, if $\hat{\beta} = (1,0)$ or $(0,1)$, then $f_x(0,0) = 0 = f_y(0,0)$.

Thus both f_x and f_y exist at $(0,0)$, but since $\lim\limits_{(x,y) \to (0,0)} \dfrac{xy}{x^2 + y^2}$ does not exist uniquely, the function is not continuous at $(0,0)$.

But, we are to remember that even when f has directional derivatives in all arbitrary directions still f may not be continuous.]

16. Approximate the change in the hypotenuse of a right angled triangle whose sides are 6cm and 8cm when the shorter side is lengthened by $\dfrac{1}{4}$cm and the longer is shortened by $\dfrac{1}{8}$cm.

17. If x increases at the rate of 2cm/sec at the instant when $x = 3$cm and $y = 1$cm, at what rate must y be changing in order that the function $2xy - 3x^2y$ shall be neither increasing nor decreasing?

18. The diameter and altitude of a can in the shape of a right circular cylinder are measured as 4cm and 6cm respectively. The possible error in each measurement is 0.1cm. Find approximately the maximum possible error in the values computed for the volume and lateral surface.

19. Let $f(x,y) = \begin{cases} \dfrac{x^3 y}{x^2 + y^2} & \text{when} \quad x^2 + y^2 \neq 0 \\ 0 & \text{when} \quad x^2 + y^2 = 0. \end{cases}$

Show that $f_{xy}(0,0) \neq f_{yx}(0,0)$.

20. Let $f(x,y) = \begin{cases} \dfrac{x^4 + y^4}{x - y} & \text{when} \quad x \neq y \\ 0 & \text{when} \quad x = y. \end{cases}$ Show that $\dfrac{\partial f}{\partial x}$ and $\dfrac{\partial f}{\partial y}$ exist at $(0,0)$, but f is not continuous at $(0,0)$.

21. If $f(x,y) = x^2 + 2y^2 - 3xy$, use mean value theorem to express $f(1,2) - f(2,-1)$ in terms of partial derivatives of f and compute θ, $(0 < \theta < 1)$.

22. If $f(x,y) = \sqrt{x + y}$, use mean value theorem to express $f(3,1) - f(1,0)$ by partial derivatives of $f(x,y)$. Compute θ, $(0 < \theta < 1)$.

23. If $f(x,y) = x^3 - 2y^3 + 3xy$, use mean value theorem to express $f(1,2) - f(2,1)$ in terms of partial derivatives of f and deduce that there exists θ, $(0 < \theta < 1)$ satisfying $30^2 + 2\theta - 2 = 0$.

24. If $f(x,y) = \sqrt{|xy|}$, prove that Taylor's series expansion about the point (x,x) is not valid in any domain which includes the origin.

[*Hints:* Clearly, $f_x(0,0) = f_y(0,0) = 0$. Also

$$f_x(x,y) = \begin{cases} \dfrac{1}{2}\sqrt{\left|\dfrac{y}{x}\right|}, & x > 0 \\ -\dfrac{1}{2}\sqrt{\left|\dfrac{y}{x}\right|}, & x < 0 \end{cases} \quad \text{and} \quad f_y(x,y) = \begin{cases} \dfrac{1}{2}\sqrt{\left|\dfrac{x}{y}\right|}, & y > 0 \\ -\dfrac{1}{2}\sqrt{\left|\dfrac{x}{y}\right|}, & y < 0. \end{cases}$$

$$\therefore \quad f_x(x,x) = f_y(x,x) = \begin{cases} \dfrac{1}{2}, & x > 0 \\[2mm] -\dfrac{1}{2}, & x < 0. \end{cases}$$

So, by first order Taylor's expansion about the point (x,x) we get,

$$f(x+h, x+h) - f(x,x) = h \left[\frac{\partial}{\partial x} + \frac{\partial}{\partial y} \right]\Big|_{(x+\theta h, x+\theta h)} f(x,y)$$

$$\text{or,} \quad |x+h| = \begin{cases} |x| + h & \text{if} \quad x + \theta h > 0 \\ |x| - h & \text{if} \quad x + \theta h < 0 \\ |x| & \text{if} \quad x + \theta h = 0 \end{cases} \tag{2.46}$$

If the said h-nbd of the point (x,x) contains the point $(0,0)$, then x and $x+h$ must be of opposite signs, i.e., either $|x+h| = x+h$, $|x| = -x$ or $|x+h| = -(x+h)$, $|x| = x$. But for these conditions the functional relationship given in (2.46) will not be valid and as a result, the expansion is not valid.]

25. Expand $x^4 + x^2 y^2 - y^4$ about the point $(1,1)$ up to terms of the second degree. Find the form of the remainder R_2.

26. Show that if f, f_x, f_y are all continuous in a domain D of (a,b), and D is large enough to contain the point $(a+h, b+k)$, within it, then for $0 < \theta < 1$,

$$f(a+h, b+k) = f(a,b) + h f_x(a+\theta h, b+\theta k) + k f_y(a+\theta h, b+\theta k).$$

If $f(x,y) = x\sqrt{x^2 + y^2}$, $a = b = -1$, $h = k = 3$, verify that the above conditions are satisfied and find the value of θ.

27. Expand $e^x \tan^{-1} y$ about $(1,1)$ up to the second degree in $(x-1)$ and $(y-1)$.

28. If $f(x,y) = f(a+ht, b+kt) = F(t)$ where a,b,h,k are constants, show that for $0 < \theta < 1$, $F^n(\theta) = \left[h\dfrac{\partial}{\partial x} + k\dfrac{\partial}{\partial y} \right]^n f(a+\theta h, y+\theta k)$.

29. If $z = f(x,y)$ and $x = a+ht$, $y = b+kt$ where a,b,h,k are constants, show that

$$\frac{d^n z}{dt^n} = \left\{ h\frac{\partial}{\partial x} + k\frac{\partial}{\partial y} \right\}^n z = h^n \frac{\partial^n z}{\partial x^n} + \begin{Bmatrix} n \\ 1 \end{Bmatrix} h^{n-1} k \frac{\partial^n z}{\partial x^{n-1} \partial y} + \cdots$$

$$+ \begin{Bmatrix} n \\ r \end{Bmatrix} h^{n-r} k^r \frac{\partial^n z}{\partial x^{n-r} \partial y^r} + \cdots + k^n \frac{\partial^n z}{\partial y^n}.$$

30. Obtain Taylor's expansion of the function $\ln\left(\dfrac{x+y}{2}\right)$, $x,y \in \mathbb{R}^+$ about the point $(1,1)$ up to and including the third degree terms.

31. With proper justification, show that

$$\cos x \cos y = 1 - \frac{1}{2}(x^2 + y^2) + \frac{1}{6}[(x^3 + 3xy^2)\sin\theta x \cos\theta y$$
$$+ (3x^2 y + y^3)\cos\theta x \sin\theta y], \quad \text{where} \quad 0 < \theta < 1.$$

32. Prove that for $0 < \theta < 1$,

$$e^{ax}\sin by = by + abxy + \frac{1}{6}[(a^3 x^3 - 3ab^2 xy^2)\sin(b\theta y)$$
$$+ (3a^2 bx^2 y - b^3 y^3)\cos(b\theta y)]e^{a\theta x}.$$

33. Prove that for sufficiently small values of $|x|, |y|$,

$$\frac{\cos x}{\cos y} \approx 1 - \frac{1}{2}(x^2 - y^2).$$

[*Hints*: Let $f(x, y) = \cos x \sec y$. Find $f_x(0,0), f_y(0,0), f_{xx}(0,0), f_{xy}(0,0)$ and $f_{yy}(0,0)$. Clearly, the assumed function and all its partial derivatives are continuous in the neighbourhood of the origin. Also, $|x|, |y|$ are sufficiently small. Hence, by applying Taylor's series expansion of $f(x, y)$ about the origin $(0,0)$ and including second degree term, we get

$$f(x, y) \approx f(0,0) + \left(x\frac{\partial}{\partial x} + y\frac{\partial}{\partial y}\right)\Bigg|_{(0,0)} f + \left(x\frac{\partial}{\partial x} + y\frac{\partial}{\partial y}\right)^2\Bigg|_{(0,0)} f$$

$$\therefore \frac{\cos x}{\cos y} \approx 1 + 0 + \frac{1}{2}(-x^2 + 0 + y^2) = 1 - \frac{1}{2}(x^2 - y^2).]$$

34. Find the expansion of $\cos(xy)$ in powers of $x - 1$ and $\left(y - \frac{\pi}{2}\right)$ up to and including third degree terms.

ANSWERS

1. (c) No, it is not continuous at $(0,0)$; **2.** (b) Not total differentiable; (c) $e^z(dx + dy + dz) + udz$; **3.** (b) No; **7.** No; **8.** Differentiable; **15.** Directional derivative does not exist at $(0,0)$ in arbitrary directions, but $f_x(0,0) = 0 = f_y(0,0)$; **16.** Hypotenuse is lengthened by $\frac{1}{20}$ cm (approximately); **17.** y is decreasing at the rate of $\frac{32}{21}$ cm/sec; **18.** Error in volume $= 1.6\pi cm^3$, error in lateral surface $= \pi cm^3$; **21.** $f(1,2) - f(2,-1) = -f_x(2 - \theta, 3\theta - 1) + 2f_y(2 - \theta, 3\theta - 1), \theta = \frac{8}{41}$; **22.** $f(3,1) - f(1,0) = 2f_x(1 + 2\theta, \theta) + f_y(1 + 2\theta, \theta), \theta = \frac{5}{12}$; **23.** $f(1,2) - f(2,1) = -f_x(2 - \theta, \theta + 1) + f_y(2 - \theta, \theta + 1)$; **25.** $1 + [6(x - 1) - (y - 1)] + R_2$, where $R_2 = [(x - 1)^2\{6(1 + \theta x - \theta)^2 + (1 + \theta y - \theta)^2\} + 4(x - 1)(y - 1)(1 + \theta x - \theta)(1 + \theta y - \theta) + (y - 1)^2\{(1 + \theta x - \theta) - 6(1 + \theta y - \theta)\}]$; **26.** $\theta = \frac{11}{6}$; **27.** $\frac{e\pi}{4} + (x - 1)\frac{e\pi}{4} + (y - 1)\frac{e}{\sqrt{2}} + \frac{1}{2}[(x - 1)^2\frac{e\pi}{4} + 2(x - 1)(y - 1)\frac{e}{\sqrt{2}} - (y - 1)^2\frac{e}{2\sqrt{2}}] + ...$; **30.** $\frac{1}{2}(x - 1) + \frac{1}{2}(y - 1) - \frac{1}{8}[(x - 1)^2 + 2(x - 1)(y - 1) + (y - 1)^2] + \frac{1}{3}[(x - 1)^3 + 3(x - 1)^2(y - 1) + 3(x - 1)(y - 1)^2 + (y - 1)^2 \left[\frac{1}{2 + (x + y - 2)\theta}\right]^3, 0 < \theta < 1;$ **34.** $-\frac{\pi}{2}(x - 1) - \left(y - \frac{\pi}{2}\right) + \frac{1}{6}\left[\frac{\pi^3}{8}(x - 1)^3 + \frac{3}{4}\pi^2(x - 1)^2 \left(y - \frac{\pi}{2}\right) + \frac{3}{2}\pi(x - 1)\left(y - \frac{\pi}{2}\right)^2 + \left(y - \frac{\pi}{2}\right)^3\right].$

Chapter 3

Functions of Several Variables: Differentiation - II

3.1 Introduction

This chapter is the continuation of the **Chapter 2** where we have discussed mainly the analytical definitions of first order and higher order partial derivatives, total differentiations, directional derivatives and the inter relationship among them and with limits and continuity of the multiple variable functions. The Schawrz's and the Young's theorems regarding the commutativity of second order partial derivatives are also taken into account there.

In this chapter we shall study the results and formulas for finding different order partial derivatives, total derivatives and differentials, Chain rule for composite functions, change of variables, homogeneous functions and Euler's theorems. Some preliminary notions are given first:

3.2 Recapitulation

Before going to study detail on different types of derivatives of functions of several variables, we first recapitulate, some of the definitions, formulae and expressions which have already been discussed in **Chapter 2**.

Partial derivatives

Let $u = f(x, y)$ be a function of two variables x and y. If we keep y as constant and vary x alone, then u is a function of x only. The derivative of u with respect to x, treating y as constant, is called *partial derivative of u with respect to x* and is denoted by one of the symbols $\dfrac{\partial u}{\partial x}, \dfrac{\partial f}{\partial x}, f_x, f_x(x, y), D_x f$ etc. Thus

$$\frac{\partial u}{\partial x} = \lim_{\Delta x \to 0} \frac{f(x + \Delta x, y) - f(x, y)}{\Delta x} = \lim_{h \to 0} \frac{f(x + h, y) - f(x, y)}{h},$$

provided the limit exist and similarly $\dfrac{\partial u}{\partial y} = \lim_{k \to 0} \dfrac{f(x, y + k) - f(x, y)}{k}$, provided the limit exist.

Similarly, if u is a function of three or more variables $x_1, x_2, x_3, ...,$ *the partial derivative of u with respect to x_1 is obtained by differentiating u with respect to*

x_1, *keeping all other variables constant and is written as* $\dfrac{\partial u}{\partial x_1}$ *which is analytically* expressed as

$$\frac{\partial u}{\partial x_1} = \lim_{\Delta x_1 \to 0} \frac{f(x_1 + \Delta x_1, x_2, x_3, ...) - f(x_1, x_2, x_3, ...)}{\Delta x_1}, \text{ if the limit exists}$$

or, putting $\Delta x_1 = h$, we get

$$\frac{\partial u}{\partial x_1} = \lim_{h \to 0} \frac{f(x_1 + h, x_2, x_3, ...) - f(x_1, x_2, x_3, ...)}{h}, \text{ if the limit exists.}$$

The expressions for other partial derivatives are obtained accordingly.

Successive partial derivatives

In general, f_x and f_y are also functions of x and y and so these can be differentiated further partially with respect to x and y. The usual notations for the second order partial derivatives with respect to x and y are $\dfrac{\partial}{\partial x}\left(\dfrac{\partial u}{\partial x}\right) = \dfrac{\partial^2 u}{\partial x^2}$ or $\dfrac{\partial^2 f}{\partial x^2}$

or f_{xx}; $\dfrac{\partial}{\partial x}\left(\dfrac{\partial u}{\partial y}\right) = \dfrac{\partial^2 u}{\partial x \partial y}$ or $\dfrac{\partial^2 f}{\partial x \partial y}$ or f_{xy}; $\dfrac{\partial}{\partial y}\left(\dfrac{\partial u}{\partial x}\right) = \dfrac{\partial^2 u}{\partial y \partial x}$ or $\dfrac{\partial^2 f}{\partial y \partial x}$ or f_{yx};

$\dfrac{\partial}{\partial y}\left(\dfrac{\partial u}{\partial y}\right) = \dfrac{\partial^2 u}{\partial y^2}$ or $\dfrac{\partial^2 f}{\partial y^2}$ or f_{yy}.

Note 3.2.1 *It is important to note that in the subscript notation, the subscripts are written in the same order in which differentiate whereas in '∂' notation the order is oppose as is seen from the notations* $f_{yx} = \dfrac{\partial}{\partial x}\left(\dfrac{\partial f}{\partial y}\right) = \dfrac{\partial^2 u}{\partial x \partial y}$ *and* $f_{xy} = \dfrac{\partial}{\partial y}\left(\dfrac{\partial f}{\partial x}\right) = \dfrac{\partial^2 u}{\partial y \partial x}$.

Note 3.2.2 *It can be easily verified that, in all ordinary cases* $\dfrac{\partial^2 u}{\partial x \partial y} = \dfrac{\partial^2 u}{\partial y \partial x}$.

Example 3.2.1 *Find the first and second partial derivatives of* $u = x^3 + x^2 y^2 + y^3$ *and show that* $\dfrac{\partial^2 u}{\partial y \partial x} = \dfrac{\partial^2 u}{\partial x \partial y}$.

Solution: Given $u = x^3 + x^2 y^2 + y^3$.

∴ $\dfrac{\partial u}{\partial x} = 3x^2 + 2xy^2$, $\dfrac{\partial u}{\partial y} = 3y^2 + 2x^2 y$.

and $\dfrac{\partial^2 u}{\partial x \partial y} = 4xy$, $\dfrac{\partial^2 u}{\partial y \partial x} = 4xy$. Hence $\dfrac{\partial^2 u}{\partial y \partial x} = \dfrac{\partial^2 u}{\partial x \partial y}$. □

Example 3.2.2 *Find the first partial derivatives* f_x *and* f_y *at* $(0,0)$ *where* $f(x,y) = x^3 + 3x^2 y + 3xy^2 + y^3$, $(x,y) \in \mathbb{R}^2$.

Solution: $f_x(x,y) = 3x^2 + 6xy + 3y^2 \Rightarrow f_x(0,0) = 0$.

Similarly, $f_y(x,y) = 3x^2 + 6xy + 3y^2 \Rightarrow f_y(0,0) = 0$. □

Example 3.2.3 *(a) If* $V = x^2 + y^2 + z^2$ *then show that* $xV_x + yV_y + zV_z = 2V$.

(b) If $V = x^2 y + y^2 z + z^2 x$ *then show that* $V_x + V_y + V_z = (x + y + z)^2$.

Solution: (a) Given $V = x^2 + y^2 + z^2$, $\therefore V_x = 2x$, $V_y = 2y$, $V_z = 2z$.
$\therefore \quad xV_x + yV_y + zV_z = 2x^2 + 2y^2 + 2z^2 = 2(x^2 + y^2 + z^2) = 2V$.
(b) $V = x^2y + y^2z + z^2x$, $\Rightarrow V_x = 2xy + z^2$, $V_y = 2yz + x^2$, $V_z = 2zx + y^2$.
$\therefore \quad V_x + V_y + V_z = x^2 + y^2 + z^2 + 2xy + 2yz + 2zx = (x + y + z)^2$. $\qquad \square$

Example 3.2.4 *From the implicit equation* $x^{\frac{2}{3}} + y^{\frac{2}{3}} = a^{\frac{2}{3}}$ *find* $\dfrac{dy}{dx}$.

Solution: We know that $\dfrac{dy}{dx}$ for the implicit equation $f(x, y) = 0$ is

$$\frac{dy}{dx} = -\frac{\frac{\partial f}{\partial x}}{\frac{\partial f}{\partial y}} = -\frac{f_x}{f_y}.$$

Here $f(x, y) \equiv x^{\frac{2}{3}} + y^{\frac{2}{3}} - a^{\frac{2}{3}} = 0$, $\therefore f_x = \dfrac{2}{3}x^{-\frac{1}{3}}$, $f_y = \dfrac{2}{3}y^{-\frac{1}{3}}$.

$$\therefore \quad \frac{dy}{dx} = -\frac{f_x}{f_y} = -\frac{\frac{2}{3}x^{-\frac{1}{3}}}{\frac{2}{3}y^{-\frac{1}{3}}} = -\left(\frac{y}{x}\right)^{\frac{1}{3}}. \qquad \square$$

3.3 Differentials and Exact (or Perfect) Differentials

The idea of differentials of a function of a single independent variable has already been achieved by us. Also in the previous chapter, we have already defined the differential of a function of several variables. Thus if $u = f(x_1, x_2, ..., x_n)$ be a function of n independent variables, then the differential of the function u is given by

$$du = \frac{\partial u}{\partial x_1}dx_1 + \frac{\partial u}{\partial x_2}dx_2 + ... + \frac{\partial u}{\partial x_n}dx_n \qquad (3.1)$$

The formula (3.1) is true whether the variables $x_1, x_2, ..., x_n$ are independent or not.

The definition of exact differentials and the condition of exactness is discussed in § 2.11 of **Chapter 2**.

3.4 Composite Functions (Functions of Functions): Chain Rule

1. For a single variable differentiable function $y = f(v)$ where v is again a single variable differentiable function, say, $v = \phi(x)$, then we have the familiar chain rule

$$\frac{dy}{dx} = \frac{dy}{dv} \cdot \frac{dv}{dx} \qquad (3.2)$$

2. Let $u = f(x, y)$ be a double variable differentiable function where x and y are itself single variable derivable functions, i.e., $x = \phi(t)$ and $y = \psi(t)$. Then u is actually a function of t alone. By substituting $x = \phi(t)$ and $y = \psi(t)$ in $u = f(x, y)$ we get $u = f(\phi(t), \psi(t))$ and then u becomes a ordinary derivable function with respect to the single independent variable t which is also known as total derivative of u with respect to t and is denoted by simply $\dfrac{du}{dt}$. Clearly, it is different from the two partial derivatives $\dfrac{\partial u}{\partial x}$ and $\dfrac{\partial u}{\partial y}$.

We now establish the following formula, an extended chain rule of the formula (3.1):

$$\frac{du}{dt} = \frac{\partial u}{\partial x}\frac{dx}{dt} + \frac{\partial u}{\partial y}\frac{dy}{dt}$$

Proof: We have $u = f(x, y)$.

Let us suppose that f_x, f_y and also $\phi'(t), \psi'(t)$ are continuous. When t changes to $t + \Delta t$, let x, y change to $x + \Delta x, y + \Delta y$ respectively. Then

$$
\begin{aligned}
u + \Delta u &= f(x + \Delta x, y + \Delta y) \\
\therefore \quad \Delta u &= f(x + \Delta x, y + \Delta y) - f(x, y) \\
&= f(x + \Delta x, y + \Delta y) - f(x, y + \Delta y) + f(x, y + \Delta y) - f(x, y) \\
\therefore \quad \frac{\Delta u}{\Delta t} &= \frac{f(x + \Delta x, y + \Delta y) - f(x, y + \Delta y)}{\Delta x} \times \frac{\Delta x}{\Delta t} \\
&\quad + \frac{f(x, y + \Delta y) - f(x, y)}{\Delta y} \times \frac{\Delta y}{\Delta t}.
\end{aligned}
$$

Taking limit $\Delta t \to 0$ we get

$$
\begin{aligned}
\frac{du}{dt} &= \lim_{\Delta t \to 0} \left\{ \frac{f(x + \Delta x, y + \Delta y) - f(x, y + \Delta y)}{\Delta x} \times \frac{\Delta x}{\Delta t} \right. \\
&\quad \left. + \frac{f(x, y + \Delta y) - f(x, y)}{\Delta y} \times \frac{\Delta y}{\Delta t} \right\}.
\end{aligned}
$$

Now, by Lagrange's Mean Value Theorem, we get

$$f(x + \Delta x, y + \Delta y) - f(x, y + \Delta y) = \Delta x f_x(x + \theta \Delta x, y + \Delta y), \quad 0 < \theta < 1$$

and $\quad f(x, y + \Delta y) - f(x, y) = \Delta y f_y(x, y + \theta' \Delta y), \ 0 < \theta' < 1$

Also when $\Delta t \to 0$, $\Delta x \to 0$, $\Delta y \to 0$ and $\dfrac{\Delta x}{\Delta t} \to \phi'(t)$, $\dfrac{\Delta y}{\Delta t} \to \psi'(t)$ and $f_x(x + \theta \Delta x, y + \Delta y) \to f_x(x, y)$, $f_y(x, y + \theta' \Delta y) \to f_y(x, y)$. Therefore we get

$$\frac{du}{dt} = f_x \frac{dx}{dt} + f_y \frac{dy}{dt} = \frac{\partial u}{\partial x}\frac{dx}{dt} + \frac{\partial u}{\partial y}\frac{du}{dt} \tag{3.3}$$

Hence the result.   □

Note 3.4.1 *As a particular case, if $u = f(x, y)$ and y is a function of x then*
$\dfrac{du}{dx} = \dfrac{\partial u}{\partial x} + \dfrac{\partial u}{\partial y}\dfrac{dy}{dx}.$

Note 3.4.2 *We can extend the formula (3.2) for functions with n independent variables, i.e., $u = f(x_1, x_2, ..., x_n)$ where $x_1 = x_1(t), x_2 = x_2(t), ..., x_n = x_n(t)$ as*

$$\frac{du}{dt} = \frac{\partial u}{\partial x_1}\frac{dx_1}{dt} + \frac{\partial u}{\partial x_2}\frac{dx_2}{dt} + ... + \frac{\partial u}{\partial x_n}\frac{dx_n}{dt} \tag{3.4}$$

3. If $u = f(\phi, \psi)$ where ϕ, ψ are again functions of two independent variables x, y, i.e., $\phi = \phi(x, y), \psi = \psi(x, y)$, then

$$\frac{\partial u}{\partial x} = \frac{\partial u}{\partial \phi}\cdot\frac{\partial \phi}{\partial x} + \frac{\partial u}{\partial \psi}\cdot\frac{\partial \psi}{\partial x}$$

$$\text{and} \quad \frac{\partial u}{\partial y} = \frac{\partial u}{\partial \phi}\cdot\frac{\partial \phi}{\partial y} + \frac{\partial u}{\partial \psi}\cdot\frac{\partial \psi}{\partial y}.$$

Proof: From the definition of differential we have

$$du = \frac{\partial u}{\partial \phi}.d\phi + \frac{\partial u}{\partial \psi}.d\psi \qquad (3.5)$$

Also $d\phi = \frac{\partial \phi}{\partial x}.dx + \frac{\partial \phi}{\partial y}.dy$ and $d\psi = \frac{\partial \psi}{\partial x}.dx + \frac{\partial \psi}{\partial y}.dy.$

When values of ϕ, ψ in terms of x, y are substituted, u becomes a function of x, y; hence

$$du = \frac{\partial u}{\partial x}.dx + \frac{\partial u}{\partial y}.dy \qquad (3.6)$$

Now, substituting the values of $d\phi, d\psi$ in (3.5), we get

$$du = \left(\frac{\partial u}{\partial \phi}.\frac{\partial \phi}{\partial x} + \frac{\partial u}{\partial \psi}.\frac{\partial \psi}{\partial x}\right) dx + \left(\frac{\partial u}{\partial \phi}.\frac{\partial \phi}{\partial y} + \frac{\partial u}{\partial \psi}.\frac{\partial \psi}{\partial y}\right) dy \qquad (3.7)$$

Comparing (3.6) and (3.7), since dx and dy are independent, we get

$$\frac{\partial u}{\partial x} = \frac{\partial u}{\partial \phi}.\frac{\partial \phi}{\partial x} + \frac{\partial u}{\partial \psi}.\frac{\partial \psi}{\partial x}$$

and $\frac{\partial u}{\partial y} = \frac{\partial u}{\partial \phi}.\frac{\partial \phi}{\partial y} + \frac{\partial u}{\partial \psi}.\frac{\partial \psi}{\partial y}.$ $\qquad \square$

Note 3.4.3 *The above results can also be established in case of more than two variables. Thus, if $u = f(\phi, \psi, \chi)$, where $\phi = \phi(x, y, z), \psi = \psi(x, y, z), \chi = \chi(x, y, z)$ and x, y, z are independent variables, then*

$$\frac{\partial u}{\partial x} = \frac{\partial u}{\partial \phi}.\frac{\partial \phi}{\partial x} + \frac{\partial u}{\partial \psi}.\frac{\partial \psi}{\partial x} + \frac{\partial u}{\partial \chi}.\frac{\partial \chi}{\partial x}$$

$$\frac{\partial u}{\partial y} = \frac{\partial u}{\partial \phi}.\frac{\partial \phi}{\partial y} + \frac{\partial u}{\partial \psi}.\frac{\partial \psi}{\partial y} + \frac{\partial u}{\partial \chi}.\frac{\partial \chi}{\partial y}$$

$$and \quad \frac{\partial u}{\partial z} = \frac{\partial u}{\partial \phi}.\frac{\partial \phi}{\partial z} + \frac{\partial u}{\partial \psi}.\frac{\partial \psi}{\partial z} + \frac{\partial u}{\partial \chi}.\frac{\partial \chi}{\partial z}.$$

Example 3.4.1 *If $u = \frac{x^2 - y^2}{x^2 + y^2}$ and $z = \sin^{-1}(\sqrt{u})$, find dz.*

Solution: We have

$$dz = \frac{1}{\sqrt{1-u}}\frac{1}{2}u^{-\frac{1}{2}}du \qquad (3.8)$$

$$and \quad du = \frac{\partial u}{\partial x}dx + \frac{\partial u}{\partial y}dy \qquad (3.9)$$

$$Now \quad \frac{\partial u}{\partial x} = \frac{2x(x^2+y^2) - 2x(x^2-y^2)}{(x^2+y^2)^2} = \frac{4xy^2}{(x^2+y^2)^2}$$

$$and \quad \frac{\partial u}{\partial y} = \frac{-2y(x^2+y^2) - 2y(x^2-y^2)}{(x^2+y^2)^2} = -\frac{4x^2y}{(x^2+y^2)^2}$$

From (3.9) $\quad du = \frac{4xy^2}{(x^2+y^2)^2}dx - \frac{4x^2y}{(x^2+y^2)^2}dy = \frac{4xy}{(x^2+y^2)^2}(ydx - xdy).$

From (3.8), $dz = \dfrac{1}{\sqrt{1-\frac{x^2-y^2}{x^2+y^2}}}.\dfrac{1}{2\sqrt{\frac{x^2-y^2}{x^2+y^2}}}.\dfrac{4xy}{(x^2+y^2)^2}(ydx - xdy)$

$$= \frac{2\sqrt{2}x}{(x^2+y^2)\sqrt{x^2-y^2}}(ydx - xdy). \qquad \square$$

Example 3.4.2 *Discuss the total differentiation of the function f defined by $f(u,v)$* $= \sqrt{|uv|}$ *where $u = x, v = x$.*

Solution: Applying the chain rule, we get

$$\frac{df}{dx} = \frac{\partial f}{\partial u} \cdot \frac{du}{dx} + \frac{\partial f}{\partial v} \cdot \frac{dv}{dx} = \frac{df}{du} + \frac{df}{dv} \quad \left(\because \frac{du}{dx} = \frac{dv}{dx} = 1 \right)$$

When $x = 0$, $u = v = 0$ and then

$$\frac{\partial f}{\partial u} = \lim_{h \to 0} \frac{f(h, 0) - f(0, 0)}{h} = 0, \quad \frac{\partial f}{\partial v} = \lim_{k \to 0} \frac{f(0, k) - f(0, 0)}{k} = 0.$$

Thus we get, $\dfrac{df}{dx} = \dfrac{df}{du} + \dfrac{df}{dv} = 0 + 0 = 0$ when $x = 0$.

But, $f(u, v) = f(x, x) = \sqrt{|x^2|} = |x|$.

$\therefore \dfrac{df}{dx} = +1$ or -1 according as x is positive or negative. Therefore at $x = 0$,

$\dfrac{df}{dx}$ does not exist.   □

Note 3.4.4 *The fallacy arises because we can not apply the chain rule due to the fact that $\sqrt{|uv|}$ is not differentiable when $u = v = 0$.*

Example 3.4.3 *If $u = \log(x^3 + y^3 + z^3 - 3xyz)$, then show that*

$i) \quad \dfrac{\partial u}{\partial x} + \dfrac{\partial u}{\partial y} + \dfrac{\partial u}{\partial z} = \dfrac{3}{x + y + z}$

$ii) \quad \dfrac{\partial^2 u}{\partial x^2} + \dfrac{\partial^2 u}{\partial y^2} + \dfrac{\partial^2 u}{\partial z^2} = -\dfrac{3}{(x + y + z)^2}.$

Solution: i) $\dfrac{\partial u}{\partial x} = \dfrac{3(x^2 - yz)}{P}, \ \dfrac{\partial u}{\partial y} = \dfrac{3(y^2 - zx)}{P}, \ \dfrac{\partial u}{\partial z} = \dfrac{3(z^2 - yx)}{P},$

where $P = x^3 + y^3 + z^3 - 3xyz = (x + y + z)(x^2 + y^2 + z^2 - xy - yz - zx).$

$$\frac{\partial u}{\partial x} + \frac{\partial u}{\partial y} + \frac{\partial u}{\partial z} = \frac{3(x^2 + y^2 + z^2 - xy - yz - zx)}{(x + y + z)(x^2 + y^2 + z^2 - xy - yz - zx)} = \frac{3}{x + y + z}.$$

Alternatively,

$$x^2 + y^2 + z^2 - xy - yz - zx = (x + y + z)(x^2 + y^2 + z^2 - xy - yz - zx)$$
$$= (x + y + z)(x + \omega y + \omega^2 z)(x + \omega^2 y + \omega z),$$

where ω is a cube root of unity.

$$\therefore \ u = \log(x + y + z) + \log(x + \omega y + \omega^2 z) + \log(x + \omega^2 y + \omega z).$$

$$\therefore \ \frac{\partial u}{\partial x} = \frac{1}{x + y + z} + \frac{1}{x + \omega y + \omega^2 z} + \frac{1}{x + \omega^2 y + \omega z} \tag{3.10}$$

$$\frac{\partial u}{\partial y} = \frac{1}{x + y + z} + \frac{\omega}{x + \omega y + \omega^2 z} + \frac{\omega^2}{x + \omega^2 y + \omega z} \tag{3.11}$$

$$\frac{\partial u}{\partial z} = \frac{1}{x + y + z} + \frac{\omega^2}{x + \omega y + \omega^2 z} + \frac{\omega}{x + \omega^2 y + \omega z} \tag{3.12}$$

$$(\because \ \omega^4 = \omega)$$

Therefore, adding (3.10), (3.11) and (3.12) we get

$$\frac{\partial u}{\partial x} + \frac{\partial u}{\partial y} + \frac{\partial u}{\partial z} = \frac{3}{x+y+z} + \frac{1+\omega+\omega^2}{x+\omega y+\omega^2 z} + \frac{1+\omega+\omega^2}{x+\omega^2 y+\omega z}$$

$$= \frac{3}{x+y+z} \quad (\because \ 1+\omega+\omega^2 = 0)$$

ii) Differentiating again (3.10), (3.11) and (3.12) partially with respect to x, y and z respectively, we get

$$\therefore \frac{\partial^2 u}{\partial x^2} = -\frac{1}{(x+y+z)^2} - \frac{1}{(x+\omega y+\omega^2 z)^2} - \frac{1}{(x+\omega^2 y+\omega z)^2} \qquad (3.13)$$

$$\frac{\partial^2 u}{\partial y^2} = -\frac{1}{(x+y+z)^2} - \frac{\omega^2}{(x+\omega y+\omega^2 z)^2} - \frac{\omega}{(x+\omega^2 y+\omega z)^2} \qquad (3.14)$$

$$\frac{\partial^2 u}{\partial z^2} = -\frac{1}{(x+y+z)^2} - \frac{\omega}{(x+\omega y+\omega^2 z)^2} - \frac{\omega^2}{(x+\omega^2 y+\omega z)^2} \qquad (3.15)$$

Now adding (3.13), (3.14) and (3.15), we get,

$$\frac{\partial^2 u}{\partial x^2} + \frac{\partial^2 u}{\partial y^2} + \frac{\partial^2 u}{\partial z^2} = -\frac{3}{(x+y+z)^2} - \frac{1+\omega^2+\omega}{(x+\omega y+\omega^2 z)^2} - \frac{1+\omega+\omega^2}{(x+\omega y+\omega^2 z)^2}$$

$$= -\frac{3}{(x+y+z)^2} \quad (\because \ 1+\omega+\omega^2 = 0). \qquad \square$$

Example 3.4.4 *Given* $u = \sin\dfrac{x}{y}, x = e^t, y = t^2$, *find* $\dfrac{du}{dt}$ *as a function of* t. *Verify your result by direct substitution.*

Solution: We have

$$\frac{du}{dt} = \frac{\partial u}{\partial x}.\frac{dx}{dt} + \frac{\partial u}{\partial y}.\frac{dy}{dt} = \left(\cos\frac{x}{y}\right).\frac{1}{y}.e^t + \left(\cos\frac{x}{y}\right).\left(-\frac{x}{y^2}\right).2t$$

$$= \left(\cos\frac{e^t}{t^2}\right).\frac{e^t}{t^2} - 2\left(\cos\frac{e^t}{t^2}\right).\frac{e^t}{t^3} = \frac{t-2}{t^3}e^t\cos\frac{e^t}{t^2}.$$

Verification: Also we have $u = \sin\dfrac{x}{y} = \sin\dfrac{e^t}{t^2}$

$$\therefore \frac{du}{dt} = \left(\cos\frac{e^t}{t^2}\right)\frac{t^2 e^t - e^t.2t}{t^4} = \frac{t-2}{t^3}e^t\cos\frac{e^t}{t^2}. \qquad \square$$

Example 3.4.5 *If* v, p *and* t *are connected by an equation then prove that*

$$\left(\frac{dp}{dt}\right)_{v=constant} \times \left(\frac{dt}{dv}\right)_{p=constant} \times \left(\frac{dv}{dp}\right)_{t=constant} = -1.$$

Solution: Let $f(v, p, t) = 0$, then $\left(\dfrac{dp}{dt}\right)_{v=constant} = -\dfrac{\frac{\partial f}{\partial t}}{\frac{\partial f}{\partial p}} = -\dfrac{f_t}{f_p}.$

Similarly, $\left(\dfrac{dt}{dv}\right)_{p=constant} = -\dfrac{f_v}{f_t}$ and $\left(\dfrac{dv}{dp}\right)_{t=constant} = -\dfrac{f_p}{f_v}.$

On multiplication, the required result follows. $\qquad \square$

Example 3.4.6 *(a) If $V = \sqrt{x^2 + y^2 + z^2}$, prove that $V_{xx} + V_{yy} + V_{zz} = \dfrac{2}{V}$.*

  (b) If $V = \dfrac{1}{\sqrt{x^2 + y^2 + z^2}}$, prove that $V_{xx} + V_{yy} + V_{zz} = 0$.

Solution: (a) Given

$$V = \sqrt{x^2 + y^2 + z^2} \quad \Rightarrow \quad V^2 = x^2 + y^2 + z^2 \tag{3.16}$$

Differentiating partially both sides of (3.16) with respect to x, we get

$$2.V.V_x = 2x \quad \Rightarrow \quad V_x = \frac{x}{V} \tag{3.17}$$

Differentiating again partially both sides of (3.17) with respect to x we get

$$V_{xx} = \frac{1}{V} - \frac{x}{V^2}V_x = \frac{1}{V} - \frac{x}{V^2}\cdot\frac{x}{V} = \frac{1}{V}\left(1 - \frac{x^2}{V^2}\right).$$

Similarly we get $V_{yy} = \dfrac{1}{V}\left(1 - \dfrac{y^2}{V^2}\right)$ and $V_{zz} = \dfrac{1}{V}\left(1 - \dfrac{z^2}{V^2}\right)$.

  Adding these values of V_{xx}, V_{yy} and V_{zz} we get

$$V_{xx} + V_{yy} + V_{zz} = \frac{1}{V}\left(3 - \frac{x^2 + y^2 + z^2}{V^2}\right) = \frac{1}{V}\left(3 - \frac{V^2}{V^2}\right) = \frac{2}{V}.$$

(b) From the given relation we get

$$V^2(x^2 + y^2 + z^2) = 1 \tag{3.18}$$

Differentiating partially both sides of (3.18) with respect to x, we get

$$2V.V_x(x^2 + y^2 + z^2) + 2xV^2 = 0 \quad \Rightarrow \quad V_x = -\frac{xV}{(x^2 + y^2 + z^2)} = -\frac{xV}{\frac{1}{V^2}} = -xV^3.$$

Differentiating this again partially both sides with respect to x we get

$$V_{xx} = -V^3 - x.3V^2V_x = -V^3 + 3xV^2(-xV^3) = -V^3 + 3x^2V^5.$$

Similarly, we get $V_{yy} = V^3 + 3y^2V^5$ and $V_{zz} = V^3 + 3z^2V^5$.

  Adding these we get,

$$V_{xx} + V_{yy} + V_{zz} = -3V^3 + 3V^5(x^2 + y^2 + z^2) = -3V^3 + 3V^5\frac{1}{V^2} = 0. \quad \square$$

Example 3.4.7 *If $u = \dfrac{x + y}{1 - xy}, v = \dfrac{x(1 - y^2) + y(1 - x^2)}{(1 + x^2)(1 + y^2)}$ then prove that $u_x v_y =$*
$u_y v_x$.

Solution: Put $x = \tan\alpha, y = \tan\beta$.

$$\therefore \quad u = \tan(\alpha + \beta) \quad \Rightarrow \quad \alpha + \beta = \tan^{-1} u$$

i.e., $\tan^{-1} x + \tan^{-1} y = \tan^{-1} u$ (3.19)

and $v = \dfrac{\tan\alpha(1 - \tan^2\beta) + \tan\beta(1 - \tan^2\alpha)}{(1 + \tan^2\alpha)(1 + \tan^2\beta)}$

$$= \frac{1}{2}\left[\frac{2\tan\alpha}{1 + \tan^2\alpha}\cdot\frac{1 - \tan^2\alpha}{1 + \tan^2\alpha} + \frac{2\tan\beta}{1 + \tan^2\beta}\cdot\frac{1 - \tan^2\beta}{1 + \tan^2\beta}\right]$$

$$= \frac{1}{2}[\sin 2\alpha\cos 2\beta + \cos 2\alpha\sin 2\beta] = \frac{1}{2}\sin 2(\alpha + \beta).$$

$$\therefore \quad \sin^{-1} 2v = 2(\tan^{-1} x + \tan^{-1} y) = 2\tan^{-1} u \quad [\text{by } (3.19)]$$

i.e., $\sin^{-1} 2v = 2\tan^{-1} u$ (3.20)

Now differentiating partially both sides of (3.20) with respect x, y respectively, we get

$$\frac{1}{\sqrt{1 - 4v^2}} v_x = \frac{1}{1 + u^2} u_x \quad \text{and} \quad \frac{1}{\sqrt{1 - 4v^2}} v_y = \frac{1}{1 + u^2} u_y$$

Dividing these we get $\dfrac{v_x}{v_y} = \dfrac{u_x}{u_y} \Rightarrow u_x v_y = u_y v_x$. $\qquad\qquad\square$

Example 3.4.8 *If $Pdx + Qdy + Rdz$ can be made a perfect differential of some function of x, y, z on multiplication by a factor, prove that*

$$P \left(\frac{\partial Q}{\partial z} - \frac{\partial R}{\partial y} \right) + Q \left(\frac{\partial R}{\partial x} - \frac{\partial P}{\partial z} \right) + R \left(\frac{\partial P}{\partial y} - \frac{\partial Q}{\partial x} \right) = 0.$$

Solution: Let u be a function of x, y and z and

$$\mu(Pdx + Qdy + Rdz) = du \tag{3.21}$$

where μ is some function of x, y and z. Also,

$$du = \frac{\partial u}{\partial x} dx + \frac{\partial u}{\partial y} dy + \frac{\partial u}{\partial z} dz \tag{3.22}$$

since u is a function of x, y and z.

Comparing (3.21) and (3.22) we get

$$\frac{\partial u}{\partial x} = \mu P \tag{3.23}$$

$$\frac{\partial u}{\partial y} = \mu Q \tag{3.24}$$

$$\frac{\partial u}{\partial z} = \mu R \tag{3.25}$$

Differentiating partially (3.23) with respect to y and (3.24) with respect to x respectively we get

$$\frac{\partial^2 u}{\partial y \partial x} = \mu \frac{\partial P}{\partial y} + P \frac{\partial \mu}{\partial y}, \quad \frac{\partial^2 u}{\partial x \partial y} = \mu \frac{\partial Q}{\partial x} + Q \frac{\partial \mu}{\partial x}$$

and assuming $\dfrac{\partial^2 u}{\partial y \partial x} = \dfrac{\partial^2 u}{\partial x \partial y}$ we get

$$\mu \frac{\partial P}{\partial y} + P \frac{\partial \mu}{\partial y} = \mu \frac{\partial Q}{\partial x} + Q \frac{\partial \mu}{\partial x} \tag{3.26}$$

Similarly, we shall get

$$\mu \frac{\partial Q}{\partial z} + Q \frac{\partial \mu}{\partial z} = \mu \frac{\partial R}{\partial y} + R \frac{\partial \mu}{\partial y} \tag{3.27}$$

$$\text{and} \quad \mu \frac{\partial R}{\partial x} + R \frac{\partial \mu}{\partial x} = \mu \frac{\partial P}{\partial z} + P \frac{\partial \mu}{\partial z} \tag{3.28}$$

Rearranging (3.26), (3.27) and (3.28), we get

$$\mu \left(\frac{\partial P}{\partial y} - \frac{\partial Q}{\partial x} \right) = Q \frac{\partial \mu}{\partial x} - P \frac{\partial \mu}{\partial x} \tag{3.29}$$

$$\mu \left(\frac{\partial Q}{\partial z} - \frac{\partial R}{\partial y} \right) = R \frac{\partial \mu}{\partial y} - Q \frac{\partial \mu}{\partial z} \tag{3.30}$$

$$\text{and} \quad \mu \left(\frac{\partial R}{\partial x} - \frac{\partial P}{\partial z} \right) = P \frac{\partial \mu}{\partial z} - R \frac{\partial \mu}{\partial x} \tag{3.31}$$

Multiplying (3.29) by R, (3.30) by P and (3.31) by Q and then adding together we get the required result.   □

Note 3.4.5 *If $P\,dx + Q\,dy + R\,dz$ is itself a perfect differential, then it can be easily deduced that* $\dfrac{\partial Q}{\partial z} - \dfrac{\partial R}{\partial y} = \dfrac{\partial R}{\partial x} - \dfrac{\partial P}{\partial z} = \dfrac{\partial P}{\partial y} - \dfrac{\partial Q}{\partial x} = 0.$

Example 3.4.9 *If $u(x, y, z)$ satisfies the equation $\dfrac{\partial^2 u}{\partial x^2} + \dfrac{\partial^2 u}{\partial y^2} + \dfrac{\partial^2 u}{\partial z^2} = 0$ then show that*

(i) $\dfrac{\partial u}{\partial x}, \dfrac{\partial u}{\partial y}$ *and* $\dfrac{\partial u}{\partial z}$ *satisfies it,*

(ii) $x\dfrac{\partial u}{\partial x} + y\dfrac{\partial u}{\partial y} + z\dfrac{\partial u}{\partial z}$ *also satisfies it.*

Solution: (i) Given that

$$\frac{\partial^2 u}{\partial x^2} + \frac{\partial^2 u}{\partial y^2} + \frac{\partial^2 u}{\partial z^2} = 0 \qquad\qquad (3.32)$$

Differentiating both sides of (3.32) partially with respect to x, we get

$$\frac{\partial^3 u}{\partial x^3} + \frac{\partial^3 u}{\partial x \partial y^2} + \frac{\partial^3 u}{\partial x \partial z^2} = 0$$

or, $\dfrac{\partial^2}{\partial x^2}\left(\dfrac{\partial u}{\partial x}\right) + \dfrac{\partial^2}{\partial y^2}\left(\dfrac{\partial u}{\partial x}\right) + \dfrac{\partial^2}{\partial z^2}\left(\dfrac{\partial u}{\partial x}\right) = 0$

showing that $\dfrac{\partial u}{\partial x}$ satisfies the given equation.

Similarly, differentiating partially both sides of (3.32) with respect to y and z in turn, we can prove that $\dfrac{\partial u}{\partial x}$ and $\dfrac{\partial u}{\partial x}$ will also satisfy the equation.

(ii) To prove this we are to show that

$$\frac{\partial^2}{\partial x^2}\left(x\frac{\partial u}{\partial x} + y\frac{\partial u}{\partial y} + z\frac{\partial u}{\partial z}\right) + \frac{\partial^2}{\partial y^2}\left(x\frac{\partial u}{\partial x} + y\frac{\partial u}{\partial y} + z\frac{\partial u}{\partial z}\right)$$
$$+ \frac{\partial^2}{\partial z^2}\left(x\frac{\partial u}{\partial x} + y\frac{\partial u}{\partial y} + z\frac{\partial u}{\partial z}\right) = 0$$

Now $\dfrac{\partial}{\partial x}\left(x\dfrac{\partial u}{\partial x} + y\dfrac{\partial u}{\partial y} + z\dfrac{\partial u}{\partial z}\right) = \dfrac{\partial u}{\partial x} + x\dfrac{\partial^2 u}{\partial x^2} + y\dfrac{\partial^2 u}{\partial x \partial y} + z\dfrac{\partial^2 u}{\partial x \partial z}.$

$\therefore\ \dfrac{\partial^2}{\partial x^2}\left(x\dfrac{\partial u}{\partial x} + y\dfrac{\partial u}{\partial y} + z\dfrac{\partial u}{\partial z}\right) = \dfrac{\partial}{\partial x}\left(\dfrac{\partial u}{\partial x} + x\dfrac{\partial^2 u}{\partial x^2} + y\dfrac{\partial^2 u}{\partial x \partial y} + z\dfrac{\partial^2 u}{\partial x \partial z}\right)$

$\qquad = \dfrac{\partial^2 u}{\partial x^2} + \dfrac{\partial^2 u}{\partial x^2} + x\dfrac{\partial^3 u}{\partial x^3} + y\dfrac{\partial^3 u}{\partial x^2 \partial y} + z\dfrac{\partial^3 u}{\partial x^2 \partial z}$

$\qquad = 2\dfrac{\partial^2 u}{\partial x^2} + x\dfrac{\partial^3 u}{\partial x^3} + y\dfrac{\partial^3 u}{\partial x^2 \partial y} + z\dfrac{\partial^3 u}{\partial x^2 \partial z} \qquad\qquad (3.33)$

Similarly we get,

$$\frac{\partial^2}{\partial y^2}\left(x\frac{\partial u}{\partial x} + y\frac{\partial u}{\partial y} + z\frac{\partial u}{\partial z}\right) = 2\frac{\partial^2 u}{\partial y^2} + x\frac{\partial^3 u}{\partial y^2 \partial x} + y\frac{\partial^3 u}{\partial y^3} + z\frac{\partial^3 u}{\partial y^2 \partial z} \qquad (3.34)$$

$$\frac{\partial^2}{\partial z^2}\left(x\frac{\partial u}{\partial x} + y\frac{\partial u}{\partial y} + z\frac{\partial u}{\partial z}\right) = 2\frac{\partial^2 u}{\partial z^2} + x\frac{\partial^3 u}{\partial z^2 \partial x} + y\frac{\partial^3 u}{\partial z^2 \partial y} + z\frac{\partial^3 u}{\partial z^3} \qquad (3.35)$$

Adding (3.33), (3.34) and (3.35) and using (3.32) we get

$$\frac{\partial^2}{\partial x^2}\left(x\frac{\partial u}{\partial x} + y\frac{\partial u}{\partial y} + z\frac{\partial u}{\partial z}\right) + \frac{\partial^2}{\partial y^2}\left(x\frac{\partial u}{\partial x} + y\frac{\partial u}{\partial y} + z\frac{\partial u}{\partial z}\right)$$

$$+ \frac{\partial^2}{\partial z^2}\left(x\frac{\partial u}{\partial x} + y\frac{\partial u}{\partial y} + z\frac{\partial u}{\partial z}\right)$$

$$= 2\left(\frac{\partial^2 u}{\partial x^2} + \frac{\partial^2 u}{\partial y^2} + \frac{\partial^2 u}{\partial z^2}\right) + x\frac{\partial}{\partial x}\left(\frac{\partial^2 u}{\partial x^2} + \frac{\partial^2 u}{\partial y^2} + \frac{\partial^2 u}{\partial z^2}\right)$$

$$+ y\frac{\partial}{\partial y}\left(\frac{\partial^2 u}{\partial x^2} + \frac{\partial^2 u}{\partial y^2} + \frac{\partial^2 u}{\partial z^2}\right) + z\frac{\partial}{\partial z}\left(\frac{\partial^2 u}{\partial x^2} + \frac{\partial^2 u}{\partial y^2} + \frac{\partial^2 u}{\partial z^2}\right)$$

$$= 2.0 + x\frac{\partial}{\partial x}(0) + y\frac{\partial}{\partial y}(0) + z\frac{\partial}{\partial z}(0) = 0.$$

Hence $x\dfrac{\partial u}{\partial x} + y\dfrac{\partial u}{\partial y} + z\dfrac{\partial u}{\partial z}$ satisfies the given relation. $\qquad\square$

Example 3.4.10 *If α, β, γ be the roots of the equation $x^3 + px^2 + qx + r = 0$, then*

$$\text{show that}\quad \begin{vmatrix} \dfrac{\partial p}{\partial \alpha} & \dfrac{\partial p}{\partial \beta} & \dfrac{\partial p}{\partial \gamma} \\[2mm] \dfrac{\partial q}{\partial \alpha} & \dfrac{\partial q}{\partial \beta} & \dfrac{\partial q}{\partial \gamma} \\[2mm] \dfrac{\partial r}{\partial \alpha} & \dfrac{\partial r}{\partial \beta} & \dfrac{\partial r}{\partial \gamma} \end{vmatrix}\quad \text{vanishes when any two of the three roots are equal.}$$

Solution: From the given cubic equation we get, $p = -(\alpha+\beta+\gamma)$, $q = \alpha\beta+\beta\gamma+\gamma\alpha$ and $r = -\alpha\beta\gamma$. Using these we get

$$\begin{vmatrix} \dfrac{\partial p}{\partial \alpha} & \dfrac{\partial p}{\partial \beta} & \dfrac{\partial p}{\partial \gamma} \\[2mm] \dfrac{\partial q}{\partial \alpha} & \dfrac{\partial q}{\partial \beta} & \dfrac{\partial q}{\partial \gamma} \\[2mm] \dfrac{\partial r}{\partial \alpha} & \dfrac{\partial r}{\partial \beta} & \dfrac{\partial r}{\partial \gamma} \end{vmatrix} = \begin{vmatrix} -1 & -1 & -1 \\ \beta+\gamma & \gamma+\alpha & \alpha+\beta \\ -\beta\gamma & -\gamma\alpha & -\alpha\beta \end{vmatrix}$$

$$= \begin{vmatrix} -1 & 0 & 0 \\ \beta+\gamma & \alpha-\beta & \beta-\gamma \\ -\beta\gamma & -(\alpha-\beta)\gamma & -(\beta-\gamma)\alpha \end{vmatrix}$$

$$= (\alpha-\beta)(\beta-\gamma)\begin{vmatrix} -1 & 0 & 0 \\ \beta+\gamma & 1 & 1 \\ -\beta\gamma & -\gamma & -\alpha \end{vmatrix} = -(\alpha-\beta)(\beta-\gamma)(\gamma-\alpha)$$

and hence, it vanishes if and only if two of the three roots are equal. $\qquad\square$

3.5   Change of Variables

In problems involving change of variables it is frequently required to transform a particular expression involving a combination of derivatives with respect to a set of variables to another set of variables by proper substitution. We restrict our derivatives up to second order only in our consideration. The higher derivatives, however, may be obtained by exactly the same method.

Problem Type 1: If $u = f(x, y)$, where $x = \phi(s, t)$ and $y = \psi(s, t)$, it is often necessary to change expressions involving $u, x, y, \dfrac{\partial u}{\partial x}, \dfrac{\partial u}{\partial y}$ etc to expressions involving $u, s, t, \dfrac{\partial u}{\partial s}, \dfrac{\partial u}{\partial t}$ etc. The necessary formula for the change of variables are easily obtained.

If t be regarded as constant, then x, y, u will be function of s alone. Therefore we get the formula

$$\frac{\partial u}{\partial s} = \frac{\partial u}{\partial x} \cdot \frac{\partial x}{\partial s} + \frac{\partial u}{\partial y} \cdot \frac{\partial y}{\partial s} \tag{3.36}$$

where the ordinary derivatives have been replaced by the partial derivatives, because, x, y are functions of two variables s and t.

Similarly, regarding s as a constant, we get

$$\frac{\partial u}{\partial t} = \frac{\partial u}{\partial x} \cdot \frac{\partial x}{\partial t} + \frac{\partial u}{\partial y} \cdot \frac{\partial y}{\partial t} \tag{3.37}$$

Solving (3.36) and (3.37) as simultaneous equations in $\dfrac{\partial u}{\partial x}$ and $\dfrac{\partial u}{\partial y}$ we get their values in terms of $u, s, t, \dfrac{\partial u}{\partial s}, \dfrac{\partial u}{\partial t}$.

The higher derivatives of u can be obtained by repeated application of the formulae (3.36) and (3.37), but here we shall consider the derivatives up to second order only.

Note 3.5.1 *If the problem be reversed, i.e., if instead of the type of the function given, we consider $v = v(s, t)$, where $s = s(x, y), t = t(x, y)$ then the transformed derivatives will be*

$$\frac{\partial v}{\partial x} = \frac{\partial v}{\partial s} \cdot \frac{\partial s}{\partial x} + \frac{\partial v}{\partial t} \cdot \frac{\partial t}{\partial x} \quad and \quad \frac{\partial v}{\partial y} = \frac{\partial v}{\partial s} \cdot \frac{\partial s}{\partial y} + \frac{\partial v}{\partial t} \cdot \frac{\partial t}{\partial y}.$$

Note 3.5.2 *In the 'change of variables' the variables x, y are called **intermediate variables** while s, t are **independent variables**.*

Note 3.5.3 *Let $u = f(x, y)$ where $x = \phi(t), y = \psi(t)$. Also let f is differentiable at (x, y) and x, y are both derivable at t then u is also derivable at t and we get*

$$\frac{du}{dt} = \frac{\partial u}{\partial x} \cdot \frac{dx}{dt} + \frac{\partial u}{\partial y} \cdot \frac{dy}{dt}.$$

Similarly if $u = u(x, y, z)$ and $x = x(t), y = y(t), z = z(t)$ and f is differentiable at (x, y, z) and x, y, z are all derivable at t, then u is also derivable at t and

$$\frac{du}{dt} = \frac{\partial u}{\partial x} \cdot \frac{dx}{dt} + \frac{\partial u}{\partial y} \cdot \frac{dy}{dt} + \frac{\partial u}{\partial z} \cdot \frac{dz}{dt}.$$

Problem Type 2: In Coordinate Geometry, *Translation, Rotation* and *Rigid body motion* (or *Orthogonal Transformation*) are frequently used to **transform from one system of Cartesian coordinates to another Cartesian system**. Also we have the equations of transformation form **plane Cartesian to plane polar coordinate system** and vice-verse as follows:

$x = r\cos\theta,\ y = r\sin\theta$ and $r = \sqrt{x^2 + y^2}$, $\theta = \tan^{-1}\dfrac{y}{x}$ where $r \geq 0$ and $-\pi < \theta \leq \pi$.

Similarly, the relations given by $x = r\sin\theta\cos\phi$, $y = r\sin\theta\sin\phi$ and $z = r\cos\theta$ where $r \geq 0$, $0 \leq \theta \leq \pi$ and $0 \leq \phi \leq 2\pi$ defines a transformation from, what is called *Spherical polar to Cartesian* and $x = r\sin\theta\cos\phi$, $y = r\sin\theta\sin\phi$ and $z = z$ give the transformation from *cylindrical coordinates to Cartesian system*.

Example 3.5.1 *If $u = F(x,y,z)$ and $z = f(x,y)$, find a formula for $\dfrac{\partial^2 u}{\partial x^2}$ in terms of the derivatives of F and that of z.*

Solution: In the expression of F, we consider x, y, z as intermediate variables while in the expression for f, we consider x, y as independent variables.

Now $\dfrac{\partial u}{\partial x} = \dfrac{\partial F}{\partial x}\cdot\dfrac{\partial x}{\partial x} + \dfrac{\partial F}{\partial y}\cdot\dfrac{\partial y}{\partial x} + \dfrac{\partial F}{\partial z}\cdot\dfrac{\partial z}{\partial x}$.

But x and y are independent variables, therefore $\dfrac{\partial y}{\partial x} = 0$. Also, $\dfrac{\partial x}{\partial x} = 1$.

$\therefore\quad \dfrac{\partial u}{\partial x} = \dfrac{\partial F}{\partial x} + \dfrac{\partial F}{\partial z}\cdot\dfrac{\partial z}{\partial x}$.

Differentiating again partially with respect to x, we get

$$\frac{\partial^2 u}{\partial x^2} = \left(\frac{\partial^2 F}{\partial x^2}\cdot\frac{\partial x}{\partial x} + \frac{\partial^2 F}{\partial x\partial y}\cdot\frac{\partial y}{\partial x} + \frac{\partial^2 F}{\partial x\partial z}\cdot\frac{\partial z}{\partial x}\right)$$
$$+ \frac{\partial z}{\partial x}\left(\frac{\partial^2 F}{\partial z\partial x}\cdot\frac{\partial x}{\partial x} + \frac{\partial^2 F}{\partial z\partial y}\cdot\frac{\partial y}{\partial x} + \frac{\partial^2 F}{\partial z^2}\cdot\frac{\partial z}{\partial x}\right) + \frac{\partial F}{\partial z}\frac{\partial^2 z}{\partial x^2}$$
$$= \frac{\partial^2 F}{\partial x^2} + 2\frac{\partial^2 F}{\partial x\partial z}\cdot\frac{\partial z}{\partial x} + \frac{\partial^2 F}{\partial x^2}\left(\frac{\partial z}{\partial x}\right)^2 + \frac{\partial F}{\partial z}\frac{\partial^2 z}{\partial x^2}. \qquad \square$$

Example 3.5.2 *Let u be a differentiable function of x and y and let $x = r\cos\theta$, $y = r\sin\theta$, where (x,y) denote rectangular Cartesian coordinates and (r,θ), the corresponding polar coordinates. Then prove the following*

(i) $\left(\dfrac{\partial u}{\partial r}\right)^2 + \dfrac{1}{r^2}\left(\dfrac{\partial r}{\partial\theta}\right)^2 = \left(\dfrac{\partial u}{\partial x}\right)^2 + \left(\dfrac{\partial u}{\partial y}\right)^2$

(ii) $\dfrac{\partial^2 u}{\partial r^2} + \dfrac{1}{r}\dfrac{\partial u}{\partial r} + \dfrac{1}{r^2}\dfrac{\partial^2 u}{\partial\theta^2} = \dfrac{\partial^2 u}{\partial x^2} + \dfrac{\partial^2 u}{\partial y^2}$.

Solution: (i) Given $x = r\cos\theta$, $y = r\sin\theta$. $\therefore\ r = \sqrt{x^2 + y^2}$, $\theta = \tan^{-1}\dfrac{y}{x}$.

Here, $\dfrac{\partial x}{\partial r} = \cos\theta$; $\quad \dfrac{\partial r}{\partial x} = \dfrac{x}{\sqrt{x^2+y^2}} = \dfrac{r\cos\theta}{r} = \cos\theta$.

$\dfrac{\partial y}{\partial r} = \sin\theta$; $\quad \dfrac{\partial r}{\partial y} = \dfrac{y}{\sqrt{x^2+y^2}} = \dfrac{r\sin\theta}{r} = \sin\theta$.

$\dfrac{\partial\theta}{\partial x} = \dfrac{1}{1+\frac{y^2}{x^2}}\left(-\dfrac{y}{x^2}\right) = -\dfrac{x^2}{x^2+y^2}\dfrac{y}{x^2} = -\dfrac{y}{x^2+y^2} = -\dfrac{\sin\theta}{r}$

and similarly $\dfrac{\partial\theta}{\partial y} = \dfrac{\cos\theta}{r}$.

Now, since u is a function of x and y and x and y are functions of r, θ, so u is also a function of r, θ and hence using chain rule, we get

$$\left. \begin{aligned} \frac{\partial u}{\partial r} &= \frac{\partial u}{\partial x} \cdot \frac{\partial x}{\partial r} + \frac{\partial u}{\partial y} \cdot \frac{\partial y}{\partial r} &&= \cos\theta \frac{\partial u}{\partial x} + \sin\theta \frac{\partial u}{\partial y} \\ \frac{\partial u}{\partial \theta} &= \frac{\partial u}{\partial x} \cdot \frac{\partial x}{\partial \theta} + \frac{\partial u}{\partial y} \cdot \frac{\partial y}{\partial \theta} &&= -r\sin\theta \frac{\partial u}{\partial x} + r\cos\theta \frac{\partial u}{\partial y} \end{aligned} \right\} \quad (3.38)$$

and similarly,

$$\left. \begin{aligned} \frac{\partial u}{\partial x} &= \frac{\partial u}{\partial r} \cdot \frac{\partial r}{\partial x} + \frac{\partial u}{\partial \theta} \cdot \frac{\partial \theta}{\partial x} &&= \cos\theta \frac{\partial u}{\partial r} - \frac{\sin\theta}{r} \frac{\partial u}{\partial \theta} \\ \frac{\partial u}{\partial y} &= \frac{\partial u}{\partial r} \cdot \frac{\partial r}{\partial y} + \frac{\partial u}{\partial \theta} \cdot \frac{\partial \theta}{\partial y} &&= \sin\theta \frac{\partial u}{\partial r} + \frac{\cos\theta}{r} \frac{\partial u}{\partial \theta} \end{aligned} \right\} \quad (3.39)$$

$$\therefore \quad \left(\frac{\partial u}{\partial r} \right)^2 + \frac{1}{r^2} \left(\frac{\partial u}{\partial \theta} \right)^2 = \left(\cos\theta \frac{\partial u}{\partial x} + \sin\theta \frac{\partial u}{\partial y} \right)^2$$

$$+ \frac{1}{r^2} \left(-r\sin\theta \frac{\partial u}{\partial x} + r\cos\theta \frac{\partial u}{\partial y} \right)^2$$

$$= \left(\frac{\partial u}{\partial x} \right)^2 (\cos^2\theta + \sin^2\theta) + \left(\frac{\partial u}{\partial y} \right)^2 (\sin^2\theta + \cos^2\theta)$$

$$= \left(\frac{\partial u}{\partial x} \right)^2 + \left(\frac{\partial u}{\partial y} \right)^2.$$

(ii) Again from the results of the set of equations (3.38), we get the following equivalent operators:

$$\left. \begin{aligned} \frac{\partial}{\partial r} &= \cos\theta \frac{\partial}{\partial x} + \sin\theta \frac{\partial}{\partial y} \\ \frac{\partial}{\partial \theta} &= -r\sin\theta \frac{\partial}{\partial x} + r\cos\theta \frac{\partial}{\partial y} \end{aligned} \right\} \quad (3.40)$$

$$\therefore \quad \frac{\partial^2 u}{\partial r^2} = \frac{\partial}{\partial r} \left(\frac{\partial u}{\partial r} \right) = \frac{\partial}{\partial r} \left(\cos\theta \frac{\partial u}{\partial x} + \sin\theta \frac{\partial u}{\partial y} \right)$$

$$= \cos\theta \frac{\partial}{\partial r} \left(\frac{\partial u}{\partial x} \right) + \sin\theta \frac{\partial}{\partial r} \left(\frac{\partial u}{\partial y} \right) \quad (\because \theta \text{ is a constant with respect to } r)$$

$$= \cos\theta \left[\cos\theta \frac{\partial}{\partial x} + \sin\theta \frac{\partial}{\partial y} \right] \left(\frac{\partial u}{\partial x} \right) + \sin\theta \left[\cos\theta \frac{\partial}{\partial x} + \sin\theta \frac{\partial}{\partial y} \right] \left(\frac{\partial u}{\partial y} \right)$$

$$\left(\text{using the operator } \frac{\partial}{\partial r} \text{ from (3.40) on } \frac{\partial u}{\partial x} \text{ and on } \frac{\partial u}{\partial y} \right)$$

$$= \left[\cos^2\theta \frac{\partial^2 u}{\partial x^2} + \cos\theta\sin\theta \frac{\partial^2 u}{\partial x \partial y} \right] + \left[\sin\theta\cos\theta \frac{\partial^2 u}{\partial x \partial y} + \sin^2\theta \frac{\partial^2 u}{\partial y^2} \right]$$

$$= \cos^2\theta \frac{\partial^2 u}{\partial x^2} + 2\cos\theta\sin\theta \frac{\partial^2 u}{\partial x \partial y} + \sin^2\theta \frac{\partial^2 u}{\partial y^2} \quad \left(\because \frac{\partial^2 u}{\partial x \partial y} = \frac{\partial^2 u}{\partial y \partial x} \right)$$

Similarly, using the operator $\dfrac{\partial}{\partial\theta}$ from (3.40) on $\dfrac{\partial u}{\partial x}$ and $\dfrac{\partial u}{\partial y}$, we get

$$\frac{\partial^2 u}{\partial\theta^2} = \frac{\partial}{\partial\theta}\left(\frac{\partial u}{\partial\theta}\right) = \frac{\partial}{\partial\theta}\left(-r\sin\theta\frac{\partial u}{\partial x} + r\cos\theta\frac{\partial u}{\partial y}\right)$$

$$= \left[-r\cos\theta\frac{\partial u}{\partial x} - r\sin\theta\frac{\partial u}{\partial y}\right] - r\sin\theta\frac{\partial}{\partial\theta}\left(\frac{\partial u}{\partial x}\right) + r\cos\theta\frac{\partial}{\partial\theta}\left(\frac{\partial u}{\partial y}\right)$$

$$= -r\left[\cos\theta\frac{\partial u}{\partial x} + \sin\theta\frac{\partial u}{\partial y}\right] - r\sin\theta\left[-r\sin\theta\frac{\partial}{\partial x} + r\cos\theta\frac{\partial}{\partial y}\right]\left(\frac{\partial u}{\partial x}\right)$$

$$\qquad\qquad + r\cos\theta\left[-r\sin\theta\frac{\partial}{\partial x} + r\cos\theta\frac{\partial}{\partial y}\right]\left(\frac{\partial u}{\partial y}\right)$$

$$= -r\frac{\partial u}{\partial r} + r^2\sin^2\theta\frac{\partial^2 u}{\partial x^2} - 2r^2\sin\theta\cos\theta\frac{\partial^2 u}{\partial x\partial y} + r^2\cos^2\theta\frac{\partial^2 u}{\partial y^2}$$

or, $$\frac{1}{r^2}\frac{\partial^2 u}{\partial\theta^2} + \frac{1}{r}\frac{\partial u}{\partial r} = \sin^2\theta\frac{\partial^2 u}{\partial y^2} - 2\sin\theta\cos\theta\frac{\partial^2 u}{\partial x\partial y} + \cos^2\theta\frac{\partial^2 u}{\partial y^2}$$

Now, adding the values of $\dfrac{\partial^2 u}{\partial r^2}$ and $\dfrac{1}{r^2}\dfrac{\partial^2 u}{\partial\theta^2} + \dfrac{1}{r}\dfrac{\partial u}{\partial r}$ we get

$$\frac{\partial^2 u}{\partial r^2} + \frac{1}{r^2}\frac{\partial^2 u}{\partial\theta^2} + \frac{1}{r}\frac{\partial u}{\partial r} = (\cos^2\theta + \sin^2\theta)\frac{\partial^2 u}{\partial x^2} + (\sin^2\theta + \cos^2\theta)\frac{\partial^2 u}{\partial y^2}$$

$$= \frac{\partial^2 u}{\partial x^2} + \frac{\partial^2 u}{\partial y^2}. \qquad\qquad \square$$

Example 3.5.3 *Show that* $f(xy, z - 2x) = 0$ *satisfies, under suitable conditions, the equation* $x\dfrac{\partial z}{\partial x} - y\dfrac{\partial z}{\partial y} = 2x$. *What are these conditions?*

Solution: Let $u = xy, v = z - 2x$, then $f(u, v) = 0$

$\therefore\ df = f_u du + f_v dv = f_u(x\,dy + y\,dx) + f_v(dz - 2dx) = 0.$

Let us take z as the dependent variable and x, y are independent variables, then we get $dz = \dfrac{\partial z}{\partial x}dx + \dfrac{\partial z}{\partial y}dy.$

$$\therefore\quad df = f_u(x\,dy + y\,dx) + f_v\left(\frac{\partial z}{\partial x}dx + \frac{\partial z}{\partial y}dy - 2dx\right) = 0$$

or, $$\left(yf_u + \left\{\frac{\partial z}{\partial x} - 2\right\}f_v\right)dx + \left(xf_u + \frac{\partial z}{\partial y}f_v\right)dy = 0$$

Since x and y are independent variables, above gives

$$yf_u + \left\{\frac{\partial z}{\partial x} - 2\right\}f_v = 0 \qquad\qquad (3.41)$$

and $$xf_u + \frac{\partial z}{\partial y}f_v = 0 \qquad\qquad (3.42)$$

From (3.42) we get $f_u = -\dfrac{1}{x}\dfrac{\partial z}{\partial y}f_v.$

$\therefore$ From (3.41) we get

$$-y\frac{1}{x}\frac{\partial z}{\partial y}f_v + \left\{\frac{\partial z}{\partial x} - 2\right\}f_v = 0 \text{ or, } -\frac{y}{x}\frac{\partial z}{\partial y} + \frac{\partial z}{\partial x} - 2 = 0$$

or, $\quad x\frac{\partial z}{\partial x} - y\frac{\partial z}{\partial y} = 2x, \quad$ provided $f_y \neq 0.$

Therefore, the result holds when f is differentiable and $f_v \neq 0$, which are the required conditions. $\qquad\square$

Example 3.5.4 *If* $F(v^2 - x^2, v^2 - y^2, v^2 - z^2) = 0$ *where* v *is a function of* x, y *and* z, *show that* $\dfrac{1}{x}\dfrac{\partial v}{\partial x} + \dfrac{1}{y}\dfrac{\partial v}{\partial y} + \dfrac{1}{z}\dfrac{\partial v}{\partial z} = \dfrac{1}{v}.$

Solution: Let $p = v^2 - x^2$, $q = v^2 - y^2$, $r = v^2 - z^2$. Then $F(p, q, r) = 0.$

Differentiating with respect to x, we get $\dfrac{\partial F}{\partial x} = 0$

or, $\quad \dfrac{\partial F}{\partial p}\cdot\dfrac{\partial p}{\partial x} + \dfrac{\partial F}{\partial q}\cdot\dfrac{\partial q}{\partial x} + \dfrac{\partial F}{\partial r}\cdot\dfrac{\partial r}{\partial x} = 0$

or, $\quad \dfrac{\partial F}{\partial p}\cdot\left(2v\dfrac{\partial v}{\partial x} - 2x\right) + \dfrac{\partial F}{\partial q}.2v\dfrac{\partial v}{\partial x} + \dfrac{\partial F}{\partial r}.2v\dfrac{\partial v}{\partial x} = 0$

or, $\quad 2v\dfrac{\partial v}{\partial x}\left(\dfrac{\partial F}{\partial p} + \dfrac{\partial F}{\partial q} + \dfrac{\partial F}{\partial r}\right) = 2x\dfrac{\partial F}{\partial p} \text{ or, } \dfrac{v}{x}\dfrac{\partial v}{\partial x}\sum\dfrac{\partial F}{\partial p} = \dfrac{\partial F}{\partial p}.$

Similarly, we get $\dfrac{v}{y}\dfrac{\partial v}{\partial y}\sum\dfrac{\partial F}{\partial p} = \dfrac{\partial F}{\partial q}$ and $\dfrac{v}{z}\dfrac{\partial v}{\partial z}\sum\dfrac{\partial F}{\partial p} = \dfrac{\partial F}{\partial r}.$

Adding all these we get

$$\left(\frac{v}{x}\frac{\partial v}{\partial x} + \frac{v}{y}\frac{\partial v}{\partial y} + \frac{v}{z}\frac{\partial v}{\partial z}\right)\sum\frac{\partial F}{\partial p} = \sum\frac{\partial F}{\partial p} \Rightarrow \frac{v}{x}\frac{\partial v}{\partial x} + \frac{v}{y}\frac{\partial v}{\partial y} + \frac{v}{z}\frac{\partial v}{\partial z} = 1$$

or, $\dfrac{1}{x}\dfrac{\partial v}{\partial x} + \dfrac{1}{y}\dfrac{\partial v}{\partial y} + \dfrac{1}{z}\dfrac{\partial v}{\partial z} = \dfrac{1}{v}.$ $\qquad\square$

Example 3.5.5 *If* $u = x\phi(x+y) + y\psi(x+y)$, *then prove that*

$$\frac{\partial^2 u}{\partial x^2} - 2\frac{\partial^2 u}{\partial x\partial y} + \frac{\partial^2 u}{\partial y^2} = 0.$$

Solution: Let $x + y = v$, then $u = x\phi(v) + y\psi(v).$

$\therefore \quad \dfrac{\partial u}{\partial x} = \phi(v) + x\phi'(v) + y\psi'(v)$

$\dfrac{\partial^2 u}{\partial x^2} = \phi'(v) + \phi'(v) + x\phi''(v) + y\psi''(v) = 2\phi'(v) + x\phi''(v) + y\psi''(v)$

and $\quad \dfrac{\partial^2 u}{\partial x\partial y} = \phi'(v) + x\phi''(v) + \psi'(v) + y\psi''(v)$

and similarly, $\dfrac{\partial^2 u}{\partial y^2} = 2\psi'(v) + x\phi''(v) + y\psi''(v).$

Adding these we get $\dfrac{\partial^2 u}{\partial x^2} - 2\dfrac{\partial^2 u}{\partial x\partial y} + \dfrac{\partial^2 u}{\partial y^2} = 2\phi'(v) + x\phi''(v) + y\psi''(v) - 2\phi'(v) -$

$2x\phi''(v) - 2\psi'(v) - y\psi''(v) + 2\psi'(v) + x\phi''(v) + y\psi''(v) = 0.$ $\qquad\square$

3.6   Homogeneous Functions

Definition 3.6.1 *A function $f(x,y)$ is said to be a homogeneous function of degree n in the variables x and y, if it can be expressed in the form $x^n \phi \left(\dfrac{y}{x} \right)$ or $y^n \phi \left(\dfrac{x}{y} \right)$.*

Example 3.6.1 $f(x,y) = x^2 + 3xy + 5y^2$ *is a homogeneous function of degree 2, for it can be written as* $f(x,y) = x^2 \left(1 + 3\dfrac{y}{x} + 5\dfrac{y^2}{x^2} \right) = x^n \phi \left(\dfrac{y}{x} \right)$.

Example 3.6.2 $f(x,y) = \sqrt{x} + \sqrt{y} + 2$ *is not a homogeneous function, because* $f(x,y) = \sqrt{x} \left(1 + \sqrt{\dfrac{y}{x}} + \dfrac{2}{\sqrt{x}} \right) \neq \phi \left(\dfrac{y}{x} \right)$ *and it can not be written in the above mentioned form in any way.*

Definition 3.6.2 *A function $u = f(x_1, x_2, ..., x_n)$ is said to be a homogeneous function of degree k in n variables $x_1, x_2, ..., x_n$, if we can write $u = f(x_1, x_2, ..., x_n)$*
$$= x_1^k \phi \left(\frac{x_2}{x_1}, \frac{x_3}{x_1}, ..., \frac{x_n}{x_1} \right) \text{ or, } x_2^k \phi \left(\frac{x_1}{x_2}, \frac{x_3}{x_2}, ..., \frac{x_n}{x_2} \right) \text{ or, } ..., ... \text{ or,}$$
$$x_n^k \phi \left(\frac{x_1}{x_n}, \frac{x_2}{x_n}, ..., \frac{x_{n-1}}{x_1} \right).$$

Definition 3.6.3 *(**Alternative form**) A function $u = f(x_1, x_2, ..., x_n)$ is said to be a homogeneous function of degree k in n variables $x_1, x_2, ..., x_n$, if we can write $f(tx_1, tx_2, ..., tx_n) = t^k f(x_1, x_2, ..., x_n)$, for all values of t, where t is independent of $x_1, x_2, ..., x_n$.*

Example 3.6.3 $f(x,y,z) = \sqrt[3]{x} + \sqrt[3]{y} + \sqrt[3]{z}$ *is a homogeneous function of degree $\dfrac{1}{3}$ in three variables x, y, z, because we write $f(tx, ty, tz) = \sqrt[3]{tx} + \sqrt[3]{ty} + \sqrt[3]{tz} = t^{\frac{1}{3}} \left(\sqrt[3]{x} + \sqrt[3]{y} + \sqrt[3]{z} \right) = t^{\frac{1}{3}} f(x,y,z)$.*

Note 3.6.1 *A polynomial homogeneous function in x, y and of degree n of the form $f(x,y) = a_0 x^n + a_1 x^{n-1} y + a_2 x^{n-2} y^2 + ... + a_r x^{n-r} y^r + ... + a_n y^n$ in which the sum of the indices of x and y in every term is n, is a polynomial homogeneous function of degree n in the two variables x, y. The above function can be expressed as $f(x,y) = x^n g \left(\dfrac{y}{x} \right)$.*

Example 3.6.4 *Examine the homogeneity of the following functions*

$$(i) \quad u = x^3 + 3x^2 y + 3xy^2 + y^3 \qquad (ii) \quad u = \sin \frac{x^2 + y^2}{xy}$$

$$(iii) \quad u = \frac{x+y}{2x-3y} \qquad\qquad (iv) \quad u = \frac{x^{\frac{1}{4}} + y^{\frac{1}{4}}}{x^{\frac{1}{5}} + y^{\frac{1}{5}}}$$

Solution: (i) Here $u = f(x,y) = x^3 \left(1 + 3\dfrac{y}{x} + 3\dfrac{y^2}{x^2} + \dfrac{y^3}{x^3} \right) = x^3 \phi \left(\dfrac{y}{x} \right)$.

So it is a homogeneous function of degree 3.

Alternatively, $f(tx, ty) = (tx)^3 + 3(tx)^2 ty + 3tx(ty)^2 + (ty)^3$

$$= t^3(x^3 + 3x^2y + 3xy^2 + y^3) = t^3 f(x, y).$$

So it is a homogeneous function of degree 3.

(ii) $u = x^0 \sin\left(\dfrac{1 + \frac{y^2}{x^2}}{\frac{y}{x}}\right) = x^0 \phi\left(\dfrac{y}{x}\right) \Rightarrow$ homogeneous function of degree 0.

(iii) $u = \dfrac{x\left(1 + \frac{y}{x}\right)}{x\left(2 - 3\frac{y}{x}\right)} = x^0 \phi\left(\dfrac{y}{x}\right)$ and is a homogeneous function of degree 0.

(iv) $u = \dfrac{x^{\frac{1}{4}}\left\{1 + \left(\frac{y}{x}\right)^{\frac{1}{4}}\right\}}{x^{\frac{1}{5}}\left\{1 + \left(\frac{y}{x}\right)^{\frac{1}{5}}\right\}} = x^{\frac{1}{20}}\phi\left(\dfrac{y}{x}\right) \Rightarrow$ homogeneous function of degree $\dfrac{1}{20}$.

$\square$

Theorem 3.6.1 *Euler's Theorem on Homogeneous Function of Two Independent Variables.*

Statement: *If* $u = f(x, y)$ *be a homogeneous function of degree n in the two variables x and y, then*

$$x\frac{\partial u}{\partial x} + y\frac{\partial u}{\partial y} = nu.$$

Proof: Since $f(x, y)$ is a homogeneous function of degree n in x, y, we can write

$$f(x, y) = x^n \phi\left(\frac{y}{x}\right) \tag{3.43}$$

Differentiating partially both sides of (3.43) with respect to x and y respectively we get

$$\frac{\partial f}{\partial x} = nx^{n-1}\phi\left(\frac{y}{x}\right) + x^n\phi'\left(\frac{y}{x}\right)\left(-\frac{y}{x^2}\right) \tag{3.44}$$

$$\text{and} \quad \frac{\partial f}{\partial y} = x^n\phi'\left(\frac{y}{x}\right)\frac{1}{x} \tag{3.45}$$

Taking $(3.44) \times x + (3.45) \times y$ we get

$$\begin{aligned} x\frac{\partial f}{\partial x} + y\frac{\partial f}{\partial y} &= nx^n\phi\left(\frac{y}{x}\right) - yx^{n-1}\phi'\left(\frac{y}{x}\right) + yx^{n-1}\phi'\left(\frac{y}{x}\right) \\ &= nx^n\phi\left(\frac{y}{x}\right) = nf(x, y) \end{aligned}$$

i.e., $x\dfrac{\partial u}{\partial x} + y\dfrac{\partial u}{\partial y} = nu.$ $\square$

Corollary 3.6.1 *If u is a homogeneous function of degree n in the two variables x and y, then* $\dfrac{\partial u}{\partial x}$ *and* $\dfrac{\partial u}{\partial x}$ *are each a homogeneous function of degree* $(n-1)$.

Proof: Since u is a homogeneous function of degree n in the two variables x and y, by Euler's theorem, we have

$$x\frac{\partial u}{\partial x} + y\frac{\partial u}{\partial y} = nu \tag{3.46}$$

Differentiating both sides of (3.46) partially with respect to x ,we get

$$\frac{\partial u}{\partial x} + x\frac{\partial^2 u}{\partial x^2} + y\frac{\partial^2 u}{\partial x \partial y} = n\frac{\partial u}{\partial x}$$

or, $\quad x\frac{\partial}{\partial x}\left(\frac{\partial u}{\partial x}\right) + y\frac{\partial}{\partial y}\left(\frac{\partial u}{\partial x}\right) = (n-1)\left(\frac{\partial u}{\partial x}\right).$

This shows that $\frac{\partial u}{\partial x}$ is a homogeneous function of degree $n-1$.

Similarly, differentiating both sides of (3.46) partially with respect to y, it can be shown that $\frac{\partial u}{\partial x}$ is also a homogeneous function of degree $n-1$. □

Corollary 3.6.2 *Euler's Theorem of Second Order*

If u is a homogeneous function of degree n in the two variables x and y, then

$$\left(x\frac{\partial}{\partial x} + y\frac{\partial}{\partial y}\right)^2 u = n(n-1)u$$

where $\left(x\frac{\partial}{\partial x} + y\frac{\partial}{\partial y}\right)^2 u = x^2\frac{\partial^2 u}{\partial x^2} + 2xy\frac{\partial^2 u}{\partial x \partial y} + y^2\frac{\partial^2 u}{\partial y^2}.$

Proof: Since u is a homogeneous function of degree n, in x and y, so by **Corollary 3.5.1** above, each of $\frac{\partial u}{\partial x}$ and $\frac{\partial u}{\partial x}$ will be a homogeneous function of degree $(n-1)$ and so by Euler's theorem we get

$$\left. \begin{array}{l} x\frac{\partial}{\partial x}\left(\frac{\partial u}{\partial x}\right) + y\frac{\partial}{\partial y}\left(\frac{\partial u}{\partial x}\right) = (n-1)\frac{\partial u}{\partial x} \\ \text{and}\quad x\frac{\partial}{\partial y}\left(\frac{\partial u}{\partial x}\right) + y\frac{\partial}{\partial y}\left(\frac{\partial u}{\partial y}\right) = (n-1)\frac{\partial u}{\partial y} \end{array} \right\}$$

Taking $\left(x\frac{\partial}{\partial x} + y\frac{\partial}{\partial y}\right)$ as an operator, above can be written as

$$\left(x\frac{\partial}{\partial x} + y\frac{\partial}{\partial y}\right)\frac{\partial u}{\partial x} = (n-1)\frac{\partial u}{\partial x} \qquad (3.47)$$

$$\text{and}\quad \left(x\frac{\partial}{\partial x} + y\frac{\partial}{\partial y}\right)\frac{\partial u}{\partial y} = (n-1)\frac{\partial u}{\partial y} \qquad (3.48)$$

Now (3.47) $\times x$ + (3.48) $\times y$ gives

$$x^2\frac{\partial^2 u}{\partial x^2} + 2xy\frac{\partial^2 u}{\partial x \partial y} + y^2\frac{\partial^2 u}{\partial y^2} = (n-1)\left(x\frac{\partial u}{\partial x} + y\frac{\partial u}{\partial y}\right) = (n-1)nu$$

or, $\quad \left(x\frac{\partial}{\partial x} + y\frac{\partial}{\partial y}\right)^2 u = n(n-1)u.$

Alternative Method (Direct Method):

Since u is a homogeneous function of degree n in the two variables x and y, we have by Euler's theorem

$$x\frac{\partial u}{\partial x} + y\frac{\partial u}{\partial y} = nu \qquad (3.49)$$

Differentiating both sides of (3.49) partially with respect to x, we get

$$\frac{\partial u}{\partial x} + x\frac{\partial^2 u}{\partial x^2} + y\frac{\partial^2 u}{\partial x \partial y} = n\frac{\partial u}{\partial x} \tag{3.50}$$

Similarly, differentiating both sides of (3.49) partially with respect to y, we get

$$x\frac{\partial^2 u}{\partial x \partial y} + \frac{\partial u}{\partial y} + y\frac{\partial^2 u}{\partial y^2} = n\frac{\partial u}{\partial y} \tag{3.51}$$

Now (3.47) $\times x$ + (3.48) $\times y$ gives

$$\left(x\frac{\partial u}{\partial x} + y\frac{\partial u}{\partial y}\right) + \left(x^2\frac{\partial^2 u}{\partial x^2} + 2xy\frac{\partial^2 u}{\partial x \partial y} + y^2\frac{\partial^2 u}{\partial y^2}\right) = n\left(x\frac{\partial u}{\partial x} + y\frac{\partial u}{\partial y}\right)$$

or, $nu + \left(x\dfrac{\partial}{\partial x} + y\dfrac{\partial}{\partial y}\right)^2 u = n.nu \Rightarrow \left(x\dfrac{\partial}{\partial x} + y\dfrac{\partial}{\partial y}\right)^2 u = n(n-1)u.$ □

Note 3.6.2 *The above result is called **Euler's theorem of second order on homogeneous function of two variables**.*

Example 3.6.5 *Verify Euler's theorem for the functions given in the **Example 3.6.4**.*

Solution: (i) The given function is a homogeneous function of degree 3. Clearly,

$$\begin{aligned} x\frac{\partial u}{\partial x} + y\frac{\partial u}{\partial y} &= x(3x^2 + 6xy + 3y^2) + y(3x^2 + 6xy + 3y^2) \\ &= 3(x^3 + 3x^2y + 3xy^2 + y^3) = 3u. \end{aligned}$$

which shows that Euler's theorem is satisfied in this case.

(ii) $u = \sin\dfrac{x^2 + y^2}{xy}$ is a homogeneous function of degree 0.

$$\text{Now}\quad \frac{\partial u}{\partial x} = \cos\frac{x^2+y^2}{xy}.\frac{2x.xy - y(x^2+y^2)}{x^2 y^2} = \frac{x^2-y^2}{x^2 y}\cos\frac{x^2+y^2}{xy}.$$

$$\text{Similarly,}\quad \frac{\partial u}{\partial y} = \frac{y^2-x^2}{xy^2}\cos\frac{x^2+y^2}{xy}$$

$$\therefore\quad x\frac{\partial u}{\partial x} + y\frac{\partial u}{\partial y} = \cos\frac{x^2+y^2}{xy}\left(\frac{x^2-y^2}{xy} + \frac{y^2-x^2}{xy}\right) = 0 = 0.u$$

Hence, Euler's theorem is verified.

(iii) Here $u = \dfrac{x+y}{2x-3y}$ is a homogeneous function of degree 0.

$$\text{Now,}\quad \frac{\partial u}{\partial x} = \frac{1.(2x-3y) - 2(x+y)}{(2x-3y)^2} = \frac{-5x}{(2x-3y)^2}$$

$$\text{and}\quad \frac{\partial u}{\partial y} = \frac{1.(2x-3y) + 3(x+y)}{(2x-3y)^2} = \frac{5x}{(2x-3y)^2}$$

$$\therefore\quad x\frac{\partial u}{\partial x} + y\frac{\partial u}{\partial y} = -\frac{5yx}{(2x-3y)^2} + \frac{5yx}{(2x-3y)^2} = 0 = 0.u.$$

Hence, Euler's theorem is verified.

(iv) Here $u = \dfrac{x^{\frac{1}{4}} + y^{\frac{1}{4}}}{x^{\frac{1}{5}} + y^{\frac{1}{5}}}$ is a homogeneous function of degree $\dfrac{1}{20}$.

Now, $\dfrac{\partial u}{\partial x} = \dfrac{\frac{1}{4}x^{-\frac{3}{4}}\left(x^{\frac{1}{5}} + y^{\frac{1}{5}}\right) - \frac{1}{5}x^{-\frac{4}{5}}\left(x^{\frac{1}{4}} + y^{\frac{1}{4}}\right)}{\left(x^{\frac{1}{5}} + y^{\frac{1}{5}}\right)^{2}}$

$= \dfrac{\frac{1}{4}x^{-\frac{11}{20}} + \frac{1}{4}x^{-\frac{3}{4}}y^{\frac{1}{5}} - \frac{1}{5}x^{-\frac{11}{20}} - \frac{1}{5}x^{-\frac{4}{5}}y^{\frac{1}{4}}}{\left(x^{\frac{1}{5}} + y^{\frac{1}{5}}\right)^{2}}$

$= \dfrac{\frac{1}{20}x^{-\frac{11}{20}} + \frac{1}{4}x^{-\frac{3}{4}}y^{\frac{1}{5}} - \frac{1}{5}x^{-\frac{4}{5}}y^{\frac{1}{4}}}{\left(x^{\frac{1}{5}} + y^{\frac{1}{5}}\right)^{2}}$

Similarly, $\dfrac{\partial u}{\partial y} = \dfrac{\frac{1}{20}y^{-\frac{11}{20}} + \frac{1}{4}y^{-\frac{3}{4}}x^{\frac{1}{5}} - \frac{1}{5}y^{-\frac{4}{5}}x^{\frac{1}{4}}}{\left(x^{\frac{1}{5}} + y^{\frac{1}{5}}\right)^{2}}$

$\therefore \quad x\dfrac{\partial u}{\partial x} + y\dfrac{\partial u}{\partial y} = \dfrac{1}{\left(x^{\frac{1}{5}} + y^{\frac{1}{5}}\right)^{2}}\left[\left(\frac{1}{20}x^{-\frac{11}{20}} + \frac{1}{4}x^{-\frac{3}{4}}y^{\frac{1}{5}} - \frac{1}{5}x^{-\frac{4}{5}}y^{\frac{1}{4}}\right)x\right.$

$\left. + \left(\frac{1}{20}y^{-\frac{11}{20}} + \frac{1}{4}y^{-\frac{3}{4}}x^{\frac{1}{5}} - \frac{1}{5}y^{-\frac{4}{5}}x^{\frac{1}{4}}\right)y\right]$

$= \dfrac{\frac{1}{20}\left(x^{\frac{9}{20}} + y^{\frac{9}{20}}\right) + \frac{1}{20}y^{\frac{1}{4}}x^{\frac{1}{5}} + \frac{1}{20}x^{\frac{1}{4}}y^{\frac{1}{5}}}{\left(x^{\frac{1}{5}} + y^{\frac{1}{5}}\right)^{2}}$

$= \dfrac{1}{20}\dfrac{x^{\frac{1}{4}}\left(x^{\frac{1}{5}} + y^{\frac{1}{5}}\right) + y^{\frac{1}{4}}\left(x^{\frac{1}{5}} + y^{\frac{1}{5}}\right)}{\left(x^{\frac{1}{5}} + y^{\frac{1}{5}}\right)^{2}}$

$= \dfrac{1}{20}\dfrac{\left(x^{\frac{1}{4}} + y^{\frac{1}{4}}\right)\left(x^{\frac{1}{5}} + y^{\frac{1}{5}}\right)}{\left(x^{\frac{1}{5}} + y^{\frac{1}{5}}\right)^{2}} = \dfrac{1}{20}\dfrac{x^{\frac{1}{4}} + y^{\frac{1}{4}}}{x^{\frac{1}{5}} + y^{\frac{1}{5}}} = \dfrac{1}{20}u.$

Hence Euler's theorem is verified for this function.

Alternative method:

Given $u = \dfrac{x^{\frac{1}{4}} + y^{\frac{1}{4}}}{x^{\frac{1}{5}} + y^{\frac{1}{5}}} = \dfrac{x^{\frac{1}{4}}\left\{1 + \left(\frac{y}{x}\right)^{\frac{1}{4}}\right\}}{x^{\frac{1}{5}}\left\{1 + \left(\frac{y}{x}\right)^{\frac{1}{5}}\right\}} = x^{\frac{1}{20}}f\left(\dfrac{y}{x}\right).$

Hence the given function is a homogeneous function of degree $\dfrac{1}{20}$.

Now as $u = \dfrac{x^{\frac{1}{4}} + y^{\frac{1}{4}}}{x^{\frac{1}{5}} + y^{\frac{1}{5}}} \quad \therefore \quad \log u = \log\left(x^{\frac{1}{4}} + y^{\frac{1}{4}}\right) - \log\left(x^{\frac{1}{5}} + y^{\frac{1}{5}}\right).$

$\therefore \dfrac{1}{u}.u_x = \dfrac{\frac{1}{4}x^{-\frac{3}{4}}}{x^{\frac{1}{4}} + y^{\frac{1}{4}}} - \dfrac{\frac{1}{5}x^{-\frac{4}{5}}}{x^{\frac{1}{5}} + y^{\frac{1}{5}}} \quad \therefore \quad xu_x = u\left[\dfrac{1}{4}\dfrac{x^{\frac{1}{4}}}{x^{\frac{1}{4}} + y^{\frac{1}{4}}} - \dfrac{1}{5}\dfrac{x^{\frac{1}{5}}}{x^{\frac{1}{5}} + y^{\frac{1}{5}}}\right].$

Similarly, we shall get $yu_y = u\left[\dfrac{1}{4}\dfrac{y^{\frac{1}{4}}}{x^{\frac{1}{4}} + y^{\frac{1}{4}}} - \dfrac{1}{5}\dfrac{y^{\frac{1}{5}}}{x^{\frac{1}{5}} + y^{\frac{1}{5}}}\right].$

So we get $xu_x + yu_y = u\left(\dfrac{1}{4} - \dfrac{1}{5}\right) = \dfrac{1}{20}u.$

This shows that Euler's theorem is satisfied. □

Now in the next theorems we discuss about the generalized form of Euler's theorem and its converse.

Theorem 3.6.2 *Euler's Theorem on Homogeneous Functions of k Independent Variables*

Statement: *If $f = f(x_1, x_2, ..., x_k)$ be a homogeneous function of degree n in k independent variables $x_1, x_2, ..., x_k$ having continuous partial derivatives, then*

$$x_1\frac{\partial f}{\partial x_1} + x_2\frac{\partial f}{\partial x_2} + ... + x_k\frac{\partial f}{\partial x_k} = nf.$$

Proof: Since $f = f(x_1, x_2, ..., x_k)$ is a homogeneous function of degree n in x_1, x_2, ..., x_k, we can write

$$f(tx_1, tx_2, ..., tx_k) = t^n f(x_1, x_2, ..., x_k) \tag{3.52}$$

for all values of t.

We put $u_1 = tx_1, u_2 = tx_2, ..., u_k = tx_k$.

Differentiating both sides of (3.52) with respect to t, we get

$$\frac{\partial f}{\partial u_1}\frac{\partial u_1}{\partial t} + \frac{\partial f}{\partial u_2}\frac{\partial u_2}{\partial t} + ... + \frac{\partial f}{\partial u_k}\frac{\partial u_k}{\partial t} = nt^{n-1}f(x_1, x_2, ..., x_k)$$

$$\text{or,} \quad x_1\frac{\partial f}{\partial u_1} + x_2\frac{\partial f}{\partial u_2} + ... + x_k\frac{\partial f}{\partial u_k} = nt^{n-1}f(x_1, x_2, ..., x_k) \tag{3.53}$$

Now put $t = 1$ in (3.53) and get

$$x_1\frac{\partial f}{\partial x_1} + x_2\frac{\partial f}{\partial x_2} + ... + x_k\frac{\partial f}{\partial x_k} = nf(x_1, x_2, ..., x_k). \quad □$$

Theorem 3.6.3 *Converse of Euler's theorem for a function of k independent variables*

Statement: *If $f = f(x_1, x_2, ..., x_k)$ admits of continuous partial derivatives satisfying the relation $x_1\dfrac{\partial f}{\partial x_1} + x_2\dfrac{\partial f}{\partial x_2} + ... + x_k\dfrac{\partial f}{\partial x_k} = nf$, where n is a positive integer, then $f(x_1, x_2, ..., x_k)$ is a homogeneous function of degree n.*

Proof: Let us put $u_1 = \dfrac{x_1}{x_k}, u_2 = \dfrac{x_2}{x_k}, ..., u_{k-1} = \dfrac{x_{k-1}}{x_k}, u_k = x_k,$

so that $x_1 = u_1 x_k, x_2 = u_2 x_k, ..., x_{k-1} = u_{k-1}x_k, x_k = u_k.$

Now $f(x_1, x_2, ..., x_k)$ can be expressed as a function of $u_1, u_2, ..., u_k$.

Let us suppose that $f(x_1, x_2, ..., x_k) = \phi(u_1, u_2, ..., u_k)$.

Then $x_1\dfrac{\partial f}{\partial x_1} = x_1\left(\dfrac{\partial \phi}{\partial u_1}\dfrac{\partial u_1}{\partial x_1} + \dfrac{\partial \phi}{\partial u_2}\dfrac{\partial u_2}{\partial x_1} + ... + \dfrac{\partial \phi}{\partial u_k}\dfrac{\partial u_k}{\partial x_1}\right)$

$= x_1\left(\dfrac{\partial \phi}{\partial u_1}\dfrac{1}{x_k} + \dfrac{\partial \phi}{\partial u_2}0 + ... + \dfrac{\partial \phi}{\partial u_k}.0\right) = \dfrac{x_1}{x_k}\dfrac{\partial \phi}{\partial u_1} = u_1\dfrac{\partial \phi}{\partial u_1}.$

Similarly, $x_2\dfrac{\partial f}{\partial x_2} = u_2\dfrac{\partial \phi}{\partial u_2}$ and so on.

Lastly, $x_k \dfrac{\partial f}{\partial x_k} = x_k \left(\dfrac{\partial \phi}{\partial u_1} \dfrac{\partial u_1}{\partial x_k} + \dfrac{\partial \phi}{\partial u_2} \dfrac{\partial u_2}{\partial x_k} + \ldots + \dfrac{\partial \phi}{\partial u_k} \dfrac{\partial u_k}{\partial x_k} \right)$

$$= x_k \left(-\dfrac{\partial \phi}{\partial u_1} \dfrac{x_1}{x_k^2} - \dfrac{\partial \phi}{\partial u_2} \dfrac{x_2}{x_k^2} + \ldots + \dfrac{\partial \phi}{\partial u_k}.1 \right).$$

$$= -u_1 \dfrac{\partial \phi}{\partial u_1} - u_2 \dfrac{\partial \phi}{\partial u_2} - \ldots - u_{k-1} \dfrac{\partial \phi}{\partial u_{k-1}} + u_k \dfrac{\partial \phi}{\partial u_k}.$$

Hence we get $x_1 \dfrac{\partial f}{\partial x_1} + x_2 \dfrac{\partial f}{\partial x_2} + \ldots + x_k \dfrac{\partial f}{\partial x_k} = u_k \dfrac{\partial \phi}{\partial u_k} = nf$

$\therefore \quad \dfrac{1}{\phi} \dfrac{\partial \phi}{\partial u_k} = \dfrac{n}{u_k} \quad [\because \phi = f]$

or, $\displaystyle \int \dfrac{\partial \phi}{\phi} = n \int \dfrac{\partial u_k}{u_k}$ + a function containing all the variables $u_1, u_2, \ldots,$

u_{k-1} except the variable x_k (i.e., free form u_k)

$$= n \int \dfrac{\partial u_k}{u_k} + \psi^*(u_1, u_2, \ldots, u_{k-1}) \text{ (say)}$$

Integrating with respect to u_k we get

$$\log \phi = n \log u_k + \psi^*(u_1, u_2, \ldots, u_{k-1})$$

For the sake of brevity we take $\psi^*(u_1, u_2, \ldots, u_{k-1}) = \log\{\psi(u_1, u_2, \ldots, u_{k-1})\}$ and then we get

$$\log \phi = n \log u_k + \psi^*(u_1, u_2, \ldots, u_{k-1}) = \log u_k^n$$
$$+ \log\{\psi(u_1, u_2, \ldots, u_{k-1})\}$$

or, $\log \phi = \log u_k^n \psi$ or, $\phi = u_k^n \psi(u_1, u_2, \ldots, u_{k-1})$

$\therefore \quad f(x_1, x_2, \ldots, x_k) = x_k^n \psi \left(\dfrac{x_1}{x_k}, \dfrac{x_2}{x_k}, \ldots, \dfrac{x_{k-1}}{x_k} \right)$

which, according to the definition of a homogeneous function, shows that $f = f(x_1, x_2, \ldots, x_k)$ is a homogeneous function of degree n. $\qquad\square$

3.7 Miscellaneous Illustrative Examples

Example 3.7.1 *If $x = r \cos \theta, y = r \sin \theta$, so that $r = \sqrt{x^2 + y^2}$ and $\theta = \tan^{-1} \dfrac{y}{x}$, show that $\dfrac{\partial x}{\partial r} \neq \dfrac{1}{\frac{\partial r}{\partial x}}$ and $\dfrac{\partial x}{\partial \theta} \neq \dfrac{1}{\frac{\partial \theta}{\partial x}}$.*

Solution: From the given relations, we get

$$\dfrac{\partial x}{\partial r} = \cos \theta; \quad \dfrac{\partial r}{\partial x} = \dfrac{1}{2\sqrt{x^2 + y^2}} 2x = \dfrac{x}{\sqrt{x^2 + y^2}} = \dfrac{r \cos \theta}{r} = \cos \theta.$$

$$\therefore \quad \dfrac{\partial x}{\partial r} \neq \dfrac{1}{\frac{\partial r}{\partial x}}.$$

Also, $\dfrac{\partial x}{\partial \theta} = -r \sin \theta; \quad \dfrac{\partial \theta}{\partial x} = \dfrac{-\frac{y}{x^2}}{1 + \frac{y^2}{x^2}} = \dfrac{-y}{x^2 + y^2} = \dfrac{-r \sin \theta}{r^2} = \dfrac{-\sin \theta}{r}.$

$$\therefore \quad \dfrac{\partial x}{\partial \theta} \neq \dfrac{1}{\frac{\partial \theta}{\partial x}}. \qquad\square$$

Note 3.7.1 *It is to be carefully noticed that, if y be a function of a single variable x, then under certain circumstances* $\dfrac{dy}{dx} = \dfrac{1}{\frac{dx}{dy}}$, *but a similar property is not true, as seen above, in case of a function of more than one variable.*

Example 3.7.2 *Prove that by the transformations* $u = x - ct, v = x + ct$, *the partial differential equation* $\dfrac{\partial^2 z}{\partial t^2} = c^2 \dfrac{\partial^2 z}{\partial x^2}$ *reduces to* $\dfrac{\partial^2 z}{\partial u \partial v} = 0$.

Hence obtain the general solution $z = \phi(x - ct) + \psi(x + ct)$, ϕ *and* ψ *being arbitrary functions.*

Solution: We consider z as a function of u and v, which, again are linear functions of the independent variables x and t.

Clearly, $x = \dfrac{u+v}{2}$, $t = \dfrac{v-u}{2c}$. Using chain rule, we write

$$\frac{\partial z}{\partial u} = \frac{\partial z}{\partial x} \cdot \frac{\partial x}{\partial u} + \frac{\partial z}{\partial t} \cdot \frac{\partial t}{\partial u} = \frac{1}{2}\frac{\partial z}{\partial x} - \frac{1}{2c}\frac{\partial z}{\partial t}$$

and

$$\frac{\partial z}{\partial v} = \frac{\partial z}{\partial x} \cdot \frac{\partial x}{\partial v} + \frac{\partial z}{\partial t} \cdot \frac{\partial t}{\partial v} = \frac{1}{2}\frac{\partial z}{\partial x} + \frac{1}{2c}\frac{\partial z}{\partial t}.$$

$$\therefore \quad \frac{\partial^2 z}{\partial u \partial v} = \frac{\partial}{\partial u}\left(\frac{\partial z}{\partial v}\right) = \frac{\partial}{\partial u}\left(\frac{1}{2}\frac{\partial z}{\partial x} + \frac{1}{2c}\frac{\partial z}{\partial t}\right)$$

$$= \frac{1}{2}\frac{\partial}{\partial u}\left(\frac{\partial z}{\partial x}\right) + \frac{1}{2c}\frac{\partial}{\partial u}\left(\frac{\partial z}{\partial t}\right)$$

$$= \frac{1}{2}\left[\frac{1}{2}\frac{\partial}{\partial x}\left(\frac{\partial z}{\partial x}\right) - \frac{1}{2c}\frac{\partial}{\partial t}\left(\frac{\partial z}{\partial x}\right)\right] + \frac{1}{2c}\left[\frac{1}{2}\frac{\partial}{\partial x}\left(\frac{\partial z}{\partial t}\right) - \frac{1}{2c}\frac{\partial}{\partial t}\left(\frac{\partial z}{\partial t}\right)\right]$$

$$\therefore \quad \frac{\partial^2 z}{\partial u \partial v} = \frac{1}{4}\left[\frac{\partial^2 z}{\partial x^2} - \frac{1}{c^2}\frac{\partial^2 z}{\partial t^2}\right].$$

Thus the equation $\dfrac{\partial^2 z}{\partial x^2} - \dfrac{1}{c^2}\dfrac{\partial^2 z}{\partial t^2} = 0$ becomes $4\dfrac{\partial^2 z}{\partial u \partial v} = 0$, i.e., $\dfrac{\partial^2 z}{\partial u \partial v} = 0$.

An Alternative Approach

The given equation is written as

$$\frac{\partial^2 z}{\partial x^2} = \frac{1}{c^2}\frac{\partial^2 z}{\partial t^2} \tag{3.54}$$

The given transformations are $u = x - ct, v = x + ct$. Then z is a function of u and v

$$\therefore \quad \frac{\partial z}{\partial x} = \frac{\partial z}{\partial u} \cdot \frac{\partial u}{\partial x} + \frac{\partial z}{\partial v} \cdot \frac{\partial v}{\partial x} = \frac{\partial z}{\partial u} + \frac{\partial z}{\partial v}.$$

Differentiating again with respect to x, we get

$$\frac{\partial^2 z}{\partial x^2} = \frac{\partial}{\partial x}\left(\frac{\partial z}{\partial u} + \frac{\partial z}{\partial v}\right) = \frac{\partial}{\partial u}\left(\frac{\partial z}{\partial u} + \frac{\partial z}{\partial v}\right)\frac{\partial u}{\partial x} + \frac{\partial}{\partial v}\left(\frac{\partial z}{\partial u} + \frac{\partial z}{\partial v}\right)\frac{\partial v}{\partial x}$$

$$= \frac{\partial^2 z}{\partial u^2} + 2\frac{\partial^2 z}{\partial u \partial v} + \frac{\partial^2 z}{\partial v^2} \tag{3.55}$$

Again, differentiating both sides of (3.54) partially with respect to t, we get

$$\frac{\partial z}{\partial t} = \frac{\partial z}{\partial u} \cdot \frac{\partial u}{\partial t} + \frac{\partial z}{\partial v} \cdot \frac{\partial v}{\partial t} = -c\frac{\partial z}{\partial u} + c\frac{\partial z}{\partial v}$$

Differentiating again with respect to t, we get

$$\frac{\partial^2 z}{\partial t^2} = \frac{\partial}{\partial t}\left(-c\frac{\partial z}{\partial u} + c\frac{\partial z}{\partial v}\right) = \frac{\partial}{\partial u}\left(-c\frac{\partial z}{\partial u} + c\frac{\partial z}{\partial v}\right)\frac{\partial u}{\partial t}$$

$$+ \frac{\partial}{\partial v}\left(-c\frac{\partial z}{\partial u} + c\frac{\partial z}{\partial v}\right)\frac{\partial v}{\partial t}$$

$$= \left(-c\frac{\partial^2 z}{\partial u^2} + c\frac{\partial^2 z}{\partial u\partial v}\right)(-c) + \left(-c\frac{\partial^2 z}{\partial u\partial v} + c\frac{\partial^2 z}{\partial v^2}\right).c$$

$$\therefore \quad \frac{\partial^2 z}{\partial t^2} = c^2\left(\frac{\partial^2 z}{\partial u^2} - 2\frac{\partial^2 z}{\partial u\partial v} + \frac{\partial^2 z}{\partial v^2}\right) \tag{3.56}$$

Now, subtracting (3.56) form (3.55), we get $\dfrac{\partial^2 z}{\partial u\partial v} = 0$ (using (3.54)), i.e., the equation $\dfrac{\partial^2 z}{\partial x^2} - \dfrac{1}{c^2}\dfrac{\partial^2 z}{\partial t^2} = 0$ reduces to $\dfrac{\partial^2 z}{\partial u\partial v} = 0$.

Second Part:

$$\because \quad \frac{\partial^2 z}{\partial u\partial v} = 0 \quad \therefore \quad \frac{\partial}{\partial u}\left(\frac{\partial z}{\partial v}\right) = 0$$

Integrating with respect to u, we get $\dfrac{\partial z}{\partial v} =$ a function which does not contain u but may contain v, i.e., $\dfrac{\partial z}{\partial v} = f(v)$.

Integrating this now with respect to v we get $z = \displaystyle\int f(v)+$ a function which does not contain v, but may contain u, i.e., we get $z = \psi(v) + \phi(u)$ as the complete solution of (3.54).

Hence replacing u and v by $x - ct$ and $x + ct$ respectively, we obtain $z = \phi(x - ct) + \psi(x + ct)$ as the complete solution of (3.54) where ϕ and ψ are any two arbitrary function of x and t in the combination of $x \quad ct$ and $x + ct$. $\qquad\square$

Note 3.7.2 *The above equation* $\dfrac{\partial^2 z}{\partial x^2} - \dfrac{1}{c^2}\dfrac{\partial^2 z}{\partial t^2} = 0$ *is known as the well-known One-dimensional wave equation in Mathematical Physics, where z is the wave disturbance and is a function of x and t only.*

Note 3.7.3 *The solution procedure which has done here is called D' Alembert's solution of one-dimensional wave equation.*

Example 3.7.3 *Let the differentiable function ϕ be such that $\phi(cx - az, cy - bz) = 0$ where a, b, c are constants and $c \neq 0$. Express z as a differentiable function of x and y. Also prove that $a\dfrac{\partial z}{\partial x} + b\dfrac{\partial z}{\partial y} = c$.*

Solution: Let us consider that

$$u = cx - az \tag{3.57}$$
$$v = cy - bz \quad \forall\ (x, y, z) \in \mathbb{R}^3 \tag{3.58}$$

Then ϕ becomes a function of two variables u and v.

Since z depends on x and y

$$\therefore \ \frac{\partial u}{\partial x} = c - a\frac{\partial z}{\partial x}, \ \frac{\partial u}{\partial y} = -a\frac{\partial z}{\partial y}, \ \frac{\partial v}{\partial x} = -b\frac{\partial z}{\partial x}, \ \frac{\partial v}{\partial y} = c - b\frac{\partial z}{\partial y}.$$

Also $\phi(u,v) = 0 \ \Rightarrow \ \frac{\partial \phi}{\partial u}\cdot\frac{\partial u}{\partial x} + \frac{\partial \phi}{\partial v}\cdot\frac{\partial v}{\partial x} = 0$ and $\frac{\partial \phi}{\partial u}\cdot\frac{\partial u}{\partial y} + \frac{\partial \phi}{\partial v}\cdot\frac{\partial v}{\partial y} = 0$.

Since $\frac{\partial \phi}{\partial u}$ and $\frac{\partial \phi}{\partial v}$ are not both zero,

$$\therefore \ \begin{vmatrix} \frac{\partial u}{\partial x} & \frac{\partial u}{\partial y} \\ \frac{\partial v}{\partial x} & \frac{\partial v}{\partial y} \end{vmatrix} = 0 \ \text{ or, } \ \frac{\partial u}{\partial x}\cdot\frac{\partial v}{\partial y} - \frac{\partial u}{\partial y}\cdot\frac{\partial v}{\partial x} = 0$$

$$\text{or, } \ \left(c - a\frac{\partial z}{\partial x}\right)\cdot\left(c - b\frac{\partial z}{\partial y}\right) - ab\frac{\partial z}{\partial x}\cdot\frac{\partial z}{\partial y} = 0$$

$$\text{or, } \ c\left(b\frac{\partial z}{\partial y} + a\frac{\partial z}{\partial x}\right) = c^2 \ \Rightarrow \ a\frac{\partial z}{\partial y} + b\frac{\partial z}{\partial x} = c \ [\because \ c \neq 0]$$

[*Second part proved*]

Now, adding (3.57) and (3.58), we get $u + v = c(x+y) - (a+b)z$ whence

$$z = \frac{(u+v) - c(x+y)}{a+b}. \hspace{3cm} \text{[\textit{First part proved}]} \quad \square$$

Example 3.7.4 *Prove that* $\frac{\partial^2 V}{\partial x^2} + \frac{\partial^2 V}{\partial y^2}$ *is an invariant for change of rectangular axes.*

Solution: Let us first consider the change of origin, i.e., shifting of the origin only. For this, the translational formulas are

$$\left. \begin{array}{l} x = x' + \alpha \\ y = y' + \beta \end{array} \right\} \Rightarrow \left. \begin{array}{l} dx = dx' \\ dy = dy', \end{array} \right\} \alpha, \beta \text{ are constants.}$$

Hence all the partial derivatives of V remain the same form.
Next, let the axes be turned through an angle θ, so that

$$\left. \begin{array}{l} x = x'\cos\theta - y'\sin\theta \\ y = x'\sin\theta + y'\cos\theta \end{array} \right\} \text{ where } \theta \text{ is a constant.}$$

We have

$$\begin{aligned} dV &= \frac{\partial V}{\partial x}dx + \frac{\partial V}{\partial y}dy \\ &= \frac{\partial V}{\partial x}(dx'\cos\theta - dy'\sin\theta) + \frac{\partial V}{\partial y}(dx'\sin\theta + dy'\cos\theta) \\ &= \left(\frac{\partial V}{\partial x}\cos\theta + \frac{\partial V}{\partial y}\sin\theta\right)dx' + \left(-\frac{\partial V}{\partial x}\sin\theta + \frac{\partial V}{\partial y}\cos\theta\right)dy' \end{aligned}$$

Also $dV = \frac{\partial V}{\partial x'}dx' + \frac{\partial V}{\partial y'}dy'$.

$$\therefore \ \frac{\partial V}{\partial x'} = \frac{\partial V}{\partial x}\cos\theta + \frac{\partial V}{\partial y}\sin\theta \text{ and } \frac{\partial V}{\partial y'} = -\frac{\partial V}{\partial x}\sin\theta + \frac{\partial V}{\partial y}\cos\theta.$$

These give
$$\frac{\partial V}{\partial x} = \frac{\partial V}{\partial x'}\cos\theta - \frac{\partial V}{\partial y'}\sin\theta$$

and
$$\frac{\partial V}{\partial y} = \frac{\partial V}{\partial x'}\sin\theta + \frac{\partial V}{\partial y'}\cos\theta. \Bigg\}$$

$$\therefore \quad \frac{\partial^2 V}{\partial x^2} = \frac{\partial}{\partial x}\left(\frac{\partial V}{\partial x}\right) = \left(\frac{\partial}{\partial x'}\cos\theta - \frac{\partial}{\partial y'}\sin\theta\right)\left(\frac{\partial V}{\partial x'}\cos\theta - \frac{\partial V}{\partial y'}\sin\theta\right)$$

$$= \frac{\partial^2 V}{\partial x'^2}\cos^2\theta - 2\frac{\partial^2 V}{\partial x'\partial y'}\sin\theta\cos\theta + \frac{\partial^2 V}{\partial y'^2}\sin^2\theta \qquad (3.59)$$

$$\frac{\partial^2 V}{\partial y^2} = \frac{\partial}{\partial y}\left(\frac{\partial V}{\partial y}\right) = \left(\frac{\partial}{\partial x'}\sin\theta + \frac{\partial}{\partial y'}\cos\theta\right)\left(\frac{\partial V}{\partial x'}\sin\theta + \frac{\partial V}{\partial y'}\cos\theta\right)$$

$$= \frac{\partial^2 V}{\partial x'^2}\sin^2\theta + 2\frac{\partial^2 V}{\partial x'\partial y'}\sin\theta\cos\theta + \frac{\partial^2 V}{\partial y'^2}\cos^2\theta \qquad (3.60)$$

Adding (3.59) and (3.60) we get $\dfrac{\partial^2 V}{\partial x^2} + \dfrac{\partial^2 V}{\partial y^2} = \dfrac{\partial^2 V}{\partial x'^2} + \dfrac{\partial^2 V}{\partial y'^2}.$

Thus the expression remains invariant. □

Example 3.7.5 *A differentiable function $f(x, y)$, when expressed in terms of new variables u and v defined by $x = \dfrac{1}{2}(u + v), y = \sqrt{uv}$ becomes $g(u, v)$. Prove that*

$$\frac{\partial^2 g}{\partial u\partial v} = \frac{1}{4}\left(\frac{\partial^2 f}{\partial x^2} + 2\frac{x}{y}\frac{\partial^2 f}{\partial x\partial y} + \frac{\partial^2 f}{\partial y^2} + \frac{1}{y}\frac{\partial f}{\partial y}\right).$$

Solution: Using chain rule we get

$$\frac{\partial g}{\partial u} = \frac{\partial f}{\partial x}\cdot\frac{\partial x}{\partial u} + \frac{\partial f}{\partial y}\cdot\frac{\partial y}{\partial u} = \frac{1}{2}\frac{\partial f}{\partial x} + \frac{\sqrt{v}}{2\sqrt{u}}\frac{\partial f}{\partial y}$$

and
$$\frac{\partial g}{\partial v} = \frac{\partial f}{\partial x}\cdot\frac{\partial x}{\partial v} + \frac{\partial f}{\partial y}\cdot\frac{\partial y}{\partial v} = \frac{1}{2}\frac{\partial f}{\partial x} + \frac{\sqrt{u}}{2\sqrt{v}}\frac{\partial f}{\partial y}.$$

In operator notation, the above results can be written as

$$\frac{\partial}{\partial u} \rightarrow \frac{1}{2}\frac{\partial}{\partial x} + \frac{1}{2}\sqrt{\frac{v}{u}}\frac{\partial}{\partial y}$$

and
$$\frac{\partial}{\partial v} \rightarrow \frac{1}{2}\frac{\partial}{\partial x} + \frac{1}{2}\sqrt{\frac{u}{v}}\frac{\partial}{\partial y}. \Bigg\}$$

$$\therefore \quad \frac{\partial^2 g}{\partial u\partial v} = \frac{\partial}{\partial u}\left(\frac{\partial g}{\partial v}\right) = \frac{\partial}{\partial u}\left(\frac{1}{2}\frac{\partial f}{\partial x} + \frac{\sqrt{u}}{2\sqrt{v}}\frac{\partial f}{\partial y}\right)$$

$$= \frac{1}{2}\frac{\partial}{\partial u}\left(\frac{\partial f}{\partial x}\right) + \frac{1}{4}\frac{1}{\sqrt{uv}}\frac{\partial f}{\partial y} + \frac{1}{2}\sqrt{\frac{u}{v}}\frac{\partial}{\partial u}\left(\frac{\partial f}{\partial y}\right)$$

$$= \frac{1}{2}\left(\frac{1}{2}\frac{\partial}{\partial x} + \frac{1}{2}\sqrt{\frac{v}{u}}\frac{\partial}{\partial y}\right)\frac{\partial f}{\partial x} + \frac{1}{4}\frac{1}{\sqrt{uv}}\frac{\partial f}{\partial y}$$

$$\qquad\qquad + \frac{1}{2}\sqrt{\frac{u}{v}}\left(\frac{1}{2}\frac{\partial}{\partial x} + \frac{1}{2}\sqrt{\frac{u}{v}}\frac{\partial}{\partial y}\right)\frac{\partial f}{\partial y}$$

$$= \frac{1}{4}\frac{\partial^2 f}{\partial x^2} + \frac{1}{4}\sqrt{\frac{v}{u}}\frac{\partial^2 f}{\partial x\partial y} + \frac{1}{4\sqrt{uv}}\frac{\partial f}{\partial y} + \frac{1}{4}\sqrt{\frac{u}{v}}\frac{\partial^2 f}{\partial x\partial y} + \frac{1}{4}\frac{\partial^2 f}{\partial y^2}$$

$$= \frac{1}{4}\left[\frac{\partial^2 f}{\partial x^2} + \frac{u+v}{\sqrt{uv}}\frac{\partial^2 f}{\partial x \partial y} + \frac{1}{\sqrt{uv}}\frac{\partial f}{\partial y} + \frac{\partial^2 f}{\partial y^2}\right]$$

$$= \frac{1}{4}\left[\frac{\partial^2 f}{\partial x^2} + \frac{2x}{y}\frac{\partial^2 f}{\partial x \partial y} + \frac{1}{y}\frac{\partial f}{\partial y} + \frac{\partial^2 f}{\partial y^2}\right].$$

[Assuming $\dfrac{\partial^2 f}{\partial x \partial y} = \dfrac{\partial^2 f}{\partial y \partial x}$ and using the conditions $u + v = 2x$, $\sqrt{uv} = y$.]   □

Example 3.7.6 *If* $\dfrac{x^2}{a^2 + u} + \dfrac{y^2}{b^2 + u} + \dfrac{z^2}{c^2 + u} = 1$, *then prove that*

$$\left(\frac{\partial u}{\partial x}\right)^2 + \left(\frac{\partial u}{\partial y}\right)^2 + \left(\frac{\partial u}{\partial z}\right)^2 = 2\left(x\frac{\partial u}{\partial x} + y\frac{\partial u}{\partial y} + z\frac{\partial u}{\partial z}\right).$$

Solution: Differentiating the given equation partially with respect to x, we get

$$\frac{2x}{a^2 + u} - \left\{\frac{x^2}{(a^2 + u)^2} + \frac{y^2}{(b^2 + u)^2} + \frac{z^2}{(c^2 + u)^2}\right\}\frac{\partial u}{\partial x} = 0$$

$$\therefore \quad \frac{\partial u}{\partial x} = \frac{1}{Q}\frac{2x}{a^2 + u} \quad \text{where } Q = \frac{x^2}{(a^2 + u)^2} + \frac{y^2}{(b^2 + u)^2} + \frac{z^2}{(c^2 + u)^2}$$

Similarly, $\quad \dfrac{\partial u}{\partial y} = \dfrac{1}{Q}\dfrac{2y}{b^2 + u}$ and $\dfrac{\partial u}{\partial z} = \dfrac{1}{Q}\dfrac{2z}{c^2 + u}$.

Squaring and adding these we get

$$\left(\frac{\partial u}{\partial x}\right)^2 + \left(\frac{\partial u}{\partial y}\right)^2 + \left(\frac{\partial u}{\partial z}\right)^2$$

$$= \frac{4}{Q^2}\left\{\frac{x^2}{(a^2 + u)^2} + \frac{y^2}{(b^2 + u)^2} + \frac{z^2}{(c^2 + u)^2}\right\} = \frac{4}{Q} \qquad (3.61)$$

Again

$$x\frac{\partial u}{\partial x} + y\frac{\partial u}{\partial y} + z\frac{\partial u}{\partial z} = \frac{2}{Q}\left\{\frac{x^2}{a^2 + u} + \frac{y^2}{b^2 + u} + \frac{z^2}{c^2 + u}\right\} = \frac{2}{Q} \qquad (3.62)$$

$(3.61) \div (3.62)$ gives

$$\left(\frac{\partial u}{\partial x}\right)^2 + \left(\frac{\partial u}{\partial y}\right)^2 + \left(\frac{\partial u}{\partial z}\right)^2 = 2\left(x\frac{\partial u}{\partial x} + y\frac{\partial u}{\partial y} + z\frac{\partial u}{\partial z}\right). \qquad □$$

Example 3.7.7 *If* $u = \dfrac{1}{\sqrt{1 - 2xy + y^2}}$, *prove that*

$$\frac{\partial}{\partial x}\left\{(1 - x^2)\frac{\partial u}{\partial x}\right\} + \frac{\partial}{\partial y}\left(y^2\frac{\partial u}{\partial y}\right) = 0.$$

Solution: We have $\dfrac{\partial u}{\partial x} = -\dfrac{1}{2}\dfrac{1}{(1 - 2xy + y^2)^{\frac{3}{2}}}(-2y) = \dfrac{y}{(1 - 2xy + y^2)^{\frac{3}{2}}}.$

$$\therefore \quad \frac{\partial}{\partial x}\left\{(1 - x^2)\frac{\partial u}{\partial x}\right\} = \frac{\partial}{\partial x}\left\{\frac{(1 - x^2)y}{(1 - 2xy + y^2)^{\frac{3}{2}}}\right\}$$

$$= \frac{-2xy}{(1 - 2xy + y^2)^{\frac{3}{2}}} + y(1 - x^2) \times \left(-\frac{3}{2}\right)(1 - 2xy + y^2)^{-\frac{5}{2}}(-2y)$$

$$= \frac{-2xy}{(1-2xy+y^2)^{\frac{3}{2}}} + \frac{3y^2(1-x^2)}{(1-2xy+y^2)^{\frac{5}{2}}}$$

$$= \frac{-2xy(1-2xy+y^2)+3y^2(1-x^2)}{(1-2xy+y^2)^{\frac{5}{2}}} \qquad (3.63)$$

Again, $\dfrac{\partial u}{\partial y} = -\dfrac{1}{2}\dfrac{1}{(1-2xy+y^2)^{\frac{3}{2}}}(2y-2x) = \dfrac{x-y}{(1-2xy+y^2)^{\frac{3}{2}}}.$

$$\therefore \frac{\partial}{\partial y}\left\{y^2\frac{\partial u}{\partial y}\right\} = \frac{\partial}{\partial y}\left\{\frac{(x-y)y^2}{(1-2xy+y^2)^{\frac{3}{2}}}\right\}$$

$$= \frac{2xy-3y^2}{(1-2xy+y^2)^{\frac{3}{2}}} + \frac{3y^2(x-y)^2}{(1-2xy+y^2)^{\frac{5}{2}}}$$

$$= \frac{(2xy-3y^2)(1-2xy+y^2)+3y^2(x-y)^2}{(1-2xy+y^2)^{\frac{5}{2}}} \qquad (3.64)$$

Adding (3.63) and (3.64) we get

$$\frac{\partial}{\partial x}\left\{(1-x^2)\frac{\partial u}{\partial x}\right\} + \frac{\partial}{\partial y}\left(y^2\frac{\partial u}{\partial y}\right)$$

$$= \frac{1}{(1-2xy+y^2)^{\frac{5}{2}}}\{-2xy(1-2xy+y^2)+3y^2(1-x^2)$$

$$+(2xy-3y^2)(1-2xy+y^2)+3y^2(x-y)^2\}$$

$$= \frac{(1-2xy+y^2)(-2xy+2xy-3y^2)+3y^2(1-x^2+x^2+y^2-2xy)}{(1-2xy+y^2)^{\frac{5}{2}}}$$

$$= \frac{-3y^2(1-2xy+y^2)+3y^2(1-x^2+x^2+y^2-2xy)}{(1-2xy+y^2)^{\frac{5}{2}}}$$

$$= \frac{-3y^2(1-2xy+y^2)+3y^2(1-2xy+y^2)}{(1-2xy+y^2)^{\frac{5}{2}}} = 0. \qquad \square$$

Example 3.7.8 *Given that z is a function of x and y and that $x = u+v, y = uv$. Prove that $\dfrac{\partial^2 z}{\partial u^2} - 2\dfrac{\partial^2 z}{\partial u\partial v} + \dfrac{\partial^2 z}{\partial v^2} = (x^2-4y)\dfrac{\partial^2 z}{\partial y^2} - 2\dfrac{\partial z}{\partial y}.$*

Solution: Using chain rule, we get

$$\frac{\partial z}{\partial u} = \frac{\partial z}{\partial x}\cdot\frac{\partial x}{\partial u} + \frac{\partial z}{\partial y}\cdot\frac{\partial y}{\partial u} = \frac{\partial z}{\partial x}.1 + \frac{\partial z}{\partial y}.v = \frac{\partial z}{\partial x} + v\frac{\partial z}{\partial y}.$$

Similarly, we can get $\dfrac{\partial u}{\partial v} = \dfrac{\partial z}{\partial x} + u\dfrac{\partial z}{\partial y}.$

So we can use the following operational symbols

$$\frac{\partial}{\partial u} \rightarrow \frac{\partial}{\partial x} + v\frac{\partial}{\partial y} \qquad (3.65)$$

and $\dfrac{\partial}{\partial v} \rightarrow \dfrac{\partial}{\partial x} + u\dfrac{\partial}{\partial y}$ \qquad (3.66)

Now $\dfrac{\partial^2 z}{\partial u^2} = \dfrac{\partial}{\partial u}\left(\dfrac{\partial z}{\partial u}\right) = \dfrac{\partial}{\partial u}\left(\dfrac{\partial z}{\partial x} + v\dfrac{\partial z}{\partial y}\right) = \dfrac{\partial}{\partial u}\left(\dfrac{\partial z}{\partial x}\right) + v\dfrac{\partial}{\partial u}\left(\dfrac{\partial z}{\partial y}\right)$

$\qquad = \left\{\dfrac{\partial}{\partial x}\left(\dfrac{\partial z}{\partial x}\right) + v\dfrac{\partial}{\partial y}\left(\dfrac{\partial z}{\partial y}\right)\right\} + v\left\{\dfrac{\partial}{\partial x}\left(\dfrac{\partial z}{\partial x}\right) + v\dfrac{\partial}{\partial y}\left(\dfrac{\partial z}{\partial y}\right)\right\}$

$\hfill$ [Using (3.65)]

$\qquad = \dfrac{\partial^2 z}{\partial x^2} + 2v\dfrac{\partial^2 z}{\partial x\partial y} + v^2\dfrac{\partial^2 z}{\partial y^2} \quad \left[\text{Assuming } \dfrac{\partial^2 z}{\partial x\partial y} = \dfrac{\partial^2 z}{\partial y\partial x}\right]$ (3.67)

Similarly, $\qquad \dfrac{\partial^2 z}{\partial v^2} = \dfrac{\partial^2 z}{\partial x^2} + 2u\dfrac{\partial^2 z}{\partial x\partial y} + u^2\dfrac{\partial^2 z}{\partial y^2}$ (3.68)

Again $\qquad \dfrac{\partial^2 z}{\partial u\partial v} = \dfrac{\partial}{\partial u}\left(\dfrac{\partial z}{\partial v}\right) = \dfrac{\partial}{\partial u}\left(\dfrac{\partial z}{\partial x} + u\dfrac{\partial z}{\partial y}\right)$

$\qquad = \dfrac{\partial}{\partial u}\left(\dfrac{\partial z}{\partial x}\right) + \dfrac{\partial z}{\partial y} + u\dfrac{\partial}{\partial u}\left(\dfrac{\partial z}{\partial y}\right)$

$\qquad = \left\{\dfrac{\partial}{\partial x}\left(\dfrac{\partial z}{\partial x}\right) + v\dfrac{\partial}{\partial y}\left(\dfrac{\partial z}{\partial x}\right)\right\} + \dfrac{\partial z}{\partial y} + u\left\{\dfrac{\partial}{\partial x}\left(\dfrac{\partial z}{\partial y}\right) + v\dfrac{\partial}{\partial y}\left(\dfrac{\partial z}{\partial y}\right)\right\}$

$\qquad = \dfrac{\partial^2 z}{\partial x^2} + v\dfrac{\partial^2 z}{\partial y\partial x} + \dfrac{\partial z}{\partial y} + u\left(\dfrac{\partial^2 z}{\partial x\partial y} + v\dfrac{\partial^2 z}{\partial y^2}\right)$

$\qquad = \dfrac{\partial^2 z}{\partial x^2} + uv\dfrac{\partial^2 z}{\partial y^2} + u\dfrac{\partial^2 z}{\partial y\partial x} + v\dfrac{\partial^2 z}{\partial x\partial y} + \dfrac{\partial z}{\partial y}$ (3.69)

Using (3.67), (3.68) and (3.69), we get

$$\dfrac{\partial^2 z}{\partial u^2} - 2\dfrac{\partial^2 z}{\partial u\partial v} + \dfrac{\partial^2 z}{\partial v^2} = \left(\dfrac{\partial^2 z}{\partial x^2} + 2v\dfrac{\partial^2 z}{\partial x\partial y} + v^2\dfrac{\partial^2 z}{\partial y^2}\right) - 2\left(\dfrac{\partial^2 z}{\partial x^2} + uv\dfrac{\partial^2 z}{\partial y^2}\right.$$

$$\left. + u\dfrac{\partial^2 z}{\partial y\partial x} + v\dfrac{\partial^2 z}{\partial x\partial y} + \dfrac{\partial z}{\partial y}\right) + \left(\dfrac{\partial^2 z}{\partial x^2} + 2u\dfrac{\partial^2 z}{\partial x\partial y} + u^2\dfrac{\partial^2 z}{\partial y^2}\right)$$

$$= -2\dfrac{\partial z}{\partial y} + \dfrac{\partial^2 z}{\partial y^2}(v^2 - 2uv + u^2) = -2\dfrac{\partial z}{\partial y} + \dfrac{\partial^2 z}{\partial y^2}\{(u+v)^2 - 4uv\}$$

$$= (x^2 - 4y)\dfrac{\partial^2 z}{\partial y^2} - 2\dfrac{\partial z}{\partial y}. \qquad\qquad \square$$

Example 3.7.9 *If $F(u,v)$ is a twice differentiable function of (u,v) and if $u = x^2 - y^2, v = 2xy$, prove that*

$$4(u^2 + v^2)\dfrac{\partial^2 F}{\partial u\partial v} + 2u\dfrac{\partial F}{\partial v} + 2v\dfrac{\partial F}{\partial u} = xy\left(\dfrac{\partial^2 F}{\partial x^2} - \dfrac{\partial^2 F}{\partial y^2}\right) + (x^2 - y^2)\dfrac{\partial^2 F}{\partial x\partial y}.$$

Solution: Using chain rule, we get

$$\dfrac{\partial F}{\partial x} = \dfrac{\partial F}{\partial u}\cdot\dfrac{\partial u}{\partial x} + \dfrac{\partial F}{\partial v}\cdot\dfrac{\partial v}{\partial x} = \dfrac{\partial F}{\partial u}(2x) + \dfrac{\partial F}{\partial v}(2y)$$

and $\qquad \dfrac{\partial F}{\partial y} = \dfrac{\partial F}{\partial u}\cdot\dfrac{\partial u}{\partial y} + \dfrac{\partial F}{\partial v}\cdot\dfrac{\partial v}{\partial y} = \dfrac{\partial F}{\partial u}(-2y) + \dfrac{\partial F}{\partial v}(2x).$

In operator notation, we get

$$\dfrac{\partial}{\partial x} \equiv 2x\dfrac{\partial}{\partial u} + 2y\dfrac{\partial}{\partial v} \quad \text{and} \quad \dfrac{\partial}{\partial y} \equiv -2y\dfrac{\partial}{\partial u} + 2x\dfrac{\partial}{\partial v}.$$

$$\frac{\partial^2 F}{\partial x^2} = \frac{\partial}{\partial x}\left(2x\frac{\partial F}{\partial u} + 2y\frac{\partial F}{\partial v}\right) = 2\frac{\partial F}{\partial u} + 2y\frac{\partial}{\partial x}\left(\frac{\partial F}{\partial v}\right) + 2x\frac{\partial}{\partial x}\left(\frac{\partial F}{\partial u}\right)$$

$$= 2\frac{\partial F}{\partial u} + 2y\left(2x\frac{\partial}{\partial u} + 2y\frac{\partial}{\partial v}\right)\frac{\partial F}{\partial v} + 2x\left(2x\frac{\partial}{\partial u} + 2y\frac{\partial}{\partial v}\right)\frac{\partial F}{\partial u}$$

$$= 2\frac{\partial F}{\partial u} + 4xy\frac{\partial^2 F}{\partial u\partial v} + 4y^2\frac{\partial^2 F}{\partial v^2} + 4x^2\frac{\partial^2 F}{\partial u^2} + 4xy\frac{\partial^2 F}{\partial v\partial u}$$

$$= 2\frac{\partial F}{\partial u} + 8xy\frac{\partial^2 F}{\partial u\partial v} + 4y^2\frac{\partial^2 F}{\partial v^2} + 4x^2\frac{\partial^2 F}{\partial u^2}$$

$$\left[\text{Assuming } \frac{\partial^2 F}{\partial u\partial v} = \frac{\partial^2 F}{\partial v\partial u}\right]$$

$$\frac{\partial^2 F}{\partial y^2} = \frac{\partial}{\partial y}\left(\frac{\partial F}{\partial y}\right) = \frac{\partial}{\partial y}\left(-2y\frac{\partial F}{\partial u} + 2x\frac{\partial F}{\partial v}\right)$$

$$= -2\frac{\partial F}{\partial u} + 2x\frac{\partial}{\partial y}\left(\frac{\partial F}{\partial v}\right) - 2y\frac{\partial}{\partial y}\left(\frac{\partial F}{\partial u}\right)$$

$$= -2\frac{\partial F}{\partial u} + 2x\left(-2y\frac{\partial}{\partial u} + 2x\frac{\partial}{\partial v}\right)\frac{\partial F}{\partial v} - 2y\left(-2y\frac{\partial}{\partial u} + 2x\frac{\partial}{\partial v}\right)\frac{\partial F}{\partial u}$$

$$= -2\frac{\partial F}{\partial u} - 8xy\frac{\partial^2 F}{\partial u\partial v} + 4x^2\frac{\partial^2 F}{\partial v^2} + 4y^2\frac{\partial^2 F}{\partial u^2}$$

and $\quad \frac{\partial^2 F}{\partial x\partial y} = \frac{\partial}{\partial x}\left(\frac{\partial F}{\partial y}\right) = \frac{\partial}{\partial x}\left(-2y\frac{\partial F}{\partial u} + 2x\frac{\partial F}{\partial v}\right)$

$$= 2\frac{\partial F}{\partial v} - 2y\frac{\partial}{\partial x}\left(\frac{\partial F}{\partial u}\right) + 2x\frac{\partial}{\partial x}\left(\frac{\partial F}{\partial v}\right)$$

$$= 2\frac{\partial F}{\partial v} - 2y\left(2x\frac{\partial}{\partial u} + 2y\frac{\partial}{\partial v}\right)\frac{\partial F}{\partial u} + 2x\left(2x\frac{\partial}{\partial u} + 2y\frac{\partial}{\partial v}\right)\frac{\partial F}{\partial v}$$

$$= 2\frac{\partial F}{\partial v} - 4xy\frac{\partial^2 F}{\partial u^2} - 4y^2\frac{\partial^2 F}{\partial v\partial u} + 4x^2\frac{\partial^2 F}{\partial u\partial v} + 4xy\frac{\partial^2 F}{\partial v^2}$$

$$= 2\frac{\partial F}{\partial v} - 4xy\frac{\partial^2 F}{\partial u^2} + 4(x^2 - y^2)\frac{\partial^2 F}{\partial v\partial u} + 4xy\frac{\partial^2 F}{\partial v^2}.$$

Now,

$$\text{R.H.S.} = xy\left(\frac{\partial^2 F}{\partial x^2} - \frac{\partial^2 F}{\partial y^2}\right) + (x^2 - y^2)\frac{\partial^2 F}{\partial x\partial y}$$

$$= xy\left[4\frac{\partial F}{\partial u} + 16xy\frac{\partial^2 F}{\partial u\partial v} + 4(y^2 - x^2)\frac{\partial^2 F}{\partial v^2} + 4(x^2 - y^2)\frac{\partial^2 F}{\partial u^2}\right]$$

$$+ (x^2 - y^2)\left[2\frac{\partial F}{\partial v} - 4xy\frac{\partial^2 F}{\partial u^2} + 4(x^2 - y^2)\frac{\partial^2 F}{\partial u\partial v} + 4xy\frac{\partial^2 F}{\partial v^2}\right]$$

$$= 4xy\frac{\partial F}{\partial u} + 2(x^2 - y^2)\frac{\partial F}{\partial v} + 4\left\{(x^2 - y^2)^2 + 4x^2 y^2\right\}\frac{\partial^2 F}{\partial u\partial v}$$

$$= 4(u^2 + v^2)\frac{\partial^2 F}{\partial u\partial v} + 2u\frac{\partial F}{\partial v} + 2v\frac{\partial F}{\partial u} = \text{L.H.S.} \qquad \square$$

Example 3.7.10 *If z be a differentiable function of x and y and $x = c\cosh u\cos v$, $y = c\sinh u\sin v$, then prove that*

$$\frac{\partial^2 z}{\partial u^2} + \frac{\partial^2 z}{\partial v^2} = \frac{1}{2}c^2(\cosh 2u - \cos 2v)\left(\frac{\partial^2 z}{\partial x^2} + \frac{\partial^2 z}{\partial y^2}\right).$$

Solution: Applying chain rule, we get

$$\frac{\partial z}{\partial u} = \frac{\partial z}{\partial x}\cdot\frac{\partial x}{\partial u} + \frac{\partial z}{\partial y}\cdot\frac{\partial y}{\partial u} = c\sinh u\cos v\frac{\partial z}{\partial x} + c\cosh u\sin v\frac{\partial z}{\partial y}$$

and $$\frac{\partial z}{\partial v} = \frac{\partial z}{\partial x}\cdot\frac{\partial x}{\partial v} + \frac{\partial z}{\partial y}\cdot\frac{\partial y}{\partial v} = -c\cosh u\sin v\frac{\partial z}{\partial x} + c\sinh u\cos v\frac{\partial z}{\partial y}.$$

Using operator notation, we get

$$\frac{\partial}{\partial u} \equiv c\sinh u\cos v\frac{\partial}{\partial x} + c\cosh u\sin v\frac{\partial}{\partial y}$$

and $$\frac{\partial}{\partial v} \equiv -c\cosh u\sin v\frac{\partial}{\partial x} + c\sinh u\cos v\frac{\partial}{\partial y}.$$

$$\therefore \frac{\partial^2 z}{\partial u^2} = \frac{\partial}{\partial u}\left(\frac{\partial z}{\partial u}\right) = \frac{\partial}{\partial u}\left[c\sinh u\cos v\frac{\partial z}{\partial x} + c\cosh u\sin v\frac{\partial z}{\partial y}\right]$$

$$= c\cosh u\cos v\frac{\partial z}{\partial x} + c\sinh u\cos v\frac{\partial}{\partial u}\left(\frac{\partial z}{\partial x}\right)$$

$$+c\sinh u\sin v\frac{\partial z}{\partial y} + c\cosh u\sin v\frac{\partial}{\partial u}\left(\frac{\partial z}{\partial y}\right)$$

$$= x\frac{\partial z}{\partial x} + y\frac{\partial z}{\partial y} + c\sinh u\cos v\left[c\sinh u\cos v\frac{\partial}{\partial x}\left(\frac{\partial z}{\partial x}\right)\right.$$

$$\left.+c\cosh u\sin v\frac{\partial}{\partial y}\left(\frac{\partial z}{\partial x}\right)\right] + c\cosh u\sin v \times$$

$$\left[c\sinh u\cos v\frac{\partial}{\partial x}\left(\frac{\partial z}{\partial y}\right) + c\cosh u\sin v\frac{\partial}{\partial y}\left(\frac{\partial z}{\partial y}\right)\right]$$

$$= x\frac{\partial z}{\partial x} + y\frac{\partial z}{\partial x} + c^2\sinh^2 u\cos^2 v\frac{\partial^2 z}{\partial x^2} + 2c^2\sinh u\cosh u\sin v\cos v\frac{\partial^2 z}{\partial x\partial y}$$

$$+c^2\cosh^2 u\sin^2 v\frac{\partial^2 z}{\partial y^2}$$

$$\frac{\partial^2 z}{\partial v^2} = \frac{\partial}{\partial v}\left(\frac{\partial z}{\partial v}\right) = \frac{\partial}{\partial v}\left[-c\cosh u\sin v\frac{\partial z}{\partial x} + c\sinh u\cos v\frac{\partial z}{\partial y}\right]$$

$$= -c\cosh u\cos v\frac{\partial z}{\partial x} - c\sinh u\sin v\frac{\partial}{\partial v}\left(\frac{\partial z}{\partial x}\right)$$

$$-c\sinh u\sin v\frac{\partial z}{\partial y} + c\cosh u\cos v\frac{\partial}{\partial v}\left(\frac{\partial z}{\partial y}\right)$$

$$= -x\frac{\partial z}{\partial x} - y\frac{\partial z}{\partial y} - c\cosh u\sin v\left[-c\cosh u\sin v\frac{\partial}{\partial x}\left(\frac{\partial z}{\partial x}\right)\right.$$

$$\left.+c\sinh u\cos v\frac{\partial}{\partial y}\left(\frac{\partial z}{\partial x}\right)\right] + c\sinh u\cos v \times$$

$$\left[-c\cosh u\sin v\frac{\partial}{\partial x}\left(\frac{\partial z}{\partial y}\right) + c\sinh u\cos v\frac{\partial}{\partial y}\left(\frac{\partial z}{\partial y}\right)\right]$$

$$= -x\frac{\partial z}{\partial x} - y\frac{\partial z}{\partial y} + c^2\cosh^2 u\sin^2 v\frac{\partial^2 z}{\partial x^2} - 2c^2\sinh u\cosh u \times$$

$$\sin v\cos v\frac{\partial^2 z}{\partial x\partial y} + c^2\sinh^2 u\cos^2 v\frac{\partial^2 z}{\partial y^2}$$

$$\therefore \frac{\partial^2 z}{\partial u^2} + \frac{\partial^2 z}{\partial v^2} = \frac{\partial^2 z}{\partial x^2} \left[c^2 \sinh^2 u \cos^2 v + c^2 \cosh^2 u \sin^2 v \right]$$

$$+ \frac{\partial^2 z}{\partial y^2} \left[c^2 \sinh^2 u \cos^2 v + c^2 \cosh^2 u \sin^2 v \right]$$

$$= (c^2 \sinh^2 u \cos^2 v + c^2 \cosh^2 u \sin^2 v) \left(\frac{\partial^2 z}{\partial x^2} + \frac{\partial^2 z}{\partial y^2} \right)$$

$$= \frac{1}{2} c^2 (2 \sinh^2 u \cos^2 v + 2 \cosh^2 u \sin^2 v) \left(\frac{\partial^2 z}{\partial x^2} + \frac{\partial^2 z}{\partial y^2} \right)$$

$$= \frac{1}{2} c^2 [(1 + \cosh 2u)(\sin^2 v) + (\cosh 2u - 1)(\cos^2 v)] \left(\frac{\partial^2 z}{\partial x^2} + \frac{\partial^2 z}{\partial y^2} \right)$$

$$= \frac{1}{2} c^2 [\cosh 2u(\sin^2 v + \cos^2 v) - (\cos^2 v - \sin^2 v)] \left(\frac{\partial^2 z}{\partial x^2} + \frac{\partial^2 z}{\partial y^2} \right)$$

$$= \frac{1}{2} c^2 (\cosh 2u - \cos 2v) \left(\frac{\partial^2 z}{\partial x^2} + \frac{\partial^2 z}{\partial y^2} \right). \qquad \square$$

Example 3.7.11 *(a) Verify Euler's theorem for the function*

$$u = ax^2 + 2hxy + by^2.$$

(b) If $u = \tan^{-1} \dfrac{x^3 + y^3}{x - y}$, then show that (i) $x \dfrac{\partial u}{\partial x} + y \dfrac{\partial u}{\partial y} = \sin 2u$ and

(ii) $x^2 \dfrac{\partial^2 u}{\partial x^2} + 2xy \dfrac{\partial^2 z}{\partial x \partial y} + y^2 \dfrac{\partial^2 u}{\partial y^2} = \sin 2u(1 - 4 \sin^2 u).$

Solution: *(a)* $u = ax^2 + 2hxy + by^2 = x^2 \left\{ a + 2h \dfrac{y}{x} + b \left(\dfrac{y}{x} \right)^2 \right\} = x^2 f \left(\dfrac{y}{x} \right).$

Hence the given function is a homogeneous function of degree 2.

Now, $xu_x + yu_y = x(2ax + 2hy) + y(2hx + 2by) = 2(ax^2 + 2hxy + by^2) = 2u,$

which shows that Euler's theorem is satisfied in this case.

(b) (i) From the given relation we get

$$\tan u = \frac{x^3 + y^3}{x - y} = \frac{x^3 \left\{ 1 + \left(\frac{y}{x} \right)^3 \right\}}{x \left\{ 1 - \frac{y}{x} \right\}} = x^2 \phi \left(\frac{y}{x} \right).$$

So, $\tan u$ is a homogeneous function of degree 2.

Let $v = \tan u$, i.e., v is a homogeneous function of degree 2. Therefore, by Euler's theorem we have

$$x \frac{\partial v}{\partial x} + y \frac{\partial v}{\partial y} = 2v \implies x \frac{\partial}{\partial x}(\tan u) + y \frac{\partial}{\partial y}(\tan u) = 2 \tan u$$

or, $\quad x \sec^2 u \dfrac{\partial u}{\partial x} + y \sec^2 u \dfrac{\partial u}{\partial y} = 2 \tan u$

or, $\quad x \dfrac{\partial u}{\partial x} + y \dfrac{\partial u}{\partial y} = \dfrac{2 \tan u}{\sec^2 u} = \sin 2u. \qquad (3.70)$

(ii) Differentiating the above relation (3.70) partially with respect to x and then multiplying by x, we get

$$x^2 \frac{\partial^2 u}{\partial x^2} + xy \frac{\partial^2 u}{\partial y \partial x} + x \frac{\partial u}{\partial x} = 2x \cos 2u \cdot \frac{\partial u}{\partial x} \qquad (3.71)$$

Again differentiating the same relation partially with respect to y and then multiplying by y, we get

$$xy\frac{\partial^2 u}{\partial x \partial y} + y\frac{\partial u}{\partial y} + y^2\frac{\partial^2 u}{\partial y^2} + = 2y\cos 2u.\frac{\partial u}{\partial y} \qquad (3.72)$$

Adding (3.71) and (3.72), we get

$$x^2\frac{\partial^2 u}{\partial x^2} + 2xy\frac{\partial^2 u}{\partial x \partial y} + y^2\frac{\partial^2 u}{\partial y^2} + \left(x\frac{\partial u}{\partial x} + y\frac{\partial u}{\partial y}\right) = 2\cos 2u\left(x\frac{\partial u}{\partial x} + y\frac{\partial u}{\partial y}\right)$$

or, $x^2\frac{\partial^2 u}{\partial x^2} + 2xy\frac{\partial^2 u}{\partial x \partial y} + y^2\frac{\partial^2 u}{\partial y^2} + \sin 2u = 2\cos 2u \sin 2u$

or, $x^2\frac{\partial^2 u}{\partial x^2} + 2xy\frac{\partial^2 u}{\partial x \partial y} + y^2\frac{\partial^2 u}{\partial y^2} = \sin 2u(2\cos 2u - 1) = \sin 2u(1 - 4\sin^2 u)$. □

Example 3.7.12 *If $u = \cos^{-1}\dfrac{x+y}{\sqrt{x} + \sqrt{y}}$, then show that*

$$x\frac{\partial u}{\partial x} + y\frac{\partial u}{\partial y} + \frac{1}{2}\cot u = 0.$$

Solution: We have $\cos u = \dfrac{x+y}{\sqrt{x} + \sqrt{y}} = \dfrac{x\left(1 + \frac{y}{x}\right)}{\sqrt{x}\left(1 + \sqrt{\frac{y}{x}}\right)} = x^{\frac{1}{2}}\phi\left(\frac{y}{x}\right) = v$ (say).

Thus $v = \cos u$ is a homogeneous function of degree $\dfrac{1}{2}$ in x and y and so using Euler's theorem, we get

$$x\frac{\partial v}{\partial x} + y\frac{\partial v}{\partial y} = \frac{1}{2}v \quad \Rightarrow \quad x\frac{\partial}{\partial x}(\cos u) + y\frac{\partial}{\partial y}(\cos u) = \frac{1}{2}\cos u$$

or, $-x\sin u\frac{\partial u}{\partial x} - y\sin u\frac{\partial u}{\partial y} = \frac{1}{2}\cos u$ or, $x\frac{\partial u}{\partial x} + y\frac{\partial u}{\partial y} = -\frac{1}{2}\cot u$.

Hence $x\frac{\partial u}{\partial x} + y\frac{\partial u}{\partial y} + \frac{1}{2}\cot u = 0$. □

Example 3.7.13 *If $u = \phi(H_n)$, where H_n is a homogeneous function of degree n in x, y, z, then show that $x\dfrac{\partial u}{\partial x} + y\dfrac{\partial u}{\partial y} + z\dfrac{\partial u}{\partial z} = n\dfrac{F(u)}{F'(u)}$ where $F(u) = H_n$.*

Solution: Given $u = \phi(H_n)$, $\therefore \dfrac{\partial u}{\partial x} = \dfrac{du}{dH_n}.\dfrac{\partial H_n}{\partial x} = \phi'(H_n)\dfrac{\partial H_n}{\partial x}$ and similarly,

we get $\dfrac{\partial u}{\partial y} = \phi'(H_n)\dfrac{\partial H_n}{\partial y}$ and $\dfrac{\partial u}{\partial z} = \phi'(H_n)\dfrac{\partial H_n}{\partial z}$.

Since H_n is a homogeneous function of degree n, so by using Euler's theorem we get $x\dfrac{\partial H_n}{\partial x} + y\dfrac{\partial H_n}{\partial y} + z\dfrac{\partial H_n}{\partial z} = nH_n$.

$$\therefore \ x\frac{\partial u}{\partial x} + y\frac{\partial u}{\partial y} + z\frac{\partial u}{\partial z} = \phi'(H_n)\left(x\frac{\partial H_n}{\partial x} + y\frac{\partial H_n}{\partial y} + z\frac{\partial H_n}{\partial z}\right) = \phi'(H_n).nH_n$$

Now, $\dfrac{du}{dH_n} = \phi'(H_n)$. Again since $H_n = F(u) \ \therefore \ \dfrac{dH_n}{du} = F'(u)$.

$$\therefore \ \phi'(H_n) = \frac{du}{dH_n} = \frac{1}{\frac{dH_n}{du}} = \frac{1}{F'(u)}.$$

So we get $x\dfrac{\partial u}{\partial x} + y\dfrac{\partial u}{\partial y} + z\dfrac{\partial u}{\partial z} = n\dfrac{F(u)}{F'(u)}.$ □

Example 3.7.14 *(i) If H is a homogeneous function of degree n in x and y and if $u = (x^2 + y^2)^{-\frac{1}{2}n}$, show that $\dfrac{\partial}{\partial x}\left(H\dfrac{\partial u}{\partial x}\right) + \dfrac{\partial}{\partial y}\left(H\dfrac{\partial u}{\partial y}\right) = 0.$*

(ii) If H is a homogeneous function of degree n in three variables x, y, z and if $u = (x^2 + y^2 + z^2)^{-\frac{1}{2}(n+1)}$, show that

$$\frac{\partial}{\partial x}\left(H\frac{\partial u}{\partial x}\right) + \frac{\partial}{\partial y}\left(H\frac{\partial u}{\partial y}\right) + \frac{\partial}{\partial y}\left(H\frac{\partial u}{\partial y}\right) = 0.$$

Solution: (i) L.H.S $= \dfrac{\partial}{\partial x}\left(H\dfrac{\partial u}{\partial x}\right) + \dfrac{\partial}{\partial y}\left(H\dfrac{\partial u}{\partial y}\right)$

$$= \frac{\partial H}{\partial x}.\frac{\partial u}{\partial x} + \frac{\partial H}{\partial y}.\frac{\partial u}{\partial y} + H\left(\frac{\partial^2 u}{\partial x^2} + \frac{\partial^2 u}{\partial y^2}\right) \tag{3.73}$$

Now, $\dfrac{\partial u}{\partial x} = -nx(x^2 + y^2)^{-\frac{1}{2}n - 1}$

$\dfrac{\partial^2 u}{\partial x^2} = -n(x^2 + y^2)^{-\frac{1}{2}n - 1} + n\left(\dfrac{1}{2}n + 1\right)(x^2 + y^2)^{-\frac{1}{2}n - 2}.2x^2$

Similarly, $\dfrac{\partial u}{\partial y} = -ny(x^2 + y^2)^{-\frac{1}{2}n - 1}$

$\dfrac{\partial^2 u}{\partial y^2} = -n(x^2 + y^2)^{-\frac{1}{2}n - 1} + n\left(\dfrac{1}{2}n + 1\right)(x^2 + y^2)^{-\frac{1}{2}n - 2}.2y^2$

$\therefore \dfrac{\partial^2 u}{\partial x^2} + \dfrac{\partial^2 u}{\partial x^2} = -2n(x^2 + y^2)^{-\frac{1}{2}n - 1} + (n^2 + 2n)(x^2 + y^2)^{-\frac{1}{2}n - 2}(x^2 + y^2)$

$$= n^2(x^2 + y^2)^{-\frac{1}{2}n - 1} \tag{3.74}$$

Again $\dfrac{\partial H}{\partial x}.\dfrac{\partial u}{\partial x} + \dfrac{\partial H}{\partial y}.\dfrac{\partial u}{\partial y}$

$= -n(x^2 + y^2)^{-\frac{1}{2}n - 1}\left(x\dfrac{\partial H}{\partial x} + y\dfrac{\partial H}{\partial x}\right)$

$$= -n(x^2 + y^2)^{-\frac{1}{2}n - 1}.nH = -n^2 H(x^2 + y^2)^{-\frac{1}{2}n - 1} \tag{3.75}$$

Hence by (3.74) and (3.75) we get,

$$\frac{\partial H}{\partial x}.\frac{\partial u}{\partial x} + \frac{\partial H}{\partial y}.\frac{\partial u}{\partial y} + H\left(\frac{\partial^2 u}{\partial x^2} + \frac{\partial^2 u}{\partial y^2}\right)$$

i.e., $\quad \dfrac{\partial}{\partial x}\left(H\dfrac{\partial u}{\partial x}\right) + \dfrac{\partial}{\partial y}\left(H\dfrac{\partial u}{\partial y}\right) = 0.$

(ii) Given $u = (x^2 + y^2 + z^2)^{-\frac{1}{2}(n+1)}$

$\therefore \dfrac{\partial u}{\partial x} = -\dfrac{1}{2}(n + 1)(x^2 + y^2 + z^2)^{-\frac{1}{2}(n+3)}.2x = -(n + 1)x(x^2 + y^2 + z^2)^{-\frac{1}{2}(n+3)}$

$$\frac{\partial^2 u}{\partial x^2} = -(n+1)(x^2+y^2+z^2)^{-\frac{1}{2}(n+3)} + (n+1)(n+3)x^2(x^2+y^2+z^2)^{-\frac{1}{2}(n+5)}.$$

Now,
$$\frac{\partial}{\partial x}\left(H\frac{\partial u}{\partial x}\right) + \frac{\partial}{\partial y}\left(H\frac{\partial u}{\partial y}\right) + \frac{\partial}{\partial z}\left(H\frac{\partial u}{\partial z}\right)$$

$$= \frac{\partial H}{\partial x}\cdot\frac{\partial u}{\partial x} + \frac{\partial H}{\partial y}\cdot\frac{\partial u}{\partial y} + \frac{\partial H}{\partial z}\cdot\frac{\partial u}{\partial z} + H\left(\frac{\partial^2 u}{\partial x^2} + \frac{\partial^2 u}{\partial y^2} + \frac{\partial^2 u}{\partial z^2}\right)$$

$$= -(n+1)(x^2+y^2+z^2)^{-\frac{1}{2}(n+3)}\left(x\frac{\partial H}{\partial x} + y\frac{\partial H}{\partial x} + z\frac{\partial H}{\partial z}\right)$$

$$\quad + H\left[-3(n+1)(x^2+y^2+z^2)^{-\frac{1}{2}(n+3)}\right.$$

$$\left.\qquad\qquad +(n+1)(n+3)(x^2+y^2+z^2)^{-\frac{1}{2}(n+3)}\right]$$

$$= -(n+1)(x^2+y^2+z^2)^{-\frac{1}{2}(n+3)}nH$$

$$\quad - (n+1)H(x^2+y^2+z^2)^{-\frac{1}{2}(n+3)}(3-n-3)$$

[by using Euler's theorem on the function H]

$$= -(n+1)nH(x^2+y^2+z^2)^{-\frac{1}{2}(n+3)}$$

$$\qquad\qquad +(n+1)nH(x^2+y^2+z^2)^{-\frac{1}{2}(n+3)} = 0. \qquad \Box$$

Example 3.7.15 *If f is a homogeneous function of degree n in the three variables x, y and z, then show that all the partial derivatives of it of order r are also homogeneous functions of degree $n-r$ where $0 \le r \le n$.*

Solution: Since f is a homogeneous function of degree n in x, y and z, we can write

$f(x,y,z) = x^n\phi(u,v)$, where $u = \dfrac{y}{x}, v = \dfrac{z}{x}$ (See **Definition 3.5.2** of homogeneous function).

$$\therefore \; f_x(x,y,z) = nx^{n-1}\phi((u,v) + x^n(\phi_u u_x + \phi_v v_x)$$

$$= x^{n-1}\left[n\phi(u,v) + x\left\{-\frac{y}{x^2}\phi_u(u,v) - \frac{z}{x^2}\phi_v(u,v)\right\}\right]$$

$$= x^{n-1}\left[n\phi(u,v) - u\phi_u(u,v) - v\phi_v(u,v)\right] = x^{n-1}\phi_1(u,v)$$

where $\phi_1(u,v) = n\phi(u,v) - u\phi_u(u,v) - v\phi_v(u,v)$.

Hence f_x is a homogeneous function of degree $(n-1)$ in x, y, z.

$$f_{yx} = (n-1)x^{n-2}\psi + x^{n-1}\{\psi_u u_y + \psi_v v_y\}$$

$$= x^{n-2}\left\{(n-1)\psi + x\left(\psi_u\cdot\frac{1}{x} + 0\right)\right\} = x^{n-2}\phi_2(u,v)$$

where $\phi_2(u,v) = (n-1)\psi + \psi_u$, a function of u and v. Hence f_{yx} is a homogeneous function of degree $(n-2)$ in x, y and z. Similarly, f_{xx}, f_{yy} can be shown to have $(n-2)$ as their degree of homogeneity.

Proceeding in the similar manner it can be shown that r-th order partial derivative of f is a homogeneous function of degree $(n-r)$. $\qquad \Box$

Example 3.7.16 *You are given a differentiable function $f(x,y)$. Prove that if the variables x and y are replaced by homogeneous linear functions of X and Y, then the obtained function $F(X,Y)$ is related with the function as follows:*

$$x\frac{\partial f}{\partial x} + y\frac{\partial f}{\partial y} = X\frac{\partial F}{\partial X} + Y\frac{\partial F}{\partial Y}.$$

Solution: Let us replace x by $aX + bY$ and y by $cX + dY$. By chain rule, we get

$$\frac{\partial F}{\partial X} = \frac{\partial F}{\partial x}\cdot\frac{\partial x}{\partial X} + \frac{\partial F}{\partial y}\cdot\frac{\partial y}{\partial X} = a\frac{\partial F}{\partial x} + c\frac{\partial F}{\partial y}$$

and
$$\frac{\partial F}{\partial Y} = \frac{\partial F}{\partial x}\cdot\frac{\partial x}{\partial Y} + \frac{\partial F}{\partial y}\cdot\frac{\partial y}{\partial Y} = b\frac{\partial F}{\partial x} + d\frac{\partial F}{\partial y}.$$

$$\therefore \quad X\frac{\partial F}{\partial X} + Y\frac{\partial F}{\partial Y} = X\left(a\frac{\partial F}{\partial x} + c\frac{\partial F}{\partial y}\right) + Y\left(b\frac{\partial F}{\partial x} + d\frac{\partial F}{\partial y}\right)$$

$$= \frac{\partial F}{\partial x}(aX + bY) + \frac{\partial F}{\partial y}(cX + dY) = x\frac{\partial F}{\partial x} + y\frac{\partial F}{\partial y}$$

$$= x\frac{\partial f}{\partial x} + y\frac{\partial f}{\partial y} \quad (\because F \text{ is obtained form } f). \qquad \square$$

Example 3.7.17 *If $x = r\cos\theta, y = r\sin\theta$, prove that*

(i) $dx^2 + dy^2 = dr^2 + r^2 d\theta^2$ and (ii) $xdy - ydx = r^2 d\theta$.

Solution: We have $dx = \cos\theta dr - r\sin\theta d\theta$ and $dy = \sin\theta dr + r\cos\theta d\theta$.

(i) Now squaring and adding these we get

$$\begin{aligned}
dx^2 + dy^2 &= (\cos\theta dr - r\sin\theta d\theta)^2 + (\sin\theta dr + r\cos\theta d\theta)^2 \\
&= dr^2(\cos^2\theta + \sin^2\theta) + r^2 d\theta^2(\sin^2\theta + \cos^2\theta) = dr^2 + r^2 d\theta^2.
\end{aligned}$$

(ii) $xdy - ydx = r\cos\theta(\sin\theta dr + r\cos\theta d\theta) - r\sin\theta(\cos\theta dr - r\sin\theta d\theta)$

$$= r^2(\cos^2\theta + \sin^2\theta)d\theta = r^2 d\theta. \qquad \square$$

Example 3.7.18 *If $u = f(x^2 + 2yz, y^2 + 2zx)$, prove that*

$$(y^2 - zx)\frac{\partial u}{\partial x} + (x^2 - yz)\frac{\partial u}{\partial y} + (z^2 - xy)\frac{\partial u}{\partial z} = 0.$$

Solution: Let $p = x^2 + 2yz, q = y^2 + 2zx$, then $u = f(p, q)$.

$$\therefore \quad \frac{\partial u}{\partial x} = \frac{\partial u}{\partial p}\cdot\frac{\partial p}{\partial x} + \frac{\partial u}{\partial q}\cdot\frac{\partial q}{\partial x} = 2x\frac{\partial u}{\partial p} + 2z\frac{\partial u}{\partial q}$$

similarly, $\quad \dfrac{\partial u}{\partial y} = 2z\dfrac{\partial u}{\partial p} + 2y\dfrac{\partial u}{\partial q}$ and $\dfrac{\partial u}{\partial z} = 2y\dfrac{\partial u}{\partial p} + 2x\dfrac{\partial u}{\partial q}$

$$\therefore \quad (y^2 - zx)\frac{\partial u}{\partial x} + (x^2 - yz)\frac{\partial u}{\partial y} + (z^2 - xy)\frac{\partial u}{\partial z}$$

$$= (y^2 - zx)\left(2x\frac{\partial u}{\partial p} + 2z\frac{\partial u}{\partial q}\right) + (x^2 - yz)\left(2z\frac{\partial u}{\partial p} + 2y\frac{\partial u}{\partial q}\right)$$

$$+ (z^2 - xy)\left(2y\frac{\partial u}{\partial p} + 2x\frac{\partial u}{\partial q}\right)$$

$$= 2\{x(y^2 - zx) + y(z^2 - xy) + z(x^2 - yz)\}\frac{\partial u}{\partial p}$$

$$+ 2\{z(y^2 - zx) + y(x^2 - zy) + x(z^2 - xy)\}\frac{\partial u}{\partial q}$$

$$= 2.0\frac{\partial u}{\partial p} + 2.0\frac{\partial u}{\partial q} = 0. \qquad \square$$

Example 3.7.19 *If $u = F(x^2 + y^2 + z^2)f(xy + yz + zx)$, prove that*

$$(y - z)\frac{\partial u}{\partial x} + (z - x)\frac{\partial u}{\partial y} + (x - y)\frac{\partial u}{\partial z} = 0.$$

Solution: Given $u = F(x^2 + y^2 + z^2) f(xy + yz + zx)$

$$\therefore \quad \frac{\partial u}{\partial x} = 2x.F'.f + (y+z)Ff'$$

where $F' = \dfrac{dF}{dR}$, $R = x^2 + y^2 + z^2$ and $f' = \dfrac{df}{dS}$, $S = xy + yz + zx$.

Similarly, $\dfrac{\partial u}{\partial y} = 2y.F'.f + (z+x)Ff'$ and $\dfrac{\partial u}{\partial z} = 2z.F'.f + (x+y)Ff'$.

$$\therefore (y-z)\frac{\partial u}{\partial x} + (z-x)\frac{\partial u}{\partial y} + (x-y)\frac{\partial u}{\partial z} = 2F'f \sum x(y-z) + Ff' \sum (y^2 - z^2) = 0.$$

□

Example 3.7.20 *If $\phi(x, y) = 0$ and $\psi(x, z) = 0$, prove that*

$$\frac{\partial \psi}{\partial x}.\frac{\partial \phi}{\partial y}.\frac{dy}{dz} = \frac{\partial \phi}{\partial x}.\frac{\partial \psi}{\partial z}.$$

Solution: We have $\dfrac{dy}{dz} = \dfrac{dy}{dx}.\dfrac{dx}{dz} - \left(-\dfrac{\phi_x}{\phi_y}\right).\left(-\dfrac{\psi_z}{\psi_x}\right)$ [See § 2.12]

$$\therefore \quad \phi_y \psi_x . \frac{dy}{dz} = \phi_x \psi_z, \text{ i.e., } \frac{\partial \psi}{\partial x}.\frac{\partial \phi}{\partial y}.\frac{dy}{dz} = \frac{\partial \phi}{\partial x}.\frac{\partial \psi}{\partial z}.$$

□

Example 3.7.21 *Let $u = \sin^{-1} \sqrt{\dfrac{x^{\frac{1}{3}} + y^{\frac{1}{3}}}{x^{\frac{1}{2}} + y^{\frac{1}{2}}}}$, prove that*

$$x^2 \frac{\partial^2 u}{\partial x^2} + 2xy \frac{\partial^2 u}{\partial x \partial y} + y^2 \frac{\partial^2 u}{\partial y^2} = \frac{\tan u}{144}(13 + \tan^2 u).$$

Solution: Here, $\sin u = \left(\dfrac{x^{\frac{1}{3}} + y^{\frac{1}{3}}}{x^{\frac{1}{2}} + y^{\frac{1}{2}}}\right)^{\frac{1}{2}} = \dfrac{x^{\frac{1}{6}}}{x^{\frac{1}{4}}}\left\{\dfrac{1 + \left(\frac{y}{x}\right)^{\frac{1}{3}}}{1 + \left(\frac{y}{x}\right)^{\frac{1}{2}}}\right\}^{\frac{1}{2}} = x^{-\frac{1}{12}}\phi\left(\dfrac{y}{x}\right)$, i.e.,

$\sin u = v$ (say) is a homogeneous function of degree $-\dfrac{1}{12}$.

Therefore, by Euler's theorem we get

$$x\frac{\partial v}{\partial x} + y\frac{\partial v}{\partial y} = -\frac{1}{12}v \text{ or, } x\frac{\partial}{\partial x}(\sin u) + y\frac{\partial}{\partial y}(\sin u) = -\frac{1}{12}(\sin u)$$

$$\text{or, } \quad x \cos u \frac{\partial u}{\partial x} + y \cos u \frac{\partial u}{\partial y} = -\frac{1}{12}\sin u$$

$$\therefore \quad x\frac{\partial u}{\partial x} + y\frac{\partial u}{\partial y} = -\frac{1}{12}\tan u \qquad (3.76)$$

Differentiating both sides of (3.76) with respect to x and then multiplying by x, we get

$$x^2 \frac{\partial^2 u}{\partial x^2} + xy\frac{\partial^2 u}{\partial x \partial y} = -\frac{1}{12}\sec^2 u.x\frac{\partial u}{\partial x} - x\frac{\partial u}{\partial x} = -x\frac{\partial u}{\partial x}\left(\frac{1}{12}\sec^2 u + 1\right) \quad (3.77)$$

Again, differentiating both sides of (3.76) with respect to y and then multiplying by y, we get

$$xy\frac{\partial^2 u}{\partial x \partial y} + y^2\frac{\partial^2 u}{\partial y^2} = -y\frac{\partial u}{\partial y}\left(\frac{1}{12}\sec^2 u + 1\right) \qquad (3.78)$$

Now adding (3.77) and (3.78) we get

$$
\begin{aligned}
x^2 \frac{\partial^2 u}{\partial x^2} + 2xy \frac{\partial^2 u}{\partial x \partial y} + y^2 \frac{\partial^2 u}{\partial y^2} &= -\left(\frac{1}{12} \sec^2 u + 1\right)\left(x \frac{\partial u}{\partial x} + y \frac{\partial u}{\partial y}\right) \\
&= -\left(\frac{1}{12} \sec^2 u + 1\right)\left(-\frac{1}{12} \tan u\right) \\
&\qquad\qquad\qquad\qquad\qquad \text{[Using (3.76)]} \\
&= \frac{\tan u}{12}\left(\frac{13}{12} + \frac{\tan^2 u}{12}\right) \\
&= \frac{\tan u}{144}(13 + \tan^2 u). \qquad \Box
\end{aligned}
$$

3.8   Exercises

1. Find f_x, f_y for the following functions $f(x, y)$:

(i) $\tan^{-1} \dfrac{y}{x}$,   (ii) $\tan^{-1} \dfrac{x^2 + y^2}{x + y}$,   (iii) $\log(x^2 + y^2)$,   (iv) $\dfrac{1}{\sqrt{x^2 + y^2}}$,

(v) $\dfrac{x^2}{a^2} + \dfrac{y^2}{b^2} = 1$,   (vi) $x^2 + xy + y^2$.

2. Find f_{xx}, f_{xy}, f_{yx} and f_{yy} from the following functions $f(x, y)$:

(i) $x^3 + x^2 y^2 + y^3$,   (ii) $e^{x^2 + xy + y^2}$,   (iii) $\log(x^2 y + xy^2)$,

(iv) $x \cos y + y \cos x$.

3. If $z(x + y) = x^2 + y^2$, show that $\left(\dfrac{\partial z}{\partial x} - \dfrac{\partial z}{\partial y} \right)^2 = 4 \left(1 - \dfrac{\partial z}{\partial x} - \dfrac{\partial z}{\partial y} \right)$.

4. If $u = f(xyz)$, show that $xu_x = yu_y = zu_z$.

5. If $u = e^{ax+by} f(ax - by)$, prove that $b \dfrac{\partial u}{\partial x} + a \dfrac{\partial u}{\partial y} = 2abu$.

6. If $u = \dfrac{u}{z} + \dfrac{z}{x} + \dfrac{x}{y}$, prove that $x \dfrac{\partial u}{\partial x} + y \dfrac{\partial u}{\partial y} + z \dfrac{\partial u}{\partial z} = 0$.

7. Show that $\dfrac{\partial^2 u}{\partial x^2} + \dfrac{\partial^2 u}{\partial y^2} = 0$, if (i) $u = \tan^{-1} a \log(x^2 + y^2) + b \tan^{-1} \dfrac{y}{x}$ and
(ii) $u = e^x(x \cos y - y \sin y)$.

8. If $u = \dfrac{x + y}{1 - xy}$ and $v = \dfrac{x(1 - y^2) + y(1 - x^2)}{(1 + x^2)(1 + y^2)}$, prove that $u_x v_y = v_x u_y$.

9. (i) If $u = z \tan^{-1} \dfrac{y}{x}$, prove that $u_{xx} + u_{yy} + u_{zz} = 0$.

(ii) If $u = 3(ax + by + cz)^2 - (x^2 + y^2 + z^2)$ and $a^2 + b^2 + c^2 = 1$, then prove that $u_{xx} + u_{yy} + u_{zz} = 0..$

(iii) If $V = ax^2 + 2hxy + by^2$, prove that
$$V_x^2 V_{yy} - 2V_x V_y V_{xy} + V_y^2 V_{xx} = 8(ab - h^2)V.$$

10. (i) Find the value of n so that the equation $V = r^n(3 \cos^2 \theta - 1)$ satisfies the relation $\dfrac{\partial}{\partial r} \left(r^2 \dfrac{\partial V}{\partial r} \right) + \dfrac{1}{\sin \theta} \dfrac{\partial}{\partial \theta} \left(\sin \theta \dfrac{\partial V}{\partial \theta} \right) = 0.$

(ii) If $\theta = t^n e^{-\frac{r}{4t}}$, find what value of n will make $\dfrac{1}{r^2} \dfrac{\partial}{\partial r} \left(r^2 \dfrac{\partial \theta}{\partial r} \right) = \dfrac{\partial \theta}{\partial t}.$

11. (i) If $u = \log r$ and $r^2 = x^2 + y^2 + z^2$, then prove that
$$r^2(u_{xx} + u_{yy} + u_{zz}) = 1.$$

(ii) If $V = r^m$ and $r^2 = x^2 + y^2 + z^2$, prove that
$$V_{xx} + V_{yy} + V_{zz} = m(m + 1)r^{m-2}.$$

12. (i) If $u = \exp(xyz)$, find the value of $\dfrac{\partial^3 u}{\partial x \partial y \partial z}$.

(ii) If $u = f(ax^2 + 2hxy + by^2), v = \phi(ax^2 + 2hxy + by^2)$ then show that
$$\dfrac{\partial}{\partial y} \left(u \dfrac{\partial v}{\partial x} \right) = \dfrac{\partial}{\partial x} \left(u \dfrac{\partial v}{\partial y} \right).$$

(iii) If $x^x y^y z^z = c$ (constant), show that at (x, y, z), where $x = y = z$, $\dfrac{\partial^2 z}{\partial x \partial y} =$
$-\{x \log_e(ex)\}^{-1}$.

13. If $V = F(x, y)$ and $x = e^u \cos t, y = e^u \sin t$, show that
$$\frac{\partial^2 V}{\partial u^2} + \frac{\partial^2 V}{\partial t^2} = e^{2u}\left(\frac{\partial^2 V}{\partial x^2} + \frac{\partial^2 V}{\partial y^2}\right).$$

14. Find $\dfrac{dy}{dx}$ in the following cases:

(i) $x^4 + x^2 y^2 + y^4 = 0$, (ii) $x^y + y^x = a^b$, (iii) $e^{xy} - 4xy = 2$,
(iv) $\log(xy) = x^2 + y^2$.

15. If $u = F(y - z, z - x, x - y)$, prove that $\dfrac{\partial u}{\partial x} + \dfrac{\partial u}{\partial y} + \dfrac{\partial u}{\partial z} = 0$.

16. If u be a function of x and y satisfying $x = \theta \cos\alpha - \phi \sin\alpha, y = \theta \sin\alpha + \phi \cos\alpha$ (α is a constant), prove that $\dfrac{\partial^2 u}{\partial x^2} + \dfrac{\partial^2 u}{\partial y^2} = \dfrac{\partial^2 u}{\partial \theta^2} + \dfrac{\partial^2 u}{\partial \phi^2}$.

17. If $x^2 = vw, y^2 = wu, z^2 = uv$ and $f(x, y, z) = \phi(u, v, w)$ then show that $xf_x + yf_y + zf_z = u\phi_u + v\phi_v + w\phi_w$.

18. If $x = u^2 - v^2$ and $y = 2uv$, then show that $u_{xx} + u_{yy} = 0$.

19. If $f(x, y, z)$ is a homogeneous function of degree n ($\neq 1$) having continuous second order partial derivatives, then show that

$$\begin{vmatrix} f_{xx} & f_{xy} & f_{xz} \\ f_{yx} & f_{yy} & f_{yz} \\ f_{zx} & f_{zy} & f_{zz} \end{vmatrix} = \frac{(n-1)^2}{z^2} \begin{vmatrix} f_{xx} & f_{xy} & f_x \\ f_{yx} & f_{yy} & f_y \\ f_x & f_y & \dfrac{nf}{n-1} \end{vmatrix}.$$

20. Let $f(x, y, z)$ be a differentiable function.

Prove that $f(x, y, z)$ is homogeneous function of degree n if and only if it satisfy $x\dfrac{\partial u}{\partial x} + y\dfrac{\partial u}{\partial y} + z\dfrac{\partial u}{\partial z} = nf$.

If $f(x, y, z)$ has the property $f(tx, t^p y, t^q z) = t^n f(x, y, z)$, where p, q, n are positive integers, prove that $xf_x + pyf_y + qzf_z = nf(x, y, z)$.

[*Hints:* For the first part see **Theorem 3.5.2** and **Theorem 3.5.3** for $k = 3$.
Second part: Let $f(x, y, z)$ be such that $f(tx, t^p y, t^q z) = t^n f(x, y, z)$. Let $u = tx, v = t^p y, w = t^q z$, then

$$f(u, v, w) = t^n f(x, y, z) \tag{3.79}$$

Differentiating both sides of (3.79) with respect to t, we get

$$f_u \frac{du}{dt} + f_v \frac{dv}{dt} + f_w \frac{dw}{dt} = nt^{n-1} f(x, y, z)$$
or, $\quad f_u.x + f_v p t^{p-1} y + f_w q t^{q-1} z = nt^{n-1} f(x, y, z)$

Now putting $t = 1$, we get $xf_x + pyf_y + qzf_z = nf(x, y, z)$.]

21. If $V = (x^2 + y^2 + z^2)^{\frac{3}{2}}$, prove that $V_{xx} + V_{yy} + V_{zz} = 12\sqrt{x^2 + y^2 + z^2}$.

22. Show that the function z defined by $F\left(\dfrac{x}{z}, \dfrac{y}{z}\right) = 0$ yields

$$x\frac{\partial u}{\partial x} + y\frac{\partial u}{\partial y} = z.$$

23. If λ, μ, ν are functions of x, y, z given by $x = \lambda + \mu + \nu, y = \lambda^2 + \mu^2 + \nu^2, z = \lambda^3 + \mu^3 + \nu^3$, prove that $\dfrac{\partial \lambda}{\partial x} = \dfrac{\mu\nu(\nu - \mu)}{(\lambda - \mu)(\mu - \nu)(\nu - \lambda)}$.

24. By the transformation $\xi = a + \alpha x + \beta y$ and $\eta = b - \beta x + \alpha y$, where a, b, α, β are all constants and $\alpha^2 + \beta^2 = 1$, the function $u(x, y)$ is transformed into the function $U(\xi, \eta)$, prove that $U_{\xi\xi}U_{\eta\eta} - U_{\xi\eta}^2 = u_{xx}u_{yy} - u_{xy}^2$.

25. If $x = r\cos\theta, y = r\sin\theta$, show that the equation $xy(u_{xx} - u_{yy}) - (x^2 - y^2)u_{xy} = 0$ reduces to $ru_{r\theta} = u_\theta$.

26. Given that f is a function of x and y and that $x = u^2v, y = uv^2$, prove that $2x^2 f_x^2 + 2y^2 f_y^2 + 5xy f_{xy} = uv f_{uv} - \dfrac{2}{3}(uf_u + vf_v)$.

27. Given that F is a differentiable function of x and y and that $x = e^u + e^{-v}, y = e^v + e^{-v}$, prove that

$$\frac{\partial^2 F}{\partial u^2} - 2\frac{\partial^2 F}{\partial u \partial v} + \frac{\partial^2 F}{\partial v^2} = x^2\frac{\partial^2 F}{\partial x^2} - 2xy\frac{\partial^2 F}{\partial x \partial y} + y^2\frac{\partial^2 F}{\partial y^2} + x\frac{\partial F}{\partial x} + y\frac{\partial F}{\partial y}.$$

28. If the variables x and y are changed to u and v, where $u = x^2 - y^2, v = 2xy$ then show that $y^2 z_{xx} + 2xy z_{xy} + x^2 z_{yy} + xz_x + yz_y = 0$ reduces to $(u^2 + v^2)z_{vv} + vz_v = 0$.

29. Applying the transformation given by $2x = e^u + e^v, 2y = e^u - e^v$, prove that $(x^2 + y^2)(z_{xx} + z_{yy}) + 4xy z_{xy} + 2xz_x + 2yz_y = 0$ is transformed into $z_{uu} + z_{vv} = 0$.

30. If the function $f(x, y)$ be transformed to $g(u, v)$ by the transformation $u = a + x\cos\theta + y\sin\theta$ and $v = b - x\sin\theta + y\cos\theta$, prove that $f_{xx}f_{yy} - f_{xy}^2 = g_{uu}g_{vv} - g_{uv}^2$.

[*Hints:* Similar to **Exercise 24**].

31. If $\Delta\phi = \dfrac{\partial^2\phi}{\partial x^2} + \dfrac{\partial^2\phi}{\partial y^2} + \dfrac{\partial^2\phi}{\partial z^2}$, prove that

$$\Delta(uv) = u\Delta v + v\Delta u + 2\left(\frac{\partial u}{\partial x}.\frac{\partial v}{\partial x} + \frac{\partial u}{\partial y}.\frac{\partial v}{\partial y} + \frac{\partial u}{\partial z}.\frac{\partial v}{\partial z}\right)$$

where $u = u(x, y, z)$ and $v = v(x, y, z)$ which admit of partial derivatives of the second order.

If $\phi(x, y, z)$ be such a function that $\Delta\phi = 0$ and $x\dfrac{\partial\phi}{\partial y} + y\dfrac{\partial\phi}{\partial y} + z\dfrac{\partial\phi}{\partial y} = nf(x, y, z)$, then apply the above result to prove

$$\Delta(r^m.\phi) = r^{m-2}[m(m + 1) + 2mn].\phi(x, y, z),$$

where $r^2 = x^2 + y^2 + z^2$, and m, n are positive integers.

32. If $u(x, y, z, t) = \dfrac{f(t + r)}{r} + \dfrac{g(t - r)}{r}$, where $r^2 = x^2 + y^2 + z^2$, prove that u satisfies the relation $u_{xx} + u_{yy} + u_{zz} = u_{tt}$.

33. If $x = r\cos\theta, y = r\sin\theta$ and V is a function of x, y possessing partial derivatives of the second order, prove that $V_{xx} + V_{yy} = \dfrac{1}{r^2}\left[r^2\dfrac{\partial}{\partial r}(rV_r) + V_{\theta\theta}\right]$.

34. If u is a function of x, y having continuous partial derivatives upto the second order and the variables x, y are changed to ξ, η by the transformation $x + y = (\xi + \eta)^n, x - y = (\xi - \eta)^n$, prove that

$$(x^2 - y^2)\left[\dfrac{\partial^2 u}{\partial x^2} - \dfrac{\partial^2 u}{\partial u^2}\right] = \dfrac{1}{n^2}(\xi^2 - \eta^2)\left[\dfrac{\partial^2 u}{\partial \xi^2} - \dfrac{\partial^2 u}{\partial \eta^2}\right].$$

35. If $V = x^2 \tan^{-1}\dfrac{y}{x} - y^2 \tan^{-1}\dfrac{x}{y}$, show that $\dfrac{\partial^2 V}{\partial x \partial y} = \dfrac{x^2 - y^2}{x^2 + y^2}$.

36. If $u = F\left(\dfrac{y-x}{xy}, \dfrac{z-x}{zx}\right)$ be a differentiable function, show that

$$x^2\dfrac{\partial u}{\partial x} + y^2\dfrac{\partial u}{\partial y} + z^2\dfrac{\partial u}{\partial z} = 0.$$

37. If z is a function of u, v where $u = x^2 - y^2 - 2xy, v = y$, then show that $(x+y)\dfrac{\partial z}{\partial x} + (x-y)\dfrac{\partial z}{\partial y} = 0$ reduces to $\dfrac{\partial z}{\partial v} = 0$.

38. If $r^2 = x^2 + y^2 + z^2$ and $v = r^3$ then prove that

$$\dfrac{1}{yz}\dfrac{\partial^2 v}{\partial y \partial z} + \dfrac{1}{zx}\dfrac{\partial^2 v}{\partial z \partial x} + \dfrac{1}{xy}\dfrac{\partial^2 v}{\partial x \partial y} = \dfrac{9}{r}.$$

39. If $V = f(r)$, where $r = \sqrt{x^2 + y^2 + z^2}$, then prove that

$$\dfrac{\partial^2 V}{\partial x^2} + \dfrac{\partial^2 V}{\partial y^2} + \dfrac{\partial^2 V}{\partial z^2} = f''(r) + \dfrac{2}{r}f'(r).$$

In particular, apply this result to solve **Exercise 11 (ii)**.

40. A function $f(x, y, z)$ is said to be homogeneous of degree n, if $f(tx, ty, tz) = t^n f(x, y, z) \ \forall \ t > 0$. Hence prove that for a homogeneous function of degree n,

$$f(x, y, z) = x^n f\left(1, \dfrac{y}{x}, \dfrac{z}{x}\right).$$

Illustrate with the following functions:

(i) $x + \dfrac{y^2}{z} + \dfrac{z^2}{y}$, (ii) $\sin^{-1}\dfrac{1}{\sqrt{x^2 + y^2 + z^2}}$.

[*Hints:* For the first part put $tx = 1$.]

41. If $f(x, y, z)$ has the property $f(tx, t^p y, t^q z) - t^n f(x, y, z) \ \forall \ t > 0$ where p, q, r are postive integers, prove that $f(x, y, z) = x^n f\left(1, \dfrac{y}{x^p}, \dfrac{z}{x^q}\right)$.

[*Hints:* Put $tx = 1$.]

42. If u is a homogeneous function of degree n in x and y and if $v = f(u)$, then show that $x\dfrac{\partial v}{\partial x} + y\dfrac{\partial v}{\partial y} = nu\dfrac{dv}{du}$.

43. If u is a homogeneous function in x, y, z of degree n having continuous second order partial derivatives and if $u = f(\xi, \eta, \zeta)$, where ξ, η, ζ are partial derivatives of u with respect to x, y, z respectively, prove that

$$\xi\dfrac{\partial u}{\partial \xi} + \eta\dfrac{\partial u}{\partial \eta} + \zeta\dfrac{\partial u}{\partial \zeta} = \dfrac{n}{n-1}u \ \ (n \neq 1).$$

44. Apply Euler's theorem to show that $x\dfrac{\partial f}{\partial x} + y\dfrac{\partial f}{\partial y} = 6f$, where $f(x,y) = x^4 y^2 \sin^{-1}\dfrac{y}{x}$.

45. If $f(x,y,z)$ is a homogeneous function of degree n in x,y,z, which satisfies Laplace's equation $\Delta f = \dfrac{\partial^2 f}{\partial x^2} + \dfrac{\partial^2 f}{\partial y^2} + \dfrac{\partial^2 f}{\partial z^2} = 0$, prove that $\Delta(r^{2m} f) = 2m(2n + 2m + 1)r^{2m-2} f$ where $r^2 = x^2 + y^2 + z^2$.

46. Prove that if V_n is a homogeneous function of degree n in x,y,z and $r^2 = x^2 + y^2 + z^2$, then $\Delta^2(r^m V_n) = m(m + 2n + 1)r^{m-2} V_n + r^m \Delta^2 V_n$, where $\Delta^2 = \dfrac{\partial^2}{\partial x^2} + \dfrac{\partial^2}{\partial y^2} + \dfrac{\partial^2}{\partial z^2}$.

Deduce that if further V_n satisfies the equation $\Delta^2 V_n = 0$, so does $r^{-2n-1} V_n$.

47. If u is a homogeneous function of degree n in x and y, then show that $(xD_x + yD_y)^2 u = n(n-1)u$, where

$$(xD_x + yD_y)^2 = x^2 D_x^2 + 2xy D_x D_y + y^2 D_y^2 = x^2 \frac{\partial^2}{\partial x^2} + 2xy \frac{\partial^2}{\partial x \partial y} + y^2 \frac{\partial^2}{\partial y^2},$$

a second order linear partial differential operator.

[*Hints:* See § 3.5 **Corollary 3.5.2**]

Remark: Generalising the above result, we can get

$$(xD_x + yD_y)^m u \ = n(n-1)(n-2)...(n-m+1)u = \underline{|m} \binom{n}{m} u$$

$$\text{where} \quad (xD_x + yD_y)^m \ = \sum_{r=0}^{m} {}^m C_r x^{m-r} y^r D_x^{m-r} D_y^r$$

$$\text{and} \quad D_x^{m-r} D_y^r \ = \frac{\partial^m}{\partial x^{m-r} \partial y^r},$$

assuming the allowance of the change in the order of differentiation.

48. If $x = r\cos\theta, y = r\sin\theta$, where r and θ are functions of t alone, prove that $x\dfrac{dy}{dt} - y\dfrac{dx}{dt} = r^2 \dfrac{d\theta}{dt}$.

49. If $\phi = \phi(x,y)$, where $x = e^u \sec v, y = e^u \tan v$, prove that

$$\left(\frac{\partial \phi}{\partial x}\right)^2 - \left(\frac{\partial \phi}{\partial y}\right)^2 = \frac{1}{e^{2u}}\left\{\left(\frac{\partial \phi}{\partial u}\right)^2 - \cos^2 v \left(\frac{\partial \phi}{\partial v}\right)^2\right\}.$$

50. If $\phi = \phi(x,y)$, where $x = e^u \sec v, y = e^u \tan v$, prove that

$$\cos u \left(\frac{\partial^2 \phi}{\partial u \partial v} - \frac{\partial \phi}{\partial u}\right) = xy\left(\frac{\partial^2 \phi}{\partial x^2} + \frac{\partial^2 \phi}{\partial y^2}\right) + (x^2 + y^2)\frac{\partial^2 \phi}{\partial x \partial y}.$$

51. If V is a function of r alone, where $r^2 = x_1^2 + x_2^2 + ... + x_n^2$, prove that

$$\frac{\partial^2 V}{\partial x^2} + \frac{\partial^2 V}{\partial x^2} + ... + \frac{\partial^2 V}{\partial x^2} = \frac{d^2 V}{dr^2} + \frac{n-1}{r}\frac{dV}{dr}.$$

52. Let $u = \sin^{-1} \dfrac{x+y}{\sqrt{x} + \sqrt{y}}$, prove that

(i) $\quad xu_x + yu_y = \dfrac{1}{2} \tan u$

(ii) $\quad x^2 u_{xx} + 2xy y_{xy} + y^2 u_{yy} = -\dfrac{1}{4} \dfrac{\sin u \cos 2u}{4 \cos^3 u}.$

53. Let $u = x\phi \left(\dfrac{y}{x}\right) + \psi \left(\dfrac{y}{x}\right)$, show that

(i) $\quad xu_x + yu_y = x\phi \left(\dfrac{y}{x}\right)$

(ii) $\quad x^2 u_{xx} + 2xy y_{xy} + y^2 u_{yy} = 0.$

Explain how can you illustrate the validity of Euler's theorem and Euler's theorem of second order respectively for this function u.

54. If V is a homogeneous function of degree n in the three variables x, y, z, prove that $V_x + V_y + V_z$ is a homogeneous function of degree $(n-1)$.

[*Hints:* By **Corollary 3.5.1**, V_x, V_y, V_z are each a homogeneous function of degree $(n-1)$, so then sum is also so.]

55. Apply Euler's theorem to find the value of $x\dfrac{\partial v}{\partial x} + y\dfrac{\partial v}{\partial y}$ at $(1,1)$ for the function $u = x\sin^{-1} \dfrac{y}{x} + y\tan^{-1} \dfrac{x}{y}$.

56. If $f(x, y, z) = \log \sin \left\{ \dfrac{\pi}{2} \dfrac{\sqrt{2x^2 + y^2 + zx}}{\sqrt[3]{x^2 + xy + 2yz + z^2}} \right\}$, find the value of $x\dfrac{\partial v}{\partial x} + y\dfrac{\partial v}{\partial y} + z\dfrac{\partial v}{\partial z}$ at $(0, 1, 2)$.

[*Hints:* Let $V(x, y, z) = \dfrac{\pi}{2} \dfrac{\sqrt{2x^2 + y^2 + zx}}{\sqrt[3]{x^2 + xy + 2yz + z^2}}$, then

$$
\begin{aligned}
V(tx, ty, tz) &= \frac{\pi}{2} \frac{\sqrt{2(tx)^2 + (ty)^2 + tz.tx}}{\sqrt[3]{(tx)^2 + tx.ty + 2.ty.tz + (tz)^2}} \\
&= \frac{\pi}{2} t^{\frac{1}{3}} \frac{\sqrt{2x^2 + y^2 + zx}}{\sqrt[3]{x^2 + xy + 2yz + z^2}} = t^{\frac{1}{3}} V(x, y, z).
\end{aligned}
$$

So, V is a homogeneous function of degree $\dfrac{1}{3}$.

Therefore, by Euler's theorem, we get

$$
x\frac{\partial V}{\partial x} + y\frac{\partial V}{\partial y} + z\frac{\partial V}{\partial z} = \frac{1}{3} V \tag{3.80}
$$

Now $f(x, y, z) = \log \sin V$, i.e., f is a function of V and V is a function of x, y, z. So using chain rule we get

$$
\frac{\partial f}{\partial x} = \frac{df}{dV} \cdot \frac{\partial V}{\partial x} = \frac{1}{\sin V} \cos V \frac{\partial V}{\partial x} = \cot V \frac{\partial V}{\partial x} \quad \therefore \frac{\partial V}{\partial x} = \tan V \frac{\partial f}{\partial x}.
$$

Similarly, $\dfrac{\partial V}{\partial y} = \tan V \dfrac{\partial f}{\partial y}$ and $\dfrac{\partial V}{\partial z} = \tan V \dfrac{\partial f}{\partial z}$.

So, from (3.80) we get

$$x \tan V \frac{\partial f}{\partial x} + y \tan V \frac{\partial f}{\partial y} + z \tan V \frac{\partial f}{\partial z} = \frac{1}{3} V$$

or, $\qquad x \frac{\partial f}{\partial x} + y \frac{\partial f}{\partial y} + z \frac{\partial f}{\partial z} = \frac{1}{3} \frac{V}{\tan V}$

$\therefore$ at $(0, 1, 2)$ $\quad x \frac{\partial f}{\partial x} + y \frac{\partial f}{\partial y} + z \frac{\partial f}{\partial z} = \frac{1}{3} \frac{V(0, 1, 2)}{\tan\{V(0, 1, 2)\}}$ (3.81)

Now, $V(0, 1, 2) = \dfrac{\pi}{2} \dfrac{\sqrt{2.0^2 + 1^2 + 2.0}}{\sqrt[3]{0^2 + 0.1 + 2.1.2 + 2^2}} = \dfrac{\pi}{2} \cdot \dfrac{1}{2} = \dfrac{\pi}{4}.$

$\therefore$ From (3.81), $\left. x \dfrac{\partial f}{\partial x} + y \dfrac{\partial f}{\partial y} + z \dfrac{\partial f}{\partial z} = \dfrac{\frac{\pi}{4}}{3 \tan \frac{\pi}{4}} = \dfrac{\pi}{12}. \right]$

57. If $u = \dfrac{x^2 y^2}{x + y}$, apply Euler's theorem to find the value of $x \dfrac{\partial f}{\partial x} + y \dfrac{\partial f}{\partial y}$ and hence deduce that $x^2 \dfrac{\partial^2 u}{\partial x^2} + 2xy \dfrac{\partial^2 u}{\partial x \partial y} + y^2 \dfrac{\partial^2 u}{\partial y^2} = 6u.$

[*Hints:* Since u is a homogeneous function of degree 3, so by Euler's theorem we get

$$x \frac{\partial u}{\partial x} + y \frac{\partial u}{\partial y} = 3u \qquad (3.82)$$

Differentiating (3.82) partially with respect to x and then multiplying by x we get

$$x^2 \frac{\partial^2 u}{\partial x^2} + xy \frac{\partial^2 u}{\partial x \partial y} + x \frac{\partial u}{\partial x} = 3x \frac{\partial u}{\partial x} \qquad (3.83)$$

Similarly, differentiating (3.82) partially with respect to y and then multiplying by y we get

$$xy \frac{\partial^2 u}{\partial x \partial y} + y^2 \frac{\partial^2 u}{\partial y^2} + y \frac{\partial u}{\partial x} = 3y \frac{\partial u}{\partial y} \qquad (3.84)$$

Now, adding (3.83) and (3.84) and using the result (3.82) we get

$$\left. x^2 \frac{\partial^2 u}{\partial x^2} + 2xy \frac{\partial^2 u}{\partial x \partial y} + y^2 \frac{\partial^2 u}{\partial y^2} = 2 \left(x \frac{\partial u}{\partial x} + y \frac{\partial u}{\partial y} \right) = 2.3u = 6u. \right]$$

Note 3.8.1 *If we apply Euler's theorem of second order directly, then we get*
$\left(x \dfrac{\partial}{\partial x} + y \dfrac{\partial}{\partial y} \right)^2 u = 3.(3 - 1)u = 6u.$ *[ref.* **Corollary 3.5.2***]*

58. If $f(x, y) = \dfrac{(x^2 + y^2)^n}{2n(2n - 1)} + x\phi \left(\dfrac{y}{x} \right) + \psi \left(\dfrac{y}{x} \right)$, apply Euler's theorem of second order on homogeneous function in x and y to prove that

$$x^2 \frac{\partial^2 u}{\partial x^2} + 2xy \frac{\partial^2 u}{\partial x \partial y} + y^2 \frac{\partial^2 u}{\partial y^2} = (x^2 + y^2)^n.$$

[*Hints:* If $f(x, y)$ is a homogeneous function of degree n in x, y, then by Euler's theorem of second order we have

$$x^2 \frac{\partial^2 u}{\partial x^2} + 2xy \frac{\partial^2 u}{\partial x \partial y} + y^2 \frac{\partial^2 u}{\partial y^2} = n(n - 1) \qquad (3.85)$$

Let $f = u + v + w$, where $u = x^{2n}\dfrac{\left\{1 + \left(\frac{y}{x}\right)^2\right\}^n}{2n(2n-1)}$ is a homogeneous function of degree $2n$, $v = x\phi\left(\dfrac{y}{x}\right)$ is a homogeneous function of degree 1 and $w = \psi\left(\dfrac{y}{x}\right) = x^0\psi\left(\dfrac{y}{x}\right)$ is a homogeneous function of degree 0. So by (3.85) we get

$$\left(x\frac{\partial}{\partial x} + y\frac{\partial}{\partial y}\right)^2 u = 2n(2n-1)u, \quad \left(x\frac{\partial}{\partial x} + y\frac{\partial}{\partial y}\right)^2 v = 0$$

and $\left(x\dfrac{\partial}{\partial x} + y\dfrac{\partial}{\partial y}\right)^2 w = 0(0-1)w = 0.$

Now, adding all these we get

$$\left(x\frac{\partial}{\partial x} + y\frac{\partial}{\partial y}\right)^2 (u + v + w) = 2n(2n-1)u + 0 + 0$$

or, $x^2\dfrac{\partial^2 u}{\partial x^2} + 2xy\dfrac{\partial^2 u}{\partial x \partial y} + y^2\dfrac{\partial^2 u}{\partial y^2} = 2n(2n-1)\dfrac{(x^2+y^2)^n}{2n(2n-1)} = (x^2+y^2)^n.$

59. If $u(x, y) = (x^2 + y^2)^{\frac{2}{3}}$, prove that $x^2 u_{xx} + 2xy u_{xy} + y^2 u_{yy} = \dfrac{4u}{9}$.

60. If $u = \dfrac{1}{x^2 - y^2}$, prove that $\left(x\dfrac{\partial}{\partial x} + y\dfrac{\partial}{\partial y}\right)^2 u = 6u$.

61. If $u = 2x^2 + 6xy + y^2$, prove that $u_x^2 u_{yy} - 2u_x u_y u_{xy} + u_y^2 u_{xx} = -48u$.

62. If $u = \dfrac{(ax^3 + by^3)^n}{3n(3n-1)} + xf\left(\dfrac{y}{x}\right)$, prove that

$$x^2 u_{xx} + 2xy u_{xy} + y^2 u_{yy} = (ax^3 + by^3)^n.$$

63. If $u = \log(x^3 + y^3 + z^3 - 3xyz)$, prove that

(i) $\dfrac{\partial u}{\partial x} + \dfrac{\partial u}{\partial y} + \dfrac{\partial u}{\partial z} = \dfrac{3}{x + y + z}$

(ii) $x\dfrac{\partial u}{\partial x} + y\dfrac{\partial u}{\partial y} + z\dfrac{\partial u}{\partial z} = 3$

(iii) $\left(\dfrac{\partial}{\partial x} + \dfrac{\partial}{\partial y} + \dfrac{\partial}{\partial z}\right)^2 u = \dfrac{9}{(x + y + z)^2}$,

(iv) $\dfrac{\partial^2 u}{\partial x^2} + \dfrac{\partial^2 u}{\partial y^2} + \dfrac{\partial^2 u}{\partial z^2} = -\dfrac{3}{(x + y + z)^2}$

[*Hints:* See **Example 3.4.3** for (i) and (ii).

(ii) Multiply (3.10) by x, (3.11) by y and (3.12) by z and then add together.

(iii) We proceed as follows:

$$\left(\frac{\partial}{\partial x}+\frac{\partial}{\partial y}+\frac{\partial}{\partial z}\right)^2 u = \left(\frac{\partial}{\partial x}+\frac{\partial}{\partial y}+\frac{\partial}{\partial z}\right)\left(\frac{\partial}{\partial x}+\frac{\partial}{\partial y}+\frac{\partial}{\partial z}\right)u$$

$$= \left(\frac{\partial}{\partial x}+\frac{\partial}{\partial y}+\frac{\partial}{\partial z}\right)\left(\frac{\partial u}{\partial x}+\frac{\partial u}{\partial y}+\frac{\partial u}{\partial z}\right)$$

$$= \left(\frac{\partial}{\partial x}+\frac{\partial}{\partial y}+\frac{\partial}{\partial z}\right)\left(\frac{3}{x+y+z}\right) \quad [\text{by } (i)]$$

$$= 3\left\{\frac{\partial}{\partial x}\left(\frac{1}{x+y+z}\right)+\frac{\partial}{\partial y}\left(\frac{1}{x+y+z}\right)+\frac{\partial}{\partial z}\left(\frac{1}{x+y+z}\right)\right\}$$

$$= -3\left[\frac{1}{(x+y+z)^2}+\frac{1}{(x+y+z)^2}+\frac{1}{(x+y+z)^2}\right] = -3.\frac{3}{(x+y+z)^2}$$

$$= \frac{9}{(x+y+z)^2}.\Big]$$

ANSWERS

1. (i) $-\frac{y}{x^2+y^2}, \frac{x}{x^2+y^2}$; **(ii)** $\frac{x^2+2xy-y^2}{(x^2+y^2)^2+(x+y)^2}, \frac{y^2+2xy-x^2}{(x^2+y^2)^2+(x+y)^2}$; **(iii)** $\frac{2x}{x^2+y^2}, \frac{2y}{x^2+y^2}$;

(iv) $-\frac{x}{(x^2+y^2)^{\frac{3}{2}}}, -\frac{y}{(x^2+y^2)^{\frac{3}{2}}}$; **(v)** $\frac{2x}{a^2}, \frac{2y}{b^2}$; **(vi)** $2x+y, 2y+x$; **2. (i)** $6x+2y^2$,

$4xy, 4xy, 6y+2x^2$; **(ii)** $e^{x^2+xy+y^2}\{(2x+y)^2+2\}, e^{x^2+xy+y^2}\{(2x+y)(x+2y)+1\}, e^{x^2+xy+y^2}\{(2x+y)(x+2y)+1\}, e^{x^2+xy+y^2}\{(x+2y)^2+2\}$; **(iii)** $-\left\{\frac{1}{x^2}+\frac{1}{(x+y)^2}\right\}$,

$-\frac{1}{(x+y)^2}, -\frac{1}{(x+y)^2}, -\left\{\frac{1}{y^2}+\frac{1}{(x+y)^2}\right\}$; **(iv)** $-y\cos x, -(\sin x+\cos x), -(\sin x+\cos x), -x\cos y$; **10. (i)** $n=2,-3$; **(ii)** $n=-1.5$; **12. (i)** $(1+3xyz+x^2y^2z^2)e^{xyz}$;

14. (i) $-\frac{x}{y}\frac{2x^2+y^2}{x^2+2y^2}$; **(ii)** $\frac{yx^{y-1}+y^x\log y}{xy^{x-1}+x^y\log x}$; **(iii)** $-\frac{y}{x}$; **(iv)** $\frac{y(2x^3-1)}{x(1-2y^3)}$; **40. (i)** Degree 1;

(ii). Degree 0; **55.** $\frac{3\pi}{4}$; **56.** $\frac{\pi}{12}$; **57.** $\frac{x^2y^2}{x^2+y^2}$.

Chapter 4

Jacobians, Functional Dependence and Implicit Functions

4.1 Introduction

A Jacobian, called after the German mathematician *Carl Guston Jacob Jacobi* (1804 - 1851), is a determinant formed by the first order partial derivatives. Jacobians play very important role in change of variables. Jacobian change of variable is a technique that can be used to solve integration problems, which would otherwise be difficult using normal techniques. In geometry, the area, volume etc. in one coordinate system is transformed to another coordinate system through Jacobians. Therefore, acquaintance with the notion of Jacobian is necessary for further development of the subject. The goal of Jacobian change of variables is to convert form a physical space defined in terms of $u(x, y)$ and $v(x, y)$. Dependence and independence can also be tested by Jacobians. Jacobians have the properties like the derivative of a function of single variable.

4.2 Change of Variables by Jacobians

Let $u_1, u_2, ..., u_n$ be n functions of n variables $x_1, x_2, ..., x_n$ possessing first order partial derivatives at every point of the common domain of definition of the functions, then the determinants

$$
\begin{vmatrix}
\dfrac{\partial u_1}{\partial x_1} & \dfrac{\partial u_1}{\partial x_2} & \cdots & \dfrac{\partial u_1}{\partial x_n} \\
\dfrac{\partial u_2}{\partial x_1} & \dfrac{\partial u_2}{\partial x_2} & \cdots & \dfrac{\partial u_2}{\partial x_n} \\
\cdots & \cdots & & \cdots \\
\dfrac{\partial u_n}{\partial x_1} & \dfrac{\partial u_n}{\partial x_2} & \cdots & \dfrac{\partial u_n}{\partial x_n}
\end{vmatrix}
$$

is called the *Jacobian* or the *Functional Determinant* of the functions $u_1, u_2, ...,$ u_n with respect to the variables $x_1, x_2, ..., x_n$ and is denoted by $\dfrac{\partial(u_1, u_2, ..., u_n)}{\partial(x_1, x_2, ..., x_n)}$ or $J\left(\dfrac{u_1, u_2, ..., u_n}{x_1, x_2, ..., x_n}\right)$ or simply by J.

Example 4.2.1 *(i) In polar coordinates, $x = r\cos\theta$, $y = r\sin\theta$, show that* $\dfrac{\partial(x,y)}{\partial(r,\theta)} = r$; *(ii) In cylindrical coordinates $x = r\cos\phi$, $y = r\sin\phi$, $z = z$, show that* $\dfrac{\partial(x,y,z)}{\partial(r,\phi,z)} = r$; *(iii) In spherical polar coordinates $x = r\sin\theta\cos\phi$, $y = r\sin\theta\sin\phi$, $z = r\cos\theta$, show that* $\dfrac{\partial(x,y,z)}{\partial(r,\theta,\phi)} = r^2\sin\theta$.

Solution: (i) $\dfrac{\partial x}{\partial r} = \cos\theta$, $\dfrac{\partial x}{\partial\theta} = -r\sin\theta$ and $\dfrac{\partial y}{\partial r} = \sin\theta$, $\dfrac{\partial y}{\partial\theta} = -r\cos\theta$

$$\therefore\quad \frac{\partial(x,y)}{\partial(r,\theta)} = \begin{vmatrix} \dfrac{\partial x}{\partial r} & \dfrac{\partial x}{\partial\theta} \\[2mm] \dfrac{\partial y}{\partial r} & \dfrac{\partial y}{\partial\theta} \end{vmatrix} = \begin{vmatrix} \cos\theta & -r\sin\theta \\ \sin\theta & r\cos\theta \end{vmatrix} = r.$$

(ii) $\dfrac{\partial x}{\partial r} = \cos\phi$, $\dfrac{\partial x}{\partial\phi} = -r\sin\phi$, $\dfrac{\partial x}{\partial z} = 0$; $\dfrac{\partial y}{\partial r} = \sin\phi$, $\dfrac{\partial y}{\partial\phi} = r\cos\phi$, $\dfrac{\partial y}{\partial z} = 0$;

and $\dfrac{\partial z}{\partial r} = 0$, $\dfrac{\partial z}{\partial\phi} = 0$, $\dfrac{\partial z}{\partial z} = 1$.

$$\therefore\quad \frac{\partial(x,y,z)}{\partial(r,\phi,z)} = \begin{vmatrix} \dfrac{\partial x}{\partial r} & \dfrac{\partial x}{\partial\phi} & \dfrac{\partial x}{\partial z} \\[2mm] \dfrac{\partial y}{\partial r} & \dfrac{\partial y}{\partial\phi} & \dfrac{\partial y}{\partial z} \\[2mm] \dfrac{\partial z}{\partial r} & \dfrac{\partial y}{\partial\phi} & \dfrac{\partial z}{\partial z} \end{vmatrix} = \begin{vmatrix} \cos\phi & -r\sin\phi & 0 \\ \sin\phi & r\cos\phi & 0 \\ 0 & 0 & 1 \end{vmatrix} = r.$$

(iii) $\dfrac{\partial x}{\partial r} = \sin\theta\cos\phi$, $\dfrac{\partial x}{\partial\theta} = r\cos\theta\cos\phi$, $\dfrac{\partial x}{\partial\phi} = -r\sin\theta\sin\phi$; $\dfrac{\partial y}{\partial r} = \sin\theta\sin\phi$, $\dfrac{\partial y}{\partial\theta} = r\cos\theta\sin\phi$, $\dfrac{\partial y}{\partial\phi} = r\sin\theta\cos\phi$; and $\dfrac{\partial z}{\partial r} = \cos\theta$, $\dfrac{\partial z}{\partial\theta} = -r\sin\theta$, $\dfrac{\partial z}{\partial\phi} = 0$.

$$\therefore\quad \frac{\partial(x,y,z)}{\partial(r,\theta,\phi)} = \begin{vmatrix} \dfrac{\partial x}{\partial r} & \dfrac{\partial x}{\partial\theta} & \dfrac{\partial x}{\partial\phi} \\[2mm] \dfrac{\partial y}{\partial r} & \dfrac{\partial y}{\partial\theta} & \dfrac{\partial y}{\partial\phi} \\[2mm] \dfrac{\partial z}{\partial r} & \dfrac{\partial y}{\partial\theta} & \dfrac{\partial z}{\partial\phi} \end{vmatrix} = \begin{vmatrix} \sin\theta\cos\phi & r\cos\theta\cos\phi & -r\sin\theta\sin\phi \\ \sin\theta\sin\phi & r\cos\theta\sin\phi & r\sin\theta\cos\phi \\ \cos\theta & -r\sin\theta & 0 \end{vmatrix}$$

$$= r^2\sin\theta. \qquad\qquad \square$$

Example 4.2.2 *If $u_1 = \cos x_1$, $u_2 = \sin x_1\cos x_2$, $u_3 = \sin x_1\sin x_2\cos x_3$, ..., $u_n = \sin x_1\sin x_2...\sin x_{n-1}\cos x_n$, find* $\dfrac{\partial(u_1,u_2,...,u_n)}{\partial(x_1,x_2,...,x_n)}$.

Solution: We have $\dfrac{\partial(u_1,u_2,...,u_n)}{\partial(x_1,x_2,...,x_n)} = \begin{vmatrix} \dfrac{\partial u_1}{\partial x_1} & \dfrac{\partial u_1}{\partial x_2} & \cdots & \dfrac{\partial u_1}{\partial x_n} \\[2mm] \dfrac{\partial u_2}{\partial x_1} & \dfrac{\partial u_2}{\partial x_2} & \cdots & \dfrac{\partial u_2}{\partial x_n} \\[2mm] \cdots & \cdots & & \cdots \\[2mm] \dfrac{\partial u_n}{\partial x_1} & \dfrac{\partial u_n}{\partial x_2} & \cdots & \dfrac{\partial u_n}{\partial x_n} \end{vmatrix}$

$$
= \begin{vmatrix}
-\sin x_1 & 0 & 0 \\
\cos x_1 \cos x_2 & -\sin x_1 \sin x_2 & 0 \\
\cos x_1 \sin x_2 \cos x_3 & \sin x_1 \cos x_2 \cos x_3 & -\sin x_1 \cos x_2 \cos x_3 \\
\cdots & \cdots & \cdots \\
& & \cdots & 0 \\
& & \cdots & 0 \\
& & \cdots & 0 \\
& & \cdots & -\sin x_1 \sin x_2 \ldots \sin x_{n-1} \sin x_n
\end{vmatrix}
$$

$$
= (-\sin x_1)(-\sin x_1 \sin x_2)(-\sin x_1 \sin x_2 \sin x_3)
$$
$$
\ldots (-\sin x_1 \sin x_2 \ldots \sin x_{n-1} \sin x_n)
$$

$$
= (-1)^n \sin^n x_1 \sin^{n-1} x_2 \sin^{n-2} x_3 \ldots \sin x_n. \qquad \square
$$

Note 4.2.1 *If* $u_1 = f_1(x_1), u_2 = f_2(x_1, x_2), \ldots, u_n = f_n(x_1, x_2, \ldots, x_n),$ *then*
$\dfrac{\partial(u_1, u_2, \ldots, u_n)}{\partial(x_1, x_2, \ldots, x_n)}$ *will always be an upper or lower triangular determinant whose*
diagonal elements are $\dfrac{\partial u_1}{\partial x_1}, \dfrac{\partial u_2}{\partial x_2}, \ldots, \dfrac{\partial u_n}{\partial x_n}.$ *Hence*

$$
\frac{\partial(u_1, u_2, \ldots, u_n)}{\partial(x_1, x_2, \ldots, x_n)} = \frac{\partial u_1}{\partial x_1} \cdot \frac{\partial u_2}{\partial x_2} \cdots \frac{\partial u_n}{\partial x_n}.
$$

4.3 Jacobian of Implicit Functions

Let the functions $u_1, u_2, \ldots, u_n$, instead of being given explicitly in terms of $x_1, x_2, \ldots, x_n$ be connected as follows

$$
\left.
\begin{aligned}
f_1(u_1, u_2, \ldots, u_n, x_1, x_2, \ldots, x_n) &= 0 \\
f_2(u_1, u_2, \ldots, u_n, x_1, x_2, \ldots, x_n) &= 0 \\
\cdots \quad \cdots \quad \cdots & \quad \cdots \quad \cdots \\
f_n(u_1, u_2, \ldots, u_n, x_1, x_2, \ldots, x_n) &= 0
\end{aligned}
\right\}
$$

Then differentiating partially with respect to $x_1, x_2, \ldots, x_n$ respectively, we get

$$
\left.
\begin{aligned}
\frac{\partial f_1}{\partial x_1} + \frac{\partial f_1}{\partial u_1} \cdot \frac{\partial u_1}{\partial x_1} + \frac{\partial f_1}{\partial u_2} \cdot \frac{\partial u_2}{\partial x_1} + \ldots + \frac{\partial f_1}{\partial u_n} \cdot \frac{\partial u_n}{\partial x_1} &= 0 \\
\frac{\partial f_1}{\partial x_2} + \frac{\partial f_1}{\partial u_1} \cdot \frac{\partial u_1}{\partial x_2} + \frac{\partial f_1}{\partial u_2} \cdot \frac{\partial u_2}{\partial x_2} + \ldots + \frac{\partial f_1}{\partial u_n} \cdot \frac{\partial u_n}{\partial x_2} &= 0 \\
\cdots \quad \cdots \quad \cdots & \quad \cdots \quad \cdots \\
\frac{\partial f_1}{\partial x_n} + \frac{\partial f_1}{\partial u_1} \cdot \frac{\partial u_1}{\partial x_n} + \frac{\partial f_1}{\partial u_2} \cdot \frac{\partial u_2}{\partial x_n} + \ldots + \frac{\partial f_1}{\partial u_n} \cdot \frac{\partial u_n}{\partial x_n} &= 0
\end{aligned}
\right\}
$$

$$
\left.
\begin{aligned}
\frac{\partial f_2}{\partial x_1} + \frac{\partial f_2}{\partial u_1} \cdot \frac{\partial u_1}{\partial x_1} + \frac{\partial f_2}{\partial u_2} \cdot \frac{\partial u_2}{\partial x_1} + \ldots + \frac{\partial f_2}{\partial u_n} \cdot \frac{\partial u_n}{\partial x_1} &= 0 \\
\cdots \quad \cdots \quad \cdots & \quad \cdots \quad \cdots
\end{aligned}
\right\}
$$

$$
\cdots \quad \cdots \quad \cdots \qquad \cdots \quad \cdots
$$

$$
\left.
\begin{aligned}
\frac{\partial f_n}{\partial x_n} + \frac{\partial f_n}{\partial u_1} \cdot \frac{\partial u_1}{\partial x_n} + \frac{\partial f_n}{\partial u_2} \cdot \frac{\partial u_2}{\partial x_n} + \ldots + \frac{\partial f_n}{\partial u_n} \cdot \frac{\partial u_n}{\partial x_n} &= 0
\end{aligned}
\right\}
$$

(4.1)

Now $\dfrac{\partial(f_1, f_2, ..., f_n)}{\partial(u_1, u_2, ..., u_n)} \times \dfrac{\partial(u_1, u_2, ..., u_n)}{\partial(x_1, x_2, ..., x_n)}$

$$= \begin{vmatrix} \dfrac{\partial f_1}{\partial u_1} & \dfrac{\partial f_1}{\partial u_2} & \cdots & \dfrac{\partial f_1}{\partial u_n} \\[2mm] \dfrac{\partial f_2}{\partial u_1} & \dfrac{\partial f_2}{\partial u_2} & \cdots & \dfrac{\partial f_2}{\partial u_n} \\[2mm] \cdots & \cdots & \cdots & \cdots \\[2mm] \dfrac{\partial f_n}{\partial u_1} & \dfrac{\partial f_n}{\partial u_2} & \cdots & \dfrac{\partial f_n}{\partial u_n} \end{vmatrix} \times \begin{vmatrix} \dfrac{\partial u_1}{\partial x_1} & \dfrac{\partial u_1}{\partial x_2} & \cdots & \dfrac{\partial u_1}{\partial x_n} \\[2mm] \dfrac{\partial u_2}{\partial x_1} & \dfrac{\partial u_2}{\partial x_2} & \cdots & \dfrac{\partial u_2}{\partial x_n} \\[2mm] \cdots & \cdots & \cdots & \cdots \\[2mm] \dfrac{\partial u_n}{\partial x_1} & \dfrac{\partial u_n}{\partial x_2} & \cdots & \dfrac{\partial u_n}{\partial x_n} \end{vmatrix}$$

$$= \begin{vmatrix} \sum\limits_{k=1}^{n} \dfrac{\partial f_1}{\partial u_k}\dfrac{\partial u_k}{\partial x_1} & \sum\limits_{k=1}^{n} \dfrac{\partial f_1}{\partial u_k}\dfrac{\partial u_k}{\partial x_2} & \cdots & \sum\limits_{k=1}^{n} \dfrac{\partial f_1}{\partial u_k}\dfrac{\partial u_k}{\partial x_n} \\[4mm] \sum\limits_{k=1}^{n} \dfrac{\partial f_2}{\partial u_k}\dfrac{\partial u_k}{\partial x_1} & \sum\limits_{k=1}^{n} \dfrac{\partial f_2}{\partial u_k}\dfrac{\partial u_k}{\partial x_2} & \cdots & \sum\limits_{k=1}^{n} \dfrac{\partial f_2}{\partial u_k}\dfrac{\partial u_k}{\partial x_n} \\[4mm] \cdots & \cdots & & \cdots \\[4mm] \sum\limits_{k=1}^{n} \dfrac{\partial f_n}{\partial u_k}\dfrac{\partial u_k}{\partial x_1} & \sum\limits_{k=1}^{n} \dfrac{\partial f_n}{\partial u_k}\dfrac{\partial u_k}{\partial x_2} & \cdots & \sum\limits_{k=1}^{n} \dfrac{\partial f_n}{\partial u_k}\dfrac{\partial u_k}{\partial x_n} \end{vmatrix}$$

(performing 'row by column' rule of multiplication of determinants)

$$= \begin{vmatrix} -\dfrac{\partial f_1}{\partial x_1} & -\dfrac{\partial f_1}{\partial x_2} & \cdots & -\dfrac{\partial f_1}{\partial x_n} \\[2mm] -\dfrac{\partial f_2}{\partial x_1} & -\dfrac{\partial f_2}{\partial x_2} & \cdots & -\dfrac{\partial f_2}{\partial x_n} \\[2mm] \cdots & \cdots & & \cdots \\[2mm] -\dfrac{\partial f_n}{\partial x_1} & -\dfrac{\partial f_n}{\partial x_2} & \cdots & -\dfrac{\partial f_n}{\partial x_n} \end{vmatrix} \quad \text{(Using relation (4.1))}$$

$$= (-1)^n \dfrac{\partial(f_1, f_2, ..., f_n)}{\partial(x_1, x_2, ..., x_n)}$$

$$\therefore \quad \dfrac{\partial(u_1, u_2, ..., u_n)}{\partial(x_1, x_2, ..., x_n)} = \dfrac{(-1)^n \frac{\partial(f_1, f_2, ..., f_n)}{\partial(x_1, x_2, ..., x_n)}}{\frac{\partial(f_1, f_2, ..., f_n)}{\partial(u_1, u_2, ..., u_n)}}.$$

4.4   Some Properties of Jacobians

Jacobians have the remarkable properties of behaving like the derivatives of functions of one variable. It also satisfies the chain rule of derivatives, which can be stated as follows:

Property 1: If $u_1, u_2, ..., u_n$ are functions of $y_1, y_2, ..., y_n$ and $y_1, y_2, ..., y_n$ are functions of $x_1, x_2, ..., x_n$, then

$$\dfrac{\partial(u_1, u_2, ..., u_n)}{\partial(x_1, x_2, ..., x_n)} = \dfrac{\partial(u_1, u_2, ..., u_n)}{\partial(y_1, y_2, ..., y_n)} \times \dfrac{\partial(y_1, y_2, ..., y_n)}{\partial(x_1, x_2, ..., x_n)}.$$

Proof: Using chain rule, we have

$$\dfrac{\partial u_1}{\partial x_1} = \dfrac{\partial u_1}{\partial y_1}\cdot\dfrac{\partial y_1}{\partial x_1} + \dfrac{\partial u_1}{\partial y_2}\cdot\dfrac{\partial y_2}{\partial x_1} + ... + \dfrac{\partial u_1}{\partial y_n}\cdot\dfrac{\partial y_n}{\partial x_1} = \sum_{k=1}^{n} \dfrac{\partial u_1}{\partial y_k}\dfrac{\partial y_k}{\partial x_1}$$

$$\frac{\partial u_1}{\partial x_2} = \frac{\partial u_1}{\partial y_1}\cdot\frac{\partial y_1}{\partial x_2} + \frac{\partial u_1}{\partial y_2}\cdot\frac{\partial y_2}{\partial x_2} + ... + \frac{\partial u_1}{\partial y_n}\cdot\frac{\partial y_n}{\partial x_2} \quad = \quad \sum_{k=1}^{n}\frac{\partial u_1}{\partial y_k}\frac{\partial y_k}{\partial x_2}$$

$$... \quad ... \quad ... \quad\quad\quad ... \quad\quad\quad ...$$

$$\frac{\partial u_1}{\partial x_n} = \frac{\partial u_1}{\partial y_1}\cdot\frac{\partial y_1}{\partial x_n} + \frac{\partial u_1}{\partial y_2}\cdot\frac{\partial y_2}{\partial x_n} + ... + \frac{\partial u_1}{\partial y_n}\cdot\frac{\partial y_n}{\partial x_n} \quad = \quad \sum_{k=1}^{n}\frac{\partial u_1}{\partial y_k}\frac{\partial y_k}{\partial x_n}$$

and similarly

$$\frac{\partial u_2}{\partial x_1} = \sum_{k=1}^{n}\frac{\partial u_2}{\partial y_k}\frac{\partial y_k}{\partial x_1},\ \frac{\partial u_2}{\partial x_1} = \sum_{k=1}^{n}\frac{\partial u_2}{\partial y_k}\frac{\partial y_k}{\partial x_1}, ..., \ \frac{\partial u_2}{\partial x_1} = \sum_{k=1}^{n}\frac{\partial u_2}{\partial y_k}\frac{\partial y_k}{\partial x_1}.$$

and so on.

In general, we get

$$\frac{\partial u_p}{\partial x_q} = \sum_{k=1}^{n}\frac{\partial u_p}{\partial y_k}\frac{\partial y_k}{\partial x_q}; \quad \text{for } p, q = 1, 2, ..., n \tag{4.2}$$

Now

$$\frac{\partial(u_1, u_2, ..., u_n)}{\partial(y_1, y_2, ..., y_n)} \times \frac{\partial(y_1, y_2, ..., y_n)}{\partial(x_1, x_2, ..., x_n)}$$

$$= \begin{vmatrix} \dfrac{\partial u_1}{\partial y_1} & \dfrac{\partial u_1}{\partial y_2} & \cdots & \dfrac{\partial u_1}{\partial y_n} \\ \dfrac{\partial u_2}{\partial y_1} & \dfrac{\partial u_2}{\partial y_2} & \cdots & \dfrac{\partial u_2}{\partial y_n} \\ \cdots & \cdots & & \cdots \\ \dfrac{\partial u_n}{\partial y_1} & \dfrac{\partial u_n}{\partial y_2} & \cdots & \dfrac{\partial u_n}{\partial y_n} \end{vmatrix} \times \begin{vmatrix} \dfrac{\partial y_1}{\partial x_1} & \dfrac{\partial y_1}{\partial x_2} & \cdots & \dfrac{\partial y_1}{\partial x_n} \\ \dfrac{\partial y_2}{\partial x_1} & \dfrac{\partial y_2}{\partial x_2} & \cdots & \dfrac{\partial y_2}{\partial x_n} \\ \cdots & \cdots & & \cdots \\ \dfrac{\partial y_n}{\partial x_1} & \dfrac{\partial y_n}{\partial x_2} & \cdots & \dfrac{\partial y_n}{\partial x_n} \end{vmatrix}$$

$$= \begin{vmatrix} \displaystyle\sum_{k=1}^{n}\frac{\partial u_1}{\partial y_k}\frac{\partial y_k}{\partial x_1} & \displaystyle\sum_{k=1}^{n}\frac{\partial u_1}{\partial y_k}\frac{\partial y_k}{\partial x_2} & \cdots & \displaystyle\sum_{k=1}^{n}\frac{\partial u_1}{\partial y_k}\frac{\partial y_k}{\partial x_n} \\ \displaystyle\sum_{k=1}^{n}\frac{\partial u_2}{\partial y_k}\frac{\partial y_k}{\partial x_1} & \displaystyle\sum_{k=1}^{n}\frac{\partial u_2}{\partial y_k}\frac{\partial y_k}{\partial x_2} & \cdots & \displaystyle\sum_{k=1}^{n}\frac{\partial u_2}{\partial y_k}\frac{\partial y_k}{\partial x_n} \\ \cdots & \cdots & & \cdots \\ \displaystyle\sum_{k=1}^{n}\frac{\partial u_n}{\partial y_k}\frac{\partial y_k}{\partial x_1} & \displaystyle\sum_{k=1}^{n}\frac{\partial u_n}{\partial y_k}\frac{\partial y_k}{\partial x_2} & \cdots & \displaystyle\sum_{k=1}^{n}\frac{\partial u_n}{\partial y_k}\frac{\partial y_k}{\partial x_n} \end{vmatrix}$$

(performing 'row by column' rule of multiplication of determinants)

$$= \begin{vmatrix} \dfrac{\partial u_1}{\partial x_1} & \dfrac{\partial u_1}{\partial x_2} & \cdots & \dfrac{\partial u_1}{\partial x_n} \\ \dfrac{\partial u_2}{\partial x_1} & \dfrac{\partial u_2}{\partial x_2} & \cdots & \dfrac{\partial u_2}{\partial x_n} \\ \cdots & \cdots & & \cdots \\ \dfrac{\partial u_n}{\partial x_1} & \dfrac{\partial u_n}{\partial x_2} & \cdots & \dfrac{\partial u_n}{\partial x_n} \end{vmatrix} \quad \text{(Using relation (4.1))}$$

$$= \frac{\partial(u_1, u_2, ..., u_n)}{\partial(x_1, x_2, ..., x_n)}. \qquad\qquad \square$$

Corollary 4.4.1 *Let* $x_r = u_r$, $r = 1, 2, ..., n$. *Then assuming the existence of inverse functions* $x_1, x_2, ..., x_n$, *i.e., assuming that the equations which define* y_1, y_2, *...,* y_n *as functions of* $x_1, x_2, ..., x_n$ *determine* $x_1, x_2, ..., x_n$ *as functions of* y_1, y_2, *...,* y_n *and noting that* $\dfrac{\partial x_i}{\partial x_j} = 1$ *for* $i = j$, $\dfrac{\partial x_i}{\partial x_j} = 0$, *for* $i \neq j$ *we find that*

$$\frac{\partial(x_1, x_2, ..., x_n)}{\partial(y_1, y_2, ..., y_n)} \cdot \frac{\partial(y_1, y_2, ..., y_n)}{\partial(x_1, x_2, ..., x_n)} = \frac{\partial(x_1, x_2, ..., x_n)}{\partial(x_1, x_2, ..., x_n)} = 1.$$

An observation: *We observe that* **Property 1** *is a particular case of* § 4.3, *where the relations* $f_1, f_2, ..., f_n$ *are to be just replaced by* $y_1, y_2, ..., y_n$.

Property 2: If $u_{m+1}, u_{m+2}, ..., u_n$ are constants with respect to $x_1, x_2, ..., x_m$ then

$$\frac{\partial(u_1, u_2, ..., u_m, ..., u_n)}{\partial(x_1, x_2, ..., x_m, ..., x_n)} = \frac{\partial(u_1, u_2, ..., u_m)}{\partial(x_1, x_2, ..., x_m)} \times \frac{\partial(u_{m+1}, u_{m+2}, ..., u_n)}{\partial(x_{m+1}, x_{m+2}, ..., x_n)}.$$

Proof: Here $\dfrac{\partial u_r}{\partial x_s} = 0$ for $r = m+1, m+2, ..., n$ and $s = 1, 2, ..., m$.

$$\therefore \quad \frac{\partial(u_1, u_2, ..., u_m, ..., u_n)}{\partial(x_1, x_2, ..., x_m, ..., x_n)}$$

$$= \begin{vmatrix} \dfrac{\partial u_1}{\partial x_1} & \dfrac{\partial u_1}{\partial x_2} & \cdots & \dfrac{\partial u_1}{\partial x_m} & \dfrac{\partial u_1}{\partial x_{m+1}} & \cdots & \dfrac{\partial u_1}{\partial x_n} \\ \dfrac{\partial u_2}{\partial x_1} & \dfrac{\partial u_2}{\partial x_2} & \cdots & \dfrac{\partial u_2}{\partial x_m} & \dfrac{\partial u_2}{\partial x_{m+1}} & \cdots & \dfrac{\partial u_2}{\partial x_n} \\ \cdots & \cdots & \cdots & \cdots & \cdots & \cdots & \cdots \\ \dfrac{\partial u_m}{\partial x_1} & \dfrac{\partial u_m}{\partial x_2} & \cdots & \dfrac{\partial u_m}{\partial x_m} & \dfrac{\partial u_m}{\partial x_{m+1}} & \cdots & \dfrac{\partial u_m}{\partial x_n} \\ \dfrac{\partial u_{m+1}}{\partial x_1} & \dfrac{\partial u_{m+1}}{\partial x_2} & \cdots & \dfrac{\partial u_{m+1}}{\partial x_m} & \dfrac{\partial u_{m+1}}{\partial x_{m+1}} & \cdots & \dfrac{\partial u_{m+1}}{\partial x_n} \\ \cdots & \cdots & \cdots & \cdots & \cdots & \cdots & \cdots \\ \dfrac{\partial u_n}{\partial x_1} & \dfrac{\partial u_n}{\partial x_2} & \cdots & \dfrac{\partial u_n}{\partial x_m} & \dfrac{\partial u_n}{\partial x_{m+1}} & \cdots & \dfrac{\partial u_n}{\partial x_n} \end{vmatrix}$$

$$= \begin{vmatrix} \dfrac{\partial u_1}{\partial x_1} & \dfrac{\partial u_1}{\partial x_2} & \cdots & \dfrac{\partial u_1}{\partial x_m} & \dfrac{\partial u_1}{\partial x_{m+1}} & \cdots & \dfrac{\partial u_1}{\partial x_n} \\ \dfrac{\partial u_2}{\partial x_1} & \dfrac{\partial u_2}{\partial x_2} & \cdots & \dfrac{\partial u_2}{\partial x_m} & \dfrac{\partial u_2}{\partial x_{m+1}} & \cdots & \dfrac{\partial u_2}{\partial x_n} \\ \cdots & \cdots & \cdots & \cdots & \cdots & \cdots & \cdots \\ \dfrac{\partial u_m}{\partial x_1} & \dfrac{\partial u_m}{\partial x_2} & \cdots & \dfrac{\partial u_m}{\partial x_m} & \dfrac{\partial u_m}{\partial x_{m+1}} & \cdots & \dfrac{\partial u_m}{\partial x_n} \\ 0 & 0 & \cdots & 0 & \dfrac{\partial u_{m+1}}{\partial x_{m+1}} & \cdots & \dfrac{\partial u_{m+1}}{\partial x_n} \\ \cdots & \cdots & \cdots & \cdots & \cdots & \cdots & \cdots \\ 0 & 0 & \cdots & 0 & \dfrac{\partial u_n}{\partial x_{m+1}} & \cdots & \dfrac{\partial u_n}{\partial x_n} \end{vmatrix}$$

$$= \begin{vmatrix} \dfrac{\partial u_1}{\partial x_1} & \dfrac{\partial u_1}{\partial x_2} & \cdots & \dfrac{\partial u_1}{\partial x_m} \\[2mm] \dfrac{\partial u_2}{\partial x_1} & \dfrac{\partial u_2}{\partial x_2} & \cdots & \dfrac{\partial u_2}{\partial x_m} \\[1mm] \cdots & \cdots & & \cdots \\[1mm] \dfrac{\partial u_m}{\partial x_1} & \dfrac{\partial u_m}{\partial x_2} & \cdots & \dfrac{\partial u_m}{\partial x_m} \end{vmatrix} \times \begin{vmatrix} \dfrac{\partial u_{m+1}}{\partial x_{m+1}} & \dfrac{\partial u_{m+1}}{\partial x_{m+2}} & \cdots & \dfrac{\partial u_{m+1}}{\partial x_n} \\[2mm] \cdots & \cdots & & \cdots \\[1mm] \dfrac{\partial u_n}{\partial x_{m+1}} & \dfrac{\partial u_n}{\partial x_{m+2}} & \cdots & \dfrac{\partial u_n}{\partial x_n} \end{vmatrix}$$

$$= \frac{\partial(u_1, u_2, ..., u_m)}{\partial(x_1, x_2, ..., x_m)} \times \frac{\partial(u_{m+1}, u_{m+2}, ..., u_n)}{\partial(x_{m+1}, x_{m+2}, ..., x_n)}. \qquad \square$$

Note 4.4.1 *If instead of $u_{m+1}, u_{m+2}, ..., u_n$ are constants with respect to $x_1, x_2, ..., x_m$, it is given that $u_1, u_2, ..., u_m$ are constants with respect to $x_{m+1}, x_{m+2}, ..., x_n$, then also the above result remains true.*

Corollary 4.4.2 *In particular, we get*

$$\frac{\partial(u_1, u_2, ..., u_m, x_{m+1}, ..., x_n)}{\partial(x_1, x_2, ..., x_m, x_{m+1}, ..., x_n)} = \frac{\partial(u_1, u_2, ..., u_m)}{\partial(x_1, x_2, ..., x_m)}.$$

Example 4.4.1 *If $J = \dfrac{\partial(u, v)}{\partial(x, y)}$ and $J' = \dfrac{\partial(x, y)}{\partial(u, v)}$, then prove that $JJ' = 1$.*

The result is stated for two variables. Verify whether the result is true for n variables also.

Solution: Let $u = f(x, y)$ and $v = g(x, y)$.

Suppose, on solving for x and y we get $x = \phi(u, v)$, $y = \psi(u, v)$. Then

$$\left. \begin{array}{rcl} \dfrac{\partial u}{\partial u} = 1 & = & \dfrac{\partial u}{\partial x} \cdot \dfrac{\partial x}{\partial u} + \dfrac{\partial u}{\partial y} \cdot \dfrac{\partial y}{\partial u} \\[3mm] \dfrac{\partial u}{\partial v} = 0 & = & \dfrac{\partial u}{\partial x} \cdot \dfrac{\partial x}{\partial v} + \dfrac{\partial u}{\partial y} \cdot \dfrac{\partial y}{\partial v} \\[3mm] \dfrac{\partial v}{\partial u} = 0 & = & \dfrac{\partial v}{\partial x} \cdot \dfrac{\partial x}{\partial u} + \dfrac{\partial v}{\partial y} \cdot \dfrac{\partial y}{\partial u} \\[3mm] \dfrac{\partial v}{\partial v} = 1 & = & \dfrac{\partial v}{\partial x} \cdot \dfrac{\partial x}{\partial v} + \dfrac{\partial v}{\partial y} \cdot \dfrac{\partial y}{\partial v} \end{array} \right\} \qquad (4.3)$$

$$\therefore \ JJ' = \begin{vmatrix} \dfrac{\partial u}{\partial x} & \dfrac{\partial u}{\partial y} \\[3mm] \dfrac{\partial v}{\partial x} & \dfrac{\partial v}{\partial y} \end{vmatrix} \times \begin{vmatrix} \dfrac{\partial x}{\partial u} & \dfrac{\partial x}{\partial v} \\[3mm] \dfrac{\partial y}{\partial u} & \dfrac{\partial y}{\partial v} \end{vmatrix}$$

$$= \begin{vmatrix} \dfrac{\partial u}{\partial x} \cdot \dfrac{\partial x}{\partial u} + \dfrac{\partial u}{\partial y} \cdot \dfrac{\partial y}{\partial u} & \dfrac{\partial u}{\partial x} \cdot \dfrac{\partial x}{\partial v} + \dfrac{\partial u}{\partial y} \cdot \dfrac{\partial y}{\partial v} \\[3mm] \dfrac{\partial v}{\partial x} \cdot \dfrac{\partial x}{\partial u} + \dfrac{\partial v}{\partial y} \cdot \dfrac{\partial y}{\partial u} & \dfrac{\partial v}{\partial x} \cdot \dfrac{\partial x}{\partial v} + \dfrac{\partial v}{\partial y} \cdot \dfrac{\partial y}{\partial v} \end{vmatrix}$$

(performing 'row by column' rule of multiplication of determinant.)

$$= \begin{vmatrix} 1 & 0 \\ 0 & 1 \end{vmatrix} = 1 \ \text{(using (4.3))}$$

2nd Part: Let $u_1 = f_1(x_1, x_2, ..., x_n)$, $u_2 = f_2(x_1, x_2, ..., x_n)$, ..., $u_n = f_n(x_1, x_2, ..., x_n)$ and $x_1 = g_1(u_1, u_2, ..., u_n)$, $x_2 = g_2(u_1, u_2, ..., u_n)$, ..., $x_n = g_n(u_1, u_2, ..., u_n)$. Then

$$\frac{\partial u_1}{\partial u_1} = 1 = \frac{\partial u_1}{\partial x_1}\cdot\frac{\partial x_1}{\partial u_1} + \frac{\partial u_1}{\partial x_2}\cdot\frac{\partial x_2}{\partial u_1} + ... + \frac{\partial u_1}{\partial x_n}\cdot\frac{\partial x_n}{\partial u_1} = \sum_{k=1}^{n}\frac{\partial u_1}{\partial x_k}\frac{\partial x_k}{\partial u_1}$$

$$\frac{\partial u_1}{\partial u_2} = 0 = \frac{\partial u_1}{\partial x_1}\cdot\frac{\partial x_1}{\partial u_2} + \frac{\partial u_1}{\partial x_2}\cdot\frac{\partial x_2}{\partial u_2} + ... + \frac{\partial u_1}{\partial x_n}\cdot\frac{\partial x_n}{\partial u_2} = \sum_{k=1}^{n}\frac{\partial u_1}{\partial x_k}\frac{\partial x_k}{\partial u_2}$$

$$... \qquad ... \qquad ... \qquad ...$$
$$... \qquad ... \qquad ... \qquad ...$$

and so on.

In general,

$$\sum_{k=1}^{n}\frac{\partial u_1}{\partial x_k}\frac{\partial x_k}{\partial u_q} = \begin{cases} 1 & \text{if} \quad p=q \\ 0 & \text{if} \quad p\neq q \end{cases}, \quad \text{for} \quad p, q = 1, 2, ..., n. \tag{4.4}$$

Now

$$\frac{\partial(u_1, u_2, ..., u_n)}{\partial(x_1, x_2, ..., x_n)} \times \frac{\partial(x_1, x_2, ..., x_n)}{\partial(u_1, u_2, ..., u_n)}$$

$$= \begin{vmatrix} \frac{\partial u_1}{\partial x_1} & \frac{\partial u_1}{\partial x_2} & \cdots & \frac{\partial u_1}{\partial x_n} \\ \frac{\partial u_2}{\partial x_1} & \frac{\partial u_2}{\partial x_2} & \cdots & \frac{\partial u_2}{\partial x_n} \\ \cdots & \cdots & \cdots & \cdots \\ \frac{\partial u_n}{\partial x_1} & \frac{\partial u_n}{\partial x_2} & \cdots & \frac{\partial u_n}{\partial x_n} \end{vmatrix} \times \begin{vmatrix} \frac{\partial x_1}{\partial u_1} & \frac{\partial x_1}{\partial u_2} & \cdots & \frac{\partial x_1}{\partial u_n} \\ \frac{\partial x_2}{\partial u_1} & \frac{\partial x_2}{\partial u_2} & \cdots & \frac{\partial x_2}{\partial u_n} \\ \cdots & \cdots & \cdots & \cdots \\ \frac{\partial x_n}{\partial u_1} & \frac{\partial x_n}{\partial u_2} & \cdots & \frac{\partial x_n}{\partial u_n} \end{vmatrix}$$

$$= \begin{vmatrix} \sum_{k=1}^{n}\frac{\partial u_1}{\partial x_k}\frac{\partial x_k}{\partial u_1} & \sum_{k=1}^{n}\frac{\partial u_1}{\partial x_k}\frac{\partial x_k}{\partial u_2} & \cdots & \sum_{k=1}^{n}\frac{\partial u_1}{\partial x_k}\frac{\partial x_k}{\partial u_n} \\ \sum_{k=1}^{n}\frac{\partial u_2}{\partial x_k}\frac{\partial x_k}{\partial u_1} & \sum_{k=1}^{n}\frac{\partial u_2}{\partial x_k}\frac{\partial x_k}{\partial u_2} & \cdots & \sum_{k=1}^{n}\frac{\partial u_2}{\partial x_k}\frac{\partial x_k}{\partial u_n} \\ \cdots & \cdots & & \cdots \\ \sum_{k=1}^{n}\frac{\partial u_n}{\partial x_k}\frac{\partial x_k}{\partial u_1} & \sum_{k=1}^{n}\frac{\partial u_n}{\partial x_k}\frac{\partial x_k}{\partial u_2} & \cdots & \sum_{k=1}^{n}\frac{\partial u_n}{\partial x_k}\frac{\partial x_k}{\partial u_n} \end{vmatrix}$$

(performing 'row by column' rule of multiplication of determinants)

$$= \begin{vmatrix} 1 & 0 & \cdots & 0 \\ 0 & 1 & \cdots & 0 \\ \cdots & \cdots & & \cdots \\ 0 & 0 & \cdots & 1 \end{vmatrix} = 1 \quad \text{(Using relation (4.1))}$$

Therefore, if we take $J = \dfrac{\partial(u_1, u_2, ..., u_n)}{\partial(x_1, x_2, ..., x_n)}$ and $J' = \dfrac{\partial(x_1, x_2, ..., x_n)}{\partial(u_1, u_2, ..., u_n)}$, then we get

$JJ' = 1$ and so the given result is generalized for any number of variables, i.e., the result is true for n variables also. $\qquad \square$

Remark: This problem is a special case of **Property 1** of § 4.4, where the existence of inverse is necessary, which has been discussed in the **Corollary 4.4.1**.

Example 4.4.2 *If u, v, w are the roots of the equation*

$$(\lambda - x)^3 + (\lambda - y)^3 + (\lambda - z)^3 = 0$$

in λ, prove that $\dfrac{\partial(u, v, w)}{\partial(x, y, z)} = -2\dfrac{(y - z)(z - x)(x - y)}{(v - w)(w - u)(u - v)}$.

Solution: Here u, v, w are the roots of the equation

$$\lambda^3 - (x + y + z)\lambda^2 + (x^2 + y^2 + z^2)\lambda - \frac{1}{3}(x^3 + y^3 + z^3) = 0$$

Let $x + y + z = p$, $x^2 + y^2 + z^2 = q$ and $\frac{1}{3}(x^3 + y^3 + z^3) = r$ (4.5)

$$\therefore \quad \sum u = u + v + w = p$$
$$\sum uv = vw + wu + uv = x^2 + y^2 + z^2 = q$$
$$\text{and} \quad uvw = \frac{1}{3}(x^3 + y^3 + z^3) = r$$

$$u + v + w = p, \quad uv + wu + vw = q \text{ and } uvw = r \quad (4.6)$$

Hence from (4.5)

$$\frac{\partial(p, q, r)}{\partial(x, y, z)} = \begin{vmatrix} \dfrac{\partial p}{\partial x} & \dfrac{\partial p}{\partial y} & \dfrac{\partial p}{\partial z} \\[2mm] \dfrac{\partial q}{\partial x} & \dfrac{\partial q}{\partial y} & \dfrac{\partial q}{\partial z} \\[2mm] \dfrac{\partial r}{\partial x} & \dfrac{\partial r}{\partial y} & \dfrac{\partial r}{\partial z} \end{vmatrix} = \begin{vmatrix} 1 & 1 & 1 \\ 2x & 2y & 2z \\ x^2 & y^2 & z^2 \end{vmatrix} = 2(y - z)(z - x)(x - y) \quad (4.7)$$

and from (4.6)

$$\frac{\partial(p, q, r)}{\partial(u, v, w)} = \begin{vmatrix} \dfrac{\partial p}{\partial u} & \dfrac{\partial p}{\partial v} & \dfrac{\partial p}{\partial w} \\[2mm] \dfrac{\partial q}{\partial u} & \dfrac{\partial q}{\partial v} & \dfrac{\partial q}{\partial w} \\[2mm] \dfrac{\partial r}{\partial u} & \dfrac{\partial r}{\partial v} & \dfrac{\partial r}{\partial w} \end{vmatrix} = \begin{vmatrix} 1 & 1 & 1 \\ v + w & w + u & u + v \\ vw & wu & uv \end{vmatrix}$$

$$= -2(v - w)(w - u)(u - v) \quad (4.8)$$

Now using **Property 1** of § 4.4, we get

$$\frac{\partial(u, v, w)}{\partial(x, y, z)} = \frac{\partial(u, v, w)}{\partial(p, q, r)} \cdot \frac{\partial(p, q, r)}{\partial(x, y, z)} = \frac{\frac{\partial(p, q, r)}{\partial(x, y, z)}}{\frac{\partial(p, q, r)}{\partial(u, v, w)}}$$

$$= -2\frac{(y - z)(z - x)(x - y)}{(v - w)(w - u)(u - v)}. \qquad \square$$

4.5   Functional Dependence

Let x and y are both functions of a real variable t, defined in a common domain, such that $f(x(t), y(t)) = 0 \ \forall \ t$ in a neighbourhood of $t = \alpha$ (say), then we say that x and y are *functionally related* and the equation $f(x, y) = 0$ defines y as a function g of x, i.e., $y = g(x)$ in a certain neighbourhood of $x = a$, where $a = x(\alpha)$.

For example, let $f(x, y) = \dfrac{x^2}{a^2} + \dfrac{y^2}{b^2} - 1$.

If we take $x = a\cos t$, $y = b\sin t \ \forall \ t \in (-\pi, \pi)$, we find that $\dfrac{x^2}{a^2} + \dfrac{y^2}{b^2} - 1$
$= \cos^2 t + \sin^2 t - 1 = 0$.

Hence x and y are functionally related and that relation is given by

$$\frac{x^2}{a^2} + \frac{y^2}{b^2} = 1.$$

The equation $f(x, y) = 0$ defines y as a function of x in the neighbourhood of some point P, where $f(P) = 0$.

Similarly, if $u = u(x, y)$, $v = v(x, y)$ be two functions defined on an open region in $\mathbb{R}^2$, then u and v are said to be *functionally related* or *functionally dependent* if we get an identity of the form $f(u, v) = 0$ which does not contain x and y explicitly and which is satisfied by all the points (x, y) in some neighbourhood of a point in the region $\mathbb{R}^2$, i.e., $f(u, v) = 0$.

For example, if $u = \dfrac{e^x}{e^y}$ and $v = x - y \ \forall \ (x, y) \in \mathbb{R}^2$ and if $f(u, v) = u - e^v$, then we find that $f(u, v) = 0 \ \forall \ (x, y) \in \mathbb{R}^2$. Also $f(u, v)$ does not contain x and y explicitly. Hence u and v are functionally related, given by $u = e^v \ \forall \ v \in \mathbb{R}$.

The case is similar for n functions u_i of n real variables x_i, i.e., when the function $f(u_1, u_2, ..., u_n)$ vanishes identically and f does not contain x_i explicitly, then $u_1, u_2, ..., u_n$ are said to be *functionally dependent*.

If $u = x + y - z$, $v = x - y + z$ and $w = x^2 + y^2 + z^2 - 2yz$ and if $f(u, v, w) = u^2 + v^2 - 2w$, then it can be easily seen that $f(u, v, w) = 0$ and $f(u, v, w)$ does not contain any of the variables x, y, z explicitly. Therefore, x, y, z are *functionally dependent*.

Note 4.5.1 *It should be noted that the Jacobian*

$$J = \frac{\partial(u, v, w)}{\partial(x, y, z)} = \begin{vmatrix} \dfrac{\partial u}{\partial x} & \dfrac{\partial u}{\partial y} & \dfrac{\partial u}{\partial z} \\ \dfrac{\partial v}{\partial x} & \dfrac{\partial v}{\partial y} & \dfrac{\partial v}{\partial z} \\ \dfrac{\partial w}{\partial x} & \dfrac{\partial w}{\partial y} & \dfrac{\partial w}{\partial z} \end{vmatrix} = \begin{vmatrix} 1 & 1 & -1 \\ 1 & -1 & 1 \\ 2x & 2(y - z) & 2(z - x) \end{vmatrix}$$

$$= 0 \ (identically).$$

Above shows that the condition for three differentiable functions u, v, w of three independent variables x, y, z possessing continuous first order partial derivatives to be functionally dependent is that the Jacobian formed by them $J = \dfrac{\partial(u, v, w)}{\partial(x, y, z)}$ vanishes identically.

The above condition is also valid for n functions of n independent variables which can be stated and proved in the form of a theorem as follows:

Theorem 4.5.1 *If $u_1, u_2, ..., u_n$ are functions of n independent variables $x_1, x_2, ..., x_n$, then the necessary and sufficient condition that there exists a functional relation among the functions is that the Jacobian*

$$J = \frac{\partial(u_1, u_2, ..., u_n)}{\partial(x_1, x_2, ..., x_n)} = 0.$$

Proof: **Condition Necessary:**

Let there exists a relation of the form

$$f(u_1, u_2, ..., u_n) = 0 \tag{4.9}$$

Differentiating partially with respect to $x_1, x_2, ..., x_n$ respectively, we get

$$\left.\begin{array}{c}
\dfrac{\partial f}{\partial u_1} \cdot \dfrac{\partial u_1}{\partial x_1} + \dfrac{\partial f}{\partial u_2} \cdot \dfrac{\partial u_2}{\partial x_1} + ... + \dfrac{\partial f}{\partial u_n} \cdot \dfrac{\partial u_n}{\partial x_1} = 0 \\[2mm]
\dfrac{\partial f}{\partial u_1} \cdot \dfrac{\partial u_1}{\partial x_2} + \dfrac{\partial f}{\partial u_2} \cdot \dfrac{\partial u_2}{\partial x_2} + ... + \dfrac{\partial f}{\partial u_n} \cdot \dfrac{\partial u_n}{\partial x_2} = 0 \\[2mm]
... \quad ... \quad ... \\[2mm]
\dfrac{\partial f}{\partial u_1} \cdot \dfrac{\partial u_1}{\partial x_n} + \dfrac{\partial f}{\partial u_2} \cdot \dfrac{\partial u_2}{\partial x_n} + ... + \dfrac{\partial f}{\partial u_n} \cdot \dfrac{\partial u_n}{\partial x_n} = 0
\end{array}\right\}$$

Since $\dfrac{\partial f}{\partial u_1}, \dfrac{\partial f}{\partial u_2}, ..., \dfrac{\partial f}{\partial u_n}$ are not all zero, eliminating $\dfrac{\partial f}{\partial u_1}, \dfrac{\partial f}{\partial u_2}, ..., \dfrac{\partial f}{\partial u_n}$ from the above set of equations, we get

$$\begin{vmatrix}
\dfrac{\partial u_1}{\partial x_1} & \dfrac{\partial u_2}{\partial x_1} & \cdots & \dfrac{\partial u_n}{\partial x_1} \\[2mm]
\dfrac{\partial u_1}{\partial x_2} & \dfrac{\partial u_2}{\partial x_2} & \cdots & \dfrac{\partial u_n}{\partial x_2} \\[2mm]
\cdots & \cdots & & \cdots \\[2mm]
\dfrac{\partial u_1}{\partial x_n} & \dfrac{\partial u_2}{\partial x_n} & \cdots & \dfrac{\partial u_n}{\partial x_n}
\end{vmatrix} = 0 \quad \text{or,} \quad
\begin{vmatrix}
\dfrac{\partial u_1}{\partial x_1} & \dfrac{\partial u_1}{\partial x_2} & \cdots & \dfrac{\partial u_1}{\partial x_n} \\[2mm]
\dfrac{\partial u_2}{\partial x_1} & \dfrac{\partial u_2}{\partial x_2} & \cdots & \dfrac{\partial u_2}{\partial x_n} \\[2mm]
\cdots & \cdots & & \cdots \\[2mm]
\dfrac{\partial u_n}{\partial x_1} & \dfrac{\partial u_n}{\partial x_2} & \cdots & \dfrac{\partial u_n}{\partial x_n}
\end{vmatrix} = 0$$

(interchanging rows and columns)

or, $\dfrac{\partial(u_1, u_2, ..., u_n)}{\partial(x_1, x_2, ..., x_n)} = 0.$

Hence the condition is necessary.

Condition Sufficient:

Let $J = \dfrac{\partial(u_1, u_2, ..., u_n)}{\partial(x_1, x_2, ..., x_n)} = 0$. Then the equations connecting the functions $u_1, u_2, ..., u_n$ and the variables $x_1, x_2, ..., x_n$ can be expressed by the process of elimination as follows:

$$\left.\begin{array}{rcl}
\phi_1(x_1, x_2, ..., x_n, u_1) & = & 0 \\
\phi_2(x_2, x_3 ..., x_n, u_1, u_2) & = & 0 \\
\phi_3(x_3, x_4 ..., x_n, u_1, u_2, u_3) & = & 0 \\
... & & ... \quad ... \\
\phi_r(x_r, x_{r+1} ..., x_n, u_1, u_2, ..., u_r) & = & 0 \\
... & & ... \quad ... \\
\phi_n(x_n, u_1, u_2, ..., u_n) & = & 0
\end{array}\right\}$$

$$\text{Now,} \quad \frac{\partial(\phi_1, \phi_2, ..., \phi_n)}{\partial(x_1, x_2, ..., x_n)} = \begin{vmatrix} \dfrac{\partial \phi_1}{\partial x_1} & \dfrac{\partial \phi_1}{\partial x_2} & \cdots & \dfrac{\partial \phi_1}{\partial x_n} \\ \dfrac{\partial \phi_2}{\partial x_1} & \dfrac{\partial \phi_2}{\partial x_2} & & \dfrac{\partial \phi_2}{\partial x_n} \\ \cdots & \cdots & \cdots & \cdots \\ \dfrac{\partial \phi_n}{\partial x_1} & \dfrac{\partial \phi_n}{\partial x_2} & \cdots & \dfrac{\partial \phi_n}{\partial x_n} \end{vmatrix}$$

$$= \begin{vmatrix} \dfrac{\partial \phi_1}{\partial x_1} & \dfrac{\partial \phi_1}{\partial x_2} & \cdots & \dfrac{\partial \phi_1}{\partial x_n} \\ 0 & \dfrac{\partial \phi_2}{\partial x_2} & \cdots & \dfrac{\partial \phi_2}{\partial x_n} \\ \cdots & \cdots & \cdots & \cdots \\ 0 & 0 & \cdots & \dfrac{\partial \phi_n}{\partial x_n} \end{vmatrix} = \frac{\partial \phi_1}{\partial x_1} \cdot \frac{\partial \phi_2}{\partial x_2} \cdots \cdot \frac{\partial \phi_n}{\partial x_n} \qquad (4.10)$$

Similarly, we get

$$\frac{\partial(\phi_1, \phi_2, ..., \phi_n)}{\partial(u_1, u_2, ..., u_n)} = \frac{\partial \phi_1}{\partial u_1} \cdot \frac{\partial \phi_2}{\partial u_2} \cdots \cdot \frac{\partial \phi_n}{\partial u_n} \qquad (4.11)$$

$$\text{Now} \quad \frac{\partial(u_1, u_2, ..., u_n)}{\partial(x_1, x_2, ..., x_n)} = (-1)^n \frac{\frac{\partial(\phi_1, \phi_2, ..., \phi_n)}{\partial(x_1, x_2, ..., x_n)}}{\frac{\partial(\phi_1, \phi_2, ..., \phi_n)}{\partial(u_1, u_2, ..., u_n)}} \quad [\text{See § 4.3}]$$

So using the given condition $\dfrac{\partial(u_1, u_2, ..., u_n)}{\partial(x_1, x_2, ..., x_n)} = 0$, we get

$$\frac{\partial(\phi_1, \phi_2, ..., \phi_n)}{\partial(x_1, x_2, ..., x_n)} \quad \text{i.e.,} \quad \frac{\partial \phi_1}{\partial x_1} \cdot \frac{\partial \phi_1}{\partial x_2} \cdots \cdot \frac{\partial \phi_1}{\partial x_n} = 0 \quad [\text{by (4.10)}]$$

$$\Rightarrow \quad \frac{\partial \phi_k}{\partial x_k} = 0 \quad \text{for some } k; 1 \le k \le n$$

i.e., $\quad \phi_k(x_{k+1}, x_{k+2}, ..., x_n, u_1, u_2, ..., u_k) = 0$

Using this in the remaining equations $\phi_{k+1} = 0, \phi_{k+2} = 0, ..., \phi_n = 0$ the variables $x_{k+1}, x_{k+2}, ..., x_n$ can be eliminated to get the desired relation among $u_1, u_2, ..., u_n$.

$\square$

4.6   Implicit Functions

If a basic definition is what we're after, an implicit function is a function in which one variable can not be expressed in terms of the other. For example, $y = 7x + 5$ is an explicit function, as y can be explicitly expressed in terms of x.

Similarly $xy = 1$ is an explicit function, as y can be expressed explicitly in terms of x, i.e., $y = \dfrac{1}{x}$.

But $x^2 y^3 + x^3 y = 0$ is not an explicit function. This is because, no amount of manipulation can explicitly define y in terms of x or x in terms of y. Thus it is an implicit function $x^2 y^2 z^2 = \sin(yz)$ is another example of implicit function.

Definition 4.6.1 *Let the values of two variables x and y be related by some equation, which can be symbolized as follows*

$$F(x, y) = 0 \qquad (4.12)$$

If the function $y = f(x)$, defined on some interval (a, b), is such that equation (4.12) becomes an identity in x when the expression $f(x)$ is substituted into it in place of y, the function $y = f(x)$ is an implicit function defined by equation (4.12).

For example, the equation

$$x^2 + y^2 - a^2 = 0 \qquad (4.13)$$

defines implicitly the following elementary functions

$$y = \sqrt{a^2 - x^2}, \quad |x| \le a \qquad (4.14)$$
$$\text{and} \quad y = -\sqrt{a^2 - x^2}, \quad |x| \le a \qquad (4.15)$$

Indeed, substitution into equation (4.13) yields the identity

$$x^2 + (a^2 - x^2) - a^2 = 0.$$

Expressions (4.14) and (4.15) were obtained by solving equation (4.13) for y. But not every implicitly defined function may be represented explicitly, i.e., in the form $y = f(x)$, where $f(x)$ is an elementary function. As for example, functions defined by the equations

$$y^6 - y - x^2 = 0 \quad \text{or,} \quad y - x - \frac{1}{4}\sin(xy) = 0$$

are not expressible in terms of elementary functions, i.e., these equations cannot be solved for y.

The equation $x^2 + y^2 = 0$ gives $y = \pm\sqrt{-x^2}$, which gives $y = 0$ for $x = 0$. But for $x \ne 0$, y cannot be obtained. So $x^2 + y^2 = 0$ is an implicit equation for $x \ne 0$ and an explicit equation for $x = 0$.

Note 4.6.1 *If a function is defined by an equation of the form $y = f(x)$, one says that the function is defined **explicitly** or is **explicit**. We observe that the terms "explicit function" and "implicit function" do not characterize the nature of the function, but merely the way it is defined. Every explicit function $y = f(x)$ may also always be represented as an implicit function $y - f(x) = 0$.*

4.7 Condition for the Existence of an Explicit Function from an Implicit Function

4.7.1 Existence Theorem (in case of two variables)

Theorem 4.7.1 *If the function $f(x, y) = 0$ be such that*
(i) $f(a, b) = 0$
(ii) its partial derivatives f_x, f_y exists and are continuous in a certain neighbourhood of (a, b) and
(iii) $f_y(a, b) \ne 0$
*then there exists a neighbourhood $(a - h, a + h; b - k, b + k)$ of (a, b) such that for every x in the interval $[a - h, a + h]$, the equation $f(x, y) = 0$ determines **uniquely** the value $y = \phi(x)$ in $[b - k, b + k]$ having following properties:*
(1) $b = \phi(a)$
(2) $f(x, \phi(x)) = 0 \forall x \in [a - h, a + h]$ and
(3) $\phi'(x)$ exists and continuous in $[a - h, a + h]$ and $\phi'(x) = -\dfrac{f_x}{f_y}$.

Note 4.7.1 *The existence and uniqueness of any solution of the equation $f(x, y) = 0$ in the neighbourhood of the point (a, b) reveals that the theorem is essentially of local character, i.e., the equation $f(x, y) = 0$ may have a different solution $y = \psi(x)$ in a different neighbourhood of (a, b). For illustration we consider the following example.*

Example 4.7.1 *We consider $f(x, y) = x^2 + y^2 - 4$ and a point $(0, 2)$. We observe that $f(0, 2) = 0$, $f_x = 2x$ and $f_y = 2y$ are always continuous and $f_y(0, 2) = 4$ ($\neq 0$).*

Therefore, all the condition of implicit function existence theorem are satisfied for the initial solution $(0, 2)$. Theorem asserts that there exists a neighbourhood of $(0, 2)$ in which y can be explicitly expressed in terms of x. There are two possible solutions of $f(x, y) = 0$ given by $y = \sqrt{4 - x^2}$ and $y = -\sqrt{4 - x^2}$ of which the former, i.e., $y = \sqrt{4 - x^2}$ is the implicit function defined in the neighbourhood of $(0, 2)$, where $|x| < 2$, $y > 0$ whereas the later solution $y = -\sqrt{4 - x^2}$ is the implicit function defined in the neighbourhood of $(0, 2)$, where $|x| < 2$, $y < 0$.

Proof of the Theorem

Existence Part: The condition (ii) states that f_x, f_y are continuous in a certain neighbourhood of the point (a, b). Let this neighbourhood be a rectangle R_1 : $[a - h_1, a + h_1; b - k_1, b + k_1]$ centered at (a, b).

Moreover, continuity of f_x and f_y imply that f is differentiable and hence continuous at R_1.

Again, since f_y is continuous at (a, b) and $f_y(a, b) \neq 0$, there exists a rectangle R_2 : $[a - h_2, a + h_2; b - k_2, b + k_2], h_2 < h_1, k_2 < k_1$, i.e., $R_1 \subset R_2$ and for every point of R_2, $f_y \neq 0$.

Since $f(a, b) = 0$ and $f_y(a, b) \neq 0$, (it is therefore either positive or negative), there exists a positive number $k < k_2$, such that $f(a, b - k)$ and $f(a, b + k)$ are of opposite signs, f being an increasing or a decreasing function at $y = b$.

Now continuity of f at (a, b) imply that $\exists\ h > 0\ (h < h_2)$ such that $\forall\ x \in [a - h, a + h]$, $f(x, b - k)$ and $f(x, b + k)$ respectively, may be as near as we please to $f(a, b - k)$ and $f(a, b + k)$ and therefore have opposite signs.

Thus, for all x in $[a - h, a + h]$, f is a continuous function of y and changes sign as y changes from $b - k$ and $b + k$. Therefore it vanishes for some y in $[b - k, b + k]$.

Thus, for each x in $[a - h, a + h]$, there is a y in $[b - k, b + k]$ for which $f(x, y) = 0$. This y is a function x, say $\phi(x)$, such that properties (1), (2) and (3) are true.

Uniqueness of the Solution: Now, we shall prove that $y = \phi(x)$ is a unique solution of $f(x, y) = 0$ in R_3 : $(a - h, a + h; b - k, b + k)$, that is $f(x, y)$ cannot be zero for more than one value of y in $[b - k, b + k]$.

Let, if possible, there be two such values y_1, y_2 in $[b - k, b + k]$ so that $f(x, y_1) = 0$, $f(x, y_2) = 0$. Also $f(x, y)$ considered as a function of a single variable y is derivable in $[b - k, b + k]$, so that applying Rolle's Theorem, we get $f_y = 0$, for a value of y between y_1 and y_2 which contradicts the fact that $f_y \neq 0$ in $R_2 \supset R_3$.

Hence our supposition is wrong and there cannot be more than one such solution.

Example 4.7.2 *Let* $f(x,y) = y^2 - yx^2 - 2x^5$. *Verify that at* $(1,-1)$

(a) $f(1,-1) = 0$,

(b) $f_y(1,-1) \neq 0$ *and*

(c) f_x, f_y *are continuous in some neighbourhood of* $(1,-1)$.

Now obtain the unique solution of $f(x,y) = 0$ *in the neighbourhood of* $(1,-1)$. *Obtain also* $\dfrac{dy}{dx}$ *at* $(1,-1)$.

Solution: (a) $f(1,-1) = (-1)^2 - (-1).1^2 - 2.(1)^5 = 1 + 1 - 2 = 0$.

(b) $f_x(x,y) = -2xy - 10x^4$ and $f_y(x,y) = 2y - x^2$ which can be easily verified to be continuous in the neighbourhood of $(1,-1)$.

(c) The two possible solutions are $y = \dfrac{x^2}{2}\left(1 \pm \sqrt{1+8x}\right)$, $x > -\dfrac{1}{8}$. Of these two solutions $y = \dfrac{x^2}{2}\left(1 - \sqrt{1+8x}\right)$, $x > -\dfrac{1}{8}$ is the unique solution of $f(x,y) = 0$ in a neighbourhood of $(1,-1)$, since $-1 = y(1)$.

The given equation is

$$f(x,y) = y^2 - yx^2 - 2x^5 = 0$$

$$\therefore \quad \frac{dy}{dx} = -\frac{f_x}{f_y} = -\frac{-2xy - 10x^4}{2y - x^2} = \frac{2xy + 10x^4}{2y - x^2}$$

$$\therefore \quad \left[\frac{dy}{dx}\right]_{(1,-1)} = \frac{2.1.(-1) + 10.1^2}{2.(-1) - 1^2} = \frac{-2+10}{-2-1} = -2\frac{2}{3}.$$

Hence the value of $\dfrac{dy}{dx} = -2\dfrac{2}{3}$ at the given point. $\qquad\square$

4.8 Generalized Form of Existence Theorem

Theorem 4.8.1 *Let* $f(x_1, x_2, ..., x_n, y)$ *be a function of* $(n+1)$ *variables* x_1, x_2, ..., x_n, y *and* $(a_1, a_2, ..., a_n, b)$ *be a point in its domain of definition such that*

(i) $f(a_1, a_2, ..., a_n, b) = 0$

(ii) *its partial derivatives with respect to all the variables exists and are continuous in the neighbourhood of the point* $(a_1, a_2, ..., a_n, b)$ *and*

(iii) $f_y(a_1, a_2, ..., a_n, b) \neq 0$,

then there exists a neighbourhood $(a_1 - h_1, a_1 + h_1; a_2 - h_2, a_2 + h_2; ...; a_n - h_n, a_n + h_n; b - k, b + k)$ *of* $(a_1, a_2, ..., a_n, b)$ *such that for every* $(x_1, x_2, ..., x_n)$ *in* $R : [a_1 - h_1, a_1 + h_1; a_2 - h_2, a_2 + h_2; ...; a_n - h_n, a_n + h_n]$ *the equation* $f(x_1, x_2, ..., x_n, y) = 0$ *determines uniquely the value* $y = \phi(x_1, x_2, ..., x_n)$ *in* $[b - k, b + k]$ *having following properties*

(1) $b = \phi(a_1, a_2, ..., a_n)$

(2) $f(x_1, x_2, ..., x_n, \phi) = 0 \ \forall \ (x_1, x_2, ..., x_n) \in R$

(3) ϕ *possesses continuous partial derivatives of the first order with respect to* $x_1, x_2, ..., x_n$ *in* R.

Proof: Similar as in case of two variables in § 4.7. $\qquad\square$

4.9   Derivatives of Implicit Functions

When the existence of the implicit function $f(x, y) = 0$ is assured, we may find $\dfrac{dy}{dx}$ from it by using the 'chain rule' of derivatives. Take $u = f(x, y)$, $y = \phi(x)$ and $x = x$. Now differentiating with respect to x, we get

$$\frac{\partial u}{\partial x} \cdot \frac{dx}{dx} + \frac{\partial u}{\partial y} \cdot \frac{dy}{dx} = 0 \quad \Rightarrow \quad \frac{dy}{dx} = -\frac{\frac{\partial u}{\partial x}}{\frac{\partial u}{\partial y}} = -\frac{f_x}{f_y} \qquad (4.16)$$

Otherwise, we may take differentials and we get

$$du = \frac{\partial u}{\partial x} . dx + \frac{\partial u}{\partial y} . dy = f_x . dx + f_y . dy = 0 \quad \Rightarrow \quad \frac{dy}{dx} = -\frac{f_x}{f_y} \qquad (4.17)$$

If the higher order partial derivatives of $f(x, y)$ are continuous, we obtain the higher derivatives of y or $\phi(x)$ by successive differentiation of (4.16), provided always that f_y is not zero. Thus we get

$$
\begin{aligned}
\frac{d^2y}{dx^2} &= -\frac{\left(f_{xx} + f_{xy}\frac{dy}{dx}\right) f_y - \left(f_{yx} + f_{yy}\frac{dy}{dx}\right) f_x}{f_y^2} \\[2mm]
&= -\frac{\left\{f_{xx}f_y + f_{xy}f_y\left(-\frac{f_x}{f_y}\right)\right\} - \left\{f_{yx}f_x + f_{yy}f_x\left(-\frac{f_x}{f_y}\right)\right\}}{f_y^2} \\[2mm]
&= -\frac{f_{xx}f_y^2 - f_xf_yf_{xy} - f_xf_yf_{xy} + f_{yy}f_x^2}{f_y^3} \\[2mm]
\therefore \quad \frac{d^2y}{dx^2} &= -\frac{f_{xx}f_y^2 - 2f_xf_yf_{xy} + f_{yy}f_x^2}{f_y^3} \quad (f_y \neq 0).
\end{aligned}
$$

Quite often we take $p = f_x, q = f_y, r = f_{xx}, s = f_{xy}$ and $t = f_{yy}$ and then

$$\frac{d^2y}{dx^2} = -\frac{1}{q^3}\left[rq^2 - 2pqs + tp^2\right], \quad (q \neq 0).$$

Theorem 4.9.1 *Let the region R contains the point (a, b, c) in its interior. Then if*

(i) $f(a, b, c) = 0$

(ii) f_x, f_y, f_z exist and all are continuous in a certain neighbourhood of (a, b, c) and

(iii) $f_z(a, b, c) \neq 0$

then there exists a neighbourhood $(a - h, a + h; b - k, b + k; c - \lambda, c + \lambda)$ in which there exists a unique differentiable function $z = \phi(x, y)$, such that

(1) $c = \phi(a, b)$

(2) $f(x, y, \phi(x, y)) = 0 \ \forall \ x \in [a - h, a + h], y \in [b - k, b + k]$ and

(3) z_x, z_y exists and continuous in $[a - h, a + h; b - k, b + k]$

Also $z_x = -\dfrac{f_x}{f_z}, z_y = -\dfrac{f_y}{f_z}$.

Proof: For existence and uniqueness of the theorem, proof is similar to § 4.9. For the derivative, taking differential, we get if $u = f(x, y, z) = 0$.

$$du = f_x . dx + f_y . dy + f_z . dz = 0 \quad \Rightarrow \quad dz = -\frac{f_x}{f_y} . dx - \frac{f_y}{f_z} . dy \qquad (4.18)$$

But when the existence of the implicit function $f(x, y, z) = 0$ is assured, we get $z = \phi(x, y)$.

$$dz = \frac{\partial \phi}{\partial x}.dx + \frac{\partial \phi}{\partial y}.dy = \frac{\partial z}{\partial x}.dx + \frac{\partial z}{\partial y}.dy \qquad (4.19)$$

$\therefore$ from (4.18) and (4.19), we get

$$\frac{\partial z}{\partial x} = -\frac{f_x}{f_z} \quad \text{(here } y \text{ is a constant)}$$

and $\quad \dfrac{\partial z}{\partial y} = -\dfrac{f_y}{f_z} \quad$ (here x is a constant).

So we get $z_x = -\dfrac{f_x}{f_z}, z_y = -\dfrac{f_y}{f_z}.$ ☐

Corollary 4.9.1 *This theorem can be generalized for $(n+1)$ variables. To solve for u as a function of n variables $x_1, x_2, ..., x_n$ from $f(x_1, x_2, ..., x_n, u) = 0$. We assume $f_u \neq 0$ and taking differential, we get*

$$-f_u.du = f_{x_1}.dx_1 + f_{x_2}.dx_2 + ... + f_{x_n}.dx_n$$

$$\Rightarrow \quad du = -\frac{f_{x_1}}{f_u}dx_1 - \frac{f_{x_2}}{f_u}dx_2 - ... - \frac{f_{x_n}}{f_u}dx_n$$

Also $\quad du = \dfrac{\partial u}{\partial x_1}.dx_1 + \dfrac{\partial u}{\partial x_2}.dx_2 + ... + \dfrac{\partial u}{\partial x_n}.dx_n$

leading to $\quad \dfrac{\partial u}{\partial x_1} = -\dfrac{f_{x_1}}{f_u}, \dfrac{\partial u}{\partial x_2} = -\dfrac{f_{x_2}}{f_u}, \quad ... \quad \dfrac{\partial u}{\partial x_n} = -\dfrac{f_{x_n}}{f_u}$

i.e., $\quad u_{x_1} = -\dfrac{f_{x_1}}{f_u}, u_{x_2} = -\dfrac{f_{x_2}}{f_u}, \quad ... \quad u_{x_n} = -\dfrac{f_{x_n}}{f_u}.$

Example 4.9.1 Let $f(x, y, z) = \dfrac{x^2}{a^2} + \dfrac{y^2}{b^2} + \dfrac{z^2}{c^2} - 3 \; \forall \; (x, y, z) \in \mathbb{R}^3$. *Show that the functional equation $f(x, y, z) = 0$ defines z as a function ϕ of x and y in a certain neighbourhood of $P(a, b, c)$ which is continuous possessing first order partial derivatives in that neighbourhood.*

Solution: Obviously
(i) $f(a, b, c) = 0$,
(ii) $f_x = \dfrac{2x}{a^2}, f_y = \dfrac{2y}{b^2}, f_z = \dfrac{2z}{c^2}$, so that all the first order partial derivatives are continuous.
(iii) $f_z(a, b, c) = \dfrac{2c}{c^2} = \dfrac{2}{c} \neq 0.$

Hence there exists a real function given by $z = \phi(x, y)$ defined in a neighbourhood of $P(x, y, z)$. (vide § 4.10)

$$f(x, y, \phi(x, y)) = 0$$

$$\phi_x = -\frac{f_x}{f_z} = -\frac{\frac{x}{a^2}}{\frac{z}{c^2}} = -\frac{xc^2}{za^2}, \quad \phi_y = -\frac{f_y}{f_z} = -\frac{\frac{y}{b^2}}{\frac{z}{c^2}} = -\frac{yc^2}{zb^2}.$$

Regarding a, b, c all positive, $z = \pm c\sqrt{3 - \dfrac{x^2}{a^2} - \dfrac{y^2}{b^2}}$ which is real.

We take $z = \phi(x,y) = c\sqrt{3 - \dfrac{x^2}{a^2} - \dfrac{y^2}{b^2}}$, so that $(a,b,c) \in \phi$.

Domain of ϕ is $\left\{(x,y) \in \mathbb{R}^2 : 3 - \dfrac{x^2}{a^2} - \dfrac{y^2}{b^2} \geq 0\right\}$, a region within and on the boundary of the ellipse $\dfrac{x^2}{a^2} + \dfrac{y^2}{b^2} = 3$.

$$\phi_x = -\frac{xc^2}{za^2} = -\frac{cx}{a^2} \cdot \frac{1}{\sqrt{3 - \frac{x^2}{a^2} - \frac{y^2}{b^2}}}.$$

so that ϕ_x is continuous everywhere within the region.

Similarly, $\phi_y = -\dfrac{cy}{b^2} \cdot \dfrac{1}{\sqrt{3 - \frac{x^2}{a^2} - \frac{y^2}{b^2}}}$ is also continuous within the same region.

$\square$

Example 4.9.2 *If $f(x,y,z) = 0$ determines each of the three variables x,y and z to be a differentiable function of the two others and if none of f_x, f_y and f_z vanishes, then show that*

$$\left(\frac{\partial x}{\partial y}\right)_z \left(\frac{\partial y}{\partial z}\right)_x \left(\frac{\partial z}{\partial x}\right)_y = -1.$$

Solution: Let $x = x(y,z)$ be the differentiable implicit function determined by the given equation $f(x,y,z) = 0$. Then we get

$$x_y = -\frac{f_y}{f_z} \quad (z \text{ is constant}: x = \text{function of } y \text{ and } z; f_x \neq 0) = \left(\frac{\partial x}{\partial y}\right)_z.$$

Similarly,

$$y_z = -\frac{f_z}{f_y} \quad (x \text{ is constant}: y = \text{function of } x \text{ and } z; f_y \neq 0) = \left(\frac{\partial y}{\partial z}\right)_x$$

$$\text{and} \quad z_x = -\frac{f_x}{f_z} \quad (y \text{ is constant}: z = \text{function of } x \text{ and } y; f_z \neq 0) = \left(\frac{\partial z}{\partial x}\right)_y$$

Hence the product $= -1$.

$\square$

Note 4.9.1 *This problem may be compared to **Example 3.3.5**.*

4.10 Implicit Functions Defined by a System of Functional Equations (Problems of Solving Two Equations are Considered)

Theorem 4.10.1 *Let the functions $f(x,y,u,v) = 0$, $g(x,y,u,v) = 0$ be two functions such that at an interior point (x_0, y_0, u_0, v_0) of a region R*
(i) $f(x_0, y_0, u_0, v_0) = 0$, $g(x_0, y_0, u_0, v_0) = 0$
(ii) all the partial derivatives $f_x, f_y, f_u, f_v, g_x, g_y, g_u, g_v$ are continuous in R
(iii) The Jacobian

$$J = \frac{\partial(f,g)}{\partial(u,v)} = \begin{vmatrix} \dfrac{\partial f}{\partial u} & \dfrac{\partial f}{\partial v} \\[2mm] \dfrac{\partial g}{\partial u} & \dfrac{\partial g}{\partial v} \end{vmatrix} \neq 0 \quad \text{at } (x_0, y_0, u_0, v_0).$$

Then, there exists a neighbourhood of (x_0, y_0) in which the differentiable functions $u = \phi(x, y)$, $v = \psi(x, y)$ exists, such that
  (1) $u_0 = \phi(x_0, y_0)$, $v_0 = \psi(x_0, y_0)$
  (2) $f(x, y, \phi(x, y), \psi(x, y)) = 0$, $g(x, y, \phi(x, y), \psi(x, y)) = 0$ and
  (3) the partial derivatives of $u = \phi(x, y)$, $v = \psi(x, y)$ are continuous functions found by solving

$$df = f_x.dx + f_y.dy + f_u.du + f_v.dv$$
$$dg = g_x.dx + g_y.dy + g_u.du + g_v.dv.$$

Observation:

Since $J \neq 0$ we can solve the last equation for du and dv. Thus

$$du = -\frac{\begin{vmatrix} f_x.dx + f_y.dy & f_v \\ g_x.dx + g_y.dy & g_v \end{vmatrix}}{\begin{vmatrix} f_u & f_v \\ g_u & g_v \end{vmatrix}} \quad \text{(using Cramer's Rule)}$$

$$= -\frac{\begin{vmatrix} f_x & f_v \\ g_x & g_v \end{vmatrix}.dx + \begin{vmatrix} f_y & f_v \\ g_y & g_v \end{vmatrix}.dy}{J} = \frac{\partial u}{\partial x}.dx + \frac{\partial u}{\partial y}.dy$$

$$\Rightarrow \quad \frac{\partial u}{\partial x} = -\frac{\begin{vmatrix} f_x & f_v \\ g_x & g_v \end{vmatrix}}{J}; \quad \frac{\partial u}{\partial y} = -\frac{\begin{vmatrix} f_y & f_v \\ g_y & g_v \end{vmatrix}}{J}.$$

Similarly,

$$dv = -\frac{1}{J}\begin{vmatrix} f_u & f_x \\ g_u & g_x \end{vmatrix} dx - \frac{1}{J}\begin{vmatrix} f_u & f_y \\ g_u & g_y \end{vmatrix} dy$$

giving $\quad \dfrac{\partial v}{\partial x} = -\dfrac{\begin{vmatrix} f_u & f_x \\ g_u & g_x \end{vmatrix}}{J}$ and $\dfrac{\partial v}{\partial y} = -\dfrac{\begin{vmatrix} f_u & f_y \\ g_u & g_y \end{vmatrix}}{J}.$

from which we can also write

$$\frac{\partial u}{\partial x} = -\frac{\frac{\partial(f,g)}{\partial(x,v)}}{\frac{\partial(f,g)}{\partial(u,v)}}, \quad \frac{\partial u}{\partial y} = -\frac{\frac{\partial(f,g)}{\partial(y,v)}}{\frac{\partial(f,g)}{\partial(u,v)}}, \quad \frac{\partial v}{\partial x} = -\frac{\frac{\partial(f,g)}{\partial(u,x)}}{\frac{\partial(f,g)}{\partial(u,v)}} \text{ and } \frac{\partial v}{\partial y} = -\frac{\frac{\partial(f,g)}{\partial(u,y)}}{\frac{\partial(f,g)}{\partial(u,v)}}.$$

Note 4.10.1 *We note that the numerator of the above formulas are obtained by just replacing u by x or v by y in $J = \dfrac{\partial(f,g)}{\partial(u,v)}$. Thus numerator of $\dfrac{\partial u}{\partial x} = \dfrac{\partial(f,g)}{\partial(x,v)}$, of $\dfrac{\partial v}{\partial x} = \dfrac{\partial(f,g)}{\partial(u,x)}$ etc.*

Note 4.10.2 *On the other hand, if we wish to solve for x and y from the functional equation $f(x, y, u, v) = g(x, y, u, v)$, then J must be replaced by $J' = \dfrac{\partial(f,g)}{\partial(x,v)} \neq 0$ at (x_0, y_0, u_0, v_0) and the respective derivatives are*

$$\frac{\partial x}{\partial u} = -\frac{\frac{\partial(f,g)}{\partial(u,y)}}{J'}, \quad \frac{\partial y}{\partial u} = -\frac{\frac{\partial(f,g)}{\partial(x,u)}}{J'}, \quad \frac{\partial x}{\partial v} = -\frac{\frac{\partial(f,g)}{\partial(v,y)}}{J'} \text{ and } \frac{\partial y}{\partial v} = -\frac{\frac{\partial(f,g)}{\partial(x,v)}}{J'}.$$

Example 4.10.1 *If u and v are determined as functions of x, y, z by the equations* $\phi(x, y, z, u, v) = 0$ *and* $\psi(x, y, z, u, v) = 0$, *find the derivatives of u and v with respect to x, y, z.*

Solution: Let $(a_1, a_2, a_3, b_1, b_2)$ be a point of the domain of definition such that $\phi(a_1, a_2, a_3, b_1, b_2) = 0 = \psi(a_1, a_2, a_3, b_1, b_2)$.

Let $\dfrac{\partial(\phi, \psi)}{\partial(u, v)} \neq 0$ at $(a_1, a_2, a_3, b_1, b_2)$.

Then at least one of $\dfrac{\partial \phi}{\partial v}$ and $\dfrac{\partial \psi}{\partial v}$ must not be equal to zero at that point. Let us take $\dfrac{\partial \phi}{\partial v} \neq 0$.

Then we know that the equation $\psi(x, y, z, u) = 0$ is satisfied by one function $v = g(x, y, z, u)$ defined in the neighbourhood of (a_1, a_2, a_3, b_1) such that $b_2 = g(a_1, a_2, a_3, b_1)$.

Replacing v by $g(x, y, z, u)$ in $\psi(x, y, z, u, v)$, we write $h(x, y, z, u) = \psi(x, y, z, u, g)$ where $h(a_1, a_2, a_3, b_1) = 0$.

Now at (a_1, a_2, a_3, b_1), we know that $\dfrac{\partial h}{\partial u} = \dfrac{\partial \psi}{\partial u} + \dfrac{\partial \psi}{\partial v} \cdot \dfrac{\partial g}{\partial u}$.

Also since $\phi(x, y, z, u, g) = 0$, we have at (a_1, a_2, a_3, b_1)

$$\frac{\partial \phi}{\partial u} + \frac{\partial \phi}{\partial v} \cdot \frac{\partial g}{\partial u} = 0 \text{ whence, we get } \frac{\partial h}{\partial u} = -\frac{\frac{\partial(\phi, \psi)}{\partial(u, v)}}{\frac{\partial \phi}{\partial v}}$$

so that $\dfrac{\partial h}{\partial u} \neq 0$ at (a_1, a_2, a_3, b_1).

Also, from the same argument, the equation $h(x, y, z, u) = 0$ is satisfied by only one function $u = \lambda_1(x, y, z)$ defined in a certain neighbourhood of (a_1, a_2, a_3) such that $b_1 = \lambda_1(a_1, a_2, a_3)$.

Replacing u by $\lambda_1(x, y, z)$ in $g(x, y, z, u)$ we get $v = g(x, y, z, \lambda_1) = \lambda_2(x, y, z)$, where $\lambda_2(a_1, a_2, a_3) = g(a_1, a_2, a_3, b_1) = b_2$.

Functions λ_1, λ_2 are continuous and possess continuous first order partial derivatives.

Applying this result on the derivation of function of functions to $\phi(x, y, z, u, v)$ and $\psi(x, y, z, u, v)$, where $u = \lambda_1(x, y, z)$ and $v = \lambda_2(x, y, z)$, we have $(\phi, \psi$ being identically zero.)

$$\frac{\partial \phi}{\partial x} + \frac{\partial \phi}{\partial u} \cdot \frac{\partial u}{\partial x} + \frac{\partial \phi}{\partial v} \cdot \frac{\partial v}{\partial x} = 0 \text{ and } \frac{\partial \psi}{\partial x} + \frac{\partial \psi}{\partial u} \cdot \frac{\partial u}{\partial x} + \frac{\partial \psi}{\partial v} \cdot \frac{\partial v}{\partial x} = 0.$$

Solving, we get $\dfrac{\partial u}{\partial x} = -\dfrac{\frac{\partial(\phi, \psi)}{\partial(x, v)}}{\frac{\partial(\phi, \psi)}{\partial(u, v)}}, \quad \dfrac{\partial v}{\partial x} = -\dfrac{\frac{\partial(\phi, \psi)}{\partial(x, u)}}{\frac{\partial(\phi, \psi)}{\partial(u, v)}}$.

Similarly, we get $\dfrac{\partial u}{\partial y}, \dfrac{\partial v}{\partial y}, \dfrac{\partial u}{\partial z}, \dfrac{\partial u}{\partial z}$. □

Example 4.10.2 *The equations*

$$\left. \begin{array}{llll} f(x, y, u, v) &=& u^2 + v^2 - x^2 - y &=& 0 \\ g(x, y, u, v) &=& u + v - x^2 + y &=& 0 \end{array} \right\} \qquad (4.20)$$

are satisfied by $x_0 = 2$, $y_0 = 1$, $u_0 = 1$, $v_0 = 2$.

Find the derivatives of u and v with respect to x and y. Also find $\dfrac{\partial x}{\partial u}, \dfrac{\partial x}{\partial v}, \dfrac{\partial y}{\partial u}$ and $\dfrac{\partial y}{\partial v}$.

Solution: Here

$$J = \begin{vmatrix} f_u & f_v \\ g_u & g_v \end{vmatrix} = \begin{vmatrix} 2u & 2v \\ 1 & 1 \end{vmatrix} = 2(u - v)$$

and $J_0 = 2(u_0 - v_0) = -2.$

Equation (4.20) can be resolved uniquely for u and v in the neighbourhood of $(x_0 = 2,\ y_0 = 1)$. Taking differential we get

$$\left.\begin{aligned} df &= 2u\,du + 2v\,dv - 2x\,dx - dy &= 0 \\ dg &= du + dv - 2x\,dx + dy &= 0 \end{aligned}\right\} \tag{4.21}$$

Solving for du and dv by using Cramer's Rule, we get

$$du = \frac{\begin{vmatrix} 2x\,dx + dy & 2v \\ 2x\,dx - dy & 1 \end{vmatrix}}{\begin{vmatrix} 2u & 2v \\ 1 & 1 \end{vmatrix}} = \frac{\begin{vmatrix} 2x & 2v \\ 2x & 1 \end{vmatrix} dx + \begin{vmatrix} 1 & 2v \\ -1 & 1 \end{vmatrix} dy}{\begin{vmatrix} 2u & 2v \\ 1 & 1 \end{vmatrix}}$$

$$= \frac{2(x - 2xv)}{2(u - v)} dx + \frac{(1 + 2v)}{2(u - v)} dy$$

$$\Rightarrow \qquad \frac{\partial u}{\partial x} = \frac{x(1 - 2v)}{(u - v)}, \ \frac{\partial u}{\partial y} = \frac{1 + 2v}{2(u - v)}$$

Similarly, $\qquad \dfrac{\partial v}{\partial x} = \dfrac{x(2u - 1)}{(u - v)}, \ \dfrac{\partial v}{\partial y} = \dfrac{-2u - 1}{2(u - v)}.$

Again, since $J' = \begin{vmatrix} f_x & f_y \\ g_x & g_y \end{vmatrix} = \begin{vmatrix} -2x & -1 \\ -2x & 1 \end{vmatrix} = -4x$ and $J_0' = -8$ (∵ $x_0 = 2$), so equation (4.20) can also be solved for x, y in the neighbourhood of $(u_0 = 1,\ v_0 = 2)$ and solving equations (4.21) for dx and dy, we shall get

$$\frac{\partial x}{\partial u} = \frac{2u + 1}{4x}, \ \frac{\partial x}{\partial v} = \frac{2v + 1}{4x}, \ \frac{\partial y}{\partial u} = \frac{2u - 1}{2} \text{ and } \frac{\partial y}{\partial v} = \frac{2v - 1}{2}. \qquad \square$$

Note 4.10.3 *From the above results we can verify that the Jacobians $\dfrac{\partial(u, v)}{\partial(x, y)}$ and $\dfrac{\partial(x, y)}{\partial(u, v)}$ are reciprocal. This is also a check on computation.*

4.11 Miscellaneous Illustrative Examples

Example 4.11.1 *If λ, μ, ν are real roots of the equation*

$$\frac{x}{a + k} + \frac{y}{b + k} + \frac{z}{c + k} = 1$$

in k, then prove that $\dfrac{\partial(x, y, z)}{\partial(\lambda, \mu, \nu)} = -\dfrac{(\mu - \nu)(\nu - \lambda)(\lambda - \mu)}{(b - c)(c - a)(a - b)}.$

Solution: Writing the given equation as a cubic equation in k (see also **Example 4.4.2**) we get

$$k^3 - (x + y + z - a - b - c)k^2 - [(b + c)x + (c + a)y + (a + b)z$$
$$-ab - bc - ca]k + abc - bcx - cay - abz = 0.$$

From this we get

$$\sum \lambda = \left(\sum x - \sum y\right), \quad \sum \lambda\mu = -\left\{\sum(b + c)x - \sum bc\right\}$$

and $\quad \lambda\mu\nu = \sum bcx - abc.$

We consider,

$$\phi_1 = \lambda + \mu + \nu - (x + y + z - a - b - c) = 0$$
$$\phi_2 = \lambda\mu + \mu\nu + \nu\lambda + [(b - c)x + (c + a)y + (a + b)z] - bc - ca - ab = 0$$
$$\phi_3 = \lambda\mu\nu - [bcx + cay + abz - abc] = 0$$

$$\therefore \quad \frac{\partial(x, y, z)}{\partial(\lambda, \mu, \nu)} = (-1)^3 \frac{\frac{\partial(\phi_1, \phi_2, \phi_3)}{\partial(\lambda, \mu, \nu)}}{\frac{\partial(\phi_1, \phi_2, \phi_3)}{\partial(x, y, z)}} \quad \text{[by \textbf{Property 1} of § 4.4].}$$

Now $\quad \dfrac{\partial(\phi_1, \phi_2, \phi_3)}{\partial(\lambda, \mu, \nu)} = \begin{vmatrix} \dfrac{\partial\phi_1}{\partial\lambda} & \dfrac{\partial\phi_2}{\partial\mu} & \dfrac{\partial\phi_1}{\partial\nu} \\[2mm] \dfrac{\partial\phi_2}{\partial\lambda} & \dfrac{\partial\phi_2}{\partial\mu} & \dfrac{\partial\phi_2}{\partial\nu} \\[2mm] \dfrac{\partial\phi_3}{\partial\lambda} & \dfrac{\partial\phi_3}{\partial\mu} & \dfrac{\partial\phi_3}{\partial\nu} \end{vmatrix} = \begin{vmatrix} 1 & 1 & 1 \\ \mu + \nu & \nu + \lambda & \lambda + \mu \\ \mu\nu & \nu\lambda & \lambda\mu \end{vmatrix}$

$$= -(\mu - \nu)(\nu - \lambda)(\lambda - \mu).$$

Similarly $\quad \dfrac{\partial(\phi_1, \phi_2, \phi_3)}{\partial(x, y, z)} = \begin{vmatrix} -1 & -1 & -1 \\ (b + c) & (c + a) & (a + b) \\ -bc & -ca & -ab \end{vmatrix}$

$$= \begin{vmatrix} 1 & 1 & 1 \\ (b + c) & (c + a) & (a + b) \\ bc & ca & ab \end{vmatrix} = -(b - c)(c - a)(a - b)$$

$$\therefore \quad \frac{\partial(x, y, z)}{\partial(\lambda, \mu, \nu)} = (-1)^3 \left\{\frac{-(\mu - \nu)(\nu - \lambda)(\lambda - \mu)}{-(b - c)(c - a)(a - b)}\right\} = -\frac{(\mu - \nu)(\nu - \lambda)(\lambda - \mu)}{(b - c)(c - a)(a - b)}. \quad \square$$

Example 4.11.2 *Let* $\xi = uf$, $\eta = vf$ *and* $\zeta = wf$, *where* f, u, v *and* w *are all functions of* x, y *and* z, *possessing first order partial derivatives in a certain region. Then show that*

$$\frac{\partial(u, v, w)}{\partial(x, y, z)} = \frac{1}{f^4} \begin{vmatrix} f & f_x & f_y & f_z \\ \xi & \xi_x & \xi_y & \xi_z \\ \eta & \eta_x & \eta_y & \eta_z \\ \zeta & \zeta_x & \zeta_y & \zeta_z \end{vmatrix}.$$

Solution: Here $u = \dfrac{\xi}{f}$ $\therefore$ $u_x = \dfrac{f\xi_x - \xi f_x}{f^2}, u_y = \dfrac{f\xi_y - \xi f_y}{f^2}, u_z = \dfrac{f\xi_z - \xi f_z}{f^2}.$

Similarly we shall get $v_x, v_y, \ldots$ etc. Now

$$
\begin{aligned}
J &= \frac{\partial(u,v,w)}{\partial(x,y,z)} = \begin{vmatrix} \frac{1}{f^2}(f\xi_x - \xi f_x) & \frac{1}{f^2}(f\xi_y - \xi f_y) & \frac{1}{f^2}(f\xi_z - \xi f_z) \\ \frac{1}{f^2}(f\eta_x - \eta f_x) & \frac{1}{f^2}(f\eta_y - \eta f_y) & \frac{1}{f^2}(f\eta_z - \eta f_z) \\ \frac{1}{f^2}(f\zeta_x - \zeta f_x) & \frac{1}{f^2}(f\zeta_y - \zeta f_y) & \frac{1}{f^2}(f\zeta_z - \zeta f_z) \end{vmatrix} \\[2mm]
&= \frac{1}{f^6} \begin{vmatrix} 1 & 0 & 0 & 0 \\ \xi & f\xi_x - \xi f_x & f\xi_y - \xi f_y & f\xi_z - \xi f_z \\ \eta & f\eta_x - \eta f_x & f\eta_y - \eta f_y & f\eta_z - \eta f_z \\ \zeta & f\zeta_x - \zeta f_x & f\zeta_y - \zeta f_y & f\zeta_z - \zeta f_z \end{vmatrix}
\end{aligned}
$$

(Bordering the determinant without changing its value.)

$$
= \frac{1}{f^6} \begin{vmatrix} 1 & f_x & f_y & f_z \\ \xi & f\xi_x & f\xi_y & f\xi_z \\ \eta & f\eta_x & f\eta_y & f\eta_z \\ \zeta & f\zeta_x & f\zeta_y & f\zeta_z \end{vmatrix} \qquad \begin{pmatrix} C'_2 = C_2 + C_1 f_x \\ C'_3 = C_3 + C_1 f_y \\ C'_4 = C_4 + C_1 f_z \end{pmatrix}
$$

$$
= \frac{1}{f^7} \begin{vmatrix} f & f_x & f_y & f_z \\ f\xi & f\xi_x & f\xi_y & f\xi_z \\ f\eta & f\eta_x & f\eta_y & f\eta_z \\ f\zeta & f\zeta_x & f\zeta_y & f\zeta_z \end{vmatrix} \qquad (C'_1 = f C_1)
$$

$$
\therefore \quad J = \frac{1}{f^4} \begin{vmatrix} f & f_x & f_y & f_z \\ \xi & \xi_x & \xi_y & \xi_z \\ \eta & \eta_x & \eta_y & \eta_z \\ \zeta & \zeta_x & \zeta_y & \zeta_z \end{vmatrix}. \qquad\qquad \square
$$

Example 4.11.3 *If* $xu = yv = zw = \sum x^2$, *prove that*

$$
\frac{\partial(u,v,w)}{\partial(x,y,z)} = \frac{x^2 y^2 z^2}{(x^2 + y^2 + z^2)^3}.
$$

Solution: Here $u = \dfrac{\sum x^2}{x} \Rightarrow u_x = \dfrac{x^2 - y^2 - z^2}{x^2},\ u_y = \dfrac{2y}{x},\ u_z = \dfrac{2z}{x}$ and similarly $v_x, v_y, \ldots$ etc. can be calculated. So, we get

$$
\begin{aligned}
J &= \frac{\partial(u,v,w)}{\partial(x,y,z)} = \begin{vmatrix} u_x & u_y & u_z \\ v_x & v_y & v_z \\ w_x & w_y & w_z \end{vmatrix} \\[2mm]
&= \begin{vmatrix} \dfrac{x^2 - y^2 - z^2}{x^2} & \dfrac{2y}{x} & \dfrac{2z}{x} \\ \dfrac{2x}{y} & \dfrac{y^2 - z^2 - x^2}{y^2} & \dfrac{2z}{y} \\ \dfrac{2x}{z} & \dfrac{2y}{z} & \dfrac{z^2 - x^2 - y^2}{z^2} \end{vmatrix}
\end{aligned}
$$

$$= \frac{1}{x^2 y^2 z^2} \begin{vmatrix} x^2 - y^2 - z^2 & 2yx & 2xz \\ 2xy & y^2 - z^2 - x^2 & 2yz \\ 2xz & 2yz & z^2 - x^2 - y^2 \end{vmatrix}$$

$$= \frac{1}{x^3 y^3 z^3} \begin{vmatrix} x(x^2 - y^2 - z^2) & 2yx^2 & 2x^2 z \\ 2xy^2 & y(y^2 - z^2 - x^2) & 2y^2 z \\ 2xz^2 & 2yz^2 & z(z^2 - x^2 - y^2) \end{vmatrix}$$

$$= \frac{1}{x^2 y^2 z^2} \begin{vmatrix} x^2 - y^2 - z^2 & 2x^2 & 2x^2 \\ 2y^2 & y^2 - z^2 - x^2 & 2y^2 \\ 2z^2 & 2z^2 & z^2 - x^2 - y^2 \end{vmatrix}$$

$$= \frac{1}{x^2 y^2 z^2} \begin{vmatrix} x^2 + y^2 + z^2 & x^2 + y^2 + z^2 & x^2 + y^2 + z^2 \\ 2y^2 & y^2 - z^2 - x^2 & 2y^2 \\ 2z^2 & 2z^2 & z^2 - x^2 - y^2 \end{vmatrix}$$

$$= \frac{x^2 + y^2 + z^2}{x^2 y^2 z^2} \begin{vmatrix} 1 & 1 & 1 \\ 2y^2 & y^2 - z^2 - x^2 & 2y^2 \\ 2z^2 & 2z^2 & z^2 - x^2 - y^2 \end{vmatrix}$$

$$= \frac{x^2 + y^2 + z^2}{x^2 y^2 z^2} \begin{vmatrix} 1 & 0 & 0 \\ 2y^2 & -(x^2 + y^2 + z^2) & 0 \\ 2z^2 & 0 & -(x^2 + y^2 + z^2) \end{vmatrix}$$

$$= \frac{(x^2 + y^2 + z^2)^3}{x^2 y^2 z^2} \begin{vmatrix} 1 & 0 & 0 \\ 2y^2 & 1 & 0 \\ 2z^2 & 0 & 1 \end{vmatrix} = \frac{(x^2 + y^2 + z^2)^3}{x^2 y^2 z^2}$$

$$\therefore \frac{\partial(u, v, w)}{\partial(x, y, z)} = \frac{(x^2 + y^2 + z^2)^3}{x^2 y^2 z^2} \quad \therefore \frac{\partial(x, y, z)}{\partial(u, v, w)} = \frac{x^2 y^2 z^2}{(x^2 + y^2 + z^2)^3}.$$

(by **Corollary 4.4.1**).   □

Example 4.11.4 *If* $u_1 = 1 - x_1, u_2 = x_1(1 - x_2), u_3 = x_1 x_2 (1 - x_3), ..., u_n = x_1 x_2 ... x_{n-1}(1 - x_n)$ *then find*

$$\frac{\partial(u_1, u_2, ..., u_n)}{\partial(x_1, x_2, ..., x_n)}.$$

Solution: Let $u_{pq} = \dfrac{\partial u_p}{\partial x_q}$. Then $u_{11} = \dfrac{\partial u_1}{\partial x_1}$.

$$u_{1r} = \frac{\partial u_1}{\partial x_r} = 0 \text{ for } r = 2, 3, ..., n.$$

$$u_{21} = \frac{\partial u_2}{\partial x_1} = 1 - x_2, u_{22} = -x_1, ..., u_{2r} = 0 \text{ for } 3 \le r \le n.$$

Similarly, $u_{33} = -x_1 x_2, u_{3r} = 0 \text{ for } 4 \le r \le n$

....

....

$$u_{ii} = \frac{\partial u_i}{\partial x_i} = -x_1 x_2 ... x_{i-1}, \; u_{ir} = 0 \text{ for } i + 1 \le r \le n.$$

Hence we get

$$\frac{\partial(u_1, u_2, ..., u_n)}{\partial(x_1, x_2, ..., x_n)} = \begin{vmatrix} u_{11} & u_{12} & u_{13} & ... & u_{1n} \\ u_{21} & u_{22} & u_{23} & ... & u_{2n} \\ u_{31} & u_{32} & u_{33} & ... & u_{3n} \\ ... & ... & ... & ... & ... \\ u_{n1} & u_{n2} & u_{n3} & ... & u_{nn} \end{vmatrix}$$

$$= \begin{vmatrix} -1 & 0 & 0 & ... & 0 \\ u_{21} & -x_1 & 0 & ... & 0 \\ u_{31} & u_{32} & -x_1 x_2 & ... & 0 \\ ... & ... & ... & ... & ... \\ u_{n1} & u_{n2} & u_{n3} & ... & -x_1 x_2 ... x_{n-1} \end{vmatrix}$$

$$= (-1)(-x_1)(-x_1 x_2)...(-x_1 x_2...x_{n-1}) = (-1)^n x_1^{n-1} x_2^{n-2} x_3^{n-3}...x_{n-1}. \qquad \square$$

Example 4.11.5 *If $f = f(x, y, z), g = g(x, y, z)$ are differentiable functions of x, y, z and $x = x(p, q), y = y(p, q), z = z(p, q)$ are differentiable functions of p, q, then prove that*

$$\begin{pmatrix} f_p & f_q \\ g_p & g_q \end{pmatrix} = \begin{pmatrix} f_x & f_y & f_z \\ g_x & g_y & g_z \end{pmatrix} \begin{pmatrix} x_p & x_q \\ y_p & y_q \\ z_p & z_q \end{pmatrix}.$$

Also prove that

$$\frac{\partial(f, g)}{\partial(p, q)} = \frac{\partial(f, g)}{\partial(x, y)} \cdot \frac{\partial(x, y)}{\partial(p, q)} + \frac{\partial(f, g)}{\partial(y, z)} \cdot \frac{\partial(y, z)}{\partial(p, q)} + \frac{\partial(f, g)}{\partial(z, x)} \cdot \frac{\partial(z, x)}{\partial(p, q)}$$

where $\dfrac{\partial(f, g)}{\partial(p, q)}$ stands for $f_p g_q - f_q g_p$ and so on.

Solution: **First Part:** We have

$$f_x x_p + f_y y_p + f_z z_p = \frac{\partial f}{\partial x} \frac{\partial x}{\partial p} + \frac{\partial f}{\partial y} \frac{\partial y}{\partial p} + \frac{\partial f}{\partial z} \frac{\partial z}{\partial p} = \frac{\partial f}{\partial p} = f_p.$$

Similarly $f_x x_q + f_y y_q + f_z z_q = f_q, \ g_x x_p + g_y y_p + g_z z_p = g_p$ and $g_x x_q + g_y y_q + g_z z_q = g_q$.

Hence $\begin{pmatrix} f_x & f_y & f_z \\ g_x & g_y & g_z \end{pmatrix} \begin{pmatrix} x_p & x_q \\ y_p & y_q \\ z_p & z_q \end{pmatrix}$

$$= \begin{pmatrix} f_x x_p + f_y y_p + f_z z_p & f_x x_q + f_y y_q + f_z z_q \\ g_x x_p + g_y y_p + g_z z_p & g_x x_q + g_y y_q + g_z z_q \end{pmatrix} = \begin{pmatrix} f_p & f_q \\ g_p & g_q \end{pmatrix}.$$

i.e., R.H.S. = L.H.S.

For the proof of next part

$$\text{L.H.S.} = \frac{\partial(f, g)}{\partial(p, q)} = \begin{vmatrix} f_p & f_q \\ g_p & g_q \end{vmatrix} = f_p g_q - f_q g_p.$$

$$\text{R.H.S.} = \frac{\partial(f, g)}{\partial(x, y)} \cdot \frac{\partial(x, y)}{\partial(p, q)} + \frac{\partial(f, g)}{\partial(y, z)} \cdot \frac{\partial(y, z)}{\partial(p, q)} + \frac{\partial(f, g)}{\partial(z, x)} \cdot \frac{\partial(z, x)}{\partial(p, q)}$$

$$= \begin{vmatrix} f_x & f_y \\ g_x & g_y \end{vmatrix} \begin{vmatrix} x_p & x_q \\ y_p & y_q \end{vmatrix} + \begin{vmatrix} f_y & f_z \\ g_y & g_z \end{vmatrix} \begin{vmatrix} y_p & y_q \\ z_p & z_q \end{vmatrix} + \begin{vmatrix} f_z & f_x \\ g_z & g_x \end{vmatrix} \begin{vmatrix} z_p & z_q \\ x_p & x_q \end{vmatrix}$$

$$= (f_x g_y - f_y g_x)(x_p y_q - x_q y_p) + (f_y g_z - f_z g_y)(y_p z_q - y_q z_p)$$
$$+ (f_z g_x - f_x g_z)(z_p x_q - z_q x_p)$$

$$= f_x g_y x_p y_q - f_x g_y x_q y_p - f_y g_x x_p y_q + f_y g_x x_q y_p + f_y g_z y_p z_q - f_y g_z z_p y_q$$
$$- f_z g_y y_p z_q + f_z g_y z_p y_q + f_z g_x z_p x_q - f_z g_x z_p x_q - f_x g_z z_p x_q + f_x g_z x_p z_q$$

$$= f_x x_p (g_y y_q + g_z z_q) - f_x x_q (g_y y_p + g_z z_p) + f_y y_p (g_x x_q + g_z z_q)$$
$$- f_y y_q (g_z z_p + g_x x_p) + f_z z_p (g_y y_q + g_x x_q) - f_z z_q (g_x x_p + g_y y_p)$$

$$= f_x x_p (g_q - g_x x_q) - f_x x_q (g_p - g_x x_p) + f_y y_p (g_q - g_y y_q)$$
$$- f_y y_q (g_p - g_y y_p) - f_z z_q (g_q - g_z z_p) + f_z z_p (g_q - g_z z_q)$$
$$[\because \ g_p = g_x x_p + g_y y_p + g_z z_p \ etc.]$$

$$= g_q (f_x x_p + f_y y_p + f_z z_p) - g_p (f_x x_q + f_y y_q + f_z z_q) - f_x x_p g_x x_q$$
$$+ f_x x_q g_x x_p - f_y y_p g_y y_q + f_y y_q g_y y_p - f_z z_p g_z z_q + f_z z_q g_z z_p$$

$$= g_q f_p - g_p f_q$$

$$\therefore \ \text{L.H.S.} = \text{R.H.S.}$$

Alternative Method for Second Part

We have $\text{L.H.S.} = \dfrac{\partial(f,g)}{\partial(p,q)} = \begin{vmatrix} f_p & f_q \\ g_p & g_q \end{vmatrix}$

$$= \begin{vmatrix} f_x x_p + f_y y_p + f_z z_p & f_x x_q + f_y y_q + f_z z_q \\ g_x x_p + g_y y_p + g_z z_p & g_x x_q + g_y y_q + g_z z_q \end{vmatrix}$$

$$= \begin{vmatrix} f_x x_p + f_y y_p & f_x x_q + f_y y_q \\ g_x x_p + g_y y_p & g_x x_q + g_y y_q \end{vmatrix} + \begin{vmatrix} f_y y_p + f_z z_p & f_y y_q + f_z z_q \\ g_y y_p + g_z z_p & g_y y_q + g_z z_q \end{vmatrix}$$
$$+ \begin{vmatrix} f_x x_p + f_z z_p & f_x x_q + f_z z_q \\ g_x x_p + g_z z_p & g_x x_q + g_z z_q \end{vmatrix}$$

$$= \begin{vmatrix} f_x & f_y \\ g_x & g_y \end{vmatrix} \begin{vmatrix} x_p & y_p \\ x_q & y_q \end{vmatrix} + \begin{vmatrix} f_y & f_z \\ g_y & g_z \end{vmatrix} \begin{vmatrix} y_p & z_p \\ y_q & z_q \end{vmatrix} + \begin{vmatrix} f_z & f_x \\ g_z & g_x \end{vmatrix} \begin{vmatrix} z_p & x_p \\ z_q & x_q \end{vmatrix}$$

$$= \begin{vmatrix} f_x & f_y \\ g_x & g_y \end{vmatrix} \begin{vmatrix} x_p & x_q \\ y_p & y_q \end{vmatrix} + \begin{vmatrix} f_y & f_z \\ g_y & g_z \end{vmatrix} \begin{vmatrix} y_p & y_q \\ z_p & z_q \end{vmatrix} + \begin{vmatrix} f_z & f_x \\ g_z & g_x \end{vmatrix} \begin{vmatrix} z_p & z_q \\ x_p & x_q \end{vmatrix}$$

$$= \dfrac{\partial(f,g)}{\partial(x,y)} \cdot \dfrac{\partial(x,y)}{\partial(p,q)} + \dfrac{\partial(f,g)}{\partial(y,z)} \cdot \dfrac{\partial(y,z)}{\partial(p,q)} + \dfrac{\partial(f,g)}{\partial(z,x)} \cdot \dfrac{\partial(z,x)}{\partial(p,q)}.$$

$$\therefore \ \text{L.H.S.} = \text{R.H.S.} \qquad \qquad \qquad \qquad \qquad \square$$

Example 4.11.6 *Show that the functions $u = x + 3y + 2z, v = 3x + 4y - 2z, w = 11x + 18y - 2z$, are not independent and find the relation among them.*

Solution: We have

$$J = \frac{\partial(u,v,w)}{\partial(x,y,z)} = \begin{vmatrix} \dfrac{\partial u}{\partial x} & \dfrac{\partial u}{\partial y} & \dfrac{\partial u}{\partial z} \\[2mm] \dfrac{\partial v}{\partial x} & \dfrac{\partial v}{\partial y} & \dfrac{\partial v}{\partial z} \\[2mm] \dfrac{\partial w}{\partial x} & \dfrac{\partial w}{\partial y} & \dfrac{\partial w}{\partial z} \end{vmatrix} = \begin{vmatrix} 1 & 3 & 2 \\ 3 & 4 & -2 \\ 11 & 18 & -2 \end{vmatrix} = \begin{vmatrix} 1 & 3 & 2 \\ 4 & 7 & 0 \\ 12 & 21 & 0 \end{vmatrix}$$

$$= 2 \times \begin{vmatrix} 4 & 7 \\ 12 & 21 \end{vmatrix} = 2 \times 4 \times 7 \times \begin{vmatrix} 1 & 3 \\ 1 & 3 \end{vmatrix} = 0.$$

Hence u, v, w are functionally related, i.e., they are not independent.

Now $2u + 3v = (2x + 6y + 4z) + (9x + 12y - 6z) = 11x + 18y - 2z = w$,

so the required relation is $w = 2u + 3v$. □

Example 4.11.7 *If* $u = \dfrac{x+y}{1-xy}$, $v = \tan^{-1} x + \tan^{-1} y$, *find* $\dfrac{\partial(u,v)}{\partial(x,y)}$. *Are* u, v *functionally related? If so, find the relation.*

Solution: We have

$$\frac{\partial(u,v)}{\partial(x,y)} = \begin{vmatrix} \dfrac{\partial u}{\partial x} & \dfrac{\partial u}{\partial y} \\[2mm] \dfrac{\partial v}{\partial x} & \dfrac{\partial v}{\partial y} \end{vmatrix} = \begin{vmatrix} \dfrac{1+y^2}{(1-xy)^2} & \dfrac{1+x^2}{(1-xy)^2} \\[3mm] \dfrac{1}{1+x^2} & \dfrac{1}{1+y^2} \end{vmatrix} = 0.$$

So u, v are functionally related.

Now, $v = \tan^{-1} x + \tan^{-1} y = \tan^{-1} \dfrac{x+y}{1-xy} = \tan^{-1} u$.

$\therefore\quad v = \tan^{-1} u$ is the required relation. □

Example 4.11.8 *If* $u = \dfrac{x}{y-z}, v = \dfrac{y}{z-x}, w = \dfrac{z}{x-y}$, *show that* $u,\ v,\ w$ *are not independent. Also find the relation.*

Solution: We have

$$\frac{\partial(u,v,w)}{\partial(x,y,z)} = \begin{vmatrix} \dfrac{\partial u}{\partial x} & \dfrac{\partial u}{\partial y} & \dfrac{\partial u}{\partial z} \\[2mm] \dfrac{\partial v}{\partial x} & \dfrac{\partial v}{\partial y} & \dfrac{\partial v}{\partial z} \\[2mm] \dfrac{\partial w}{\partial x} & \dfrac{\partial w}{\partial y} & \dfrac{\partial w}{\partial z} \end{vmatrix} = \begin{vmatrix} \dfrac{1}{y-z} & \dfrac{-x}{(y-z)^2} & \dfrac{x}{(y-z)^2} \\[3mm] \dfrac{y}{(z-x)^2} & \dfrac{1}{z-x} & \dfrac{-y}{(z-x)^2} \\[3mm] \dfrac{-z}{(x-y)^2} & \dfrac{z}{(x-y)^2} & \dfrac{1}{x-y} \end{vmatrix}$$

$$= \frac{1}{(x-y)^2(y-z)^2(z-x)^2} \begin{vmatrix} y-z & -x & x \\ y & z-x & -y \\ -z & z & x-y \end{vmatrix}$$

$$= \frac{1}{(x-y)^2(y-z)^2(z-x)^2} \begin{vmatrix} 0 & -2z & 2y \\ y & z-x & -y \\ -z & z & x-y \end{vmatrix} \quad (R_1' \to R_1 - \overline{R_2 + R_3})$$

$$= \frac{2}{(x-y)^2(y-z)^2(z-x)^2} [z\{y(x-y) - yz\} + y\{yz + z(z-x)\}]$$

$$= \frac{2yz}{(x-y)^2(y-z)^2(z-x)^2} [x - y - z + y + z - x] = 0.$$

Hence u, v, w are functionally related.

Now $\quad uv + vw + wu = \dfrac{x}{y-z}\dfrac{y}{z-x} + \dfrac{y}{z-x}\dfrac{z}{x-y} + \dfrac{z}{x-y}\dfrac{x}{y-z}$

$$= \frac{xy(x-y) + yz(y-z) + zx(z-x)}{(x-y)(y-z)(z-x)} = -\frac{(x-y)(y-z)(z-x)}{(x-y)(y-z)(z-x)} = -1$$

i.e., $uv + vw + wu + 1 = 0$ is the required relation. ☐

Example 4.11.9 *Show that the expression*

$$ax^2 + by^2 + cz^2 + 2fyz + 2gzx + 2hxy$$

can be expressed as a product of two linear factors if $\begin{vmatrix} a & h & g \\ h & b & f \\ g & f & c \end{vmatrix} = 0.$

Solution: Let $u = ax^2 + by^2 + cz^2 + 2fyz + 2gzx + 2hxy = vw$ where $v = lx + my + nz$ and $w = l'x + m'y + n'z$.

Since u, v, w are functionally related, we have

$$\frac{\partial(u, v, w)}{\partial(x, y, z)} = \begin{vmatrix} \dfrac{\partial u}{\partial x} & \dfrac{\partial u}{\partial y} & \dfrac{\partial u}{\partial z} \\ \dfrac{\partial v}{\partial x} & \dfrac{\partial v}{\partial y} & \dfrac{\partial v}{\partial z} \\ \dfrac{\partial w}{\partial x} & \dfrac{\partial w}{\partial y} & \dfrac{\partial w}{\partial z} \end{vmatrix} = 0$$

or, $\begin{vmatrix} ax + hy + gz & hx + by + fz & gx + fy + cz \\ l & m & n \\ l' & m' & n' \end{vmatrix} = 0$

Since x, y, z are independent, the coefficients of x, y, z in the above determinant must separately vanish and then we get

$$\begin{aligned} a(mn' - m'n) + h(nl' - n'l) + g(lm' - l'm) &= 0 \\ h(mn' - m'n) + b(nl' - n'l) + f(lm' - l'm) &= 0 \\ g(mn' - m'n) + f(nl' - n'l) + c(lm' - l'm) &= 0. \end{aligned}$$

Eliminating $(mn' - m'n), (nl' - n'l)$ and $(lm' - l'm)$ form the above equations we get $\begin{vmatrix} a & h & g \\ h & b & f \\ g & f & c \end{vmatrix} = 0.$ ☐

Example 4.11.10 *Prove that* $2xy - \log_e xy = 2$ *determines y as a unique function of x near the point $(1, 1)$ and find* $\dfrac{dy}{dx}$ *at $(1, 1)$.*

Solution: Here $f(x, y) = 2xy - \log_e xy - 2$, $\therefore f_x = 2y - \dfrac{1}{x}$, $f_y = 2y - \dfrac{1}{y}$. Clearly, f, f_x, f_y are continuous in neighbourhood of $(1, 1)$.

Now $f(1, 1) = 0, f_x(1, 1) = 1 \neq 0, f_y(1, 1) = 1 \neq 0$.

Hence $f(x, y) = 0$ determines y as a unique function of x near the point $(1, 1)$. [See **Theorem 4.7.1**].

Now $\left(\dfrac{dy}{dx}\right)_{(1,1)} = -\dfrac{(f_x)_{(1,1)}}{(f_y)_{(1,1)}} = -1.$ ☐

Example 4.11.11 *Show that the function* $f(x, y, z) = xyz + x + y - z$ *defines an implicit function* $z = g(x, y)$ *near* $(0, 0, 0)$ *and find* $g_x(0, 0, 0), \ g_y(0, 0, 0)$.

Solution: Here $f(x,y,z) = xyz + x + y - z$, $\therefore$ $f_x = yz + 1$, $f_y = zx + 1$ and $f_z = xy - 1$ and hence f, f_x, f_y, f_z are clearly continuous in any neighbourhood of $(0,0,0)$. Now $f(0,0,0) = 0$, $f_x(0,0,0) \neq 0$, $f_y(0,0,0) \neq 0$ and $f_z(0,0,0) \neq 0$.

Hence by the existence theorem of three variables (See § 4.8), $f(x,y,z)$ determine z as a function x,y in the neighbourhood of $(0,0,0)$.

Solving for z, we get $z = \dfrac{x+y}{1-xy} = g(x,y)$ and we get

$$g_x(0,0,0) = -\frac{f_x(0,0,0)}{f_z(0,0,0)} = -\frac{-1}{-1} = 1, \; g_y(0,0,0) = -\frac{f_y(0,0,0)}{f_z(0,0,0)} = 1. \qquad \square$$

Example 4.11.12 *If $z = f(x,y)$, $p = \dfrac{\partial z}{\partial x}$, $q = \dfrac{\partial z}{\partial y}$, prove that $p, q, (px + qy - z)$ can be expressed in terms of one of them, if $z_{xx}z_{yy} - z_{xy}^2 = 0$.*

Solution: Let $u = p$, $v = q$, $w = px + qy - z$.

u, v, w can be expressed in terms of one of them when they are functionally related, i.e.,

$$\text{if} \quad \frac{\partial(u,v,w)}{\partial(x,y,z)} = 0, \text{ i.e.,}$$

$$\text{if} \quad \begin{vmatrix} u_x & u_y & u_z \\ v_x & v_y & v_z \\ w_x & w_y & w_z \end{vmatrix} = 0 \Rightarrow \begin{vmatrix} z_{xx} & z_{xy} & 0 \\ z_{yx} & z_{yy} & 0 \\ w_x & w_y & w_z \end{vmatrix} = 0$$

$$\Rightarrow \quad z_{xx}z_{yy} - z_{xy}^2 = 0$$

(assuming $z_{xx}z_{yy}$ and we have $w_z = -1 \neq 0$.) $\qquad \square$

Example 4.11.13 *If $y_r = \dfrac{u_r}{u}$, $r = 1, 2, ..., n$ and if u and u_r are functions of the n independent variables $x_1, x_2, ..., x_n$, prove that*

$$\frac{\partial(y_1, y_2, ..., y_n)}{\partial(x_1, x_2, ..., x_n)} = \frac{1}{u^{n+1}} \begin{vmatrix} u & \dfrac{\partial u}{\partial x_1} & \dfrac{\partial u}{\partial x_2} & \cdots & \dfrac{\partial u}{\partial x_n} \\ u_1 & \dfrac{\partial u_1}{\partial x_1} & \dfrac{\partial u_1}{\partial x_2} & \cdots & \dfrac{\partial u_1}{\partial x_n} \\ u_2 & \dfrac{\partial u_2}{\partial x_1} & \dfrac{\partial u_2}{\partial x_2} & \cdots & \dfrac{\partial u_2}{\partial x_n} \\ \cdots & \cdots & \cdots & \cdots & \cdots \\ u_n & \dfrac{\partial u_n}{\partial x_1} & \dfrac{\partial u_n}{\partial x_2} & \cdots & \dfrac{\partial u_n}{\partial x_n} \end{vmatrix}.$$

Solution: Given $y_r = \dfrac{u_r}{u}$, $r = 1, 2, ..., n$

$$\therefore \quad \frac{\partial y_r}{\partial x_s} = \frac{1}{u}\frac{\partial u_r}{\partial u_s} - \frac{u_r}{u^2}\frac{\partial u}{\partial x_s}$$

$$\therefore \quad \frac{\partial(y_1, y_2, ..., y_n)}{\partial(x_1, x_2, ..., x_n)}$$

$$= \begin{vmatrix} \dfrac{1}{u}\dfrac{\partial u_1}{\partial x_1} - \dfrac{u_1}{u^2}\dfrac{\partial u}{\partial x_1} & \dfrac{1}{u}\dfrac{\partial u_1}{\partial x_2} - \dfrac{u_1}{u^2}\dfrac{\partial u}{\partial x_2} & \cdots & \dfrac{1}{u}\dfrac{\partial u_1}{\partial x_n} - \dfrac{u_1}{u^2}\dfrac{\partial u}{\partial x_n} \\[2ex] \dfrac{1}{u}\dfrac{\partial u_2}{\partial x_1} - \dfrac{u_2}{u^2}\dfrac{\partial u}{\partial x_1} & \dfrac{1}{u}\dfrac{\partial u_2}{\partial x_2} - \dfrac{u_2}{u^2}\dfrac{\partial u}{\partial x_2} & \cdots & \dfrac{1}{u}\dfrac{\partial u_2}{\partial x_n} - \dfrac{u_2}{u^2}\dfrac{\partial u}{\partial x_n} \\[1ex] \cdots & \cdots & \cdots & \cdots \\[1ex] \dfrac{1}{u}\dfrac{\partial u_n}{\partial x_1} - \dfrac{u_n}{u^2}\dfrac{\partial u}{\partial x_1} & \dfrac{1}{u}\dfrac{\partial u_n}{\partial x_2} - \dfrac{u_n}{u^2}\dfrac{\partial u}{\partial x_2} & \cdots & \dfrac{1}{u}\dfrac{\partial u_n}{\partial x_n} - \dfrac{u_n}{u^2}\dfrac{\partial u}{\partial x_n} \end{vmatrix}$$

$$= \frac{1}{u^n}\begin{vmatrix} 1 & 0 & \cdots & 0 \\[1ex] u_1 & \dfrac{\partial u_1}{\partial x_1} - \dfrac{u_1}{u}\dfrac{\partial u}{\partial x_1} & \cdots & \dfrac{\partial u_1}{\partial x_n} - \dfrac{u_1}{u}\dfrac{\partial u}{\partial x_n} \\[2ex] u_2 & \dfrac{\partial u_2}{\partial x_1} - \dfrac{u_1}{u}\dfrac{\partial u}{\partial x_1} & \cdots & \dfrac{\partial u_1}{\partial x_n} - \dfrac{u_1}{u}\dfrac{\partial u}{\partial x_n} \\[1ex] \cdots & \cdots & \cdots & \cdots \\[1ex] u_n & \dfrac{\partial u_n}{\partial x_1} - \dfrac{u_n}{u}\dfrac{\partial u}{\partial x_1} & \cdots & \dfrac{\partial u_n}{\partial x_n} - \dfrac{u_n}{u}\dfrac{\partial u}{\partial x_n} \end{vmatrix}$$

(Taking out $\dfrac{1}{u}$ form each column and bordering the determinant.)

$$= \frac{1}{u^n}\begin{vmatrix} 1 & \dfrac{1}{u}\dfrac{\partial u}{\partial x_1} & \dfrac{1}{u}\dfrac{\partial u}{\partial x_2} & \cdots & \dfrac{1}{u}\dfrac{\partial u}{\partial x_n} \\[2ex] u_1 & \dfrac{\partial u_1}{\partial x_1} & \dfrac{\partial u_1}{\partial x_2} & \cdots & \dfrac{\partial u_1}{\partial x_n} \\[2ex] u_2 & \dfrac{\partial u_2}{\partial x_1} & \dfrac{\partial u_2}{\partial x_2} & \cdots & \dfrac{\partial u_2}{\partial x_n} \\[1ex] \cdots & \cdots & \cdots & \cdots & \cdots \\[1ex] u_n & \dfrac{\partial u_n}{\partial x_1} & \dfrac{\partial u_n}{\partial x_2} & \cdots & \dfrac{\partial u_n}{\partial x_n} \end{vmatrix} \qquad \begin{pmatrix} C_2' \to C_2 + \dfrac{1}{u}\dfrac{\partial u}{\partial x_1} \\[1ex] C_3' \to C_3 + \dfrac{1}{u}\dfrac{\partial u}{\partial x_2} \\[1ex] \cdots \\[1ex] C_n' \to C_n + \dfrac{1}{u}\dfrac{\partial u}{\partial x_n} \end{pmatrix}$$

$$= \frac{1}{u^{n+1}}\begin{vmatrix} u & \dfrac{\partial u}{\partial x_1} & \dfrac{\partial u}{\partial x_2} & \cdots & \dfrac{\partial u}{\partial x_n} \\[2ex] u_1 & \dfrac{\partial u_1}{\partial x_1} & \dfrac{\partial u_1}{\partial x_2} & \cdots & \dfrac{\partial u_1}{\partial x_n} \\[2ex] u_2 & \dfrac{\partial u_2}{\partial x_1} & \dfrac{\partial u_2}{\partial x_2} & \cdots & \dfrac{\partial u_2}{\partial x_n} \\[1ex] \cdots & \cdots & \cdots & \cdots & \cdots \\[1ex] u_n & \dfrac{\partial u_n}{\partial x_1} & \dfrac{\partial u_n}{\partial x_2} & \cdots & \dfrac{\partial u_n}{\partial x_n} \end{vmatrix} .$$

$\square$

4.12 Exercises

1. If $u = \dfrac{yz}{x}, v = \dfrac{zx}{y}$ and $w = \dfrac{xy}{z}$, find $\dfrac{\partial(x, y, z)}{\partial(u, v, w)}$.

2. If $u = x + y + z, uv = y + z$ and $uvw = z$, prove that $\dfrac{\partial(x, y, z)}{\partial(u, v, w)} = u^2 v$.

3. Let $u = \dfrac{x}{\sqrt{1 - r^2}}, v = \dfrac{y}{\sqrt{1 - r^2}}, w = \dfrac{z}{\sqrt{1 - r^2}}$ where $r^2 = x^2 + y^2 + z^2$. Prove that $\dfrac{\partial(u, v, w)}{\partial(x, y, z)} = \dfrac{1}{(1 - r^2)^{\frac{5}{2}}}$.

4. If γ, θ, ϕ are roots (real) of the following equation in p

$$\frac{x}{\alpha_1 + p} + \frac{y}{\alpha_2 + p} + \frac{z}{\alpha_3 + p} = 1,$$

find the Jacobian $\dfrac{\partial(x, y, z)}{\partial(\gamma, \theta, \phi)}$.

[*Hints: Proceed as in **Example 4.11.1**.*]

5. If $x = c \cos u \cosh v, y = c \sin u \sinh v$, prove that $\dfrac{\partial(x, y)}{\partial(u, v)} = \dfrac{c^2}{2}(\cos 2u - \cosh 2v)$.

6. If $u = \cos x, v = \sin x \cos y, w = \sin x \sin y \cos z$, then show that

$$\frac{\partial(x, y, z)}{\partial(u, v, w)} = (-1)^3 \sin^3 x \sin^2 y \sin z.$$

[*Hints: Proceed as in **Example 4.2.2**.*]

7. .Prove that the Jacobian

$$\frac{\partial(y_1, y_2, ..., y_n)}{\partial(x_1, x_2, ..., x_n)} = x_1^{n-1} x_2^{n-2} . x_3^{n-3} ... x_{n-1}$$

if $y_1 = x_1(1 - x_1), y_2 = x_1 x_2(1 - x_3), y_3 = x_1 x_2 x_3(1 - x_4), ..., y_{n-1} = x_1 x_2 ... x_{n-1}(1 - x_n)$

[*Hints: We have*

$$J = \frac{\partial(y_1, y_2, ..., y_n)}{\partial(x_1, x_2, ..., x_n)} = \begin{vmatrix} \dfrac{\partial y_1}{\partial x_1} & \dfrac{\partial y_1}{\partial x_2} & \cdots & \dfrac{\partial y_1}{\partial x_n} \\ \dfrac{\partial y_2}{\partial x_1} & \dfrac{\partial y_2}{\partial x_2} & \cdots & \dfrac{\partial y_2}{\partial x_n} \\ \cdots & \cdots & & \cdots \\ \dfrac{\partial y_n}{\partial x_1} & \dfrac{\partial y_n}{\partial x_2} & \cdots & \dfrac{\partial y_n}{\partial x_n} \end{vmatrix}$$

$$= \begin{vmatrix} 1-x_2 & -x_1 & 0 \\ x_2(1-x_3) & x_1(1-x_3) & -x_1x_2 \\ x_2x_3(1-x_4) & x_1x_3(1-x_4) & x_1x_2(1-x_4) \\ \cdots & \cdots & \cdots \\ x_2x_3...x_{n-1}(1-x_n) & x_1x_3x_4...x_{n-1}(1-x_n) & x_1x_2x_4...x_{n-1}(1-x_n) \\ x_2x_3...x_{n-1}x_n & x_1x_3x_4...x_{n-1}x_n & x_1x_2x_4...x_{n-1}x_n \end{vmatrix}$$

$$\begin{matrix} \cdots & 0 \\ \cdots & 0 \\ \cdots & 0 \\ \cdots & \cdots \\ \cdots & -x_1x_2...x_{n-1} \\ \cdots & x_1x_2...x_{n-1} \end{matrix}$$

$$= \begin{vmatrix} 1-x_2 & -x_1 & 0 \\ x_2(1-x_3) & x_1(1-x_3) & -x_1x_2 \\ x_2x_3(1-x_4) & x_1x_3(1-x_4) & x_1x_2(1-x_4) \\ \cdots & \cdots & \cdots \\ x_2x_3...x_{n-1}(1-x_n) & x_1x_3x_4...x_{n-1}(1-x_n) & x_1x_2x_4...x_{n-1}(1-x_n) \\ 1 & 0 & 0 \end{vmatrix}$$

$$\begin{matrix} \cdots & 0 \\ \cdots & 0 \\ \cdots & 0 \\ \cdots & \cdots \\ \cdots & -x_1x_2...x_{n-1} \\ \cdots & 0 \end{matrix} \qquad (R'_n \rightarrow R_1 + R_2 + ... + R_n)$$

$$= (-1)^{n-1}[(-x_1)(-x_1x_2)(-x_1x_2x_3)...(-x_1x_2...x_{n-1})]$$

(Expanding in terms of the elements of the last row)

$$= (-1)^{n-1}(-1)^{n-1}x_1^{n-1}x_2^{n-2}.x_3^{n-3}...x_{n-1}]$$

8. If $u^3 = xyz, \dfrac{1}{v} = \dfrac{1}{x} + \dfrac{1}{y} + \dfrac{1}{z}$ and $w^2 = x^2 + y^2 + z^2$, prove that

$$\frac{\partial(u, v, w)}{\partial(x, y, z)} = \frac{v(y-z)(z-x)(x-y)(x+y+z)}{3u^2w(yz + zx + xy)}.$$

9. Show that the function $u = x+y+z$, $v = xy+yz+zx$, $w = x^3+y^3+z^3-3xyz$ are functionally related. Find the relation.

[*Hints: Show that* $\dfrac{\partial(u, v, w)}{\partial(x, y, z)} = 0$; *Required relation is* $u^3 = 3uv + w$.]

10. If $f(0) = 0, f'(0) = \dfrac{1}{1+x^2}$, prove without using the method of integration

that $f(x) + f(y) = f\left(\dfrac{x+y}{1-xy}\right).$

[*Hints: Let* $u = f(x) + f(y), v = \dfrac{x+y}{1-xy}$

$$\therefore \quad J = \frac{\partial(u,v)}{\partial(x,y)} = \begin{vmatrix} \dfrac{\partial u}{\partial x} & \dfrac{\partial u}{\partial y} \\ \dfrac{\partial v}{\partial x} & \dfrac{\partial v}{\partial y} \end{vmatrix} = \begin{vmatrix} f'(x) & f'(y) \\ \dfrac{1+y^2}{(1-xy)^2} & \dfrac{1+x^2}{(1-xy)^2} \end{vmatrix}$$

$$= \begin{vmatrix} \dfrac{1}{1+x^2} & \dfrac{1}{1+y^2} \\ \dfrac{1+y^2}{(1-xy)^2} & \dfrac{1+x^2}{(1-xy)^2} \end{vmatrix} = 0.$$

Hence u and v are connected by a functional relation, which is given by $u = \phi(v)$.

$$\therefore \quad f(x) + f(y) = \phi\left(\frac{x+y}{1-xy}\right)$$

When $y = 0$, using $f(0) = 0$, we get $f(x) = \phi(x)$ and so we get

$$f(x) + f(y) = f\left(\frac{x+y}{1-xy}\right)\bigg].$$

11. Let $u = \dfrac{x+y}{1-xy}, v = \dfrac{(x+y)(1-xy)}{(1+x^2)(1+y^2)}$. Find $\dfrac{\partial(u,v)}{\partial(x,y)}$. Are they functionally related? If so, find the relationship.

12. Show that the functions $u = x+y-z, v = x-y+z, w = x^2+(y-z)^2$ are dependent. Find the relation.

13. Show that the functions $u = x^2+y^2+z^2, v = x+y+z, w = xy+yz+zx$ are functionally related and find the relation between them.

14. Show that $ax^2+2hxy+by^2$ and $Ax^2+2Hxy+By^2$ are independent unless $\dfrac{a}{A} = \dfrac{h}{H} = \dfrac{b}{B}$.

15. State the theorem on existence and uniqueness of an implicit function. Deduce that $x^2 + xy + y^2 = 1$ defines a function of x in some neighbourhood of $(1,0)$. Also find the first derivative of the function so defined at $x = 1$.

[*Hints: First part of* § 4.8]

16. If $u = f(x,y)$ and $v = g(x,y)$ have continuous partial derivatives in a region R of xy-plane, show that a necessary and sufficient condition that they satisfy a functional relation $F(u,v) = 0$ is that the Jacobian $\dfrac{\partial(u,v)}{\partial(x,y)} = 0$.

[*Hints: See **Theorem 4.5.1**.*]

17. What do you mean by a implicit function $y = \phi(x)$, defined by $F(x,y) = 0$ near (a,b)? Verify implicit function theorem for $y^3 \cos x + y^2 \sin^2 x - 7 = 0$ near $\left(\dfrac{\pi}{3}, 2\right)$.

18. Examine the following equations for the existence of unique implicit function near the points indicated and verify by direct calculation. Also find the first derivatives of the solutions whenever these exists

    (a) $x^2 + y^2 - 1 = 0, (0,1)$

    (b) $xy \sin x + \cos y = 0, \left(0, \dfrac{\pi}{2}\right)$

    (c) $y^3 \cos x + y^2 \sin^2 x = 7, \left(\dfrac{\pi}{3}, 2\right)$

    (d) $y^4 + x^2 y^2 - 2x^5 = 0, (1,1)$

    (e) $x^2 y^2 + x^2 + y^2 - 1 = 0, (0,1)$.

19. Examine the equation $f(x,y) = x^2 + x^3 y + y^2$ for the existence of unique implicit function at $(0,0)$.

20. Write the conditions so that the function $f(x,y) = 0$ does define an implicit function. Show that the equation $y^2 - yx^2 - 2x^5 = 0$ determines uniquely implicit function in the neighbourhood of the point $(1,-1)$. Also find the first order derivative of the function.

    [*Hints:* First part: **Definition** § 4.6; Second part: **Example 4.7.2**]

21. State the implicit function theorem for a function of two variables. Show that $2xy - \log xy = 2e - 1$ determines y uniquely as a function of x near the point $(1,e)$ and find $\dfrac{dy}{dx}$ at $(1,e)$.

    [*Hints:* $\left(\dfrac{dy}{dx}\right)_{(1,e)} = \left(-\dfrac{2y - \frac{1}{x}}{2x - \frac{1}{y}}\right)_{(1,e)} = -\dfrac{2e-1}{2-1} = 1 - 2e.$]

22. If $F(x,y,z) = x^2 + y^2 + z^2 - 1$, then show that the equation $F = 0$ defines z as a function of x and y in a neighbourhood of any point $P(a,b,c)$ for which $F(P) = 0$. Show also that $z_x = \dfrac{x}{z}$ and $z_y = \dfrac{y}{z}$.

23. Given $f(x + y) = \dfrac{f(x) + f(y)}{1 - f(x)f(y)}$ where $f(x)f(y) \neq 1$, x and y are independent, $f(t)$ is a differentiable function of t and $f(0) = 0$. Using the property of Jacobian, show that $f(t) = \tan \alpha t$ where α is a constant.

24. Prove that the functions $\phi_1 = x + y + z + t, \phi_2 = x^2 + y^2 + z^2 + t^2, \phi_3 = x^3 + y^3 + z^3 + t^3, \phi_4 = xyz + xyt + xzt + yzt$ are dependent. Obtain the relation $\phi_1^3 - 3\phi_1\phi_2 + 2\phi_3 - 6\phi_4 = 0$.

25. If $u = 3x + 2y - z, v = x - 2y + z, w = x(x + 2y - z)$, then show that u, v and w are connected by a functional equation and find the equation.

    [*Hints:* $2y - z = u - 3x = x - v \Rightarrow 4x = u + v$]

ANSWERS

1. $\frac{1}{4}$; **4.** $\dfrac{\partial(x,y,z)}{\partial(\gamma,\theta,\phi)} = -\dfrac{(\theta-\phi)(\phi-\gamma)(\gamma-\theta)}{(\alpha_2-\alpha_3)(\alpha_3-\alpha_1)(\alpha_1-\alpha_2)}$; **9.** $u^3 = 3uv + w$; **11.** $J = 0$, Yes, $v = \dfrac{u}{1+u^2}$; **12.** $(u+v)^2 + (u-v)^2 = 4w$; **13.** $v^2 = u + 2w$; **15.** $\left(\dfrac{dy}{dx}\right)_{(1,0)} = -2$ [Note that at $x = 1, y = 0$.]; **18.** (a) $\left(\dfrac{dy}{dx}\right)_{(0,1)} = 0$; (b). $\left(\dfrac{dy}{dx}\right)_{(0,\frac{\pi}{2})} = 0$; (c). $\left(\dfrac{dy}{dx}\right)_{(\frac{\pi}{3},2)} = \dfrac{2\sqrt{3}}{9}$; (d) $\left(\dfrac{dy}{dx}\right)_{(1,1)} = \left(-\dfrac{2xy^2 - 10x^4}{4y^3 + 2x^2 y}\right)_{(1,1)} = \dfrac{4}{3}$; **19.** $f(x,y) = 0$ does not determine uniquely x as a function of y or y as a function of x in the neighbourhood of $(0,0)$; **20.** $\left(\dfrac{dy}{dx}\right)_{(1,-1)} = -2\frac{2}{3}$; **21.** $\left(\dfrac{dy}{dx}\right)_{(1,e)} = 1 - 2e$; **25.** $u^2 - v^2 = 8w$.

Chapter 5

Extrema of Functions of Several Variables

5.1 Introduction

In the calculus of one variable the study of local and global extrema have been done in details. This chapter deals with the same for multi-variable functions. This study is of great importance not only in the field of theoretical Mathematics but also in various branches of applied Mathematical areas including Physics, Engineering and Technologies as well as in Economics and Financial sectors. Very recently the development of optimization theory are being used in several big financial concerns and famous companies where profits and losses are controlled by applying the theories of extrema.

The concept of maximum or minimum values of the functions of several variables, definition of extreme point, unconstrained and constrained optimization problems along with Lagrange Multiplier method are to be discussed in the present chapter.

5.2 Maxima and Minima of Functions of Two Variables

Definition 5.2.1 *A function $f(x, y)$ is said to have a **maximum** or **minimum** at $x = a, y = b$, according as $f(a + h, b + k) <$ or $> f(a, b)$ for all positive as well as negative small values of h and k.*

*In other words, if $f(a+h, b+k) - f(a, b)$ always **maintains the same sign** for all small values of h, k and if this sign is **negative**, then $f(a, b)$ is a **maximum**. If this sign is **positive**, $f(a, b)$ is a **minimum**.*

Example 5.2.1 *The function $f(x, y) = (x-1)^2 + (y-2)^2 - 1$ attains a minimum at $x = 1, y = 2$, i.e., at the point $(1, 2)$. Indeed $f(1, 2) = -1$ and since $(x-1)^2 + (y-2)^2$ is always positive for $x \neq 1, y \neq 2$, it follows that $(x - 1)^2 + (y - 2)^2 - 1 > -1$ that is $f(x, y) > f(1, 2)$.*

*If we write $z = (x - 1)^2 + (y - 2)^2 - 1$ then we have the following geometrical analogy. See **Figure** 5.1.*

Here $z = (x - 1)^2 + (y - 2)^2 - 1$ represents the lower hemispherical surface with its lowest point at $(0, 0, -1)$.

179

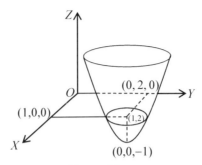

Figure 5.1

Note 5.2.1 *Considering $z = f(x, y)$ as a surface,* **maximum value** *of z occurs at the* **top** *of an elevation (e.g., a dome) from which the surface descends in every direction (see* **Example 5.2.2**) *and a* **minimum value** *occurs at the* **bottom** *of a depression (e.g., a bow) from which the surface ascends in every direction. Some times the maximum or minimum value may form a* **ridge** *such that the surface descends or ascends in all direction except that of the ridge. Besides these, we have such a point of the surface, where tha tangent plane is horizontal and the surface falls for displacement in certain directions and rises for displacement in other directions. Such a point is called a* **saddle point.**

Note 5.2.2 *A maximum or minimum value of a function is called an* **extreme value.**

Note 5.2.3 *The above definitions of maximum or minimum at any point (a, b) are also called a* **local maximum** *or* **local minimum** *at that point.*

Example 5.2.2 *For $a > 0$, the function $z = \sqrt{a^2 - x^2 - y^2}$ represents the upper hemispherical surface with its highest point at $(0, 0, 0)$.*

Example 5.2.3 *If $f(x, y) = |xy|$, then f is not an extreme at O. For although $f(O) = f(0) = 0$, yet there are points P other than O such that $f(P) = 0$.*

Also $f(P) - f(O)$ does not always maintain its sign and some times it vanishes.

Again, when $f(x, y) = xy$, then also f has no extreme at O. For although $f(O) = 0$, yet

$$f(P_1) = f(1, 0) = 0 = f(O)$$
$$f(P_2) = f(1, 1) = 1 > 0 = f(O)$$
$$f(P_3) = f(1, -1) = -1 < 0 = f(O)$$

showing that the sign of the difference $f(P) - f(O)$ is not maintained.

5.3 The Necessary Conditions for Extreme Values of a Function of Two Variables

Theorem 5.3.1 *If a function $f(x, y)$ be a maximum or a minimum at (a, b) and if the partial derivatives f_x, f_y exist at the point (a, b) then $f_x(a, b) = 0$ and $f_y(a, b) = 0$.*

Proof: If $f(a, b)$ is a maximum or minimum value of $f(x, y)$, then clearly $f(x, b)$, which is a function of a single variable x only, is also a maximum or a minimum at $x = a$, therefore if $f_x(a, b)$ exists it must be equal to zero. Similarly, we have $f_y(a, b) = 0$. $\qquad\qquad\qquad\qquad\qquad\qquad\qquad\qquad\qquad\qquad\qquad\qquad$ □

5.4   Sufficient Conditions for Maximum or Minimum

Theorem 5.4.1 *If the function $f(x, y)$ has continuous partial derivatives of first and second order in the neighbourhood of the point (a, b) and if $f_x(a, b) = 0$, $f_y(a, b) = 0$, then*

(i) *f is maximum at (a, b) if $f_{xx}(a, b).f_{yy}(a, b) - f_{xy}^2(a, b) > 0$ and $f_{xx}(a, b) < 0$*

(ii) *f is minimum at (a, b) if $f_{xx}(a, b).f_{yy}(a, b) - f_{xy}^2(a, b) > 0$ and $f_{xx}(a, b) > 0$*

(iii) *if $f_{xx}(a, b).f_{yy}(a, b) - f_{xy}^2(a, b) < 0$ then f is neither a maximum nor a minimum at (a, b)*

(iv) *if $f_{xx}(a, b).f_{yy}(a, b) - f_{xy}^2(a, b) = 0$, no conclusion can be drawn and in this case, further analysis is required.*

Proof: Expanding $f(x, y)$ using Taylor's theorem in the neighbourhood of (a, b) up to third order, we get

$$f(a + h, b + k) = f(a, b) + \left(h\frac{\partial}{\partial x} + k\frac{\partial}{\partial y} \right) f(a, b)$$

$$+ \frac{1}{2!} \left(h\frac{\partial}{\partial x} + k\frac{\partial}{\partial y} \right)^2 f(a, b) + R_3$$

$$\text{where } R_3 = \frac{1}{3!} \left(h\frac{\partial}{\partial x} + k\frac{\partial}{\partial y} \right)^3 f(a + \theta h, b + \theta k)$$

or, $\qquad f(a + h, b + k) - f(a, b) = \frac{1}{2}(Ah^2 + 2Bhk + Ck^2) + R_3$

where $A = f_{xx}(a, b)$, $B = f_{xy}(a, b)$ and $C = f_{yy}(a, b)$.

From this we see that for sufficiently small values of h, k, sign of $f(a + h, b + k) - f(a, b)$ is same as the sign of $Ah^2 + 2Bhk + Ck^2$.

Now, $\quad \frac{1}{2}(Ah^2 + 2Bhk + Ck^2) = \frac{1}{2A}[A^2h^2 + 2Ahk + B^2k^2 + (AC - B^2)k^2]$

$$= \frac{1}{2A} \left[(Ah + Bk)^2 + (AC - B^2)k^2 \right]$$

This shows that

(i) if $AC - B^2 > 0$ and $A < 0$, then $f(a + h, b + k) - f(a, b) < 0$ implies $f(a, b)$ is maximum.

(ii) if $AC - B^2 > 0$ and $A > 0$, then $f(a + h, b + k) - f(a, b) > 0$ implies $f(a, b)$ is minimum.

(iii) if $AC - B^2 < 0$ and $A \neq 0$, then

$$f(a + h, b + k) - f(a, b)$$

$$= \frac{1}{2A}[\text{a positive quantity} + \text{a negative quantity}]$$

$$= \frac{1}{2A}[\text{a positive or a negative quantity}].$$

So $f(a,b)$ is neither a maximum nor a minimum value of $f(x,y)$. Similar results occur when $C \neq 0$.

Again, when $A = 0, C = 0$, then $f(a+h, b+k) - f(a,b) = Bhk$ which does not keep the same sign in the neighbourhood of (a,b). Hence $f(a,b)$ is neither a maximum nor a minimum value of $f(x,y)$.

(iv) if $AC - B^2 = 0$ and $A \neq 0$, then $f(a+h, b+k) - f(a,b) = R_3$ when $Ah + Bk = 0$. So the sign of $f(a+h, b+k) - f(a,b)$ depends on sign of R_3 and hence no definite conclusion on the extreme value of $f(x,y)$ at (a,b) can be made.

Again, if $AC - B^2 = 0$ and $A = 0$, then $B = 0$ and

$$f(a+h, b+k) - f(a,b) = \frac{1}{2}Ck^2 + R_3 = R_3 \text{ if } k = 0.$$

Hence, again, no definite conclusion on the extreme value of $f(x,y)$ at (a,b) can be drawn.   □

5.5 Stationary Value

Definition 5.5.1 *$f(a,b)$ is said to be a stationary value of $f(x,y)$, if $f_x(a,b) = 0$, $f_y(a,b) = 0$, i.e., the function is stationary at the point (a,b). The point (a,b) is said to be the stationary point.*

Note 5.5.1 *Every extreme value is a stationary value but the converse may not be true.*

5.6 Working Rule to Find the Maximum and Minimum of a Function

1. For Two Variables

Step 1: Find $p = \dfrac{\partial f}{\partial x}$ and $q = \dfrac{\partial f}{\partial y}$ and equate each to zero. Solve these as simultaneous equations in x and y. Let $(a,b), (c,d), \dots$ be the pairs of values.

Step 2: Calculate the values of $r = \dfrac{\partial^2 f}{\partial x^2}$, $s = \dfrac{\partial^2 f}{\partial x \partial y}$ and $t = \dfrac{\partial^2 f}{\partial y^2}$ for each pair of values.

Step 3: (i) if $rt - s^2 > 0$ and $r < 0$ at (a,b), $f(a,b)$ is a maximum value.

(ii) if $rt - s^2 > 0$ and $r > 0$ at (a,b), $f(a,b)$ is a minimum value.

(iii) if $rt - s^2 < 0$ at (a,b), $f(a,b)$ is not an extreme value, i.e., (a,b) is a saddle point.

(iv) if $rt - s^2 = 0$ at (a,b), the case is doubtful and needs further investigation.

Similarly, examine the other pairs of values one by one.

2. For Three Variables

Let $F(x, y, z)$ be a function of three independent variables x, y and z.

$$\text{For extremum,} \quad dF = 0 \quad \Rightarrow \quad \frac{\partial F}{\partial x} = \frac{\partial F}{\partial y} = \frac{\partial F}{\partial z} = 0 \tag{5.1}$$

and

$$
\begin{aligned}
d^2 F &= \left(dx\frac{\partial}{\partial x} + dy\frac{\partial}{\partial y} + dz\frac{\partial}{\partial z} \right)^2 F \\
&= \frac{\partial^2 F}{\partial x^2} dx^2 + \frac{\partial^2 F}{\partial y^2} dy^2 + \frac{\partial^2 F}{\partial z^2} dz^2 + 2\frac{\partial^2 F}{\partial x \partial y} dxdy \\
&\qquad\qquad +2\frac{\partial^2 F}{\partial y \partial z} dydz + 2\frac{\partial^2 F}{\partial z \partial x} dzdx \\
&= F_{xx}(dx)^2 + F_{yy}(dy)^2 + F_{zz}(dz)^2 + 2F_{xy}dxdy \\
&\qquad\qquad +2F_{yz}dydz + 2F_{zx}dzdx \tag{5.2}
\end{aligned}
$$

(See **Note 2.9.1**)

Let a set of values of x, y, z obtained by solving the system of equations (5.1) be a, b, c. The values of triads (a, b, c) will give the stationary values of $F(x, y, z)$.

The extreme point (a, b, c) will be a maximum point or a minimum point according as $d^2 F < 0$ or $d^2 F > 0$.

The point (a, b, c) will neither be a maximum nor a minimum if $d^2 F$ does not keep the same sign in the neighbourhood of this point.

Further investigation is to be required if $d^2 F$ keeps the same sign but vanishes at some points in the neighbourhood of stationary point (a, b, c).

To discuss maximum and minimum values at (a, b, c), we find the following six partial derivatives of second order

$$
A = F_{xx} = \frac{\partial^2 F}{\partial x^2}, B = F_{yy} = \frac{\partial^2 F}{\partial y^2}, C = F_{zz} = \frac{\partial^2 F}{\partial z^2}, F = F_{yz} = \frac{\partial^2 F}{\partial y \partial z},
$$

$$
G = F_{zx} = \frac{\partial^2 F}{\partial z \partial x}, H = F_{xy} = \frac{\partial^2 F}{\partial x \partial y}.
$$

The matrix of the quadratic form shown in (5.2) is given by

$$
\begin{bmatrix} F_{xx} & F_{xy} & F_{xz} \\ F_{yx} & F_{yy} & F_{yz} \\ F_{zx} & F_{zy} & F_{zz} \end{bmatrix} = \begin{bmatrix} A & H & G \\ H & B & F \\ G & F & C \end{bmatrix}.
$$

Now form the theory of **Quadratic Forms** that this form, i.e., $d^2 F$ is *positive definite*, if and only if the three principal minors A, $\begin{vmatrix} A & H \\ H & B \end{vmatrix}$ and $\begin{vmatrix} A & H & G \\ H & B & F \\ G & F & C \end{vmatrix}$

are all *positive*, $d^2 F$ is always *negative*, if and only if the signs of above three minors are alternatively *negative* and *positive*.

On the other hand, if the signs of the above three minors *neither be all positive nor having alternatively negative and positive* at (a, b, c), the function $F(x, y, z)$ is *neither a maximum nor a minimum thereat*.

Note 5.6.1 *To find the maximum and minimum of the function at stationary point, in practice, it is sufficient to find the value of a second order partial derivative of the function with respect to **any of the independent variables**. The value of the function is then **maximum** or **minimum** according as the value of this second order partial derivative at the stationary point under consideration is **negative** or **positive**.*

Example 5.6.1 *Examine the following function for extreme values*
$$u = f(x, y) = 2(x - y)^2 - x^4 - y^4.$$

Solution: For stationary point,
$$f_x = 4(x - y) - 4x^3 = 0 \text{ and } f_y = -4(x - y) - 4y^3 = 0$$
$$\Rightarrow \quad x^3 + y^3 = 0, \text{ i.e., } (x + y)(x^2 - xy + y^2) = 0$$
or, $x + y = 0$, $x^2 - xy + y^2 = 0$.

Considering $x - y - x^3 = 0$ and $x + y = 0$ we get
$$(x, y) \equiv (0, 0), \left(\sqrt{2}, -\sqrt{2}\right) \text{ and } \left(-\sqrt{2}, \sqrt{2}\right).$$

Again, considering $x - y - x^3 = 0$ and $x^2 - xy + y^2 = 0$, we get $(0, 0)$ is the only real solution.

Now, $r = f_{xx} = 4 - 12x^2, s = f_{xy} = -4, t = f_{yy} = 4 - 12y^2$

$\therefore$ At $(0, 0)$, $r = 4$, $s = -4$, $t = 4$. So,

for $(0, 0)$, $rt - s^2 = 0$ and accordingly this case needs further consideration and for $\left(\sqrt{2}, -\sqrt{2}\right)$, $r = 4 - 12.2 = -20$, $s = -4$, $t = 4 - 12.2 = -20$, so that $rt - s^2 = (-20)(-20) - (-4)^2 > 0$ with $r < 0$. Therefore, $f\left(\sqrt{2}, -\sqrt{2}\right)$ is a maximum value.

Similarly, $f\left(-\sqrt{2}, \sqrt{2}\right)$ is also a maximum value.

Now we consider the case at $(0, 0)$, where $f(0, 0) = 0$.
For point $(x, 0)$ along x-axis the value of the function is $2x^2 - x^4 = x^2(2 - x^2) > 0$ for points in the immediate neighbourhood of the origin.
But for points along $y = x$, the value of the function is $-2x^4 < 0$. Thus every neighbourhood of $(0, 0)$ there are points where the function assumes positive values, i.e., $> f(0, 0)$ and there are points where the function assumes negative values, i.e., $< f(0, 0)$. Hence $f(0, 0)$ is neither a maximum value nor a minimum value. □

Example 5.6.2 *Show that the function $f : \mathbb{R}^2 \to \mathbb{R}$ given by $f(x, y) = 2x^4 - 3x^2y + y^2$ for all $(x, y) \in \mathbb{R}^2$ has neither maximum nor minimum at $(0, 0)$ though $f_{xx}f_{yy} - f_{xy}^2 = 0$ at $(0, 0)$.*

Solution: Here $f_x = 8x^3 - 6xy$, $f_y = 2y - 3x^2$. Clearly, at $(0, 0)$, $f = 0$, $f_x = 0$ and $f_y = 0$.

$\therefore$ $r = f_{xx} = 24x^2 - 6y$, $s = f_{xy} = -6x$, $t = f_{yy} = 2$.

Therefore, at $(0, 0)$, $f_{xx}f_{yy} - f_{xy}^2 = 0$. So the case needs further investigation.

Now, from $(x^2 - y)(2x^2 - y)$, we see that
$$f(x, y) \quad > \quad 0 \text{ for } y < 0 \text{ or } x^2 > y > 0$$
$$< \quad 0 \text{ for } 0 < \frac{y}{2} < x^2 < y$$

and $f(0,0) = 0$. Thus $f(x,y)$ is positive as well as negative in the neighbourhood of $(0,0)$. So $f(x,y)$ is neither maximum nor minimum at $(0,0)$. $\qquad\square$

Example 5.6.3 *Show that minimum value of* $u = xy + c^3\left(\dfrac{1}{x} + \dfrac{1}{y}\right)$ *is* $3c^2$.

Solution: Here, $u_x = y - \dfrac{c^3}{x^2}$, $u_y = x - \dfrac{c^3}{y^2}$. Solving $y - \dfrac{c^3}{x^2} = 0$ and $x - \dfrac{c^3}{y^2} = 0$, we get $x^2 y = y^2 x = c^3 \Rightarrow x = y$.

Hence the stationary point of the given function is (c,c).

Now $r = u_{xx} = \dfrac{2c^3}{x^3}$, $s = u_{xy} = 1$ and $t = u_{yy} = \dfrac{2c^3}{y^3}$.

Therefore at (c,c), $rt - s^2 = 2.2 - 1^2 = 3 > 0$ and $r = 2 > 0$. So $u(x,y)$ is minimum at (c,c) and $u_{min} = c^2 + c^3\left(\dfrac{1}{c} + \dfrac{1}{c}\right) = 3c^2$. $\qquad\square$

Example 5.6.4 *Find the maximum or minimum value of* $f(x,y) = x^2 + xy + y^2 + ax + by$ *and determine whether the value you get is maximum or minimum.*

Solution: Here $f_x = 2x + y + a$, $f_y = x + 2y + b$

$\therefore$ Solving $f_x = 0$ and $f_y = 0$, we get $x = \dfrac{1}{3}(b - 2a)$, $y = \dfrac{1}{3}(a - 2b)$.

Now $r = f_{xx} = 2$, $s = f_{xy} = 1$ and $t = f_{yy} = 2$.

Since $rt - s^2 = 4 - 1 = 3 > 0$ and $r > 0$, therefore $f(x,y)$ is minimum at $x = \dfrac{1}{3}(b - 2a)$, $y = \dfrac{1}{3}(a - 2b)$ and the minimum value of $f(x,y)$ is

$$f_{min} = \frac{1}{9}(b - 2a)^2 + \frac{1}{9}(b - 2a)(a - 2a) + \frac{1}{9}(a - 2b)^2 + \frac{a}{3}(b - 2a) + \frac{b}{3}(a - 2b)$$

$$= \frac{1}{3}(ab - a^2 - b^2).$$ $\qquad\square$

Example 5.6.5 *Show that* $u = x^2 y^4 (1 - x - y)^6$ *defined in the region* $x \geq 0, y \geq 0, x + y \leq 1$ *attains its greatest at an interior point of the region and find this greatest values.*

Solution: Here $u_x = 2xy^4(1 - x - y)^6 - 6x^2 y^4(1 - x - y)^5$
$$= 2xy^4(1 - x - y)^5(1 - 4x - 4y).$$
Similarly, $u_y = 4x^2 y^3(1 - x - y)^6 - 6x^2 y^4(1 - x - y)^5$
$$= 2x^2 y^3(1 - x - y)^5(2 - 2x - 5y).$$

$\therefore$ $u_x = 0$ and $u_y = 0$ give $x = 0$, $y = 0$, $x + y = 1$, $4x + 4y = 1$, $2x + 5y = 2$. Solving these equations we get the following set of points as solutions of $u_x = 0$ and $u_y = 0$: $(0,0), (0,1), \left(0, \dfrac{2}{5}\right), (1,0), \left(\dfrac{1}{4}, 0\right), \left(\dfrac{1}{6}, \dfrac{1}{3}\right)$ of which the last one, i.e., $\left(\dfrac{1}{6}, \dfrac{1}{3}\right)$ is an interior point of the given region. So we are to test this point only for extremum.

Now $\quad r = u_{xx} \quad = 2y^4(1 - x - y)^6(1 - 4x - y)$
$$- 10xy^4(1 - x - y)^4(1 - 4x - y) - 8xy^4(1 - x - y)^5$$

$\therefore \quad (r)_{\left(\frac{1}{6}, \frac{1}{3}\right)} \quad = -8\frac{1}{6} \cdot \frac{1}{3^4} \cdot \frac{1}{2^5} = -\frac{1}{6.3^4.2^2}$

$$\left[\because \; 1 - 4x - y = 0 \text{ and } 1 - x - y = 1 - \frac{1}{6} - \frac{1}{3} = \frac{1}{2}\right]$$

$s = u_{xy} \quad = 8xy^3(1 - x - y)^5(1 - 4x - y) - 10xy^4(1 - x - y)^4 \times$
$$(1 - 4x - y) - 2xy^4(1 - x - y)^5$$

$\therefore \quad (s)_{\left(\frac{1}{6}, \frac{1}{3}\right)} \quad = -2\frac{1}{6} \cdot \frac{1}{3^4} \cdot \frac{1}{2^5} = -\frac{1}{3^5.2^5}$

$t = u_{yy} \quad = 6x^2y^2(1 - x - y)^5(2 - 2x - 5y) - 10x^2y^3(1 - x - y)^4 \times$
$$(2 - 2x - 5y) - 10x^2y^3(1 - x - y)^5$$

$\therefore \quad (t)_{\left(\frac{1}{6}, \frac{1}{3}\right)} \quad - -\frac{10}{6^2.3^3} \cdot \frac{1}{2^5} = -\frac{5}{6^2.3^3.2^4} \quad \left[\because 2 - 2x - 5y = 2 - 2.\frac{1}{6} - 5.\frac{1}{3} = 0\right]$

$(rt - s^2)_{\left(\frac{1}{6}, \frac{1}{3}\right)} \quad = \quad \frac{5}{6^3.3^7.2^6} - \frac{1}{3^{10}.2^{10}} = \frac{5}{6^9.2} - \frac{1}{6^{10}}$

$$= \quad \frac{15}{6^{10}} - \frac{1}{6^{10}} = \text{ positive, i.e., } > 0 \text{ and } r < 0$$

$\therefore \; u$ is maximum at $\left(\frac{1}{6}, \frac{1}{3}\right)$ and the maximum value of u is

$$(u)_{max} = \frac{1}{6^2} \cdot \frac{1}{3^4}\left(1 - \frac{1}{6} - \frac{1}{3}\right)^6 = \frac{1}{36 \times 81} \times \frac{1}{2^6} = \frac{1}{36 \times 81 \times 64}. \qquad \square$$

Example 5.6.6 *Show that the function*

$$F(x, y, z) = (x + y + z)^3 - 3(x + y + y) - 24xyz + a^3$$

has minimum at $(1, 1, 1)$ *and maximum at* $(-1, -1, -1)$.

Solution: Given $F(x, y, z) = (x + y + z)^3 - 3(x + y + y) - 24xyz + a^3$.

$$\left. \begin{array}{rcl} \therefore \quad \dfrac{\partial F}{\partial x} & = & 3(x + y + z)^2 - 3 - 24yz \\[3mm] \dfrac{\partial F}{\partial y} & = & 3(x + y + z)^2 - 3 - 24zx \\[3mm] \text{and} \quad \dfrac{\partial F}{\partial z} & = & 3(x + y + z)^2 - 3 - 24xy \end{array} \right\} \qquad (5.3)$$

For stationary points, we have

$$\frac{\partial F}{\partial x} = 0, \; \frac{\partial F}{\partial y} = 0 \text{ and } \frac{\partial F}{\partial z} = 0 \qquad (5.4)$$

The system of equations (5.4) are satisfied when $x = y = z$.
Putting $y = z$ and $z = x$ in the first equation of the set (5.3), we get

$$27x^2 - 3 - 24x^2 = 0 \quad \Rightarrow \quad x = \pm 1 \quad \Rightarrow \quad x = y = z = 1 \text{ and } x = y = z = -1.$$

Therefore, the stationary points of F are given by $(1,1,1)$ and $(-1,-1,-1)$.

Now, the six second order partial derivatives are given by

$$A = F_{xx} = \frac{\partial^2 F}{\partial x^2} = 6(x+y+z), \ B = F_{yy} = \frac{\partial^2 F}{\partial y^2} = 6(x+y+z),$$

$$C = F_{zz} = \frac{\partial^2 F}{\partial z^2} = 6(x+y+z), \ F = F_{yz} = \frac{\partial^2 F}{\partial y \partial z} = 6(x+y+z) - 24x,$$

$$G = F_{zx} = \frac{\partial^2 F}{\partial z \partial x} = 6(x+y+z) - 24y \text{ and}$$

$$H = F_{xy} = \frac{\partial^2 F}{\partial x \partial y} = 6(x+y+z) - 24z.$$

So, at the stationary point $(1,1,1)$, we get $A = 18, B = 18, C = 18, F = -6, G = -6$ and $H = -6$. So, for this point we have

$$A > 0, \ \begin{vmatrix} A & H \\ H & B \end{vmatrix} = \begin{vmatrix} 18 & -6 \\ -6 & 18 \end{vmatrix} = 288 > 0$$

and $\begin{vmatrix} A & H & G \\ H & B & F \\ G & F & C \end{vmatrix} = \begin{vmatrix} 18 & -6 & -6 \\ -6 & 18 & -6 \\ -6 & -6 & 18 \end{vmatrix} = 6^3 \begin{vmatrix} 3 & -1 & -1 \\ -1 & 3 & -1 \\ -1 & -1 & 3 \end{vmatrix} = 3456 > 0$

showing that all the three principal minors A, $\begin{vmatrix} A & H \\ H & B \end{vmatrix}$ and $\begin{vmatrix} A & H & G \\ H & B & F \\ G & F & C \end{vmatrix}$ are

positive. As a consequence, $F(x,y,z)$ becomes minimum at $(1,1,1)$.

Again, at the stationary point $(-1,-1,-1)$, we have $A = -18, B = -18, C = -18, F = 6, G = 6$ and $H = 6$.

$\therefore$ At the point $(-1,-1,-1)$, we have $A < 0$.

$$\begin{vmatrix} A & H \\ H & B \end{vmatrix} = \begin{vmatrix} 18 & -6 \\ -6 & 18 \end{vmatrix} = 288 > 0 \text{ and}$$

$$\begin{vmatrix} A & H & G \\ H & B & F \\ G & F & C \end{vmatrix} = \begin{vmatrix} -18 & 6 & 6 \\ 6 & -18 & 6 \\ 6 & 6 & -18 \end{vmatrix} = 6^3 \begin{vmatrix} -3 & 1 & 1 \\ 1 & -3 & 1 \\ 1 & 1 & -3 \end{vmatrix} = -3456 < 0.$$

Hence, the above three expressions are alternately negative and positive, showing that the function is maximum at the point $(-1,-1,-1)$. $\quad\square$

An Important Observation

At $(1,1,1)$, we have $f_{xx} = f_{yy} = f_{zz} = 18$ and $f_{xy} = f_{yz} = f_{zx} = -6$, so,

$$d^2 F = 18(dx^2 + dy^2 + dz^2) - 12(dxdy + dydz + dzdx)$$

[From equation (5.2)]

$$= 6[(dx^2 + dy^2 + dz^2) + (dx - dy)^2 + (dy - dz)^2 + (dz - dx)^2]$$

which is *positive* for all values of dx, dy and dz and does not vanish for $(dx, dy, dz) \neq (0,0,0)$ ensuring that the function is minimum at $(1,1,1)$.

Again at $(-1,-1,-1)$, we have $f_{xx} = f_{yy} = f_{zz} = -18$ and $f_{xy} = f_{yz} = f_{zx} = 6$

$$\therefore \ d^2 F = -18(dx^2 + dy^2 + dz^2) + 12(dxdy + dydz + dzdx)$$

$$= -6[(dx^2 + dy^2 + dz^2) + (dx - dy)^2 + (dy - dz)^2 + (dz - dx)^2]$$

which is *negative* for all values of dx, dy and dz and never vanishes except $(dx, dy, dz) = (0, 0, 0)$. Therefore, the function is maximum at the point $(-1, -1, -1)$.

Example 5.6.7 *Examine for extreme values of the function*

$$u = 2xyz + x^2 + y^2 + z^2.$$

Solution: Here $u_x = 2yz + 2x$, $u_y = 2xz + 2y$, $u_z = 2xy + 2z$.

For stationary points $u_x = 0$, $u_y = 0$, $u_z = 0$, which implies

$$yz + x = 0, \ zx + y = 0 \text{ and } xy + z = 0$$
$$\Rightarrow \quad xyz + x^2 = 0, \ yzx + y^2 = 0 \text{ and } zxy + z^2 = 0$$
$$\Rightarrow \quad x^2 = y^2 = z^2 = -xyz$$
$$\Rightarrow \quad x = y = z \text{ or } x = -y, \ y = -z.$$

Now $x = y = z \Rightarrow x = 0, x = -1 \quad \therefore \quad (x, y, z) = (0, 0, 0), (-1, -1, -1)$ and $y = -x, z = x \Rightarrow -x^2 + x = 0 \Rightarrow x = 0, x = 1$

$\therefore \ x = 1, y = -1, z = 1 \quad \therefore \quad (x, y, z) = (1, -1, 1)$.

Now the six second order partial derivatives are given by $A = u_{xx} = 2$, $B = u_{yy} = 2$, $C = u_{zz} = 2$, $F = u_{yz} = 2x$, $G = u_{zx} = 2y$, $H = u_{xy} = 2z$.

Therefore, the principal minors at $(0, 0, 0)$ are

$$A = 2 > 0, \quad \begin{vmatrix} A & H \\ H & B \end{vmatrix} = \begin{vmatrix} 2 & 0 \\ 0 & 2 \end{vmatrix} = 4 > 0$$

and

$$\begin{vmatrix} A & H & G \\ H & B & F \\ G & F & C \end{vmatrix} = \begin{vmatrix} 2 & 0 & 0 \\ 0 & 2 & 0 \\ 0 & 0 & 2 \end{vmatrix} = 8 > 0.$$

Thus the three minors are all positive. $\therefore \ d^2u > 0$.

Hence $u(x, y, z)$ has minimum value at $(0, 0, 0)$.

At the point $(-1, -1, -1)$, we have

$$A = 2 > 0, \quad \begin{vmatrix} A & H \\ H & B \end{vmatrix} = \begin{vmatrix} 2 & -2 \\ -2 & 2 \end{vmatrix} = 4 - 4 = 0$$

and

$$\begin{vmatrix} A & H & G \\ H & B & F \\ G & F & C \end{vmatrix} = \begin{vmatrix} 2 & -2 & -2 \\ -2 & 2 & -2 \\ -2 & -2 & 2 \end{vmatrix} = 8 \begin{vmatrix} 1 & -1 & -1 \\ -1 & 1 & -1 \\ -1 & -1 & 1 \end{vmatrix} = -32 < 0.$$

Above results show that it does not follow any rule of positive definite or negative definite of d^2u. Therefore $u(x, y, z)$ has neither a maximum nor a minimum at $(-1, -1, -1)$.

Similarly, it can be shown that the result is also true for the points $(-1, 1, 1)$, $(1, -1, 1)$ and $(1, 1, -1)$ So the function possesses only the extreme point $(0, 0, 0)$, which is maximum.   □

5.7 Constrained Optimization

Mathematically, the term *Optimization* means the best allocation of the values of the independent variable/variables of a function for which the function values will be maximized or minimized, i.e., an optimization problem, in general, consists of minimizing or maximizing a function of several variables by systematic selection of values of the independent variables from the specified domain and calculating the function values. The function which is to be maximized or minimized (i.e., optimized) is called *Objective function* and the corresponding independent variables are called *Decision Variables*. If the objective function be optimized with respect to some extra constrained relations then the problem is called *Constrained Optimization*.

In this treatise, we consider problems of constrained extrema with equality constrained relations only and the problems with inequality constrained relations are beyond the scope of our discussion.

5.7.1 Constrained extrema of a function having two independent variables

To find the maximum or minimum value of $f(x,y)$ subject to the given constrained relation $\phi(x,y) = 0$

Let us suppose that $u = f(x,y)$ be a differentiable function in a given domain. We have the constrained relation as

$$\phi(x,y) = 0 \tag{5.5}$$

If $\phi_y \neq 0$, we can solve the relation (5.5) for y. Let $y = \psi(x)$ be such solution. Then the given function $u = f(x,y)$ is reduced to a function of single variable x alone as

$$u = f(x, \psi(x)) \tag{5.6}$$

Differentiating (5.6) with respect to x, we get

$$\frac{du}{dx} = \frac{\partial u}{\partial x} + \frac{\partial u}{\partial \psi}\cdot\frac{d\psi}{dx}$$

For extremum value $\dfrac{du}{dx} = 0 \Rightarrow \dfrac{\partial u}{\partial x} + \dfrac{\partial u}{\partial \psi}\cdot\dfrac{d\psi}{dx} = 0$

or, $\dfrac{\partial u}{\partial x} + \dfrac{\partial u}{\partial y}\cdot\dfrac{dy}{dx}$ (5.7)

Again form $\phi(x,y) = 0$, we have

$$\frac{\partial \phi}{\partial x} + \frac{\partial \phi}{\partial y}\cdot\frac{d\psi}{dx} = 0 \tag{5.8}$$

Eliminating $\dfrac{dy}{dx}$ from (5.7) and (5.8), we get $\dfrac{u_x}{\phi_x} = \dfrac{u_y}{\phi_y}$ or $u_x\phi_y - u_y\phi_x = 0$

or, $\dfrac{\partial(f,\phi)}{\partial(x,y)} = 0$ (5.9)

Equations (5.5) and (5.9) will now give the system of values of x and y for which $f(x,y)$ may be a maximum or a minimum.

Example 5.7.1 *Find the extreme value of $x^2 + y^2$ when $x - y + 2 = 0$.*

Solution: Here $u = f(x,y) = x^2 + y^2$, $\phi(x,y) = x - y + 2$.

$\phi = 0$ gives $y = x + 2 = \psi(x)$.

$y = \psi(x)$ gives $u = f(x, \psi(x)) = x^2 + (x + 2)^2 = 2x^2 + 4x + 4$.

For extrema, $\dfrac{du}{dx} = 0 \Rightarrow 4x + 4 = 0$ i.e., $x = -1$.

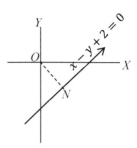

Figure 5.2

Substituting in $\phi = 0$ we get $y = 1$. Thus $(-1, 1)$ is the only stationary point. The same stationary point could also be obtained by solving $\phi = 0$ and $\dfrac{\partial(f, \phi)}{\partial(x, y)} = 0$, i.e., by solving

$x - y + 2 = 0$ and $\begin{vmatrix} 2x & 2y \\ 1 & -1 \end{vmatrix} = 0$, or, by solving $x - y + 2 = 0$ and $x + y = 0$.

So, extreme value of u at $(-1, 1) = (u)_{(-1, 1)} = (-1)^2 + 1^2 = 2$.   □

Note 5.7.1 *We observe that $f(x, y) = x^2 + y^2$ represents the square of the distance of a line from the origin and for this problem the line is restricted to move on $x - y + 2 = 0$. Also we are to remember in mind that the distance of any line from a point would be minimum when it is perpendicular.*

Now, the distance from the origin $O(0, 0)$ to $\phi = 0$, i.e., $x - y + 2 = 0$ is
$$\frac{2}{\sqrt{1^2 + 1^2}} = \sqrt{2}.$$

So, the above extremum point is a minimum point and we get $(u)_{min} = (\sqrt{2})^2 = 2$.

5.7.2 Constrained extrema of a function having three independent variables

To find the maximum or minimum value of $f(x, y, z)$ subject to the given constrained relation $\phi(x, y, z) = 0$

Let us consider that $u = f(x, y, z)$ be a differentiable function in a given domain of definition. We have the constrained relation as

$$\phi(x, y, z) = 0 \tag{5.10}$$

Solving for z from (5.10) we suppose that $z = \psi(x, y)$. The problem is then reduced to an unconstrained extreme value problem of the function

$$u = f(x, y, \psi(x, y)) \tag{5.11}$$

For extrema, we must have $u_x = 0, u_y = 0$ separately, which gives

$$\frac{\partial u}{\partial x} + \frac{\partial u}{\partial z} \cdot \frac{\partial \psi}{\partial x} = 0 \tag{5.12}$$

and $\quad \dfrac{\partial u}{\partial y} + \dfrac{\partial u}{\partial z} \cdot \dfrac{\partial \psi}{\partial y} = 0 \tag{5.13}$

Also form (5.10) we get

$$\frac{\partial \phi}{\partial x} + \frac{\partial \phi}{\partial z}.\frac{\partial \psi}{\partial x} = 0 \tag{5.14}$$

and $$\frac{\partial \phi}{\partial y} + \frac{\partial \phi}{\partial z}.\frac{\partial \psi}{\partial y} = 0 \tag{5.15}$$

Eliminating $\dfrac{\partial \psi}{\partial x}$ from (5.12) and (5.14) we get

$$\frac{\frac{\partial u}{\partial x}}{\frac{\partial \phi}{\partial x}} = \frac{\frac{\partial u}{\partial z}}{\frac{\partial \phi}{\partial z}}, \text{ i.e., } u_x \phi_z - \phi_x u_z = 0, \text{ or, } \frac{\partial(f, \phi)}{\partial(x, z)} = 0 \tag{5.16}$$

Similarly, eliminating $\dfrac{\partial \psi}{\partial y}$ from (5.13) and (5.15) we get

$$\frac{\partial(f, \phi)}{\partial(y, z)} = 0 \tag{5.17}$$

Equations (5.10), (5.16) and (5.17) now give a system of values of x, y, z for which the given function is extremum.

Example 5.7.2 *Find the shortest distance of the plane given by $2x + 3y + 4z = 12$ form the origin.*

Solution: Let δ be the distance of any point $P(x, y, z)$ from the origin $O(0, 0, 0)$, then $\delta^2 = x^2 + y^2 + z^2$. If $u = f(x, y, z) = \delta^2 = x^2 + y^2 + z^2$, then minimum value of u will give the minimum value of δ.

The subsidiary equation given here is

$$\phi(x, y, z) = 2x + 3y + 4z - 12 = 0$$

$$\therefore \quad J_1 = \frac{\partial(f, \phi)}{\partial(x, z)} = \begin{vmatrix} 2x & 2z \\ 2 & 4 \end{vmatrix} = 0 \Rightarrow 2x - z = 0$$

and $$J_2 = \frac{\partial(f, \phi)}{\partial(y, z)} = \begin{vmatrix} 2y & 2z \\ 3 & 4 \end{vmatrix} = 0 \Rightarrow 4y - 3z = 0.$$

Therefore, the stationary points are obtained by solving the above two equations which are $\left.\begin{array}{c} 2x + 0.y - z = 0 \\ 0.x + 4y - 3z = 0 \end{array}\right\}$ giving $\dfrac{x}{4} = \dfrac{y}{6} = \dfrac{z}{8}$, i.e., $\dfrac{x}{2} = \dfrac{y}{3} = \dfrac{z}{4} = k$ (say). Therefore, there is only one stationary point $P(2k, 3k, 4k)$.

$$\therefore \quad (f)_{min} = f(P) = k^2(2^2 + 3^2 + 4^2) = 29k^2$$
$$\therefore \quad \delta = |k|\sqrt{29}.$$

Now form the equation of the plane $2.2k + 3.3k + 4.4k = 12 \Rightarrow k(29) = 12$, i.e., $|k| = \dfrac{12}{29}$.

So, we get $(f)_{min} = (\delta) = |k|\sqrt{29} = \dfrac{12}{29}.\sqrt{29} = \dfrac{12}{\sqrt{29}}$ which is the perpendicular distance of the plane form O. $\qquad \square$

Note 5.7.2 *Actually, geometrically u represents the distance of $P(x, y, z)$ from the origin O. Therefore, $u_{min} = 0$ but u_{max} is not finite. Now when it is asked to find out the minimum distance of the origin from the plane constrained to move* on $ax + by + cz + d = 0$, then the variable point P is not free to move anywhere in space but is restricted to move on the given plane, which is called the subsidiary condition. So, the minimum value of u subject to the subsidiary condition,

$$(u)_{min} = \frac{|d|}{\sqrt{a^2 + b^2 + c^2}}.$$

Note 5.7.3 *There are two distinct approaches for determining the stationary points of a function subject to subsidiary condition or conditions.*

　　1. *Reduction of the number of independent variables using the subsidiary condition or conditions* [**Examples 5.7.1** *and* **5.7.2**].

　　2. *Lagrange's Method of undetermined multipliers* [*Discussed in* § 5.8 *and illustrated in* **Examples 5.8.1** *and* **5.8.2**].

5.8　Lagrange's Method of Undetermined Multipliers

Lagrange, a great French mathematician, has given a more systematic and uniform method of finding the stationary points of a function subject to a number of constraints.

　　Let us consider the following function of n independent variables

$$u = f(x_1, x_2, ..., x_n) \tag{5.18}$$

connected by m number of constrained equations

$$\left.\begin{array}{rcl} g_1(x_1, x_2, ..., x_n) & = & 0 \\ g_2(x_1, x_2, ..., x_n) & = & 0 \\ ... \ ... \ ... & & \\ g_m(x_1, x_2, ..., x_n) & = & 0 \end{array}\right\} \tag{5.19}$$

　　There are m equations in n variables $(m < n)$ in (5.19), so only $n - m$ variables are independent.

　　For an extremum value of the function u, we get $du = 0$. Hence from (5.18) and also from the set of constrained equations (5.19) we have

$$\left.\begin{array}{rcccc} du & = & \dfrac{\partial u}{\partial x_1}dx_1 + \dfrac{\partial u}{\partial x_2}dx_2 + ... + \dfrac{\partial u}{\partial x_n}dx_n & = & 0 \\[3mm] dg_1 & = & \dfrac{\partial g_1}{\partial x_1}dx_1 + \dfrac{\partial g_1}{\partial x_2}dx_2 + ... + \dfrac{\partial g_1}{\partial x_n}dx_n & = & 0 \\[3mm] dg_2 & = & \dfrac{\partial g_2}{\partial x_1}dx_1 + \dfrac{\partial g_2}{\partial x_2}dx_2 + ... + \dfrac{\partial g_2}{\partial x_n}dx_n & = & 0 \\[1mm] & & ... \ ... \ ... & & \\[1mm] dg_m & = & \dfrac{\partial g_m}{\partial x_1}dx_1 + \dfrac{\partial g_m}{\partial x_2}dx_2 + ... + \dfrac{\partial g_m}{\partial x_n}dx_n & = & 0 \end{array}\right\} \tag{5.20}$$

　　Multiplying the equations in (5.20) by $1, \lambda_1, \lambda_2, ..., \lambda_m$ respectively and adding all together, we get

$$f_1 dx_1 + f_2 dx_2 + ... + f_n dx_n = 0 \tag{5.21}$$

where $f_k = \dfrac{\partial u}{\partial x_k} + \lambda_1 \dfrac{\partial g_1}{\partial x_k} + \lambda_2 \dfrac{\partial g_2}{\partial x_k} + ... + \lambda_m \dfrac{\partial g_m}{\partial x_k}, \ k = 1, 2, ..., m.$

The m quantities $\lambda_1, \lambda_2, ..., \lambda_m$ are at our choice. Let us choose these quantities in such a way that they satisfy the following m equations

$$f_1 = 0, \ f_2 = 0, ..., f_m = 0 \tag{5.22}$$

Then the equation (5.21) becomes

$$f_{m+1}dx_{m+1} + f_{m+2}dx_{m+2} + ... + f_n dx_n = 0 \tag{5.23}$$

Now, it has already been seen that $n - m$ variables are independent. To be specific, let $x_{m+1}, x_{m+2}, ..., x_n$ are these independent variables and therefore, there coefficients must separately become zero, i.e., we get $n - m$ extra equations

$$f_{m+1} = 0, \ f_{m+2} = 0, ..., f_n = 0 \tag{5.24}$$

Thus form (5.19), (5.22) and (5.24) we have $n + m$ equations

$$g_1 = 0, g_2 = 0, ..., g_m = 0; \ f_1 = 0, f_2 = 0, ..., f_n = 0$$

From these equations we can find the m number of multipliers $\lambda_1, \lambda_2, ..., \lambda_m$ and the n number of variables $x_1, x_2, ..., x_n$ for which u becomes extremum.

Note 5.8.1 *Considering the general case when $u = f(x_1, x_2, ..., x_n)$ is a function of n independent variables, if the point $(x_1, x_2, ..., x_n)$ is a stationary point, then this point will be a point of minimum if $d^2 F > 0$ and a point of maximum if $d^2 F < 0$, where $F = f + \lambda_1 g_1 + \lambda_2 g_2 + ... + \lambda_m g_m$.*

Example 5.8.1 *If $u = x^2 + y^2$ and $ax^2 + 2hxy + by^2 = 1$, find the exteme value of u.*

Solution: Here $du = 2xdx + 2ydy = 0$.

Let $f(x, y) = ax^2 + 2hxy + by^2 - 1$.

Then $df = 2axdx + 2h(ydx + xdy) + 2bydy = 0$

i.e., we get the following two equations

$$xdx + ydy = 0 \tag{5.25}$$
$$(ax + hy)dx + (hx + by)dy = 0 \tag{5.26}$$

Multiplying (5.26) by λ and adding with (5.25) and then equating the coefficients of dx and dy, we get

$$x + \lambda(ax + hy) = 0 \tag{5.27}$$
$$y + \lambda(hx + by) = 0 \tag{5.28}$$

Now (5.27) $\times x$ + (5.28) $\times y$ gives

$$x^2 + y^2 + \lambda(ax^2 + 2hxy + by^2) = 0 \ \text{ or, } u + \lambda = 0$$
$$(\because ax^2 + 2hxy + by^2 = 1)$$

or, $\lambda = -u$.

Therefore, from (5.23) and (5.24), for extremum we get

$$x - u(ax + hy) = 0 \ \text{ and } y - u(hx + by) = 0$$

or, $(1 - au)x - huy = 0$ and $-hux + (1 - bu)y = 0$

$\therefore \quad \dfrac{1 - au}{-hu} = \dfrac{-hu}{1 - bu}$

or, $(h^2 - ab)u^2 + (a + b)u - 1 = 0$.

$\therefore\ u = \dfrac{-(a+b) \pm \sqrt{(a+b)^2 + 4(h^2 - ab)}}{2(h^2 - ab)}$ is the extreme value of u.   □

Example 5.8.2 *Find the maximum and minimum value of $x^m y^n z^p$ subject to the condition $\dfrac{a}{x} + \dfrac{b}{y} + \dfrac{c}{z} = 1$.*

Solution: Let $u = x^m y^n z^p$ and $f = \dfrac{a}{x} + \dfrac{b}{y} + \dfrac{c}{z} - 1$.

Taking logarithmic differentiation of u, we get

$$\frac{1}{u} du = \frac{m}{x} dx + \frac{n}{y} dy + \frac{p}{z} dz.$$

Also, differentiating f we get

$$df = -\frac{a}{x^2} dx - \frac{b}{y^2} dy - \frac{c}{z^2} dz.$$

Thus we get the following two equations for extrema

$$\frac{m}{x} dx + \frac{n}{y} dy + \frac{p}{z} dz = 0 \tag{5.29}$$

and $$\frac{a}{x^2} dx + \frac{b}{y^2} dy + \frac{c}{z^2} dz = 0 \tag{5.30}$$

Now $(5.29) + (5.30) \times \lambda$ gives,

$$\left(\frac{m}{x} + \frac{\lambda a}{x^2}\right) dx + \left(\frac{n}{y} + \frac{\lambda b}{y^2}\right) dy + \left(\frac{p}{z} + \frac{\lambda c}{z^2}\right) dz = 0$$

Equating the coefficients of dx, dy and dz, we get

$$m + \frac{\lambda a}{x} = 0,\ n + \frac{\lambda b}{y} = 0,\ p + \frac{\lambda c}{z} = 0 \tag{5.31}$$

Therefore, adding these we get,

$$(m + n + p) + \lambda \left(\frac{a}{x} + \frac{b}{y} + \frac{c}{z}\right) = 0$$

or, $(m + n + p) + \lambda = 0 \ \Rightarrow\ \lambda = -(m + n + p).$

Therefore from (5.31),

$$\frac{mx}{a} = \frac{ny}{b} = \frac{pz}{c} = (m + n + p)$$

$$\therefore\ x = \frac{a(m + n + p)}{m},\, y = \frac{b(m + n + p)}{n},\, z = \frac{c(m + n + p)}{p} \tag{5.32}$$

Therefore, maximum or minimum value of u

$$= \frac{a^m}{m^m} \cdot \frac{b^n}{n^n} \cdot \frac{c^p}{p^p} \cdot (m + n + p)^{m+n+p}.$$ □

5.9 Miscellaneous Illustrative Examples

Example 5.9.1 *Find the minimum value of $x^2 + y^2 + z^2$, subject to the condition $2x + 3y + 5z = 30$.*

Solution: The objective function and the subsidiary condition are

$$u = x^2 + y^2 + z^2 \tag{5.33}$$
$$2x + 3y + 5z = 30 \tag{5.34}$$

For extrema of u, we have $du = 0$

$$\therefore \ 2xdx + 2ydy + 2zdz = 0 \ \Rightarrow \ xdx + ydy + zdz = 0 \tag{5.35}$$

Taking differentials form (5.34), we get

$$2dx + 3dy + 5dz = 0 \tag{5.36}$$

Now, $(5.35) \times 1 + (5.36) \times \lambda$ gives

$$(x + 2\lambda)dx + (y + 3\lambda)dy + (z + 5\lambda)dz = 0.$$

Equating coefficients of dx, dy and dz to zero, we get

$$x + 2\lambda = 0, \quad y + 3\lambda = 0, \quad z + 5\lambda = 0.$$

From these equations we get,

$$\frac{x}{2} = \frac{y}{3} = \frac{z}{5} = \frac{2x + 3y + 5z}{2.2 + 3.3 + 5.5} = \frac{30}{38} = \frac{15}{19}$$

$$\therefore \quad x = \frac{30}{19}, \ y = \frac{45}{19}, \ z = \frac{75}{19}.$$

From the equation $x + 2\lambda = 0$, we get $\lambda = -\dfrac{x}{2} = -\dfrac{15}{19}$.

$\therefore \ u$ has an extremum at $\left(\dfrac{30}{19}, \dfrac{45}{19}, \dfrac{75}{19} \right)$.

From equation (5.34) one of the variable, say z can be expressed as a function of two independent variables x and y and we write

$$z = 6 - \frac{2}{5}x - \frac{3}{5}y$$

$$\therefore \quad \frac{\partial z}{\partial x} = -\frac{2}{5}, \frac{\partial z}{\partial y} = -\frac{3}{5}$$

Now $\dfrac{\partial u}{\partial x} = 2x + 2z\dfrac{\partial z}{\partial x} = 2x - \dfrac{4}{5}z = p, \quad \dfrac{\partial u}{\partial y} = 2y + 2z\dfrac{\partial z}{\partial y} = 2y - \dfrac{6}{5}z = q$

$$\frac{\partial^2 u}{\partial x^2} = 2 - \frac{4}{5}\frac{\partial z}{\partial x} = 2 + \frac{8}{25} = \frac{58}{25} = r$$

$$\frac{\partial^2 u}{\partial x \partial y} = -\frac{6}{5}\frac{\partial z}{\partial x} = \left(-\frac{6}{5} \right)\left(-\frac{2}{5} \right) = \frac{12}{25} = s$$

and $\dfrac{\partial^2 u}{\partial y^2} = 2 - \dfrac{6}{5}\left(-\dfrac{3}{5} \right) = 2 + \dfrac{18}{25} = \dfrac{68}{25} = t$

$$\therefore \quad rt - s^2 = \frac{58}{25} \times \frac{68}{25} - \left(\frac{12}{25}\right)^2 > 0.$$

Therefore, $(rt - s^2) > 0$ at $\left(\frac{30}{19}, \frac{45}{19}, \frac{75}{19}\right)$. Also $r > 0$ thereat. Hence u is

minimum herein and $u_{min} = \dfrac{30^2 + 45^2 + 75^2}{19^2} = \dfrac{8550}{361}.$ ☐

Example 5.9.2 *Find the minimum value of $x^2 + y^2 + z^2$ having given that $xy + yz + zx = 12$.*

Solution: Let us consider that

$$u = x^2 + y^2 + z^2 \tag{5.37}$$

It is also given that

$$xy + yz + zx = 12 \tag{5.38}$$

Taking differentials of both sides of (5.37), we get

$$du = 2xdx + 2ydy + 2zdz \tag{5.39}$$

For minimum u, we should have

$$du = 0 \;\Rightarrow\; xdx + ydy + zdz = 0 \tag{5.40}$$

Taking differentials of both sides of (5.38) we get

$$zdy + ydz + xdz + zdx + ydx + xdy = 0$$
$$\text{or,} \quad (y + z)dx + (z + x)dy + (x + y)dz = 0 \tag{5.41}$$

Now $(5.40) \times 1 + (5.41) \times \lambda$ gives

$$\{x + \lambda(y + z)\}dx + \{y + \lambda(z + x)\}dy + \{z + \lambda(x + y)\}dz = 0.$$

Equating coefficients of dx, dy and dz to zero, we get

$$x + \lambda(y + z) = 0 \tag{5.42}$$
$$y + \lambda(z + x) = 0 \tag{5.43}$$
$$z + \lambda(x + y) = 0 \tag{5.44}$$

Adding these we get, $\quad (x + y + z)(1 + 2\lambda) = 0.$

Therefore $1 + 2\lambda = 0 \;(\because\; x + y + z \neq 0) \;\Rightarrow\; \lambda = -\dfrac{1}{2}.$

Putting this value of λ in $(5.42), (5.43)$ and (5.44), we get

$$\left.\begin{array}{c} 2x - y - z = 0 \\ -x + 2y - z = 0 \\ -x - y + 2y = 0 \end{array}\right\} \tag{5.45}$$

By cross multiplication between the first two relations of (5.45), we get

$$\frac{x}{1 + 2} = \frac{y}{1 + 2} = \frac{z}{4 - 1} \;\Rightarrow\; x = y = z.$$

From (5.38), we get $x^2 + x^2 + x^2 = 12$ whence $x = \pm 2$.

Thus $x = y = z = 2$ or $x = y = z = -2$.

So u may be a minimum at the points $(2, 2, 2)$, $(-2, -2, -2)$.

Since there are three variables and one constraint, two variables must be independent, say x, y and one variable, say z is dependent upon them.

From (5.38) we have $z(x + y) + xy - 12 = 0$, so that

$$\frac{\partial}{\partial x}\{z(x + y) + xy - 12\} = 0 \tag{5.46}$$

and $\quad \dfrac{\partial}{\partial y}\{z(x + y) + xy - 12\} = 0 \tag{5.47}$

From (5.42) we get $(x + y)\dfrac{\partial z}{\partial x} + z.1 + y = 0$ or, $\dfrac{\partial z}{\partial x} = -\dfrac{y + z}{x + y}$.

Similarly, from (5.43) we get $\dfrac{\partial z}{\partial y} = -\dfrac{x + z}{x + y}$.

Now, $\dfrac{\partial u}{\partial x} = 2x + 2z\dfrac{\partial z}{\partial x}$ $\left(\because y \text{ is independent of } x, \dfrac{\partial y}{\partial x} = 0\right)$

$$= 2x - 2z\frac{y + z}{x + y} \quad \left(\text{putting the value of } \frac{\partial z}{\partial x}\right).$$

$r = \dfrac{\partial^2 u}{\partial x^2}$ $\quad = 2 - 2\dfrac{(x + y)\left(y\frac{\partial z}{\partial x} + 2z\frac{\partial z}{\partial x}\right) - z(y + z).1}{(x + y)^2}$

$$= 2 - 2\frac{(x + y)(y + 2z)\left(-\frac{y + z}{x + y}\right) - z(y + z).1}{(x + y)^2}$$

$$= 2 + \frac{(y + z)(y + 3z)}{(x + y)^2}.$$

$s = \dfrac{\partial^2 u}{\partial x \partial y} = \dfrac{\partial}{\partial x}\left(\dfrac{\partial u}{\partial y}\right) = \dfrac{\partial}{\partial x}\left(2y - 2z\dfrac{z + x}{x + y}\right)$

$$\left[\because \frac{\partial u}{\partial y} = 2y + 2z\frac{\partial z}{\partial x} = 2y - 2z\frac{z + x}{z + y}\right]$$

$$= 0 - 2\frac{(x + y)\left(2z\frac{\partial z}{\partial x} + z + x\frac{\partial z}{\partial x}\right) - z(z + x).1}{(x + y)^2}$$

$$= -2\frac{(x + y)(2z + x)\left(-\frac{y + z}{x + y}\right) + z(x + y - z - x)}{(x + y)^2}$$

$$= -2\frac{-(y + z)(2z + x) + z(y - z)}{(x + y)^2} = 2\frac{(y + z)(2z + x) - z(y - z)}{(x + y)^2}$$

and $\quad t \quad = \dfrac{\partial^2 u}{\partial y^2} = 2 - 2\dfrac{(x + y)(2z + x)\frac{\partial z}{\partial y} - (z + x).z.1}{(x + y)^2}$

$$= 2 - 2\frac{(x + y)(2z + x)\left(-\frac{z + x}{x + y}\right) - z(z + x)}{(x + y)^2}$$

$$= 2 + \frac{2(x + z)(x + 3z)}{(x + y)^2}.$$

At the points $(2, 2, 2)$ and $(-2, -2, -2)$, we have

$$r = 2 + 2\frac{y^2 + 4yz + 3z^2}{(x+y)^2} = 2 + 2\frac{4 + 4.4 + 3.4}{16} = \frac{32 + 64}{16} = 6 > 0$$

$$s = 2\frac{2yz + 2z^2 + xy + zx - zy + y^2}{(x+y)^2} = 2.\frac{8 + 8 + 4 + 4 - 4 + 4}{16} = 3$$

and $\quad t = 2 + 2\frac{x^2 + 4xz + 3z^2}{(x+y)^2} = 2 + 2\frac{4 + 4.4 + 3.4}{16} = \frac{32 + 64}{16} = 6$

$$\therefore \quad rt - s^2 = 36 - 9 = 27 > 0.$$

Hence u, i.e., $x^2 + y^2 + z^2$ is a minimum at the points $(2, 2, 2)$ and $(-2, -2, -2)$ and the minimum value at each is $4 + 4 + 4 = 12$. $\qquad\square$

Example 5.9.3 *Find out the extremum of* $f(x, y, z) = x^2 + y^2 + z^2 - xy - yz - zx$, *subject to the condition* $\phi(x, y, z) \equiv x^2 + y^2 + z^2 - 2x + 2y + 6z + 9 = 0$.

Solution: Let $F(x, y, z) = f(x, y, z) + \lambda\phi(x, y, z)$

$$= x^2 + y^2 + z^2 - xy - yz - zx + \lambda(x^2 + y^2 + z^2 - 2x + 2y + 6z + 9)$$

$$\therefore \quad \frac{\partial F}{\partial x} = 2x - y - z + \lambda(2x - 2)$$

$$\frac{\partial F}{\partial y} = 2y - z - x + \lambda(2y + 2)$$

$$\frac{\partial F}{\partial z} = 2z - x - y + \lambda(2z + 6)$$

For extrema, $dF = 0 \quad\Rightarrow\quad \dfrac{\partial F}{\partial x} = \dfrac{\partial F}{\partial y} = \dfrac{\partial F}{\partial z} = 0$

$$\Rightarrow \quad 2x - y - z + \lambda(2x - 2) = 0 \tag{5.48}$$
$$2y - z - x + \lambda(2y + 2) = 0 \tag{5.49}$$
$$2z - x - y + \lambda(2z + 6) = 0 \tag{5.50}$$

From $(5.48), (5.49)$ and (5.50), we get

$$\frac{2x - y - z}{2(x - 1)} = \frac{2y - z - x}{2(y + 1)} = \frac{2z - x - y}{2(z + 3)} = -\lambda = \frac{0}{2(x + y + z) + 6}$$

$$\Rightarrow \quad \lambda = 0, \; x + y + z \neq -3.$$

Also $2x - y - z = 0, 2y - z - x = 0$ and $2z - x - y = 0 \;\Rightarrow\; x = y = z$.

The system has infinite number of solutions of the form (c, c, c), $c \neq -1$ and the stationary point is (c, c, c), $c \neq -1$.

So, $F_x = 2x - y - z, F_y = 2y - z - x, F_z = 2z - x - y, F_{xx} = 2, F_{yy} = 2, F_{zz} = 2, F_{xy} = F_{yx} = -1, F_{zy} = F_{yz} = -1, F_{xz} = F_{zx} = -1$.

Now, for all the stationary point (c, c, c), $c \neq -1$,

$$d^2F = \sum \frac{\partial^2 f}{\partial x^2}(dx)^2 + 2\sum \frac{\partial^2 f}{\partial x \partial y}dxdy$$

$$= 2(dx)^2 + 2(dy)^2 + 2(dz)^2 - 2dxdy - 2dydz - 2dxdz$$

$$= (dx - dy)^2 + (dy - dz)^2 + (dz - dx)^2 > 0.$$

$\therefore$ F is minimum at (c, c, c), $c \neq -1$ which implies f is minimum at (c, c, c), $c \neq -1$.

$\qquad\square$

Example 5.9.4 *(i) Show that the volume of the greatest rectangular parallelepiped that can be inscribed in the ellipsoid* $\dfrac{x^2}{a^2} + \dfrac{y^2}{b^2} + \dfrac{z^2}{c^2} = 1$ *is* $\dfrac{8abc}{3\sqrt{3}}$.

(ii) Divide the number 27 into three parts x, y, z such that $2yz + 3zx + 4xy$ is minimum.

Solution: (i) Let the coordinates of the corners of the rectangular parallelepiped inscribed in the ellipsoid

$$\frac{x^2}{a^2} + \frac{y^2}{b^2} + \frac{z^2}{c^2} = 1 \tag{5.51}$$

be $(x, y, z), (x, -y, z), (x, y, -z)$ and $(-x, y, z)$.

Then the lengths of the sides of the parallelepiped are $2x, 2y, 2z$. The volume of the parallelepiped is given by, say,

$$u = 8xyz \tag{5.52}$$

Now the problem is to maximize u subject to the condition (5.51).

From (5.52), we get $du = 8(yz\,dx + zx\,dy + xy\,dz)$.

For maximum $du = 0 \Rightarrow yz\,dx + zx\,dy + xy\,dz = 0$ $\tag{5.53}$

From (5.51), we get $\dfrac{x}{a^2}dx + \dfrac{y}{b^2}dy + \dfrac{z}{c^2}dz = 0$ $\tag{5.54}$

Now $(5.49) + \lambda \times (5.50)$ gives,

$$\left(yz + \lambda\frac{x}{a^2}\right)dx + \left(zx + \lambda\frac{y}{b^2}\right)dy + \left(xy + \lambda\frac{z}{c^2}\right)dz = 0.$$

Equating the coefficients of dx, dy, dz of the above relation to zero, we get

$$yz + \lambda\frac{x}{a^2} = 0 \quad \Rightarrow \quad xyz = -\lambda\frac{x^2}{a^2} \tag{5.55}$$

$$zx + \lambda\frac{y}{b^2} = 0 \quad \Rightarrow \quad xyz = -\lambda\frac{y^2}{b^2} \tag{5.56}$$

$$xy + \lambda\frac{z}{c^2} = 0 \quad \Rightarrow \quad xyz = -\lambda\frac{z^2}{c^2} \tag{5.57}$$

Adding the relations $(5.51), (5.52)$ and (5.53) we get

$$3xyz = -\lambda\left(\frac{x^2}{a^2} + \frac{y^2}{b^2} + \frac{z^2}{c^2} =\right) = -\lambda \; [\text{Using } (5.51)]$$

$$\therefore \; \lambda = -3xyz \tag{5.58}$$

So we get $x = \pm\dfrac{a}{\sqrt{3}}$ [by using (5.55) and (5.58)]

$$y = \pm\frac{b}{\sqrt{3}} \quad [\text{by using (5.56) and (5.58)}]$$

and $z = \pm\dfrac{c}{\sqrt{3}}$ [by using (5.57) and (5.58)]

From (5.58), we get $\lambda = -\dfrac{abc}{\sqrt{3}}$ or, $\dfrac{abc}{\sqrt{3}}$ according as xyz is positive or negative.

Since there are three variables and one constraint, there are two independent variables which we assume to be x, y and consider z as a function of both x, y.

Now, we have $\dfrac{\partial x}{\partial y} = 0, \dfrac{\partial y}{\partial x} = 0$ and from (5.55), we get

$$z^2 = c^2 \left(1 - \frac{x^2}{a^2} - \frac{y^2}{b^2} \right)$$

$$\therefore \quad \frac{\partial z}{\partial x} = -\frac{c^2}{a^2} \cdot \frac{x}{z}, \quad \left[\because \frac{\partial y}{\partial x} = 0 \right]; \quad \frac{\partial z}{\partial y} = -\frac{c^2}{b^2} \cdot \frac{y}{z}, \quad \left[\because \frac{\partial x}{\partial y} = 0 \right]$$

Now, $\dfrac{1}{8} \dfrac{\partial u}{\partial x} = yz + xy \dfrac{\partial z}{\partial x} = 0 \quad \left[\because \dfrac{\partial y}{\partial x} = 0 \right]$

$$= yz - \frac{c^2}{a^2} \frac{x^2 y}{z} \quad \left[\because \frac{\partial z}{\partial x} = -\frac{c^2}{a^2} \frac{x}{z} \right]$$

$$
\begin{aligned}
\frac{1}{8} \frac{\partial^2 u}{\partial x^2} &= y \frac{\partial z}{\partial x} - \frac{c^2}{a^2} \left\{ \frac{2xy}{z} + \frac{(-1)x^2 y}{z^2} \frac{\partial z}{\partial x} \right\} \\
&= -\frac{c^2}{a^2} \frac{xy}{z} - \frac{c^2}{a^2} \left\{ \frac{2xy}{z} + \frac{c^2}{a^2} \frac{x^3 y}{z^3} \right\} \\
&= -\frac{c^2}{a^2} \frac{3xyz}{z^2} - \frac{c^4}{a^4} \frac{x^2 xyz}{z^4}
\end{aligned}
\tag{5.59}
$$

$\dfrac{1}{8} \dfrac{\partial u}{\partial y} = xz + xy \dfrac{\partial z}{\partial y} \quad \left[\because \dfrac{\partial z}{\partial y} = 0 \right] = xz - \frac{c^2}{b^2} \cdot \frac{xy^2}{z} \quad \left[\because \frac{\partial z}{\partial y} = -\frac{c^2}{b^2} \frac{y}{z} \right].$

$$
\begin{aligned}
\frac{1}{8} \frac{\partial^2 u}{\partial y^2} &= x \frac{\partial z}{\partial y} - \frac{c^2}{b^2} \left\{ \frac{2xy}{z} + \frac{(-1)xy^2}{z^2} \frac{\partial z}{\partial y} \right\} \\
&= -\frac{c^2}{b^2} \frac{xy}{z} - \frac{c^2}{b^2} \left\{ \frac{2xy}{z} + \frac{c^2}{b^2} \frac{xy^3}{z^3} \right\} \\
&= -\frac{c^2}{b^2} \frac{3xyz}{z^2} - \frac{c^4}{b^4} \frac{y^2 xyz}{z^4}
\end{aligned}
\tag{5.60}
$$

$$
\begin{aligned}
\frac{1}{8} \frac{\partial^2 u}{\partial x \partial y} &= \frac{1}{8} \frac{\partial}{\partial x} \left(\frac{\partial u}{\partial y} \right) = \frac{\partial}{\partial x} \left(xz - \frac{c^2}{b^2} \frac{xy^2}{z} \right) \\
&= z + x \frac{\partial z}{\partial x} - \frac{c^2}{b^2} \left\{ \frac{y^2}{z} + \frac{xy^2(-1)}{z^2} \frac{\partial z}{\partial x} \right\} \\
&= z - \frac{c^2}{a^2} \frac{x^2}{z} - \frac{c^2}{b^2} \frac{y^2}{z} - \frac{c^4}{a^2 b^2} \cdot \frac{x^2 y^2}{z^3}
\end{aligned}
\tag{5.61}
$$

When xyz is positive, we have

$$x = \frac{a}{\sqrt{3}}, \ y = \frac{b}{\sqrt{3}}, \ z = \frac{c}{\sqrt{3}}$$

$$\text{or,} \quad x = \frac{a}{\sqrt{3}}, \ y = -\frac{b}{\sqrt{3}}, \ z = -\frac{c}{\sqrt{3}}$$

$$\text{or,} \quad x = -\frac{a}{\sqrt{3}}, \ y = -\frac{b}{\sqrt{3}}, \ z = \frac{c}{\sqrt{3}}$$

$$\text{or,} \quad x = -\frac{a}{\sqrt{3}}, \ y = \frac{b}{\sqrt{3}}, \ z = -\frac{c}{\sqrt{3}}$$

So, when xyz are positive, we have

$$\frac{1}{8}\frac{\partial^2 u}{\partial x^2} = -\frac{c^2}{a^2}\frac{3abc}{\sqrt{3}\sqrt{3}\sqrt{3}}\cdot\frac{3}{c^2} - \frac{c^4}{a^4}\frac{a^2}{3}\frac{3.3}{c^4}\frac{abc}{\sqrt{3}\sqrt{3}\sqrt{3}} \quad \text{[from (5.59)]}$$

$$= -\frac{3bc}{a\sqrt{3}} - \frac{bc}{a\sqrt{3}} = -\frac{4bc}{a\sqrt{3}}$$

i.e., $\quad \dfrac{\partial^2 u}{\partial x^2} = -\dfrac{8.4bc}{a\sqrt{3}} < 0$

$$\frac{1}{8}\frac{\partial^2 u}{\partial y^2} = -\frac{c^2}{b^2}\frac{3abc}{\sqrt{3}\sqrt{3}\sqrt{3}}\cdot\frac{3}{c^2} - \frac{c^4}{b^4}\frac{b^2}{3}\frac{3.3}{c^4}\frac{abc}{\sqrt{3}\sqrt{3}\sqrt{3}} \quad \text{[from (5.60)]}$$

$$= -\frac{3ac}{b\sqrt{3}} - \frac{ac}{b\sqrt{3}} = -\frac{4ac}{b\sqrt{3}}$$

i.e., $\quad \dfrac{\partial^2 u}{\partial y^2} = -\dfrac{8.4ac}{b\sqrt{3}}$

$$\frac{1}{8}\frac{\partial^2 u}{\partial x\partial y} = \left(\pm\frac{c}{\sqrt{3}}\right) - \frac{c^2}{a^2}\frac{a^2}{3}\left(\pm\frac{\sqrt{3}}{c}\right) - \frac{c^2}{b^2}\frac{b^2}{3}\left(\pm\frac{\sqrt{3}}{c}\right)$$

$$- \frac{c^4}{a^2b^2}\cdot\frac{a^2}{3}\frac{b^2}{3}\left(\pm\frac{3\sqrt{3}}{c^3}\right) \quad \text{[from (5.61)]}$$

$$= \pm\frac{c}{\sqrt{3}} \mp \frac{c}{\sqrt{3}} \mp \frac{c}{\sqrt{3}} \mp \frac{c}{\sqrt{3}} = \mp\frac{2c}{\sqrt{3}}$$

i.e., $\quad \dfrac{\partial^2 u}{\partial x\partial y} = \mp\dfrac{8.2c}{\sqrt{3}}$

$$\therefore \frac{\partial^2 u}{\partial x^2}\frac{\partial^2 u}{\partial x^2} - \left(\frac{\partial^2 u}{\partial x\partial y}\right)^2 = \frac{64.16abc^2}{ab.3} - \frac{64.4c^2}{3} = \frac{64.16c^2}{3} - \frac{64.4c^2}{3} > 0.$$

So, u is maximum in this case and the maximum value is $\dfrac{8ab}{3\sqrt{3}}$.

When xyz is negative, we have

$$x = -\frac{a}{\sqrt{3}},\ y = -\frac{b}{\sqrt{3}},\ z = -\frac{c}{\sqrt{3}} \quad \text{or,}\ x = \frac{a}{\sqrt{3}},\ y = \frac{b}{\sqrt{3}},\ z = -\frac{c}{\sqrt{3}}$$

or, $\quad x = -\frac{a}{\sqrt{3}},\ y = \frac{b}{\sqrt{3}},\ z = \frac{c}{\sqrt{3}} \quad \text{or,}\ x = -\frac{a}{\sqrt{3}},\ y = \frac{b}{\sqrt{3}},\ z = \frac{c}{\sqrt{3}}$

So, when xyz is negative, we have

$$\frac{1}{8}\frac{\partial^2 u}{\partial x^2} = -\frac{c^2}{a^2}\frac{3(-abc)}{a^2\sqrt{3}\sqrt{3}\sqrt{3}}\cdot\frac{3}{c^2} - \frac{c^4}{a^4}\frac{a^2}{3}\frac{3.3}{a^2}\frac{(-abc)}{\sqrt{3}\sqrt{3}\sqrt{3}} = \frac{4bc}{a\sqrt{3}} > 0,$$

so that u is not maximum.

Thus the maximum value will be $\dfrac{8abc}{3\sqrt{3}}$.

Alternative Method:

Volume of the rectangular parallelepiped inscribed in the given ellipsoid is $8xyz$. Let $V = 8xyz$.

Subject to the condition $\quad \dfrac{x^2}{a^2} + \dfrac{y^2}{b^2} + \dfrac{z^2}{c^2} - 1 = 0 \qquad (5.62)$

Let us define the Lagrangian function F by

$$F = 8xyz + \lambda\left(\frac{x^2}{a^2} + \frac{y^2}{b^2} + \frac{z^2}{c^2} - 1\right)$$

$$\therefore \; dF = \left(8yz + \frac{2\lambda x}{a^2}\right)dx + \left(8xz + \frac{2\lambda y}{b^2}\right)dy + \left(8xy + \frac{2\lambda z}{c^2}\right)dz.$$

For extremum, we have $dF = 0 \;\Rightarrow\; \dfrac{\partial F}{\partial x} = \dfrac{\partial F}{\partial y} = \dfrac{\partial F}{\partial z} = 0$

$$\Rightarrow \; 8yz + 2\lambda\frac{x}{a^2} = 0 \;\Rightarrow\; 4yz + \lambda\frac{x}{a^2} = 0 \tag{5.63}$$

$$8zx + 2\lambda\frac{y}{b^2} = 0 \;\Rightarrow\; 4zx + \lambda\frac{y}{b^2} = 0 \tag{5.64}$$

$$8xy + 2\lambda\frac{z}{c^2} = 0 \;\Rightarrow\; 4xy + \lambda\frac{z}{c^2} = 0 \tag{5.65}$$

Taking $x \times (5.63) + y \times (5.64) + z \times (5.65)$ and using (5.62) we get

$$12xyz + \lambda\left(\frac{x^2}{a^2} + \frac{y^2}{b^2} + \frac{z^2}{c^2}\right) = 0 \;\Rightarrow\; 12xyz + \lambda = 0 \;\Rightarrow\; \lambda = -12xyz.$$

From (5.63), $4yz - \dfrac{12x^2 yz}{a^2} = 0 \;\Rightarrow\; x^2 = \dfrac{a^2}{3} \;\Rightarrow\; x = \dfrac{a}{\sqrt{3}}$ and similarly, we get

$$y = \frac{b}{\sqrt{3}}, z = \frac{c}{\sqrt{3}} \;\therefore\; \lambda = -\frac{12abc}{\sqrt{3}\sqrt{3}\sqrt{3}} = -\frac{4abc}{\sqrt{3}}$$

$$\begin{aligned}
d^2 F &= \frac{2\lambda}{a^2}(dx)^2 + \frac{2\lambda}{b^2}(dy)^2 + \frac{2\lambda}{c^2}(dz)^2 + 16zdxdy + 16ydzdx + 16xdydz \\
&= -\frac{8abc}{\sqrt{3}}\left[\frac{(dx)^2}{a^2} + \frac{(dy)^2}{b^2} + \frac{(dz)^2}{c^2}\right] \\
&\qquad\qquad + \frac{16}{\sqrt{3}}(cdxdy + bdzdx + adxdy) \tag{5.66}
\end{aligned}$$

Now, differentiating (5.62), we get $\dfrac{2x}{a^2}dx + \dfrac{2y}{b^2}dy + \dfrac{2z}{c^2}dz = 0.$

At the point $\left(\dfrac{a}{\sqrt{3}}, \dfrac{b}{\sqrt{3}}, \dfrac{c}{\sqrt{3}}\right)$, we get $\dfrac{dx}{a} + \dfrac{dy}{b} + \dfrac{dz}{c} = 0.$

Squaring we get

$$\frac{(dx)^2}{a^2} + \frac{(dy)^2}{b^2} + \frac{(dz)^2}{c^2} + \frac{2dxdy}{ab} + \frac{2dydz}{bc} + \frac{2dzdx}{ca} = 0$$

$$\text{or, } \; -abc\left[\frac{(dx)^2}{a^2} + \frac{(dy)^2}{b^2} + \frac{(dz)^2}{c^2}\right]$$

$$= 2cdxdy + 2adydz + 2bdzdy \tag{5.67}$$

$$\begin{aligned}
\therefore\; d^2 F &= -\frac{16abc}{2\sqrt{3}}\left[\frac{(dx)^2}{a^2} + \frac{(dy)^2}{b^2} + \frac{(dz)^2}{c^2}\right] \\
&\qquad\qquad - \frac{8abc}{\sqrt{3}}\left[\frac{(dx)^2}{a^2} + \frac{(dy)^2}{b^2} + \frac{(dz)^2}{c^2}\right] \\
&= -\frac{16abc}{\sqrt{3}}\left[\frac{(dx)^2}{a^2} + \frac{(dy)^2}{b^2} + \frac{(dz)^2}{c^2}\right] < 0.
\end{aligned}$$

Hence $\left(\dfrac{a}{\sqrt{3}}, \dfrac{b}{\sqrt{3}}, \dfrac{c}{\sqrt{3}}\right)$ is a point of maximum and its maximum value is

$$(V)_{max} = \frac{8abc}{3\sqrt{3}}.$$

(ii) Let us consider

$$u = 2yz + 3zx + 4xy \qquad (5.68)$$

where $\quad x + y + z = 27 \qquad (5.69)$

Here, we have three variables and one given constraint. So, two variables are independent and the third variable is dependent on the former two. Without loss of generality, let us assume that x, y are to be independent and z to be a dependent function on x and y.

Taking differentials of both sides of (5.68), we get

$$\begin{aligned} du &= 2zdy + 2ydz + 3xdz + 3zdx + 4ydx + 4xdy \\ &= (3z + 4y)dx + (2z + 4x)dy + (2y + 3x)dz \end{aligned}$$

For extremum of u, we have $du = 0$

$$\Rightarrow \quad (3z + 4y)dx + (2z + 4x)dy + (2y + 3x)dz = 0 \qquad (5.70)$$

Again from (5.69), taking differential we get

$$dx + dy + dz = 0 \qquad (5.71)$$

Now, $1 \times (5.70) + \lambda \times (5.71)$ gives

$$(3z + 4y + \lambda)dx + (2z + 4x + \lambda)dy + (2y + 3x + \lambda)dz = 0.$$

Equating to zero, the coefficients of dx, dy and dz of the above relation, we get

$$3z + 4y + \lambda = 0 \qquad (5.72)$$
$$2z + 4x + \lambda = 0 \qquad (5.73)$$
$$\text{and} \quad 2y + 3x + \lambda = 0 \qquad (5.74)$$

which implies $\quad 3z + 4y = 2z + 4x = 2y + 3x \ (= -\lambda).$

Again, from these we get

$$4x - 4y - z = 0 \quad \text{(using (5.72) and (5.73))} \qquad (5.75)$$
$$x - 2y + 2z = 0 \quad \text{(using (5.73) and (5.74))} \qquad (5.76)$$
$$\text{and} \quad 3x - 2y - 3z = 0 \quad \text{(using (5.72) and (5.74))} \qquad (5.77)$$

From (5.75) and (5.76), by cross multiplication, we get

$$\frac{x}{-8-2} = \frac{y}{-1-8} = \frac{z}{-8+4}$$

i.e., $\quad \dfrac{x}{-10} = \dfrac{y}{-9} = \dfrac{z}{-4} = \dfrac{x+y+z}{-23} = \dfrac{-27}{23}$

$\therefore \quad x = \dfrac{270}{23}, \ y = \dfrac{243}{23}, \ z = \dfrac{108}{23}.$

We see that these values of x, y, z satisfy (5.77).

From (5.74), we get $\lambda = -\left(2.\dfrac{243}{23} + 3.\dfrac{270}{23}\right) = -\dfrac{1296}{23}$.

From (5.69), we have $z = 27 - x - y \Rightarrow \dfrac{\partial z}{\partial x} = -1, \ \dfrac{\partial z}{\partial y} = -1$.

Now

$$\dfrac{\partial u}{\partial x} = 2y\dfrac{\partial u}{\partial x} + 3z + 3x\dfrac{\partial z}{\partial x} + 4y = -2y + 3z - 3x + 4y$$

$$= -3x - 6y + 3z$$

$$\dfrac{\partial^2 u}{\partial x^2} = -3 + 3\dfrac{\partial z}{\partial x} = -3 - 3 = -6 < 0$$

$$\dfrac{\partial u}{\partial y} = 2y\dfrac{\partial z}{\partial y} + 2z + 3x\dfrac{\partial z}{\partial y} + 4x = 2z - 2y - 3x + 4x$$

$$= x - 2y + 2z$$

$$\dfrac{\partial^2 u}{\partial y^2} = -2 + 2\dfrac{\partial z}{\partial y} = -2 - 2 = -4$$

$$\dfrac{\partial^2 u}{\partial x \partial y} = 1 - 2\dfrac{\partial z}{\partial x} = 1 - 2 = -1.$$

At $\left(x = \dfrac{270}{23}, y = \dfrac{243}{23}, z = \dfrac{108}{23}\right)$, we get $\dfrac{\partial^2 u}{\partial x^2} . \dfrac{\partial^2 u}{\partial x^2} - \left(\dfrac{\partial^2 u}{\partial x \partial y}\right)^2$

$= (-6)(-4) - (-1)^2 = 23 > 0$. So, u is maximum here and so $2yz + 3zx + 4xy$ is

maximum for $x = \dfrac{270}{23}, y = \dfrac{243}{23}, z = \dfrac{108}{23}$.   □

Example 5.9.5 *Determine the maximum and minimum values of $7x^2 + 8xy + y^2$ when $x^2 + y^2 = 1$.*

Solution: Let us consider

$$u = 7x^2 + 8xy + y^2 \tag{5.78}$$

Given condition $\qquad x^2 + y^2 = 1 \tag{5.79}$

Taking differentials of both sides of (5.78), we get

$$du = 14x\,dx + 8y\,dx + 8x\,dy + 2y\,dy = 2(7x + 4y)dx + 2(4x + y)dy$$

For extremum, we have $du = 0 \Rightarrow (7x + 4y)dx + (4x + y)dy = 0 \tag{5.80}$

Again taking differentials of both sides of (5.79), we get

$$x\,dx + y\,dy = 0 \tag{5.81}$$

$1 \times (5.80) + \lambda \times (5.81)$ gives $(7x + 4y + \lambda x)dx + (4x + y + \lambda y)dy = 0$

Equating to zero the coefficients of dx and dy, we get

$$7x + 4y + \lambda x = 0 \ \Rightarrow \ 7 + 4\dfrac{y}{x} = -\lambda \tag{5.82}$$

and $\quad 4x + y + \lambda y = 0 \ \Rightarrow \ 1 + 4\dfrac{x}{y} = -\lambda \tag{5.83}$

From (5.82) and (5.83), we get

$$7 + 4\frac{y}{x} = 1 + 4\frac{x}{y} \quad \Rightarrow \quad \frac{y}{x} - \frac{x}{y} = -\frac{3}{2} \tag{5.84}$$

or, $\dfrac{y^2}{x^2} + \dfrac{x^2}{y^2} - 2 = \dfrac{9}{4}$ or, $\dfrac{y^2}{x^2} + \dfrac{x^2}{y^2} + 2 = \dfrac{9}{4} + 4 = \dfrac{25}{4}$

or, $\left(\dfrac{y}{x} + \dfrac{x}{y}\right)^2 = \dfrac{25}{4}$ or, $\dfrac{(x^2 + y^2)^2}{x^2 y^2} = \dfrac{25}{4}$

or, $\dfrac{1}{x^2 y^2} = \dfrac{25}{4}$ [Using (5.79)] $\quad \therefore \quad x^2 = \dfrac{4}{25 y^2}$ \tag{5.85}

Substituting this value of x in (5.79), we get

$$\frac{4}{25 y^2} + y^2 = 1 \quad \text{or, } 25 y^4 - 25 y^2 + 4 = 0$$

or, $(5 y^2 - 4)(5 y^2 - 1) = 0$

So either $5 y^2 - 4 = 0 \Rightarrow y = \pm \dfrac{2}{\sqrt{5}}$ or, $5 y^2 - 1 = 0 \Rightarrow y = \pm \dfrac{1}{\sqrt{5}}$.

When $y = \pm \dfrac{2}{\sqrt{5}}$, we get $x^2 = \dfrac{4}{25} \cdot \dfrac{5}{4}$ [from (5.81)], i.e., $x = \pm \dfrac{1}{\sqrt{5}}$.

But $x = \dfrac{1}{\sqrt{5}}, y = \dfrac{2}{\sqrt{5}}$ and $x = -\dfrac{1}{\sqrt{5}}, y = -\dfrac{2}{\sqrt{5}}$ do not satisfy (5.84).

Hence $x = \dfrac{1}{\sqrt{5}}, y = -\dfrac{2}{\sqrt{5}}$ and $x = -\dfrac{1}{\sqrt{5}}, y = \dfrac{2}{\sqrt{5}}$.

Substituting $x = \dfrac{1}{\sqrt{5}}, y = -\dfrac{2}{\sqrt{5}}$ in (5.82), we get $-\lambda = 7 - 4 \times 2$, i.e., $\lambda = 1$

and substituting $x = -\dfrac{1}{\sqrt{5}}, y = \dfrac{2}{\sqrt{5}}$ in (5.79), we get $-\lambda = 1 - 4.\dfrac{1}{2}$, i.e., $\lambda = 1$.

When $y = \pm \dfrac{1}{\sqrt{5}}$, we get $x^2 = \dfrac{4}{25}.5$ [from (5.85)], i.e., $x = \pm \dfrac{2}{\sqrt{5}}$.

But $x = \dfrac{2}{\sqrt{5}}, y = -\dfrac{1}{\sqrt{5}}$ and $x = -\dfrac{2}{\sqrt{5}}, y = \dfrac{1}{\sqrt{5}}$ do not satisfy (5.84).

Hence $x = \dfrac{2}{\sqrt{5}}, y = \dfrac{1}{\sqrt{5}}$ and $x = -\dfrac{2}{\sqrt{5}}, y = -\dfrac{1}{\sqrt{5}}$.

Substituting $x = \dfrac{2}{\sqrt{5}}, y = \dfrac{2}{\sqrt{5}}$ in (5.78), we get $\lambda = 7 + 4.\dfrac{1}{\sqrt{5}}.\dfrac{\sqrt{5}}{2} = 9$ and

substituting $x = -\dfrac{2}{\sqrt{5}}, y = -\dfrac{1}{\sqrt{5}}$ in (5.83), we get $\lambda = 1 + 4.\dfrac{2}{\sqrt{5}}.\sqrt{5} = 9$.

Since there are two variables and one constraint, only one variable will be independent which we assume to be x and then y is treated as a function of x.

From (5.79), $2x + 2y\dfrac{dy}{dx} = 0 \quad \therefore \quad \dfrac{dy}{dx} = -\dfrac{x}{y}$ \tag{5.86}

From (5.78) $u = 6x^2 + 8xy + x^2 + y^2 = 6x^2 + 8xy + 1$ [by (5.79)]

$$\therefore \quad \frac{du}{dx} = 12x + 8y + 8x\frac{dy}{dx} = 12x + 8y - 8\frac{x^2}{y} \quad \text{[by (5.86)]}$$

$$\therefore \quad \frac{d^2u}{dx^2} = 12 + 8\frac{dy}{dx} - \frac{y.16x - 8x^2\frac{dy}{dx}}{y^2} = 12 - \frac{8x}{y} - \frac{16x}{y} - 8\frac{x^3}{y^3}$$

[by using (5.86)]

$$= 12 - \frac{24x}{y} - 8\frac{x^3}{y^3}.$$

When $x = \dfrac{1}{\sqrt{5}}, y = -\dfrac{2}{\sqrt{5}}$ or $x = -\dfrac{1}{\sqrt{5}}, y = \dfrac{2}{\sqrt{5}}$, we get $\dfrac{d^2u}{dx^2} > 0$ and when $x = \dfrac{2}{\sqrt{5}}, y = \dfrac{1}{\sqrt{5}}$ or $x = -\dfrac{2}{\sqrt{5}}, y = -\dfrac{1}{\sqrt{5}}$, we get $\dfrac{d^2u}{dx^2} < 0$.

Thus $7x^2 + 8xy + y^2$ is minimum at the points $\left(\dfrac{1}{\sqrt{5}}, -\dfrac{2}{\sqrt{5}}\right)$ and $\left(-\dfrac{1}{\sqrt{5}}, \dfrac{2}{\sqrt{5}}\right)$ and is maximum at the points $\left(\dfrac{2}{\sqrt{5}}, \dfrac{1}{\sqrt{5}}\right)$ and $\left(-\dfrac{2}{\sqrt{5}}, -\dfrac{1}{\sqrt{5}}\right)$, the minimum and maximum values being $7.\dfrac{1}{5} - 16\dfrac{1}{5} + \dfrac{4}{5} = -1$ and $7.\dfrac{4}{5} + \dfrac{16}{5} + \dfrac{1}{5} = 9$ respectively.

Alternative Method:

$$\text{Here} \qquad u = 7x^2 + 8xy + y^2 \qquad\qquad (5.87)$$

$$\text{Subject to} \qquad x^2 + y^2 = 1 \qquad\qquad (5.88)$$

From (5.88) we have $y = \pm\sqrt{1 - x^2}$.

Let us first take $y = \sqrt{1 - x^2}$ \qquad\qquad (5.89)

From (5.87), we get

$$u = 7x^2 + 8x\sqrt{1 - x^2} + (1 - x^2) \quad \text{[Using (5.89)]}$$

$$= 6x^2 + 8x\sqrt{1 - x^2} + 1.$$

$$\therefore \quad \frac{du}{dx} = 12x + 8\sqrt{1 - x^2} + 8x.\frac{1}{2\sqrt{1 - x^2}}(-2x)$$

$$= 12x + 8\sqrt{1 - x^2} - \frac{8x^2}{\sqrt{1 - x^2}} \qquad\qquad (5.90)$$

$$= \frac{1}{\sqrt{1 - x^2}}\left[12x\sqrt{1 - x^2} + 8(1 - x^2) - 8x^2\right]$$

$$= \frac{1}{\sqrt{1 - x^2}}\left[12x\sqrt{1 - x^2} + 8 - 16x^2\right] \qquad\qquad (5.91)$$

$$\frac{d^2u}{dx^2} = 12 + 8.\frac{1}{2}\frac{1}{\sqrt{1 - x^2}}(-2x) - \frac{16x\sqrt{1 - x^2} - 8x^2\frac{1}{2\sqrt{1-x^2}}(-2x)}{1 - x^2}$$

[Using (5.90)]

$$= 12 - \frac{8x}{\sqrt{1 - x^2}} - \frac{16x(1 - x^2) + 8x^3}{(1 - x^2)\sqrt{1 - x^2}}$$

$$= 12 - \frac{8x}{\sqrt{1-x^2}} - \frac{16x}{\sqrt{1-x^2}} - \frac{8x^3}{(1-x^2)\sqrt{1-x^2}}$$

$$= 12 - \frac{24x}{\sqrt{1-x^2}} - \frac{8x^3}{(1-x^2)\sqrt{1-x^2}} \tag{5.92}$$

For extreme value we have

$$\frac{du}{dx} = 0 \quad \Rightarrow \quad 3x\sqrt{1-x^2} + 2 - 4x^2 = 0 \text{ [from (5.91)]}$$

or, $2 - 4x^2 = -3x\sqrt{1-x^2}$ (5.93)

$$\therefore \quad 4 - 16x^2 + 16x^4 = 9x^2(1-x^2) \Rightarrow 25x^4 - 25x^2 + 4 = 0$$
or, $(5x^2 - 1)(5x^2 - 4) = 0$

So, either $5x^2 - 1 = 0 \Rightarrow x = \pm\frac{1}{\sqrt{5}}$ or, $5x^2 - 4 = 0 \Rightarrow x = \pm\frac{2}{\sqrt{5}}$.

Now, $x = \frac{1}{\sqrt{5}}$, $y = -\frac{2}{\sqrt{5}}$ do not satisfy (5.89). Therefore, we get $x = -\frac{1}{\sqrt{5}}$, $y = \frac{2}{\sqrt{5}}$.

When $x = \frac{1}{\sqrt{5}}$, we get $y = \sqrt{1 - \frac{1}{5}} = \frac{2}{\sqrt{5}}$ [Using (5.89)]

and $\frac{d^2u}{dx^2} = 12 + \frac{24}{\sqrt{5}\sqrt{1-\frac{1}{5}}} + \frac{8.1}{5\sqrt{5}\sqrt{1-\frac{1}{5}}} > 0$ [from (5.92)]

So, u is minimum in this case.

When $x = \frac{2}{\sqrt{5}}$, we get $y = \sqrt{1 - \frac{4}{5}} = \frac{1}{\sqrt{5}}$ [Using (5.89)]

and $\frac{d^2u}{dx^2} = 12 - \frac{24.2}{\sqrt{5}\sqrt{1-\frac{4}{5}}} - \frac{8.8}{\sqrt{5}\left(1-\frac{4}{5}\right)^{\frac{3}{2}}} < 0$ [from (5.92)]

So, u is maximum in this case.

Next, we take $y = -\sqrt{1-x^2}$ (5.94)

From (5.87), we get

$$u = 7x^2 - 8x\sqrt{1-x^2} + (1-x^2) \text{ [Using (5.94)]}$$
$$= 6x^2 - 8x\sqrt{1-x^2} + 1.$$
$$\therefore \frac{du}{dx} = 12x - 8\sqrt{1-x^2} + 8x.\frac{1}{2\sqrt{1-x^2}}(-2x)$$
$$= 12x - 8\sqrt{1-x^2} - \frac{8x^2}{\sqrt{1-x^2}} \tag{5.95}$$
$$= \frac{1}{\sqrt{1-x^2}}\left[12x\sqrt{1-x^2} - 8(1-x^2) + 8x^2\right]$$
$$= \frac{1}{\sqrt{1-x^2}}\left[12x\sqrt{1-x^2} - 8 + 16x^2\right] \tag{5.96}$$

$$\frac{d^2 u}{dx^2} = 12 - 8 \cdot \frac{1}{2} \frac{1}{\sqrt{1-x^2}}(-2x) + \frac{16x\sqrt{1-x^2} - 8x^2 \frac{1}{2\sqrt{1-x^2}}(-2x)}{1-x^2}$$

[Using (5.95)]

$$= 12 + \frac{8x}{\sqrt{1-x^2}} + \frac{16x(1-x^2) + 8x^3}{(1-x^2)\sqrt{1-x^2}}$$

$$= 12 + \frac{24x}{\sqrt{1-x^2}} - \frac{8x^3}{(1-x^2)\sqrt{1-x^2}} \tag{5.97}$$

For extremum, $\dfrac{du}{dx} = 0 \;\Rightarrow\; 3x\sqrt{1-x^2} + 2(1-x^2) + 2x^2 = 0$

or, $\quad 2 - 4x^2 = 3x\sqrt{1-x^2}$ \hfill (5.98)

$\therefore \quad 4 - 16x^2 + 16x^4 = 9x^2(1-x^2) \Rightarrow 25x^4 - 25x^2 + 4 = 0$

or, $\quad (5x^2 - 1)(5x^2 - 4) = 0$ giving $x = \pm\dfrac{1}{\sqrt{5}}$ or $x = \pm\dfrac{2}{\sqrt{5}}$ as before.

Now, $x = -\dfrac{1}{\sqrt{5}}$, $x = \dfrac{2}{\sqrt{5}}$ do not satisfy (5.98). Therefore, we get $x = \dfrac{1}{\sqrt{5}}$, $y = -\dfrac{2}{\sqrt{5}}$.

For $x = \dfrac{1}{\sqrt{5}}$, we get $y = -\sqrt{1 - \dfrac{1}{5}} = -\dfrac{2}{\sqrt{5}}$ [Using (5.94)]

and $\dfrac{d^2 u}{dx^2} = 12 + \dfrac{24.1}{\sqrt{5}\sqrt{1 - \frac{1}{5}}} - \dfrac{8.1}{5\sqrt{5}\left(1 - \frac{4}{5}\right)^{\frac{3}{2}}} > 0$ [from (5.97)] showing that

u is minimum in this case.

For $x = -\dfrac{2}{\sqrt{5}}$, we get $y = -\sqrt{1 - \dfrac{4}{5}} = -\dfrac{1}{\sqrt{5}}$

and $\dfrac{d^2 u}{dx^2} = 12 - \dfrac{48}{\sqrt{5}\sqrt{1 - \frac{4}{5}}} - \dfrac{8.8}{\sqrt{5}\left(1 - \frac{4}{5}\right)^{\frac{3}{2}}} < 0$ [from (5.92)] Showing that u

is maximum in this case.

Considering all the above cases, we conclude the following: $7x^2 + 8xy + y^2$ is minimum at the points $\left(\dfrac{1}{\sqrt{5}}, -\dfrac{2}{\sqrt{5}}\right)$ and $\left(-\dfrac{1}{\sqrt{5}}, \dfrac{2}{\sqrt{5}}\right)$ whereas its maximum value attains at the points $\left(\dfrac{2}{\sqrt{5}}, \dfrac{1}{\sqrt{5}}\right)$ and $\left(-\dfrac{2}{\sqrt{5}}, -\dfrac{1}{\sqrt{5}}\right)$, the minimum and maximum values being -1 and 9 respectively as before. $\qquad\square$

5.10 Exercises

1. Examine the followings for extremum values:

   (i) $x^4 + y^4 - 2x^2 + 4xy - 2y^2$.   (ii) $x^2 + y^2 + (x + y + 1)^2$.
   (iii) $y^2 + 2x^2 y + 2x^4$.   (iv) $y^2 + x^2 y + 2x^4$.
   (v) $x^3 + y^3 - 3axy$.   (vi) $x^2 y^2 - 5x^2 - 8xy - 5y^2$.

2. Find all the extreme values of the function $x^3 + y^3 - 63(x + y) + 12xy$.

3. Show that the function $f(x, y) = x^2 + 2xy + y^2 + x^3 + y^3 + x^7$ has neither a maximum nor a minimum at the origin.

4. Show that the function $(x - y)^4 + (x - 2)^2$ has a minimum at $(2, 2)$.

5. Find the maximum value of $\{\sin x \sin y \sin(x + y)\}$ where $0 \le x \le \pi, 0 \le y \le \pi, 0 \le (x + y) \le \pi$.

   [*Hints:* Take $f(x, y) = \sin x \sin y \sin(x + y)$, then $f_x = f_y = 0 \Rightarrow \dfrac{\sin(x + y)}{\cos(x + y)} =$

   $- \tan x = - \tan y$, i.e., $x = y$ as $0 \le (x + y) \le \pi$.

   Now, from $\dfrac{\sin 2x}{\cos 2x} = \dfrac{\sin x}{\cos x}$, we get $\sin 3x = 0$, i.e., $x = \dfrac{\pi}{3}$ as $0 \le x \le \pi$. So,

   $\left(\dfrac{\pi}{3}, \dfrac{\pi}{3}\right)$ is the stationary point of the given function $f(x, y)$.

   Now, check that at the point $\left(\dfrac{\pi}{3}, \dfrac{\pi}{3}\right)$, $f_{xx} \cdot f_{yy} - f_{xy}^2 > 0$ and $f_{xx} < 0$. Hence

   $f(x, y)$ is minimum at $\left(\dfrac{\pi}{3}, \dfrac{\pi}{3}\right)$ and its minimum value $= \dfrac{3\sqrt{3}}{8}$.]

6. Find the maximum value of $x^2 y^3 z^4$ subject to the condition $x + y + z = 18$.

7. Find the shortest distance from the origin to the hyperbola $x^2 + 8xy + 7y^2 = 225$, $z = 0$.

8. Find the minimum distance of the point $(1, 2, 3)$ from the plane $x + y - 4z = 9$.

9. Prove that of all rectangular parallelepiped of same volume, the cube has the least surface area.

10. Find the maximum value of $f(x, y) = \sin \dfrac{x}{2} \sin \dfrac{y}{2} \sin \dfrac{x + y}{2}$ defined on the triangular region bounded by the coordinate axes and the line $x + y = 2\pi$.

11. Use Lagrange Multiplier Method to show that if $f(x, y, z) = x^2 + y^2 + 2z^2$ where $xyz^2 = 1$, the four points $(1, 1, \pm 1)$ and $(-1, -1, \pm 1)$ are stationary points of f and each of the points f attains a minimum.

12. Which point of the sphere $x^2 + y^2 + z^2 = 1$ is at a minimum distance from the point $(2, 1, 3)$?

13. Use Lagrange Multiplier Method to find the shortest distance from $(0, b)$ to the parabola $x^2 = 4y$.

14. If x, y, z are the angles of a triangle then prove that the following functions (i) $f(x, y, z) = \sin x \sin y \sin z$ and (ii) $f(x, y, z) = \cos x \cos y \cos z$ are both maximum at $x = y = z = \dfrac{\pi}{3}$.

15. Find the maximum and minimum values of $x^2 + y^2 + z^2$ subject to the conditions $\dfrac{x^2}{4} + \dfrac{y^2}{5} + \dfrac{z^2}{25} = 1$ and $x + y - z = 0$.

16. Use the method of Lagrange Multiplier to find the minimum of $x^2 + y^2 + z^2$ where $2x + 3y + 6z = 12$.

Multivariate Calculus

17. Examine the function $f(x, y) = y^2 + 4xy + 3x^2 + x^3$ for extreme values.

18. Discuss the extreme values of u where (i) $u = x^2 + y^2 + z^2 + x - 2z - xy$ and (ii) $u = 2a^3xy - 3ax^2y - ay^2 + x^3y + xy^3$.

ANSWER

1. (i) Maximum at $(\sqrt{2}, -\sqrt{2})$ and $(-\sqrt{2}, \sqrt{2})$; (ii) Minimum at $\left(-\frac{1}{3}, \frac{1}{3}\right)$; (iii) Minimum at $(0, 0)$; (iv) Neither a maximum nor a minimum point; (v) Maximum at $(0, 0)$ if $a < 0$ and minimum at $(0, 0)$ if $a > 0$; (vi) Maximum at $(0, 0)$ but no minimum points. **2.** Minimum at $(3, 3)$, maximum at $(-7, -7)$, neither a maximum nor a minimum at $(5, -1)$ and at $(-1, 5)$; **5.** $\frac{3\sqrt{3}}{8}$; **6.** $4^2.6^3.8^4$; **7.** 25; **8.** $\sqrt{18}$; **12.** $\left(-\frac{2}{\sqrt{14}}, -\frac{1}{\sqrt{14}}, -\frac{3}{\sqrt{14}}\right)$; **13.** 5; **15.** Maximum value is 10, minimum value is $\frac{75}{17}$; **16.** Minimum value is $\frac{144}{49}$; **17.** Minimum at $\left(\frac{2}{3}, -\frac{4}{3}\right)$.

Chapter 6

Multiple Integrals

6.1 Introduction

In this chapter we shall study integrals of functions of several variables involving multiple integrals. The motivations connected with these integrals are much the same as with simple integrals of functions of one variable which we have studied earlier. The chapter deals mainly with two types of integrals in $\mathbb{R}^2$ and $\mathbb{R}^3$, *viz.*, (i) *Double integrals* and (ii) *Triple integrals*. These integrals have important applications to geometry and mechanics.

6.2 Double Integrals

The definite integral $\int_a^b f(x)dx$ is defined as the limit of the sum $\sum_{i=1}^{n} f(x_i)\delta x_i$,

where $n \to \infty$ and each of the lengths $\delta x_1, \delta x_2, ...$ tends to zero [*ref.* § 0.3.4]. A double integral is its counterpart in two dimensions.

Let $f(x,y)$ be a function of the two independent variables x and y defined at each point in the finite region R of the xy plane. Let the region R be divided into n elementary areas $\delta A_1, \delta A_2, ..., \delta A_n$ of which (x_r, y_r) be a point within the r-th elementary area δA_r. Now we consider the sum

$$f(x_1, y_1)\delta A_1 + f(x_2, y_2)\delta A_2 + ... + f(x_n, y_n)\delta A_n \text{ i.e., the sum } \sum_{r=1}^{n} f(x_r, y_r)\delta x_r.$$

The limit of this sum, if it exists, as the number of sub divisions increases indefinitely and area of each sub division decreases to zero is defined to be the double integral of $f(x, y)$ over the region R and is written as $\iint_R f(x, y)dA$. Thus

$$\iint_R f(x, y)dA = \lim_{\substack{n \to \infty \\ \delta A_r \to 0}} \sum_{r=1}^{n} f(x_r, y_r)\delta A_r \tag{6.1}$$

Definition 6.2.1 *Let the region R be divided into the rectangular partitions and dx be the length of a sub rectangle and dy be its width, so that $dxdy$ is an ele-*

ment of area in certain coordinates, then the integral $\iint f(x,y)dA$ *is written as*

$\iint_R f(x,y)dA$ *and is called the double integral of* $f(x,y)$ *over the region* R.

Note 6.2.1 *In the definition of double integral, if we put* $f(x,y) = 1$, *we have the area* A *of the region bounded by the curves* $y = f(x), y = \phi(x)$, *the straight lines* $x = a, x = b$ *respectively,*

$$i.e., \quad A = \int_{x=a}^{x=b} \int_{y=f(x)}^{y=\phi(x)} dy dx. \quad [See \ \S \ 7.10]$$

6.3   Properties of Double Integral

Analogous to the properties of ordinary single integral, the double integral also obey the following properties:

Property 1: When the region of integration R is partitioned into two parts, say R_1 and R_2, then

$$\iint_R f(x,y)dxdy = \iint_{R_1} f(x,y)dxdy + \iint_{R_2} f(x,y)dxdy.$$

The result is true when the region is divided into three or more sub regions.

Property 2: The double integral of an algebraic sum of a finite number of functions is equal to the algebraic sum of double integrals, taken for each terms separately. Thus

$$\iint_R [f_1(x,y) + f_2(x,y) + ...]dxdy = \iint_R f_1(x,y)dxdy + \iint_R f_2(x,y)dxdy + ...$$

Property 3: A constant factor may be taken outside the integral sign. Thus

$$\iint_R mf(x,y)dxdy = m \iint_R f(x,y)dxdy, \quad m \text{ is a constant.}$$

6.4   Evaluation of Double Integrals

The utility of double integrals would be limited if it were required to take limits of sums to evaluate them. However, there is another method of evaluating double integrals.

For purposes of evaluation of the integral in (6.1) is expressed as the repeated integral

$$\int_{x_1}^{x_2} \int_{y_1}^{y_2} f(x,y)dydx.$$

Its value is found as follows:

Case 1: When y_1, y_2 are functions of x and x_1, x_2 are constants, $f(x,y)$ is first integrated with respect to y, keeping x fixed between limits y_1, y_2 and then the resulting expression is integrated with respect to x within the limits x_1, x_2, i.e.,

$$I = \int_{x_1}^{x_2} \left[\int_{y_1}^{y_2} f(x,y)dy \right] dx,$$

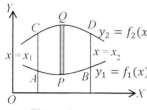

Figure 6.1

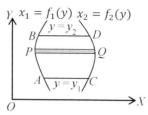

Figure 6.2

where integration is carried from the inner to the outer rectangle. This is illustrated as follows:

In **Figure** 6.1, let $y_1 = f_1(x)$ and $y_2 = f_2(x)$ be the two curves represented by AB and CD respectively. Let PQ is a vertical strip of width dx. Then the inner rectangle integral means that the integration is along one edge of the strip PQ from P to Q, x remaining constant, while the outer rectangle integral corresponds to the sliding of the edge form AC to BD.

Thus the whole region of integration is the area $ABDC$.

Case 2: When x_1, x_2 are functions of y and y_1, y_2 are constants, $f(x, y)$ is first integrated with regard to x, keeping y fixed, within the limits x_1, x_2 and the resulting expression is integrated with respect to y between the limits y_1, y_2, i.e.,

$$I = \int_{y_1}^{y_2} \left[\int_{x_1}^{x_2} f(x, y) dx \right] dy.$$

This is geometrically illustrated in **Figure** 6.2.

Here AB and CD are the curves $x_1 = f_1(y)$ and $x_2 = f_2(y)$, PQ is a horizontal strip of width dy.

The inner rectangle indicates that the integration is taken along one edge of this strip from P to Q while the outer rectangle corresponds to the sliding of this edge from AC to BD.

Thus the whole region of integration is the area $ABDC$.

Case 3: When both pairs of the limits are constants, the region of integration is the rectangle $ABDC$ as shown in **Figure** 6.3.

Over a rectangular region R, as shown in **Figure** 6.3, if the region R be given by the inequalities $x_1 \le x \le x_2, y_1 \le y \le y_2$, then the double integral

$$\iint_R f(x, y) dx dy = \int_{x_1}^{x_2} \int_{y_1}^{y_2} f(x, y) dy dx$$

$$= \int_{x_1}^{x_2} \left[\int_{y_1}^{y_2} f(x, y) dy \right] dx = I_1 \text{ (say)}.$$

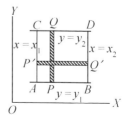

Figure 6.3

Here, in I_1, we first integrate along the strip PQ and then slide it from AC to BD.

Also, we can get

$$\iint_R f(x, y) dx dy = \int_{y_1}^{y_2} \left[\int_{x_1}^{x_2} f(x, y) dx \right] dy = I_2 \text{ (say)},$$

where the integration is taken first along the strip $P'Q'$ and then slide it from AB to CD.

In this case, obviously $I_1 = I_2$.

Thus for constant limits, it hardly matters whether we first integrate with respect to x and then with respect to y and vice-verse.

Note 6.4.1 *Calculations of a double integral **over a rectangle** are called **Iterated or Repeated Integral**.*

6.5   Evaluation Procedure of Double Integrals

1. While evaluating double integrals, first integrate with respect to the variable having variable limits and treating the other variable as constant and then integrate with respect to variate with constant limits. We have already seen, this may be done in two different ways: either by formula given in **Case 1** or by **Case 2** of § 6.4. Depending upon the type of region R or the integrand in each specific case, we choose one of the formula to calculate the double integral. The following notations are also sometimes used

$$I = \int_{x_1}^{x_2} \left[\int_{y_1}^{y_2} f(x,y) dy \right] dx = \int_{x_1}^{x_2} dx \int_{y_1}^{y_2} f(x,y) dy$$

$$\text{or,} \quad I = \int_{y_1}^{y_2} \left[\int_{x_1}^{x_2} f(x,y) dx \right] dy = \int_{y_1}^{y_2} dy \int_{x_1}^{x_2} f(x,y) dx.$$

2. In case the limits of integration of both the variables are constants, it is not matter whether we first integrate with respect to x and then with respect to y or vice-verse. Thus we have

$$I = \int_{x_1}^{x_2} \int_{y_1}^{y_2} f(x,y) dy dx$$

$$= \int_{x_1}^{x_2} \left[\int_{y_1}^{y_2} f(x,y) dy \right] dx = \int_{x_1}^{x_2} \left[\int_{y_1}^{y_2} f(x,y) dy \right] dx.$$

Example 6.5.1 *Evaluate* $\displaystyle\int_0^5 \int_0^{x^2} x(x^2 + y^2) dy dx.$

Solution: Here $\displaystyle I = \int_0^5 dx \int_0^{x^2} x(x^2+y^2) dy = \int_0^5 \left[x^3 y + x.\frac{y^3}{3} \right]_0^{x^2} dx$

$$= \int_0^5 \left(x^3.x^2 + x.\frac{x^6}{3} \right) dx = \int_0^5 \left(x^5 + \frac{x^7}{3} \right) dx = \left[\frac{x^6}{6} + \frac{x^8}{24} \right]_0^5$$

$$= \left[\frac{5^6}{6} + \frac{5^8}{24} \right] = 5^6 \left(\frac{1}{6} + \frac{5^2}{24} \right) = 18880.2 \text{ (approximate).} \qquad \square$$

Example 6.5.2 *Show that* $\displaystyle\int_1^2 \int_0^{\frac{y}{2}} y \, dy dx = \int_1^2 \int_0^{\frac{x}{2}} x \, dy dx = \frac{7}{6}.$

Solution: We have $\displaystyle\int_1^2 \int_0^{\frac{y}{2}} y \, dy dx = \int_1^2 y \left[\int_0^{\frac{y}{2}} dx \right] dy = \int_1^2 y \left[x \right]_0^{\frac{y}{2}} dy$

$$= \int_1^2 y.\frac{y}{2} dy = \frac{1}{2} \int_1^2 y^2 dy = \frac{1}{2} \left[\frac{y^3}{3} \right]_1^2 = \frac{1}{2} \left(\frac{8}{3} - \frac{1}{3} \right) = \frac{7}{6}$$

and $\displaystyle\int_1^2\int_0^{\frac{x}{2}} x\,dydx = \int_1^2 x\left[\int_0^{\frac{x}{2}} dy\right]dx = \int_1^2 x\,[y]_0^{\frac{x}{2}}\,dx$

$$= \int_1^2 x\cdot\frac{x}{2}dx = \frac{1}{2}\int_1^2 x^2 dx = \frac{1}{2}\left[\frac{x^3}{3}\right]_1^2 = \frac{1}{2}\left(\frac{2^3}{3} - \frac{1^3}{3}\right) = \frac{7}{6}.$$

Hence the result obtained. □

Example 6.5.3 *Evaluate* $\displaystyle\int_0^{3/2}\int_0^1 (4 - x^2 - y^2)dxdy.$

Solution: We have $I = \displaystyle\int_0^{\frac{3}{2}}\left[\int_0^1 (4 - x^2 - y^2)dx\right]dy$

$$= \int_0^{\frac{3}{2}}\left[4x - \frac{x^3}{3} - y^2 x\right]_0^1 dy = \int_0^{\frac{3}{2}}\left(4 - \frac{1}{3} - y^2\right)dy = \int_0^{\frac{3}{2}}\left(\frac{11}{3} - y^2\right)dy$$

$$= \left[\frac{11}{3}y - \frac{y^3}{3}\right]_0^{\frac{3}{2}} = \frac{11}{3}\cdot\frac{3}{2} - \frac{27}{8}\cdot\frac{1}{3} = \frac{11}{2} - \frac{9}{8} = \frac{44-9}{8} = \frac{35}{8}.$$ □

Example 6.5.4 *Evaluate the double integral* $\displaystyle\iint_R (4 - x^2 - y^2)dxdy$ *if the region*
R is bounded by the straight lines $x = 0, x = 1, y = 0$ *and* $y = \dfrac{3}{2}.$

Solution: Clearly, the bounded region of integration R is a rectangle bounded by
the straight lines $x = 0, x = 1, y = 0$ and $y = \dfrac{3}{2}.$

Therefore, the required double integral $= \displaystyle\iint_R (4 - x^2 - y^2)dxdy$

$$= \int_0^{3/2}\int_0^1 (4 - x^2 - y^2)dxdy = \int_0^{\frac{3}{2}}\left[\int_0^1 (4 - x^2 - y^2)dx\right]dy = \frac{35}{8}.$$

[following the **Example 6.5.3**]. □

Note 6.5.1 *In the above example the limits of integration can easily be determined*
from the given data without drawing the geometrical figure.

Example 6.5.5 *Evaluate the double integral of* $f(x, y) = 1 + x + y$ *over a region*
bounded by the lines $y = -x, x = \sqrt{y}, y = 2$ *and* $y = 0.$

Solution: Required integral is

$$\int_0^2 dy\left[\int_{-y}^{\sqrt{y}}(1 + x + y)dx\right] = \int_0^2 dy\left[x + xy + \frac{x^2}{2}\right]_{-y}^{\sqrt{y}}$$

$$= \int_0^2\left[\left(\sqrt{y} + y\sqrt{y} + \frac{y}{2}\right) - \left(-y - y^2 + \frac{y^2}{2}\right)\right]dy$$

$$= \int_0^2\left[\sqrt{y} + \frac{3y}{2} + y\sqrt{y} + \frac{y^2}{2}\right]dy = \left[\frac{2}{3}y^{\frac{3}{2}} + \frac{3}{4}y^2 + \frac{2}{5}y^{\frac{5}{2}} + \frac{y^3}{6}\right]_0^2$$

$$= \frac{2}{3}\cdot 2\sqrt{2} + 3 + \frac{2}{5}\cdot 4\sqrt{2} + \frac{8}{6} = \frac{4\sqrt{2}}{3} + \frac{8\sqrt{2}}{5} + 3 + \frac{4}{3} = \frac{44\sqrt{2}}{15} + \frac{13}{3}.$$ □

Note 6.5.2 *Though the region of integration is non rectangular here we still have not drawn the geometrical figure to determine the limits of integration because the limits can be obtained directly from the given data.*

Example 6.5.6 *Evaluate* $\iint_R (x^2 + y^2)\, dxdy$ *where* R *is the region bounded by* $x = 0, y = 0$ *and* $x + y = 1$.

Solution: From the equation $x + y = 1$ we have $x = 1$ for $y = 0$ and for the positive quadrant x varies from 0 to 1 and for y, it varies from $y = 0$ to $y = 1 - x$. Therefore, the limits of integration for the region R can be expressed as $0 \leq x \leq 1, 0 \leq y \leq 1 - x$.

Hence the required integral =

$$\int_0^1 \int_0^{1-x} (x^2 + y^2)\, dxdy = \int_0^1 dx \int_0^{1-x} (x^2 + y^2)\, dy$$

$$= \int_0^1 dx \left[x^2 y + \frac{y^3}{3} \right]_0^{1-x} = \int_0^1 \left[x^2 (1 - x) + \frac{(1 - x)^3}{3} \right] dx$$

$$= \frac{1}{3} \int_0^1 \left[(3x^2 - 3x^3) + (1 - 3x + 3x^2 - x^3) \right] dx$$

$$= \frac{1}{3} \int_0^1 (1 - 3x + 6x^2 - 4x^3) dx = \frac{1}{3} \left[x - \frac{3}{2} x^2 + 6 \frac{x^3}{3} - 4 \frac{x^4}{4} \right]_0^1$$

$$= \frac{1}{3} \left[1 - \frac{3}{2} + \frac{6}{3} - 1 \right] = \frac{1}{3} \left(2 - \frac{3}{2} \right) = \frac{1}{6}. \qquad \square$$

Example 6.5.7 *Evaluate* $\iint_D xy\, dxdy$, *where* D *is the domain bounded by the x-axis, ordinate* $x = 2a$ *and the curve* $x^2 = 4ay$.

Solution: The line $x = 2a$ and the parabola $x^2 = 4ay$ intersect at $L(2a, a)$. **Figure 6.4** shows that the domain D is the area OML. Integrating first over the vertical strip PQ, i.e., with respect to y form $P(y = 0)$ to $Q\left(y = \frac{x^2}{4a} \right)$ on the parabola and then with respect to x from $x = 0$ to $x = 2a$, we get

$$\iint_D xy\, dxdy = \int_0^{2a} dx \left[\int_0^{\frac{x^2}{4a}} xy\, dy \right] = \int_0^{2a} x \left[\frac{y^2}{2} \right]_0^{\frac{x^2}{4a}} dx$$

$$= \int_0^{2a} \frac{x^5}{32a^2}\, dx = \frac{1}{32a^2} \left[\frac{x^6}{6} \right]_0^{2a} = \frac{1}{32a^2} \cdot \frac{64a^6}{6} = \frac{1}{3} a^4.$$

Alternative Method:

Here, we first integrate over the horizontal strip $P'Q'$, (ref. **Figure** 6.5) i.e., with respect to x form $P'(x = 2\sqrt{ay})$ on the parabola to $Q'(x = 2a)$ and then with respect to y for $y = 0$ to $y = a$, we get

$$\iint_D xy\, dxdy = \int_0^a dy \left[\int_{2\sqrt{ay}}^{2a} xy\, dx \right] = \int_0^a y \left[\frac{x^2}{2} \right]_{2\sqrt{ay}}^{2a} dy$$

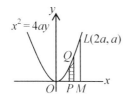

Figure 6.4

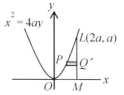

Figure 6.5

$$= \int_0^a y \left(\frac{4a^2}{2} - \frac{4ay}{2} \right) dy = 2a \int_0^a (ay - y^2) \, dy = 2a \left[\frac{ay^2}{2} - \frac{y^3}{3} \right]_0^a = \frac{1}{3} a^4. \qquad \square$$

Example 6.5.8 *Evaluate* $\iint xy(x+y) \, dxdy$ *over the region between* $y = x^2$ *and* $y = x$.

Solution: The parabola $y = x^2$ and the line $y = x$ intersect at the point $(0,0)$ and $(1,1)$, when $x = 0, y = 0$ and $x = 1, y = 1$. (see **Figure** 6.6)
So the area of integration for x is $x = 0$ to $x = 1$ and for y from x^2 to x. Therefore, given integral is

$$\int_0^1 dx \int_{x^2}^x x(yx + y^2) dy$$

$$= \int_0^1 x \left[x\frac{y^2}{2} + \frac{y^3}{3} \right]_{x^2}^x dx$$

$$= \int_0^1 x \left[x\frac{x^2}{2} + \frac{x^3}{3} - x\frac{x^4}{2} - \frac{x^6}{3} \right] dx$$

$$= \int_0^1 \left(\frac{x^4}{2} + \frac{x^4}{3} - \frac{x^6}{2} - \frac{x^7}{3} \right) dx$$

$$= \int_0^1 \left[\frac{5}{6}x^4 - \frac{1}{2}x^6 - \frac{1}{3}x^7 \right] dx = \left[\frac{5}{6}\frac{x^5}{5} - \frac{1}{2}\frac{x^7}{7} - \frac{1}{3}\frac{x^8}{8} \right]_0^1 = \frac{1}{6} - \frac{1}{14} - \frac{1}{24}$$

$$= \frac{3}{24} - \frac{1}{14} = \frac{3}{56}. \qquad \square$$

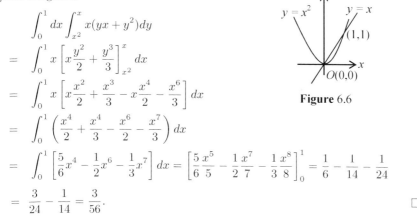

Figure 6.6

6.6 Double Integrals in Polar Coordinates

Let us consider a function $f(r, \theta)$ of polar coordinates (r, θ) over a region R, whose boundary is also given in polar coordinates. Similarly, as defined the double integral in plane Cartesian coordinates we divide the region R into n elementary areas $\delta A_1, \delta A_2, ..., \delta A_n$ of which (r_i, θ_i) is a point within the i-th elementary area δA_i.

Then the double integral $\iint_R f(r, \theta) dA$ is given by

$$\iint_R f(r, \theta) dA = \lim_{\substack{n \to \infty \\ \delta A_i \to 0}} \sum_{i=1}^n f(r_i, \theta_i) \, \delta A_i,$$

provided the limit exits.

6.7 Evaluation of Double Integrals in Polar Coordinates

To evaluate $\int_{\theta_1}^{\theta_2} \int_{r_1}^{r_2} f(r,\theta) \, dr d\theta$, we first integrate with respect to r between limits $r = r_1$ and $r = r_2$, keeping θ fixed and the resulting expression is integrated with respect to θ from $\theta = \theta_1$ to $\theta = \theta_2$. In this integral r_1, r_2 are functions of θ and θ_1, θ_2 are constants. The following **Figure** 6.7 illustrates the process geometrically.

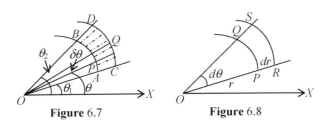

Figure 6.7   **Figure** 6.8

Here AB and CD are the curves $r_1 = f_1(\theta)$ and $r_2 = f_2(\theta)$ bounded by the lines $\theta = \theta_1$ to $\theta = \theta_2$. PQ is a wedge of angular thickness $\delta\theta$. Then $\int_{r_1}^{r_2} f(r,\theta) \, dr$ indicates that the integration is along PQ from P to Q while the integration with respect to θ corresponding to the turning of PQ from AC to BD. Therefore the whole region of integration is the area $ACDB$. The order of integration may however be changed with appropriate changes in the limits.

Remark: In case of Cartesian coordinates when the integral $\iint_R f(x,y) \, dA$ is expressed in the form of repeated integrals, dA represents the area of the rectangle with sides dx and dy and hence $dA = dxdy$.

If the radius vectors of OP and OR be r and $r + \delta r$ (vide. **Figure** 6.8) respectively and $\angle POQ = d\theta$, $PQ = r \, d\theta$ (as PQ and RS are arcs of circles), then

$$dA = PQ.PR = rd\theta.dr = rd\theta dr.$$

Example 6.7.1 *Evaluate* $\int_0^{\frac{\pi}{2}} \int_0^{\sin\theta} r \, d\theta dr.$

Solution: Required integral is $\int_0^{\frac{\pi}{2}} d\theta \int_0^{\sin\theta} r \, dr = \int_0^{\frac{\pi}{2}} \left[\frac{1}{2} r^2\right]_0^{\sin\theta} d\theta$

$$= \frac{1}{2} \int_0^{\frac{\pi}{2}} \sin^2\theta \, d\theta = \frac{1}{4} \int_0^{\frac{\pi}{2}} (1 - \cos 2\theta) \, d\theta = \frac{1}{4} \left[\theta - \frac{1}{2}\sin 2\theta\right]_0^{\frac{\pi}{2}}$$

$$= \frac{1}{4} \left[\frac{1}{2}\pi\right] = \frac{\pi}{8}. \qquad \square$$

Example 6.7.2 *Evaluate* $\int_0^{\frac{\pi}{2}} \int_0^{2a\cos\theta} r^2 \sin\theta \cos\theta \, d\theta dr.$

Solution: Given integral is $\int_0^{\frac{\pi}{2}} \sin\theta \cos\theta \, d\theta \int_0^{2a\cos\theta} r^2 \, dr$

$$= \int_0^{\frac{\pi}{2}} \sin\theta \cos\theta \left[\frac{r^3}{3}\right]_0^{2a\cos\theta} d\theta = \frac{1}{3}\int_0^{\frac{\pi}{2}} \sin\theta \cos\theta.(2a\cos\theta)^3 d\theta$$

$$= \frac{8a^3}{3}\int_0^{\frac{\pi}{2}} \sin\theta \cos^4\theta \ d\theta = -\frac{8a^3}{3}\int_0^{\frac{\pi}{2}} \cos^4\theta \ d(\cos\theta)$$

$$= -\frac{8a^3}{3}\left[\frac{\cos^5\theta}{5}\right]_0^{\frac{\pi}{2}} = -\frac{8a^3}{3}\left[0 - \frac{1}{5}\right] = \frac{8a^3}{15}. \qquad \square$$

Example 6.7.3 *Integrate*

(i) $\displaystyle\iint r^2 \sin\theta \ drd\theta$ *over upper half of the circle* $r = 2a\cos\theta$

(ii) $\displaystyle\iint \frac{rd\theta dr}{\sqrt{a^2 + r^2}}$ *over one loop of the lemniscate* $r^2 = a^2 \cos 2\theta$.

Solution: (i) The region R is the upper half of the circle $r = 2a\cos\theta$ whose centre is at $r = a, \theta = 0$.

$$\therefore I = \int_0^{\frac{\pi}{2}} \sin\theta \ d\theta \int_0^{2a\cos\theta} r^2 \ dr = \int_0^{\frac{\pi}{2}} \sin\theta \ d\theta \left[\frac{r^3}{3}\right]_0^{2a\cos\theta}$$

$$= \frac{8a^3}{3}\int_0^{\frac{\pi}{2}} \sin\theta \cos^3\theta \ d\theta = -\frac{8a^3}{3}\int_0^{\frac{\pi}{2}} \cos^3\theta \ d(\cos\theta)$$

$$= -\frac{8a^3}{3}\left[\frac{\cos^4\theta}{4}\right]_0^{\frac{\pi}{2}} = \frac{2a^3}{3}.$$

(ii) The given curve is a lemniscate which is symmetrical about the initial line since by putting $-\theta$ for θ, its equation (i.e., $r^2 = a^2 \cos 2\theta$) remains unchanged. Also, the curve passes through the pole.

The straight lines $\theta = \pm\dfrac{\pi}{4}$ are tangents at pole since by putting $r = 0$, we get

$\cos 2\theta = 0 \ \Rightarrow \ \theta = \pm\dfrac{\pi}{4}.$

For different values of θ, the corresponding values of r are given as follows:

θ	0	$\dfrac{\pi}{6}$	$\dfrac{\pi}{4}$	$\dfrac{\pi}{2}$	$\dfrac{3\pi}{4}$	$\dfrac{5\pi}{6}$	π
r^2	a^2	$\dfrac{a^2}{2}$	0	$-a^2$	0	$\dfrac{a^2}{2}$	a^2
r	$\pm a$	$\pm\dfrac{a}{\sqrt{2}}$	0	imaginary	0	$\pm\dfrac{a}{\sqrt{2}}$	$\pm a$

It is clear from the above table that no part of the curve lies between $\theta = \dfrac{\pi}{4}$ and $\theta = \dfrac{3\pi}{4}$.

Combining the above facts, the curve is traced as shown in the **Figure** 6.9 and we notice that

(a) In lemniscate, there are two loops.

(b) When $-\dfrac{\pi}{4} < \theta < \dfrac{\pi}{4}$

  or $\dfrac{3\pi}{4} < \theta < \dfrac{5\pi}{4}$, r is real.

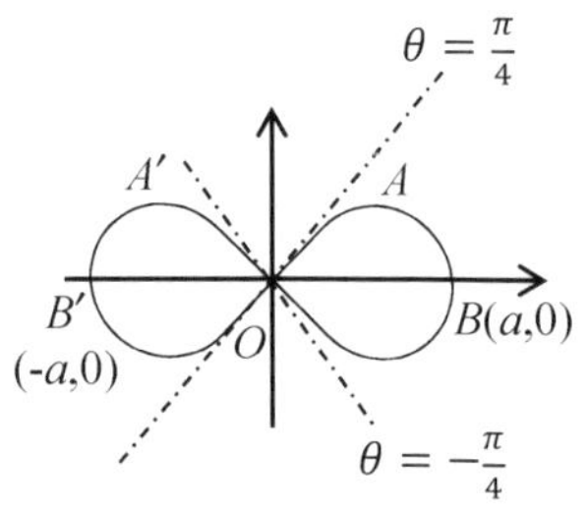

Figure 6.9

Now we want to evaluate the given integral over the right loop of the lemniscate. Therefore

$$\iint \frac{r\,d\theta dr}{\sqrt{a^2+r^2}} = \int_{-\frac{\pi}{4}}^{\frac{\pi}{4}} d\theta \int_{-a\sqrt{\cos 2\theta}}^{a\sqrt{\cos 2\theta}} \frac{r}{\sqrt{a^2+r^2}}\,dr$$

($\because r^2 = a^2 \cos 2\theta \therefore r = +a\sqrt{\cos 2\theta}$, so r varies form $-a\sqrt{\cos 2\theta}$ to $a\sqrt{\cos 2\theta}$.)

Now since, there is a symmetry about x-axis, we should evaluate the double integral over half of the right loop as follows:

$$\iint \frac{r\,d\theta dr}{\sqrt{a^2+r^2}} = \int_0^{\frac{\pi}{4}} d\theta \int_0^{a\sqrt{\cos 2\theta}} \frac{r}{\sqrt{a^2+r^2}}\,dr$$

$$= \frac{1}{2}\int_0^{\frac{\pi}{4}} d\theta \int_0^{a\sqrt{\cos 2\theta}} \frac{2r}{\sqrt{a^2+r^2}}\,dr = \frac{1}{2}\int_0^{\frac{\pi}{4}} d\theta \int_0^{a\sqrt{\cos 2\theta}} \frac{d(a^2+r^2)}{\sqrt{a^2+r^2}}$$

$$= \frac{1}{2}\int_0^{\frac{\pi}{4}} d\theta \left[2\sqrt{a^2+r^2}\right]_0^{a\sqrt{\cos 2\theta}} = \int_0^{\frac{\pi}{4}} \left(\sqrt{a^2+a^2\cos 2\theta} - \sqrt{a^2}\right) d\theta$$

$$= \int_0^{\frac{\pi}{4}} \left(\sqrt{2}.a\cos\theta - a\right) d\theta = \sqrt{2}a \int_0^{\frac{\pi}{4}} \cos\theta\, d\theta - a\int_0^{\frac{\pi}{4}} d\theta$$

$$= \sqrt{2}.a\,[\sin\theta]_0^{\frac{\pi}{4}} - a\,[\theta]_0^{\frac{\pi}{4}} = \sqrt{2}.a\frac{1}{\sqrt{2}} - a\frac{\pi}{4} = a - a\frac{\pi}{4}.$$

Therefore, value of the double integral over the complete right loop

$$= 2\left(a - a\frac{\pi}{4}\right) = 2a\left(1 - \frac{\pi}{4}\right).$$ □

Note 6.7.1 *While evaluating the double integral* $\int_{-\frac{\pi}{4}}^{\frac{\pi}{4}} d\theta \int_{-a\sqrt{\cos 2\theta}}^{a\sqrt{\cos 2\theta}} \dfrac{r\,dr}{\sqrt{a^2+r^2}}$ *we*

must be careful about the fact that the function $\left(\dfrac{r}{\sqrt{a^2+r^2}}\right)$ *is an odd function resulting its value equal to zero. Therefore we must not calculate the double integral over the complete loop.*

Example 6.7.4 *Calculate* $\displaystyle\iint r^3\,drd\theta$ *over the area included between the circles* $r = 2\sin\theta$ *and* $r = 4\sin\theta$.

Solution: Given circles are $r = 2\sin\theta$ and $r = 4\sin\theta$ which are shown in **Figure** 6.10. The shaded area between these circles is the region of integration.

To calculate the value of the double integration, we first integrate with respect to r, then its limits are from $P(r = 2\sin\theta)$ to $Q(r = 4\sin\theta)$ and to cover the whole region varies form 0 to π. Thus the required integral

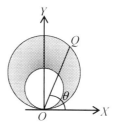

Figure 6.10

$$\iint r^3 \, drd\theta = \int_{\theta=0}^{\pi} d\theta \int_{r=2\sin\theta}^{4\sin\theta} r^3 \, dr$$

$$= \int_{\theta=0}^{\pi} d\theta \left[\frac{r^4}{4}\right]_{r=2\sin\theta}^{4\sin\theta}$$

$$= \frac{1}{4}\int_0^{\pi} \left[(4\sin\theta)^4 - (2\sin\theta)^4\right] d\theta = \frac{1}{4}\int_0^{\pi} \sin^4\theta\left(4^4 - 2^4\right) d\theta$$

$$= 60 \int_0^{\pi} \sin^4\theta d\theta = 120 \times \frac{3}{4}.\frac{1}{2}.\frac{\pi}{2} = 22.5\pi. \qquad \square$$

6.8 Change of Order of Integration

In a double integral with variable limits, the change of order of integration changes the limits of integration. While doing so, sometimes it is required to split up the region of integration and the given integral is expressed as the sum of a number of double integrals with changed limits. For finding the new limits it may require to draw a rough sketch of the region of integration. We need to change the order of integration to make the evaluation of a double integration easier.

The following examples will make the idea clear.

Example 6.8.1 *Change the order of integration in the following integral*

$$I = \int_{-a}^{a}\int_{0}^{\sqrt{a^2-y^2}} f(x,y) \, dxdy.$$

Solution: The integral is equivalent to $I = \int_{-a}^{a} dy \int_{0}^{\sqrt{a^2-y^2}} f(x,y) \, dx.$

The elementary strip considered here is, parallel to x axis like PQ which extends form $x = 0$ to $x = \sqrt{a^2 - y^2}$, i.e., to the circle $x^2 + y^2 = a^2$. This strip slides from $y = -a$ to $y = a$. The semi-circular area $BPOB'RAQSB$ in the positive side of the x-axis is, therefore, the region of integration.

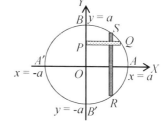

Figure 6.11

If we want to change the order of integration, we first integrate with respect to y along a vertical strip RS which extends from R [where $y = -\sqrt{a^2 - x^2}$] to S [where $y = \sqrt{a^2 - x^2}$].

To cover the same region, we, therefore next integrate with respect to x from $x = 0$ to $x = a$ and thus we get

$$I = \int_0^a dx \int_{-\sqrt{a^2-x^2}}^{\sqrt{a^2-x^2}} f(x,y)dy = \int_0^a \int_{-\sqrt{a^2-x^2}}^{\sqrt{a^2-x^2}} f(x,y) \, dxdy. \qquad \square$$

Example 6.8.2 *Change the order of integration in* $I = \int_0^1 dy \int_y^1 e^{x^2}\, dx$. *and hence*
evaluate it.

Solution: Here the x values for $0 \le y \le 1$ range from y to 1. The region of
integration OAB is bounded by $y = 0, x = 1$ and $x = y$ as shown in **Figure** 6.12
(a).

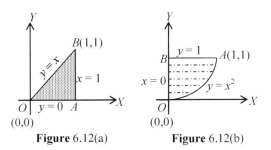

| **Figure** 6.12(a) | **Figure** 6.12(b) |

By changing the order of integration, we get

$$I = \int_0^1 dx \int_0^x e^{x^2}\, dy = \int_0^1 e^{x^2}[y]_0^x dx = \int_0^1 xe^{x^2}\, dx = \left[\frac{1}{2}e^{x^2}\right]_0^1 = \frac{e-1}{2}. \qquad \square$$

Remark: We see that $\int e^{x^2}\, dx$ is difficult to calculate in terms of our standard
integrals. That's why we needed changing the order of integration.

Example 6.8.3 *Show that*

$$\int_0^1 \left\{\int_0^1 \frac{x^2 - y^2}{x^2 + y^2} dy\right\} dx = \int_0^1 \left\{\int_0^1 \frac{x^2 - y^2}{x^2 + y^2} dx\right\} dy.$$

Solution: L.H.S. $= \int_0^1 \left\{\int_0^1 \frac{x^2 - y^2}{x^2 + y^2} dy\right\} dx = \int_0^1 dx \int_0^1 \left(\frac{2x^2}{x^2 + y^2} - 1\right) dy$

$$= \int_0^1 \left(\int_0^1 \frac{2x^2}{x^2 + y^2} dy - \int_0^1 dy\right) dx = \int_0^1 \left(2x \tan^{-1}\frac{1}{x} - 1\right) dx$$

$$= \left[\tan^{-1}\frac{1}{x}.x^2\right]_0^1 - \int_0^1 \frac{1}{1 + \frac{1}{x^2}}\left(-\frac{1}{x^2}\right) x^2 dx - \int_0^1 dx$$

$$= \frac{\pi}{4} + \int_0^1 \frac{x^2}{x^2 + 1} dx - 1 = \frac{\pi}{4} + \int_0^1 dx - \int_0^1 \frac{1}{x^2 + 1} dx - 1$$

$$= \frac{\pi}{4} + 1 - \left[\tan^{-1} x\right]_0^1 - 1 = \frac{\pi}{4} - \frac{\pi}{4} = 0.$$

and similarly R.H.S. $= \int_0^1 \left\{\int_0^1 \frac{x^2 - y^2}{x^2 + y^2} dx\right\} dy$

$$= \int_0^1 dy \int_0^1 \left(1 - \frac{2y^2}{x^2 + y^2}\right) dx = \int_0^1 \left(1 - 2y \tan^{-1}\frac{1}{y}\right) dy = 0.$$

Hence $\int_0^1 \left\{\int_0^1 \frac{x^2 - y^2}{x^2 + y^2} dy\right\} dx = \int_0^1 \left\{\int_0^1 \frac{x^2 - y^2}{x^2 + y^2} dx\right\} dy.$ $\qquad \square$

Note 6.8.1 *The above example therefore shows that change in the order of integration is permissible here.*

Example 6.8.4 *Prove that*

$$\int_0^1 \left\{ \int_0^1 \frac{x-y}{(x+y)^3} \, dy \right\} dx \neq \int_0^1 \left\{ \int_0^1 \frac{x-y}{(x+y)^3} \, dx \right\} dy.$$

Solution: Here the integrand $f(x,y) = \dfrac{x-y}{(x+y)^3}$ is bounded over the region $0 \leq x \leq 1, 0 \leq y \leq 1$, except at the origin $(0,0)$, which is a point of discontinuity. Also, for a fixed point $x \neq 0, f(x,y)$ is a bounded function of y and this is same for a fixed $y \neq 0$.

Let for $x \neq 0$, $\phi(x) = \int_0^1 \frac{x-y}{(x+y)^3} \, dy = \int_0^1 \left\{ \frac{2x}{(x+y)^3} - \frac{1}{(x+y)^2} \right\} dy$

$$= \left[\frac{-x}{(x+y)^2} + \frac{1}{x+y} \right]_0^1 = \left[-\frac{x}{(x+1)^2} + \frac{1}{x+1} - \left(-\frac{1}{x} + \frac{1}{x} \right) \right]$$

$$= \frac{-x+x+1}{(x+1)^2} = \frac{1}{(x+1)^2}.$$

$\phi(0) = \int_0^1 \left(-\frac{1}{y^2} \right) dy$, which doesn't exist.

But $\int_0^1 \left\{ \int_0^1 \frac{x-y}{(x+y)^3} \, dy \right\} dx = \int_0^1 \phi(x) dx = \int_0^1 \frac{dx}{(1+x)^2} = \left[-\frac{1}{x+1} \right]_0^1 = \frac{1}{2}.$

Again let $y \neq 0$, then $\psi(y) = \int_0^1 f(x,y) dx = \int_0^1 \frac{x-y}{(x+y)^3} \, dx$

$$= \int_0^1 \left[\frac{1}{(x+y)^2} - \frac{2y}{(x+y)^3} \right] dx = -\frac{1}{(1+y)^2}$$

and $\psi(0) = \int_0^1 \left(-\frac{1}{x^2} \right) dy$, which doesn't exist.

But $\int_0^1 \left\{ \int_0^1 f(x,y) \, dy \right\} dx = \int_0^1 \frac{1}{(1+y)^2} \, dy = -\frac{1}{2}.$

From these the result follows. □

Note 6.8.2 *The above example shows that change in the order of integration may not always give the same result.*

Note 6.8.3 *In the above case, where such a change in the order of integration is not permissible, we say that the double integral $\iint_R f(x,y) \, dxdy$ fails to exist.*

Note 6.8.4 *If a double integral exists, then the two repeated integrals can not exists without being equal. However, if the double integral does not exist, nothing can be said about the repeated integrals; they may or may not exist.*

Sometimes, it may also happen that one of the repeated integrals exists or even both exist and be equal, yet the double integral may not exist, i.e., the existence of one or both the repeated integrals gives no guarantee for the existence of the double integral. However, if the two repeated integrals exist but are unequal, the double integral cannot exist.

6.9   Change in the Variable in a Double Integral

Sometimes a double integral can be more easily evaluated by changing the variables and expressing the functions and the integrals in terms of new coordinates.

We hereby state only the procedure in working out problems:-

Let us suppose that we require to transform the double integral

$$\iint_R f(x,y)\ dxdy$$ to the variables u and v by the substitution $x = \phi(u,v),\ y = \psi(u,v)$. Then it can be shown that

$$\iint_R f(x,y)\ dxdy = \iint_{R'} f\left\{\phi(u,v), \psi(u,v)\right\}.|J|\ dudv$$

where R' is the region of the transformed integral in (u,v)-plane corresponding to the region R in the (x,y)-plane and J is their Jacobian given by

$$J = \frac{\partial(x,y)}{\partial(u,v)} = \begin{vmatrix} \dfrac{\partial x}{\partial u} & \dfrac{\partial y}{\partial u} \\ \dfrac{\partial x}{\partial v} & \dfrac{\partial y}{\partial v} \end{vmatrix} = x_u y_v - x_v y_u \neq 0.$$

Note 6.9.1 *For completeness we add that the transformation formula remains valid even if the value of the Jacobian J vanishes without however changing its sign at a finite number of isolated points.*

Example 6.9.1 *Evaluate* $\displaystyle\iint x^2 y^2\ dxdy$ *extended over the (a) region* $x \geq 0, y \geq 0, x^2 + y^2 \leq 1$ *and (b) circle* $x^2 + y^2 \leq 1$.

Solution: (a) Here the region of integration R is the positive quadrant of the circle $x^2 + y^2 = 1$. Here $f(x,y) = x^2 y^2$.

Let us use the transformation $x = r\cos\theta,\ y = r\sin\theta$.

$$\therefore\ J = \frac{\partial(x,y)}{\partial(r,\theta)} = \begin{vmatrix} \dfrac{\partial x}{\partial r} & \dfrac{\partial x}{\partial \theta} \\ \dfrac{\partial y}{\partial r} & \dfrac{\partial y}{\partial \theta} \end{vmatrix} = \begin{vmatrix} \cos\theta & -r\sin\theta \\ \sin\theta & r\sin\theta \end{vmatrix} = r(\sin^2\theta + \cos^2\theta) = r.$$

$$\begin{aligned} \therefore\ I &= \iint_{x^2+y^2\leq 1} x^2 y^2\ dxdy = \iint_{x^2+y^2\leq 1} f(x,y)dxdy \\ &= \iint f(r\cos\theta, r\sin\theta)|J|drd\theta = \iint r^2\cos^2\theta.r^2\sin^2\theta.rdrd\theta \\ &= \iint r^4\cos^2\theta\sin^2\theta.rdrd\theta = \int_0^{\frac{\pi}{2}}\cos^2\theta\sin^2\theta d\theta\int_0^1 r^5.dr \\ &= \frac{1}{6}\int_0^{\frac{\pi}{2}}(\cos\theta\sin\theta)^2 d\theta = \frac{1}{6}\int_0^{\frac{\pi}{2}}\frac{1}{4}(\sin^2\theta)^2 d\theta \\ &= \frac{1}{48}\int_0^{\frac{\pi}{2}}(1-\cos 4\theta)d\theta = \frac{1}{48}\left[\theta - \frac{\sin 4\theta}{4}\right]_0^{\frac{\pi}{2}} = \frac{1}{48}.\frac{\pi}{2} = \frac{\pi}{96}. \end{aligned}$$

(*b*) Using the same transformation as in (*a*) and proceeding as before we get in this case, required integral

$$= \int_0^{2\pi} \cos^2\theta \sin^2\theta \, d\theta \int_0^1 r^5 . dr = \frac{1}{6} \int_0^{2\pi} (\cos\theta \sin\theta)^2 d\theta$$

$$= \frac{1}{6} \int_0^{2\pi} \frac{1}{4}(\sin^2\theta)^2 d\theta = \frac{1}{48} \int_0^{2\pi} (1 - \cos 4\theta) d\theta = \frac{1}{48} \left[\theta - \frac{\sin 4\theta}{4} \right]_0^{2\pi}$$

$$= \frac{1}{48}.2\pi = \frac{\pi}{24}. \qquad \square$$

Example 6.9.2 *Evaluate* $I = \int_0^1 dx \int_0^x \sqrt{x^2 + y^2} \, dy$ *by transforming to polar co-ordinates.*

Solution: We see that the given integral is $I = \iint_R \sqrt{x^2 + y^2} \, dxdy$ where R is the $\triangle OAB$ bounded by $y = 0, y = x, x = 1$ [See **Figure** 6.13].

Let us use the transformation $x = r\cos\theta$, $y = r\sin\theta$, whereby Jacobian $J = r$ and so $I = \iint_{R'} r.r \, drd\theta$ where R is mapped into R', where $r\sin\theta = 0$ $r\sin\theta = r\cos\theta, r\cos\theta = 1$ giving $r = \sec\theta$ and θ changes for 0 to $\frac{\pi}{4}$.

$$\therefore \ I = \int_0^{\frac{\pi}{4}} \left\{ \int_0^{\sec\theta} r^2 dr \right\} d\theta = \frac{1}{3} \int_0^{\frac{\pi}{4}} \sec^3\theta \, d\theta$$

$$= \frac{1}{6} \left\{ \sqrt{2} + \log(1 + \sqrt{2}) \right\}.$$

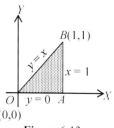

Figure 6.13

$\square$

Example 6.9.3 *Compute the double integral* $\iint_R (y - x) \, dxdy$ *where R is the region in the xy-plane bounded by the straight lines $y = x + 1, y = x - 3, y = -\frac{1}{3}x + \frac{7}{3}, y = -\frac{1}{3}x + 5$.*

Solution: It would be difficult to compute this double integral directly. However, a simple change of variables permits reducing this integral to one over a rectangle whose sides are parallel to the coordinate axes. For this, we get

$$u = y - x, \ v = y + \frac{1}{3}x \qquad (6.2)$$

Then the given straight lines are transformed, respectively, into the straight lines $u = 1, u = -3, v = \frac{7}{3}, v = 5$ in the uv-plane. Consequently the given region R is transformed into the rectangular region R' as shown in **Figure** 6.14.

It remains to compute the Jacobian of the transformation. To do this, we express x and y in terms of u and v by solving the system of equation (6.2) as

$$x = -\frac{3}{4}u + \frac{3}{4}v, \ y = \frac{1}{4}u + \frac{3}{4}v.$$

$$\therefore \quad J = \frac{\partial(x,y)}{\partial(u,v)} = \begin{vmatrix} \dfrac{\partial x}{\partial u} & \dfrac{\partial x}{\partial v} \\[2mm] \dfrac{\partial y}{\partial u} & \dfrac{\partial y}{\partial v} \end{vmatrix} = \begin{vmatrix} -\dfrac{3}{4} & \dfrac{3}{4} \\[2mm] \dfrac{1}{4} & \dfrac{3}{4} \end{vmatrix} = -\frac{9}{16} - \frac{3}{16} = -\frac{3}{4}.$$

Therefore, we get

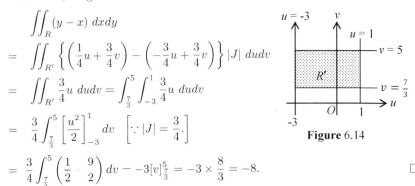

$$\iint_R (y-x)\,dxdy$$

$$= \iint_{R'} \left\{ \left(\frac{1}{4}u + \frac{3}{4}v \right) - \left(-\frac{3}{4}u + \frac{3}{4}v \right) \right\} |J|\,dudv$$

$$= \iint_{R'} \frac{3}{4}u\,dudv = \int_{\frac{7}{3}}^{5} \int_{-3}^{1} \frac{3}{4}u\,dudv$$

$$= \frac{3}{4} \int_{\frac{7}{3}}^{5} \left[\frac{u^2}{2} \right]_{-3}^{1} dv \quad \left[\because |J| = \frac{3}{4}. \right]$$

$$= \frac{3}{4} \int_{\frac{7}{3}}^{5} \left(\frac{1}{2} - \frac{9}{2} \right) dv - -3[v]_{\frac{7}{3}}^{5} = -3 \times \frac{8}{3} = -8. \qquad \square$$

Figure 6.14

Example 6.9.4 *Evaluate* $\displaystyle\iint \sqrt{4a^2 - x^2 - y^2}\,dxdy$ *taken over the upper half of the circle* $x^2 + y^2 - 2ax = 0$.

Solution: $\displaystyle\iint_R \sqrt{4a^2 - x^2 - y^2}\,dxdy$, where the region R is the upper half of the circle $x^2 + y^2 - 2ax = 0$.

Using the transformation $x = r\cos\theta, y = r\sin\theta$, whence $J = r$, we get

$\displaystyle\iint_{R'} \sqrt{4a^2 - r^2}\,r\,drd\theta$, where R' is the upper half of the circle $r^2 = 2ar\cos\theta$ or $r = 2a\cos\theta$.

The limits of r will be 0 to $2a\cos\theta$ and those of θ will be 0 to $\dfrac{\pi}{2}$.

$$\therefore \quad I = \int_0^{\frac{\pi}{2}} d\theta \int_0^{2a\cos\theta} \sqrt{4a^2 - r^2}\,r\,dr = \int_0^{\frac{\pi}{2}} d\theta \left[-\frac{1}{3} \left(4a^2 - r^2 \right)^{\frac{3}{2}} \right]_0^{2a\cos\theta}$$

$$= -\frac{1}{3} \int_0^{\frac{\pi}{2}} d\theta \left[\left(4a^2 - 4a^2\cos^2\theta \right)^{\frac{3}{2}} - \left(4a^2 \right)^{\frac{3}{2}} \right]$$

$$= -\frac{1}{3} \int_0^{\frac{\pi}{2}} \left(8a^3\sin^3\theta - 8a^3 \right) d\theta = -\frac{8}{3}a^3 \int_0^{\frac{\pi}{2}} \left(\sin^3\theta - 1 \right) d\theta$$

$$= -\frac{8}{3}a^3 \left[-\cos\theta + \frac{1}{3}\cos^3\theta - \theta \right]_0^{\frac{\pi}{2}} = \frac{4a^3}{9}(3\pi - 4). \qquad \square$$

6.10 Triple Integral or Integral over a Volume

We have introduced the notion of double integral in § 6.2. Now we are going to define the integral of a function of three independent variables, the so called *Volume Integral*, also known as *Triple Integral* in $\mathbb{R}^3$.

Let $f(x,y,z)$ be a single-valued function of the independent variables x, y, z in finite region V. We divide the region V into n sub regions $\delta V_1, \delta V_2, ..., \delta V_n$. Let

P be any point on the boundary on inside. Now we take a point in each part and form the sum

$$
\begin{aligned}
S_n &= f(x_1, y_1, z_1)\delta V_1 + f(x_2, y_2, z_2)\delta V_2 + \ldots + f(x_n, y_n, z_n)\delta V_n \\
&= \sum_{r=1}^{n} f(x_r, y_r, z_r)\delta V_r.
\end{aligned}
$$

The limit of this sum, if it exists, as the number of sub divisions increases indefinitely and volume of each sub division decreases is defined as the *triple integral* of $f(x, y, z)$ over the region V and is written as $\iiint_V f(x, y, z)\ dV$. Thus

$$
\iiint_V f(x, y, z)\ dV = \lim_{\substack{n \to \infty \\ \delta V_r \to 0}} \sum_{r=1}^{n} f(x_r, y_r, z_r)\ \delta V_r \tag{6.3}
$$

Remark 1: Triple integral can be utilised in evaluating a number of physical quantities. As an example, if $f(x, y, z) = \rho$ is considered the volume density of distribution of a substance over the domain R then we find $\iiint_V f(x, y, z)\ dV$ and putting $f(x, y, z) = \rho$, we get

$$
\text{Mass} = \iiint_V \rho\ dV.
$$

Remark 2: If the integrand $f(x, y, z) = 1$, then the triple integral over the domain R expresses the *volume* of R, i.e.,

$$
\text{Volume } V = \iiint_R dx\,dy\,dz.
$$

6.11   Evaluation of Triple Integrals

6.11.1   Triple integration over a parallelepiped

Let $f(x, y, z)$ be a bounded and integrable function of three independent variables x, y, z over the rectangular parallelepiped bounded by the planes $x = a, x = b, y = c, y = d, z = e, z = f$. This region will be bounded by $R = \{a, b; c, d; e, f\}$. Then the triple integral $\iiint_V f(x, y, z)\ dx\,dy\,dz$ is given by

$$
\begin{aligned}
\iiint_V f(x, y, z)\ dx\,dy\,dz &= \int_a^b \int_c^d \int_e^f f(x, y, z)dx\,dy\,dz \\
&= \int_a^b dx \int_c^d dy \int_e^f f(x, y, z)dz \tag{6.4}
\end{aligned}
$$

Here, we integrate first with respect to z keeping x and y constant and then the remaining integration is done as in case of double integrals (discussed in § 6.4 (4)), i.e., integration with respect to y regarding x as a constant and then with respect to x.

Note 6.11.1 *An integral of the form* $\iint_{R_1} dxdy \int_e^f f(x, y, z)\, dz$ *or*

$\iint_{R_1} \left[\int_e^f f(x, y, z)\, dz \right] dxdy$, *where* $R_1 = [a, b; c, d]$ *is the projection of* $R =$

$[a, b; c, d; e, f]$ *on the xy-plane is known as* **Iterated Integral**.

Note 6.11.2 *If* $f(x, y, z)$ *is continuous, all the integrals of the form*

$$\iiint_R f(x, y, z)\, dxdydz, \int_a^b dx \iint_{R_{yz}} f(x, y, z)dydz, \int_c^d dy \iint_{R_{zx}} f(x, y, z)dzdx,$$

$$\int_a^b \int_c^d dxdy \int_e^f f(x, y, z)dz, \int_a^b dx \int_c^d dy \int_e^f f(x, y, z)dz \text{ etc. all exist, so that the}$$

integral of a continuous function defined on a rectangle can be evaluated by **iterated integrals** *where iteration can be taken in any order.*

6.11.2 Triple integral over any finite region

Let A be a finite region bounded by a given surface. We construct a rectangular parallelepiped R enclosing A completely and define a function $g(x, y, z)$ over R as follows:

$$g(x, y, z) = \begin{cases} f(x, y, z), & \text{at all points of } A \\ 0, & \text{outside } A. \end{cases}$$

Then the function $f(x, y, z)$ is said to be integrable over A, if $g(x, y, z)$ is integrable over R and we have $\iiint_A f(x, y, z)\, dxdydz = \iiint_R g(x, y, z)\, dxdydz.$

6.11.3 Evaluation of triple integral over the region A

Let the region A be bounded by the surfaces $z = u(x, y), z = v(x, y); y = \phi(x), y = \psi(x); x = a, y = b$. If $f(x, y, z)$ be a continuous function in A then

$$\iiint_A f(x, y, z)\, dxdydz = \int_a^b dx \int_{\phi(x)}^{\psi(x)} dy \int_u^v f(x, y, z)\, dz.$$

Thus, for example, $\iiint f(x, y, z)\, dxdydz$ over the region bounded by the sphere

$x^2 + y^2 + z^2 = a^2$ is equivalent to $\int_{-a}^a dx \int_{-\sqrt{a^2-x^2}}^{\sqrt{a^2-x^2}} dy \int_{-\sqrt{a^2-x^2-y^2}}^{\sqrt{a^2-x^2-y^2}} f(x, y, z)\, dz.$

Example 6.11.1 *Evaluate* $\int_0^{3a} \int_0^{2a} \int_0^a (x + y + z)\, dxdydz.$

Solution: $I = \int_0^{3a} dz \int_0^{2a} dy \int_0^a (x + y + z)\, dx$

$$= \int_0^{3a} dz \int_0^{2a} dy \left[\frac{x^2}{2} + yx + zx \right]_0^a = \int_0^{3a} dz \int_0^{2a} \left(\frac{a^2}{2} + ay + az \right) dy$$

$$= \int_0^{3a} dz \left[\frac{a^2}{2}y + \frac{ay^2}{2} + azy \right]_0^{2a} = \int_0^{3a} \left(\frac{2a^3}{2} + \frac{4a^3}{2} + 2a^2 z \right) dz$$

$$= \int_0^{3a} (3a^3 + 2a^2 z)\, dz = \left[3a^3 z + 2a^2 \frac{z^2}{2} \right]_0^{3a} = 9a^4 + 9a^4 = 18a^4. \qquad \square$$

Example 6.11.2 *Evaluate* $\int_0^4 \int_0^{2\sqrt{z}} \int_0^{\sqrt{4z-x^2}} dz dx dy.$

Solution: $I = \int_0^4 dz \int_0^{2\sqrt{z}} dx \int_0^{\sqrt{4z-x^2}} dy = \int_0^4 dz \int_0^{2\sqrt{z}} dx\, [y]_0^{\sqrt{4z-x^2}}$

$= \int_0^4 dz \int_0^{2\sqrt{z}} \sqrt{4z-x^2} dx = \int_0^4 \left[\frac{x}{2}\sqrt{4z-x^2} + \frac{4z}{2}\sin^{-1}\frac{x}{2\sqrt{z}} \right]_0^{2\sqrt{z}} dz$

$= \int_0^4 \left[0 + \frac{4z}{2}\sin^{-1}\frac{2\sqrt{z}}{2\sqrt{z}} \right] dz = \int_0^4 2z\frac{\pi}{2} dz = \pi \int_0^4 z dz = \pi \left[\frac{z^2}{2} \right]_0^4 = 8\pi.$ □

Example 6.11.3 *Evaluate* $\int_{x=0}^1 \int_{y=0}^{\sqrt{1-x^2}} \int_{z=0}^{\sqrt{1-x^2-y^2}} xyz\, dxdydz.$

Solution: $I = \int_0^1 dx \int_0^{\sqrt{1-x^2}} dy \int_0^{\sqrt{1-x^2-y^2}} xyz\, dz$

$= \int_0^1 dx \int_0^{\sqrt{1-x^2}} dy \left[xy\frac{z^2}{2} \right]_0^{\sqrt{1-x^2-y^2}} = \frac{1}{2}\int_0^1 dx \int_0^{\sqrt{1-x^2}} xy(1-x^2-y^2)dy$

$= \frac{1}{2}\int_0^1 dx \int_0^{\sqrt{1-x^2}} x\{y(1-x^2)-y^3\}dy = \frac{1}{2}\int_0^1 \left[x(1-x^2)\frac{y^2}{2} - x\frac{y^4}{4} \right]_0^{\sqrt{1-x^2}}$

$= \frac{1}{8}\int_0^1 2x\left\{ (1-x^2)(1-x^2) - (1-x^2)^2 \right\} dx = \frac{1}{8}\int_0^1 x(1-x^2)^2 dx$

$= \frac{1}{8}\int_0^1 x(1-2x^2+x^4)dx = \frac{1}{8}\int_0^1 (x-2x^3+x^5)dx = \frac{1}{8}\left[\frac{x^2}{2} - 2\frac{x^4}{4} + \frac{x^6}{6} \right]_0^1$

$= \frac{1}{8}\left[\frac{1}{2} - \frac{1}{2} + \frac{1}{6} \right] = \frac{1}{48}.$ □

Example 6.11.4 *Compute the integral* $\iiint_R xyz\, dxdydz$ *over a domain bounded by* $x=0, y=0, z=0, x+y+z=1.$

Solution: The domain is regular and bounded, above and below by $z = 1-x-y$ and $z = 0$. Its projection R' on the xy-plane is a triangle bounded by $x=0, y=0, y=1-x.$

$\therefore \iiint_R xyz\, dxdydz = \iint_{R'} \left[\int_0^{1-x-y} xyz\, dz \right] dxdy$

$= \int_0^1 dx \int_0^{1-x} xy\left[\frac{z^2}{2} \right]_0^{1-x-y} dy = \frac{1}{2}\int_0^1 dx \int_0^{1-x} xy(1-x-y)^2 dy$

$= \frac{1}{2}\int_0^1 xdx \int_0^{1-x} y(1-x-y)^2 dy$

$= \frac{1}{2}\int_0^1 xdx \int_0^{1-x} y(1+x^2+y^2-2x+2xy-2y)dy$

$= \frac{1}{2}\int_0^1 xdx \int_0^{1-x} (y+x^2y+y^3-2xy+2xy^2-2y^2)dy$

$$= \frac{1}{2}\int_0^1 x\,dx\left[\frac{y^2}{2} + x^2\frac{y^2}{2} + \frac{y^4}{4} - 2x\frac{y^2}{2} + 2x\frac{y^3}{3} - 2\frac{y^3}{3}\right]_0^{1-x}$$

$$= \frac{1}{2}\int_0^1 x\left[\frac{y^2}{2}(1 - 2x + x^2) - \frac{2y^3}{3}(1 - x) + \frac{y^4}{4}\right]_0^{1-x}$$

$$= \frac{1}{2}\int_0^1 x\left[\frac{(1-x)^4}{2} - \frac{2}{3}(1-x)^4 + \frac{1}{4}(1-x)^4\right]dx$$

$$= \frac{1}{2}\int_0^1 \frac{x}{12}(1-x)^4 dx = -\frac{1}{24}\int_1^0 (1-p)p^4 dp \;\;\text{[by putting } 1 - x = p]$$

$$= -\frac{1}{24}\left[\frac{p^5}{5} - \frac{p^6}{6}\right]_1^0 = \frac{1}{720}. \qquad\qquad \square$$

Note 6.11.3 *The above may be clear from the following geometrical explanation:*
*At first we consider a vertical strip MN from the plane z = 0 (i.e., the xy-plane) to the plane z = 1 − x − y (i.e., the ABC-plane), giving the limits of the variable z as z = 0 to z = 1 − x − y (See **Figure** 6.15 (a)). Then we move the variable strip so as to cover the volume, so that M covers the plane triangular area OAB, i.e., the variable y runs from 0 to 1 − x and finally x runs from 0 to 1.*

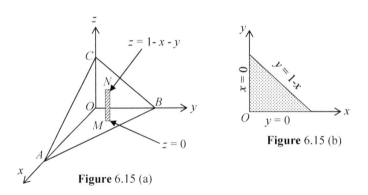

Figure 6.15 (a)

Figure 6.15 (b)

Example 6.11.5 *Evaluate* $\displaystyle\iiint \frac{dx\,dy\,dz}{(1 + x + y + z)^3}$ *extended over the tetrahedron bounded by the planes* $x = 0, y = 0, z = 0, x + y + z = 1.$

Solution: Here the region of integration is a tetrahedron in which z varies from 0 to $1 - x - y$, y varies form 0 to $1 - x$ and x varies from 0 to 1 and so we get the given integral

$$
\begin{aligned}
I &= \int_0^1 dx\int_0^{1-x} dy\int_0^{1-x-y} \frac{dz}{(1 + x + y + z)^3}\\
&= \int_0^1 dx\int_0^{1-x} dy\left[-\frac{1}{2}\frac{1}{(1 + x + y + z)^2}\right]_0^{1-x-y}\\
&= \frac{1}{2}\int_0^1 dx\int_0^{1-x}\left[\frac{1}{(1 + x + y)^2} - \frac{1}{4}\right]dy = \frac{1}{2}\int_0^1 dx\left[-\frac{1}{1 + x + y} - \frac{y}{4}\right]_0^{1-x}\\
&= \frac{1}{2}\int_0^1 dx\left(-\frac{1}{2} + \frac{1}{1 + x} - \frac{1-x}{4} + 0\right) = \frac{1}{2}\int_0^1 \left(\frac{1}{1 + x} + \frac{x}{4} - \frac{3}{4}\right)dx
\end{aligned}
$$

$$= \frac{1}{2}\left[\log_e(1+x) + \frac{x^2}{8} - \frac{3}{4}x\right]_0^1 = \frac{1}{2}\left[\log_e 2 - \log_e 1 + \frac{1}{8} - 0 - \frac{3}{4} + 0\right]$$

$$= \frac{1}{16}(\log_e 2 - 5) = \frac{1}{16}\left(8\log_e 2^8 - \log_e e^5\right) = \frac{1}{16}\log_e\left(\frac{256}{e^5}\right). \qquad \square$$

Remark: The above two examples **6.11.4** and **6.11.5** may also be stated as follows:

Example 6.11.6 *Compute the threefold iterated integral of the function $f(x,y,z)$*
$= xyz$ *or* $\dfrac{1}{(1+x+y+z)^3}$ *over the region R bounded by the planes $x = 0, y = 0, z = 0, x+y+z = 1$.*

Solution: We observe that the domain is regular, it is bounded above and below by the planes $z = 0$ and $z = 1 - x - y$ and is projected on the xy-plane into a regular plane domain D, which is a triangle bounded by the straight lines $x = 0, y = 0, y = 1 - x$ [**Figure** 6.15 (b)]. Therefore the threefold iterated integral is computed as follows:

$$I = \iint_D \left[\int_0^{1-x-y} f(x,y,z)dz\right]dS.$$

Setting up the limits in the twofold iterated integral over the domain D, we get $I = \displaystyle\int_0^1 \left\{\int_0^{1-x}\left[\int_0^{1-x-y} f(x,y,z)dz\right]dy\right\}dx.$

Now for complete solution we have to proceed like previous examples. $\qquad \square$

Example 6.11.7 *Compute the volume of the ellipsoid $\dfrac{x^2}{a^2} + \dfrac{y^2}{b^2} + \dfrac{z^2}{c^2} = 1$.*

Solution: The ellipsoid is bounded below by the surface
$z = -c\sqrt{1 - \dfrac{x^2}{a^2} - \dfrac{y^2}{b^2}}$ and above by the surface $z = c\sqrt{1 - \dfrac{x^2}{a^2} - \dfrac{y^2}{b^2}}$.

The projection of this ellipsoid on the xy-plane is an ellipse $\dfrac{x^2}{a^2} + \dfrac{y^2}{b^2} = 1$. Hence reducing the computation of volume to that of a threefold iterated integral, we get

$$V = \int_{-a}^a \left[\int_{-b\sqrt{1-\frac{x^2}{a^2}}}^{b\sqrt{1-\frac{x^2}{a^2}}} \left\{\int_{-c\sqrt{1-\frac{x^2}{a^2}-\frac{y^2}{b^2}}}^{c\sqrt{1-\frac{x^2}{a^2}-\frac{y^2}{b^2}}} dz\right\}dy\right]dx \text{ [See Remark of § 6.10]}$$

$$= 2c\int_{-a}^a \left[\int_{-b\sqrt{1-\frac{x^2}{a^2}}}^{b\sqrt{1-\frac{x^2}{a^2}}} \sqrt{1 - \frac{x^2}{a^2} - \frac{y^2}{b^2}}\, dy\right]dx.$$

While computing the integral within the third bracket, x is held constant. We substitute $y = b \sqrt{1 - \dfrac{x^2}{a^2}} \, \sin\theta$ $\quad \therefore \quad dy = b \sqrt{1 - \dfrac{x^2}{a^2}} \, \cos\theta \, d\theta.$

when $y = -b \sqrt{1 - \dfrac{x^2}{a^2}}$ then $\sin\theta = -1 \;\Rightarrow\; \theta = -\dfrac{\pi}{2}$

and when $y = b \sqrt{1 - \dfrac{x^2}{a^2}}$ then $\sin\theta = 1 \;\Rightarrow\; \theta = \dfrac{\pi}{2}$

$$\therefore \quad V = 2c \int_{-a}^{a} \left[\int_{-\frac{\pi}{2}}^{\frac{\pi}{2}} \sqrt{\left(1 - \frac{x^2}{a^2}\right) - \left(1 - \frac{x^2}{a^2}\right) \sin^2\theta} \right.$$

$$\left. \times b\sqrt{1 - \frac{x^2}{a^2}} \, \cos\theta \, d\theta \right] dx$$

$$= 2cb \int_{-a}^{a} \left[\left(1 - \frac{x^2}{a^2}\right) \int_{-\frac{\pi}{2}}^{\frac{\pi}{2}} \cos^2\theta \, d\theta \right] dx$$

$$= \frac{cb\pi}{a^2} \int_{-a}^{a} (a^2 - x^2) \, dx = \frac{4}{3}\pi abc.$$

Hence we get the volume of the ellipsoid $= V = \dfrac{4}{3}\pi abc.$ $\qquad\square$

Note 6.11.4 *If $a = b = c = r$ (say) $=$ radius of a sphere, then the volume of the sphere is $V = \dfrac{4}{3}\pi r^3$.*

Example 6.11.8 *Evaluate $\displaystyle\iiint_V zy^2 \, dx\,dy\,dz$, where V is the region bounded between the xy-plane and the sphere $x^2 + y^2 + z^2 = 1$.*

Solution: Here the column parallel to z-axis is bounded by the plane $z = 0$ and the surface of the sphere $x^2 + y^2 + z^2 = 1$, i.e., $z = \sqrt{1 - x^2 - y^2}$.

The region R in which the volume V stands is the area of the circle of intersection of the sphere $x^2 + y^2 + z^2 = a^2$ by the plane $z = 0$ (i.e., xy-plane). So the region R is the circle $x^2 + y^2 = 1$. [See **Figure** 6.16]

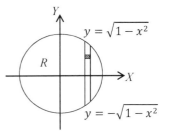

Therefore the limits of integration are $-\sqrt{1 - x^2}$ to $\sqrt{1 - x^2}$ for y and -1 to 1 for x and hence

$$I = \int_{-1}^{1} dx \int_{-\sqrt{1-x^2}}^{\sqrt{1-x^2}} dy \int_{0}^{\sqrt{1-x^2-y^2}} y^2 z \, dz$$

Figure 6.16

$$= \int_{-1}^{1} dx \int_{-\sqrt{1-x^2}}^{\sqrt{1-x^2}} dy \left[y^2 \frac{z^2}{2} \right]_{0}^{\sqrt{1-x^2-y^2}}$$

$$= \frac{1}{2} \int_{-1}^{1} dx \int_{-\sqrt{1-x^2}}^{\sqrt{1-x^2}} y^2 (1 - x^2 - y^2) dy$$

$$= \frac{1}{2} \int_{-1}^{1} dx \int_{-\sqrt{1-x^2}}^{\sqrt{1-x^2}} (y^2 - y^2 x^2 - y^4) dy$$

$$= \frac{1}{2} \int_{-1}^{1} dx \left[\frac{y^3}{3} - x^2 \frac{y^3}{3} - \frac{y^5}{5} \right]_{-\sqrt{1-x^2}}^{\sqrt{1-x^2}}$$

$$= \frac{1}{2} \int_{-1}^{1} \frac{1 \times 2}{15} \left[5(1-x^2)^{\frac{3}{2}} - 5x^2(1-x^2)^{\frac{3}{2}} - 3(1-x^2)^{\frac{5}{2}} \right] dx$$

$$= \frac{1}{15} \int_{-1}^{1} \left[5(1-x^2)^{\frac{3}{2}}(1-x^2) - 3(1-x^2)^{\frac{5}{2}} \right] dx$$

$$= \frac{1}{15} \int_{-1}^{1} 2(1-x^2)^{\frac{5}{2}} dx = \frac{2}{15} \int_{-1}^{1} (1-x^2)^{\frac{5}{2}} dx$$

$$= \frac{4}{15} \int_{0}^{1} (1-x^2)^2 \sqrt{1-x^2} dx$$

$$= \frac{4}{15} \int_{0}^{\frac{\pi}{2}} (1-\sin^2 \theta)^2 \cos\theta . \cos\theta \; d\theta \qquad [\text{Putting } x = \sin\theta]$$

$$= \frac{4}{15} \int_{0}^{\frac{\pi}{2}} \cos^6 \theta \; d\theta = \frac{4}{15} . \frac{5}{6} . \frac{3}{4} . \frac{1}{2} . \frac{\pi}{2} = \frac{\pi}{24}. \qquad \Box$$

6.12 Change of Variables in Triple Integration

6.12.1 In General Transformation

Let a closed region R in three dimensional space with the variables x, y, z be mapped on of region R' with the variables u, v, w by a one to one transformation whose Jacobian $J = \dfrac{\partial(x, y, z)}{\partial(u, v, w)}$ $(\neq 0)$. Then the transformation formula for the transformation $x = \phi_1(u, v, w), y = \phi_2(u, v, w), z = \phi_3(u, v, w)$ is given by

$$\iiint_R f(x, y, z) \; dxdydz = \iiint_{R'} f(\phi_1, \phi_2, \phi_3)|J| \; dudvdw.$$

Note 6.12.1 *As in case of change in the variables in Double Integral (See **Note 6.9.1**) here also we remark that this transformation formula remains valid if the Jacobian vanishes without, however, changing its sign at a finite number of isolated points of the region.*

6.12.2 Triple Integral in Cylindrical Coordinates

If a triple integral of the function $f(x, y, z)$ is given in rectangular Cartesian coordinates x, y, z it can readily be changed to a triple integral in Cylindrical coordinates by substituting $x = r\cos\theta, y = r\sin\theta, z = z$ and we get

$$\iiint_R f(x, y, z) \; dxdydz = \iiint_{R'} F(r, \theta, z)r \; drd\theta dz,$$

where $f(r\cos\theta, r\sin\theta, z) = F(r, \theta, z)$.

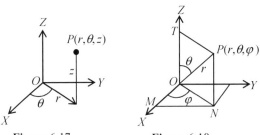

Figure 6.17   **Figure** 6.18

Note that, in this case

$$J = \frac{\partial(x,y,z)}{\partial(r,\theta,z)} = \begin{vmatrix} \dfrac{\partial x}{\partial r} & \dfrac{\partial x}{\partial \theta} & \dfrac{\partial x}{\partial z} \\ \dfrac{\partial y}{\partial r} & \dfrac{\partial y}{\partial \theta} & \dfrac{\partial y}{\partial z} \\ \dfrac{\partial z}{\partial r} & \dfrac{\partial y}{\partial \theta} & \dfrac{\partial z}{\partial z} \end{vmatrix} = \begin{vmatrix} \cos\theta & -r\sin\theta & 0 \\ \sin\theta & r\cos\theta & 0 \\ 0 & 0 & 1 \end{vmatrix} = r, \text{ i.e., } |J| = r.$$

6.12.3 Triple Integral in Spherical Polar Coordinates

For transformation into spherical polar coordinates in three dimensional space $x = r\sin\theta\cos\phi, y = r\sin\theta\sin\phi, z = r\cos\theta$ in which ϕ ranges form 0 to 2π, θ from 0 to π and r from 0 to $+\infty$. And in this case

$$J = \frac{\partial(x,y,z)}{\partial(r,\theta,\phi)} = \begin{vmatrix} \dfrac{\partial x}{\partial r} & \dfrac{\partial x}{\partial \theta} & \dfrac{\partial x}{\partial \phi} \\ \dfrac{\partial y}{\partial r} & \dfrac{\partial y}{\partial \theta} & \dfrac{\partial y}{\partial \phi} \\ \dfrac{\partial z}{\partial r} & \dfrac{\partial y}{\partial \theta} & \dfrac{\partial z}{\partial \phi} \end{vmatrix}$$

$$= \begin{vmatrix} \sin\theta\cos\phi & r\cos\theta\cos\phi & -r\sin\theta\sin\phi \\ \sin\theta\sin\phi & r\cos\theta\sin\phi & r\sin\theta\cos\phi \\ \cos\theta & -r\sin\theta & 0 \end{vmatrix} = r^2\sin\theta.$$

Therefore, we get

$$\iiint_R f(x,y,z)\ dxdydz$$

$$= \iiint_{R'} f(r\sin\theta\cos\phi, r\sin\theta\sin\phi, r\cos\theta)|J|\ drd\theta d\phi$$

$$= \iiint_{R'} f(r\sin\theta\cos\phi, r\sin\theta\sin\phi, r\cos\theta)r^2\sin\theta\ drd\theta d\phi.$$

Example 6.12.1 *Find the volume of the portion of the sphere $x^2 + y^2 + z^2 = a^2$ lying inside the cylinder $x^2 + y^2 = ax$.*

Solution: Let us use the transformation $x = r\cos\theta, y = r\sin\theta, z = z$ for changing the coordinates to cylindrical coordinates, where Jacobian $J = \dfrac{\partial(x,y,z)}{\partial(r,\theta,z)} = r$ (See § 6.12.2).

Then the equation of the sphere becomes $r^2 + z^2 = a^2$ and that of the cylinder becomes $r = a\cos\theta$.

The volume of the portion of the given sphere inside the given cylinder is twice the volume shown shaded in **Figure** 6.19 and here z varies from 0 to $\sqrt{a^2 - r^2}$, r varies from 0 to $a\cos\theta$ and θ varies form 0 to π. Hence the required volume

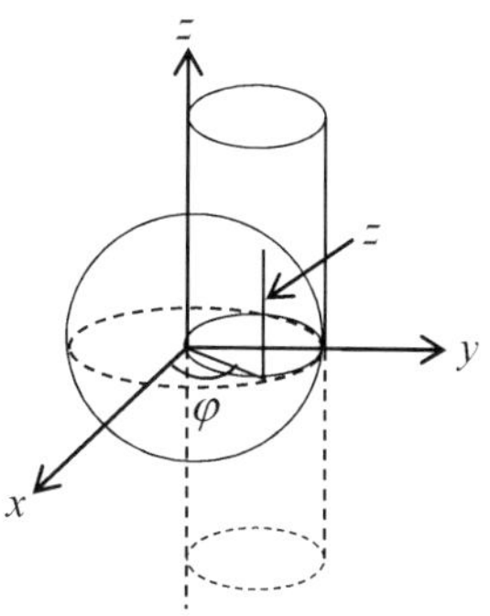

Figure 6.19

$$= 2\int_0^\pi d\theta \int_0^{a\cos\theta} r\,dr \int_0^{\sqrt{a^2-r^2}} dz$$

$$= 2\int_0^\pi d\theta \int_0^{a\cos\theta} r\sqrt{a^2 - r^2}\,dr$$

$$= 2\int_0^\pi \left[-\frac{1}{3}(a^2 - r^2)^{\frac{3}{2}}\right]_0^{a\cos\theta} d\theta = 2\int_0^\pi \left[-\frac{1}{3}(a^2 - a^2\cos^2\theta)^{\frac{3}{2}} + \frac{1}{3}a^3\right] d\theta$$

$$= \frac{2}{3}a^3 \int_0^\pi (1 - \sin^3\theta)d\theta = \frac{2}{3}a^3 \int_0^\pi \left(1 - \frac{3}{4}\sin\theta + \frac{1}{4}\sin 3\theta\right) d\theta$$

$$= \frac{2}{3}a^3 \left[\theta + \frac{3}{4}\cos\theta - \frac{1}{4}\frac{\cos 3\theta}{3}\right]_0^\pi = \frac{2}{3}a^3 \left[\pi - \frac{3}{4} + \frac{1}{12} - 0 - \frac{3}{4} + \frac{1}{12}\right]$$

$$= \frac{2}{9}a^3(3\pi - 4). \qquad \square$$

Example 6.12.2 *Evaluate using suitable transformation*

$$\iiint \sqrt{\frac{1 - x^2 - y^2 - z^2}{1 + x^2 + y^2 + z^2}}\,dxdydz$$

over the positive octant of the sphere $x^2 + y^2 + z^2 \leq 1$.

Solution: Here the region of integration R is the positive octant of the sphere $x^2 + y^2 + z^2 = 1$.

We change the coordinates to the spherical polar coordinates by the transformation $x = r\sin\theta\cos\phi, y = r\sin\theta\sin\phi, z = r\cos\theta$, where $J = r^2\sin\theta$ (See § 6.12.3) and we get

$$I = \iiint_{R'} \sqrt{\frac{1 - r^2}{1 + r^2}}\, r^2 \sin\theta\, drd\theta d\phi = \iiint_{R'} \frac{1 - r^2}{\sqrt{1 - r^4}}\, r^2 \sin\theta\, drd\theta d\phi$$

where the region R transforms to R', r varies from 0 to 1, θ from 0 to $\frac{\pi}{2}$ and ϕ from 0 to $\frac{\pi}{2}$, i.e.,

$$I = \int_0^{\frac{\pi}{2}} d\phi \int_0^{\frac{\pi}{2}} \sin\theta d\theta \int_0^1 \frac{(1 - r^2)r^2}{\sqrt{1 - r^4}}dr$$

$$= [\phi]_0^{\frac{\pi}{2}} [-\cos\theta]_0^{\frac{\pi}{2}} \left[\int_0^1 \frac{r^2}{\sqrt{1 - r^4}}dr - \int_0^1 \frac{r^4}{\sqrt{1 - r^4}}dr\right]$$

Put $r^2 = \sin t$, i.e., $2rdr = \cos t\,dt$, we get

$$
\begin{aligned}
I &= \frac{\pi}{2}\left[\frac{1}{2}\int_0^{\frac{\pi}{2}}\frac{\sqrt{\sin t}}{\cos t}\cos t\,dt - \frac{1}{2}\int_0^{\frac{\pi}{2}}\frac{\sin^{\frac{3}{2}}t}{\cos t}\cos t\,dt\right] \\
&= \frac{\pi}{4}\left[\int_0^{\frac{\pi}{2}}\sin^{\frac{1}{2}}\cos^0 t\,dt - \int_0^{\frac{\pi}{2}}\sin^{\frac{3}{2}}\cos^0 t\,dt\right] \\
&= \frac{\pi}{4}\left[\frac{1}{2}B\left(\frac{\frac{1}{2}+1}{2},\frac{0+1}{2}\right) - \frac{1}{2}B\left(\frac{\frac{3}{2}+1}{2},\frac{0+1}{2}\right)\right] \\
&= \frac{\pi}{8}\left[B\left(\frac{3}{4},\frac{1}{2}\right) - B\left(\frac{5}{4},\frac{1}{2}\right)\right]
\end{aligned}
$$

where $B(m,n) = \int_0^1 x^{m-1}(1-x)^{n-1}dx = \int_0^{\frac{\pi}{2}}\sin^{2m-2}\theta\cos^{2n-2}\theta\,d\theta$ (Putting $x = \sin\theta$) is known as the First Eulerian Integral or Beta Function. □

6.13   Differentiation under the Sign of Integration

If a function $f(x,\alpha)$ of two variables x and α (called a parameter), be integrated with respect to x between the limits a and b, then $\int_a^b f(x,\alpha)\,dx$ is a function of α, $F(\alpha)$ say. To find the derivative of $F(\alpha)$, when it exists, it may be difficult or even impossible to first evaluate the integral and then to find the derivative. Such types of problems are solved by the following rule.

Statement: *If $F(\alpha) = \int_a^b f(x,\alpha)\,dx$, where $f(x,\alpha)$ is continuous in $a < x < b$ and a,b are functions of α, α being a parameter, then prove that*

$$
\frac{dF}{d\alpha} = \int_a^b \frac{\partial f(x,\alpha)}{\partial \alpha}dx + \frac{\partial b}{\partial \alpha}\cdot f(b,\alpha) - \frac{\partial a}{\partial \alpha}\cdot f(a,\alpha).
$$

Proof: Let $f(x,\alpha)$ and $f_\alpha(x,\alpha)$ be both continuous in the region R (say), $\{a < x < b; c < \alpha < d\}$, where a,b are functions of α, a parameter and differentiable with respect to α in our chosen interval $c < \alpha < d$ and $F(\alpha) = \int_a^b f(x,\alpha)\,dx$ is differentiable in (c,d).

We have

$$
\begin{aligned}
F(\alpha+\delta\alpha) - F(\alpha) &= \int_{a(\alpha)}^{b(\alpha)}[f(x,\alpha+\delta\alpha) - f(x,\alpha)]\,dx \\
&\quad - \int_{a(\alpha)}^{a(\alpha+\delta\alpha)}f(x,\alpha+\delta\alpha)dx + \int_{b(\alpha)}^{b(\alpha+\delta\alpha)}f(x,\alpha+\delta\alpha)dx \quad (6.5)
\end{aligned}
$$

Since from the first mean value theorem of Integral Calculus, if $f(x)$ be continuous in (a,b), there exists a ξ, where

$$
\int_a^b f(x)dx = (b-a)f(\xi), \quad a < \xi < b
$$

Applying the same argument to the last two integrals of (6.5) and dividing both the sides by $\delta\alpha$, we get

$$\frac{F(\alpha+\delta\alpha)-F(\alpha)}{\delta\alpha} = \int_a^b f_\alpha(x,\alpha+\theta\delta\alpha)dx$$

$$-\frac{a(\alpha+\delta\alpha)-a(\alpha)}{\delta\alpha}f(\xi_1,\alpha+\delta\alpha) + \frac{b(\alpha+\delta\alpha)-b(\alpha)}{\delta\alpha}f(\xi_2,\alpha+\delta\alpha)$$

where $0 < \theta < 1$ and $a(\alpha) < \xi_1 < a(\alpha+\delta\alpha)$, $b(\alpha) < \xi_2 < b(\alpha+\delta\alpha)$.

Now considering limit as $\delta\alpha \to 0$, we see

$$\int_{a(\alpha)}^{b(\alpha)} f_\alpha(x,\alpha+\theta\delta\alpha)dx \to \int_{a(\alpha)}^{b(\alpha)} f_\alpha(x,\alpha)dx, \ \xi_1 \to a \text{ and } \xi_2 \to b.$$

Therefore, since $F(\alpha), a(\alpha), b(\alpha)$ are differentiable with respect to α, we get

$$\frac{dF}{d\alpha} = \int_a^b \frac{\partial f(x,\alpha)}{\partial\alpha}\,dx + \frac{\partial b}{\partial\alpha}f(b,\alpha) - \frac{\partial a}{\partial\alpha}f(a,\alpha) \qquad (6.6)$$

which is the desired result. $\qquad\qquad\square$

Note 6.13.1 *If instead of being function of α, a and b are **constants** then the above result reduces to*

$$\frac{dF}{d\alpha} = \int_a^b \frac{\partial f(x,\alpha)}{\partial\alpha}\,dx \qquad (6.7)$$

Note 6.13.2 *The result deduced above is known as **differentiation under the sign of integration** because in the R.H.S., integrand is differentiation under the integral sign.*

Note 6.13.3 *The above result is also called **Leibnitz's Rule**. Thus Leibnitz's rule enables us to derive from the value of a simple definite integral, the value of another definite integral which it may otherwise be difficult or even impossible to evaluate. Result (6.6) is called Leibnitz's rule for **variable limits of integration** whereas result (6.7) is that of for **constant limits**.*

Note 6.13.4 *The rule for differentiation under the sign of integration of an indefinite integral is the same as for definite integral.*

Corollary 6.13.1 *If $f(x,y)$ be continuous on the rectangle $S = \{f(x,y)|a \le x \le b; c \le y \le d\}$, then*

$$\int_a^b \left\{\int_c^d f(x,y)dx\right\}dy = \int_c^d \left\{\int_a^b f(x,y)dy\right\}dx.$$

Proof: Since $f(x,y)$ is a continuous function of (x,y) in S, it is also a continuous function of x in $[a,b]$ and a continuous function of y in $[c,d]$. Also since $f(x,y)$ is continuous in S, $\int_a^b f(x,y)dx$ and $\int_c^d f(x,y)dy$ are both continuous in S. Hence both the repeated integrals exist.

Let us now define $g(x,z) = \int_c^z f(x,y)dy$, $x \in [a,b]$. Then by fundamental the-
orem of calculus $\dfrac{\partial g}{\partial z} = f(x,z)$, which is continuous. Let $G(z) = \int_a^b g(x,z)dx$, $z \in$
$[c,d]$.

Then by Leibnitz's rule, we get $G(z) = \int_a^b \dfrac{\partial g(x,z)}{\partial z}\, dx = \int_a^b f(x,z)\, dx$, which
is again continuous. So we get

$$\int_c^d \left\{ \int_a^b f(x,z)dx \right\} dz = \int_c^d G'(z)dz = G(d) - G(c),$$

by the fundamental theorem of definite integral

$$\therefore \quad \int_c^d \left\{ \int_a^b f(x,z)dy \right\} dz = G(d) = \int_a^b \left\{ \int_c^d f(x,y)dy \right\} dx.$$

Hence the result. $\qquad\qquad\qquad\qquad\qquad\qquad\qquad\qquad\qquad\qquad\qquad \square$

Remark: From the above **Corollary 2.13.1** we say that "Under suitable condition, the derivative of the integral and the integral of the derivative are equal and
consequently the two repeated integrals are equal for continuous functions."

Example 6.13.1 *Using differentiation under the sign of integration with proper
justification, prove that* $\int_0^\pi \dfrac{\log(1 + a\cos x)}{\cos x}dx = \pi \sin^{-1} a$, $|a| < 1$.

Solution: Let $f(x,a) = \dfrac{\log(1 + a\cos x)}{\cos x}$, $\dfrac{\partial f}{\partial a} = \dfrac{1}{1 + a\cos x}$.

and $\quad F(a) = \int_0^\pi \dfrac{\log(1 + a\cos x)}{\cos x}dx$, $|a| < 1 \qquad\qquad (6.8)$

Clearly $f(x,a)$ is continuous on $\left\{0 \le x \le \dfrac{\pi}{2}, |a| < 1\right\}$ except at the point $x = \dfrac{\pi}{2}$

and $\lim\limits_{x \to \frac{\pi}{2}} f(x,a) = \lim\limits_{x \to \frac{\pi}{2}} \dfrac{\log(1 + a\cos x)}{\cos x} = \lim\limits_{z \to 0} \dfrac{\log(1 + a\sin z)}{\sin z} = a$ (by putting $z = \dfrac{\pi}{2} - x$.)

So at $x = \dfrac{\pi}{2}$, it has a removable discontinuity.

Also $\dfrac{\partial f}{\partial a} = \dfrac{1}{1 + a\cos x}$ is continuous on $0 \le x \le \dfrac{\pi}{2}$.

Therefore, we can apply Leibnitz's rule and differentiate (6.8) under the integral
sign to get

$$
\begin{aligned}
F'(a) &= \int_0^\pi \frac{\partial}{\partial a}\left(\frac{\log(1 + a\cos x)}{\cos x}\right)dx = \int_0^\pi \frac{1}{1 + a\cos x}dx \\
&= \lim_{\varepsilon \to 0+} \int_0^{\pi - \varepsilon} \frac{\sec^2 \frac{x}{2}}{(1 + a\cos x)\sec^2 \frac{x}{2}}dx \\
&= \lim_{\varepsilon \to 0+} \int_0^{\pi - \varepsilon} \frac{\sec^2 \frac{x}{2}}{(1 + a) + (1 - a)\tan^2 \frac{x}{2}}dx \\
&= \lim_{\lambda \to \infty} \int_0^\lambda \frac{2\, dz}{(1 + a) + (1 - a)z^2} \quad \left[\text{putting } \tan\frac{x}{2} = z\right]
\end{aligned}
$$

$$= \frac{2}{1-a} \lim_{\lambda \to \infty} \int_0^\lambda \frac{dz}{z^2 + \left(\frac{1+a}{1-a}\right)}$$

$$= \lim_{\lambda \to \infty} \frac{2}{1-a} \sqrt{\frac{1-a}{1+a}} \left[\tan^{-1} \sqrt{\frac{1-a}{1+a}} z \right]_0^\lambda = \frac{\pi}{\sqrt{1-a^2}}, \quad |a| < 1.$$

Integrating, $F(a) = \pi \int \frac{da}{\sqrt{1-a^2}} = \pi \sin^{-1} a + c \qquad (6.9)$

To find the value of the constant of integration c, we put $a = 0$ in (6.9) giving $F(0) = 0 = c$. So we get, from (6.8) and (6.9),

$$F(a) = \int_0^\pi \frac{\log(1 + a\cos x)}{\cos x} dx = \pi \sin^{-1} a, \quad |a| < 1.$$

Alternative Method:

Let $F(a) = \int_0^\pi \frac{\log(1 + a\cos x)}{\cos x} dx \qquad (6.10)$

Differentiating both sides of (6.10) with respect to a and using Leibnitz's rule of differentiation under integral sign for constant limits we get

$$\frac{dF}{da} = \int_0^\pi \frac{\partial}{\partial a} \left(\frac{\log(1 + a\cos x)}{\cos x} \right) dx \quad [\text{See } \textbf{Note 6.13.1} \text{ and } \textbf{6.13.3}]$$

$$= \int_0^\pi \frac{1}{\cos x} \cdot \frac{1}{1 + a\cos x} \cdot \cos x \; dx = \int_0^\pi \frac{dx}{1 + a\cos x}$$

$$= \frac{1}{\sqrt{1-a^2}} \left[\cos^{-1} \frac{a + \cos x}{1 + a\cos x} \right]_0^\pi$$

$$\left[\text{Using formula } \int \frac{dx}{a + b\cos x} = \frac{1}{\sqrt{a^2 - b^2}} \cos^{-1} \frac{b + a\cos x}{a + b\cos x}, \right.$$

$$\left. \text{if } b^2 < a^2 \text{ and since here } |a| < 1 \Rightarrow a^2 < 1 \right]$$

$$= \frac{1}{\sqrt{1-a^2}} \left[\cos^{-1} \frac{a-1}{1-a} - \cos^{-1} \frac{a+1}{1+a} \right]$$

$$= \frac{1}{\sqrt{1-a^2}} [\cos^{-1}(-1) - \cos^{-1}(1)] = \frac{\pi - 0}{\sqrt{1-a^2}} = \frac{\pi}{\sqrt{1-a^2}}.$$

Integrating we get

$$F(a) = \int \frac{\pi}{\sqrt{1-a^2}} da = \pi \sin^{-1} a + c \qquad (6.11)$$

where c is an arbitrary constant.

Putting $a = 0$ in (6.10) we get $F(0) = 0$.

Next putting $a = 0$ and $F(0) = 0$ in (6.11) we get $c = 0$. Hence (6.11) reduces to $F(a) = \pi \sin^{-1} a$.

or, $F(a) = \int_0^\pi \frac{\log(1 + a\cos x)}{\cos x} dx = \pi \sin^{-1} a, \quad |a| < 1.$ $\qquad \square$

Remarks: The results obtained in § 6.6 and § 6.7 may not be applicable in the case of an improper integral. The validity of the result to an improper integral requires further investigation. However, we shall omit this investigation in our discussion throughout this chapter. Accordingly, whenever we are going to deal with any improper integral, we shall assume that the necessary conditions for validity of the results are satisfied as done in the alternative method of solution above **Example 6.13.1**.

Example 6.13.2 *Applying the rule of differentiation under the sign of integration,*
show that $\int_0^{\frac{\pi}{2}} \frac{\log(1 + \cos\alpha\cos x)}{\cos x} dx = \frac{1}{2}\left(\frac{\pi}{4} - \alpha^2\right).$

Solution: We have

$$F(\alpha) = \int_0^{\frac{\pi}{2}} \frac{\log(1 + \cos\alpha\cos x)}{\cos x} dx \qquad (6.12)$$

Differentiating both sides of (6.12) with respect to α and using Leibnitz's rule of differentiation under integral sign for constant limits (discussed in **Note 6.13.1** and **6.13.3**), we get

$$\begin{aligned}
\frac{dF}{d\alpha} &= \int_0^{\frac{\pi}{2}} \frac{1}{\cos x}\left(\frac{-\sin\alpha\cos x}{1 + \cos\alpha\cos x}\right) dx = -\int_0^{\frac{\pi}{2}} \frac{\sin\alpha}{1 + \cos\alpha\cos x} dx \\
&= -\sin\alpha \int_0^{\frac{\pi}{2}} \frac{\sec^2\frac{x}{2}}{(1 + \cos\alpha\cos x)\sec^2\frac{x}{2}} dx \\
&= -\sin\alpha \int_0^{\frac{\pi}{2}} \frac{\sec^2\frac{x}{2}}{(1 + \cos\alpha) + (1 - \cos\alpha)\tan^2\frac{x}{2}} dx \\
&= -\frac{\sin\alpha}{1 - \cos\alpha} \int_0^1 \frac{2\,dz}{z^2 + \frac{1+\cos\alpha}{1-\cos\alpha}} \quad \left(\text{putting } \tan\frac{x}{2} = z\right) \\
&= -\frac{2.2.\sin\frac{\alpha}{2}\cos\frac{\alpha}{2}}{2\sin^2\frac{\alpha}{2}} \int_0^1 \frac{dz}{z^2 + \cot^2\frac{\alpha}{2}} = -2\cot\frac{\alpha}{2}\frac{1}{\cot\frac{\alpha}{2}}\left[\tan^{-1}\frac{z}{\cot\frac{\alpha}{2}}\right]_0^1 \\
&= -2\left[\tan^{-1}\tan\frac{\alpha}{2} - \tan^{-1}0\right] = -\alpha.
\end{aligned}$$

$$\therefore \; F(\alpha) = -\int \alpha\,d\alpha = -\frac{\alpha^2}{2} + c \qquad (6.13)$$

From (6.12) and using (6.13) at $\alpha = \frac{\pi}{2}$, $F = 0 = -\frac{1}{2}\frac{\pi^2}{4} + c \Rightarrow c = \frac{\pi^2}{8}.$

$$\therefore \int_0^{\frac{\pi}{2}} \frac{\log(1 + \cos\alpha\cos x)}{\cos x} dx = -\frac{1}{2}\alpha^2 + \frac{\pi^2}{8} = \frac{1}{2}\left(\frac{\pi}{4} - \alpha^2\right).$$

Alternative Method:

$$\text{Let } F(\alpha) = \int_0^{\frac{\pi}{2}} \frac{\log(1 + \cos\alpha\cos x)}{\cos x} dx \qquad (6.14)$$

We assume that the Leibnitz's rule is applicable here. Now differentiating both sides of (6.14) with respect to α and using Leibnitz's rule of differentiation under integral sign for constant limits, we get

$$\frac{dF}{d\alpha} = \int_0^{\frac{\pi}{2}} \frac{\partial}{\partial \alpha} \left[\frac{\log(1 + \cos\alpha\cos x)}{\cos x} \right] dx = -\sin\alpha \int_0^{\frac{\pi}{2}} \frac{dx}{1 + \cos\alpha\cos x}$$

$$= -\sin\alpha \left[\frac{1}{\sqrt{1 - \cos^2\alpha}} \cos^{-1} \left(\frac{\cos\alpha + \cos x}{1 + \cos\alpha\cos x} \right) \right]_0^{\frac{\pi}{2}}$$

$$\left[\text{Using the result } \int \frac{dx}{a + b\cos x} = \frac{1}{\sqrt{a^2 - b^2}} \cos^{-1} \frac{b + a\cos x}{a + b\cos x}, \right.$$

$$\left. \text{if } b^2 < a^2 \text{ and noticing that here } \cos^2\alpha < 1. \right]$$

$$= -(\cos^{-1}\cos\alpha - \cos^{-1}1) = -\alpha.$$

Integrating we get, $F(\alpha) = -\dfrac{\alpha^2}{2} + c$ \hfill (6.15)

where c is an arbitrary constant.

Now, putting $\alpha = \dfrac{\pi}{2}$ in (6.14) we get $F\left(\dfrac{\pi}{2}\right) = 0$. Also putting $\alpha = \dfrac{\pi}{2}$ in (6.15)

we get $F\left(\dfrac{\pi}{2}\right) = -\dfrac{\pi^2}{8} + c$, so that we have $-\dfrac{\pi^2}{8} + c = 0 \Rightarrow c = \dfrac{\pi^2}{8}$.

Therefore (6.15) yields, $F(\alpha) = -\dfrac{\alpha^2}{2} + \dfrac{\pi^2}{8} = \dfrac{1}{2}\left(\dfrac{\pi^2}{4} - \alpha^2\right)$

i.e., $\displaystyle\int_0^{\frac{\pi}{2}} \frac{\log(1 + \cos\alpha\cos x)}{\cos x} dx = \frac{1}{2}\left(\frac{\pi}{4} - \alpha^2\right).$ \hfill $\square$

6.14 Miscellaneous Illustrative Examples

Example 6.14.1 *Evaluate* $\displaystyle\iint_R \left(1 - \frac{x^2}{a^2} - \frac{y^2}{b^2}\right) dxdy$ *where R consists of the points*

in the positive quadrant of the ellipse $\dfrac{x^2}{a^2} + \dfrac{y^2}{b^2} = 1$.

Solution: Let us use the transformation $x = aX$, $y = bY$.

$$\therefore \quad J = \frac{\partial(x,y)}{\partial(X,Y)} = \begin{vmatrix} \dfrac{\partial x}{\partial X} & \dfrac{\partial x}{\partial Y} \\[2mm] \dfrac{\partial y}{\partial X} & \dfrac{\partial y}{\partial Y} \end{vmatrix} = \begin{vmatrix} a & 0 \\ 0 & b \end{vmatrix} = ab.$$

Now R transforms to R' which is the positive quadrant of the circle $X^2 + Y^2 = 1$.

$$\therefore \ I = \iint_{R'} \left(1 - X^2 - Y^2\right) ab \, dXdY = ab \iint_{R'} \left(1 - X^2 - Y^2\right) dXdY.$$

Next we use the transformation into polar coordinates given by $X = r\cos\theta$, $Y = r\sin\theta$, where

$$J = \frac{\partial(X,Y)}{\partial(r,\theta)} = \begin{vmatrix} \dfrac{\partial X}{\partial r} & \dfrac{\partial X}{\partial \theta} \\[2mm] \dfrac{\partial Y}{\partial r} & \dfrac{\partial Y}{\partial \theta} \end{vmatrix} = \begin{vmatrix} \cos\theta & -r\sin\theta \\ \sin\theta & r\cos\theta \end{vmatrix} = r.$$

$$\therefore \ I = ab \int_{r=0}^1 \int_{\theta=0}^{\frac{\pi}{2}} (1 - r^2) r \, drd\theta = ab \int_{r=0}^1 (1 - r^2) r \, dr \int_{\theta=0}^{\frac{\pi}{2}} d\theta$$

$$= \frac{\pi ab}{2} \int_0^1 (r - r^3) \, dr = \frac{\pi ab}{2} \left[\frac{r^2}{2} - \frac{r^4}{4} \right]_0^1 = \frac{\pi ab}{2} \left(\frac{1}{2} - \frac{1}{4} \right) = \frac{\pi ab}{8}.$$

Alternatively,

We put $x = ar\cos\theta,\ y = br\sin\theta$

$$\therefore\ J = \frac{\partial(x,y)}{\partial(r,\theta)} = \begin{vmatrix} \frac{\partial x}{\partial r} & \frac{\partial x}{\partial \theta} \\ \frac{\partial y}{\partial r} & \frac{\partial y}{\partial \theta} \end{vmatrix} = \begin{vmatrix} a\cos\theta & -ar\sin\theta \\ b\sin\theta & br\cos\theta \end{vmatrix} = rab.$$

$$\therefore\ I = \iint_{R'} (1 - r^2) abr\ drd\theta, \text{ where } R' \text{ is given by } 0 \le r \le 1, 0 \le \theta \le \frac{\pi}{2}$$

$$= ab \int_0^{\frac{\pi}{2}} d\theta \int_0^1 r(1 - r^2) dr = \frac{\pi ab}{2}\left(\frac{1}{2} - \frac{1}{4}\right) = \frac{\pi ab}{8}. \qquad \square$$

Example 6.14.2 *Evaluate* $\iint_R (x^2 + y^2) dxdy$

(i) *Over the region enclosed by the triangle having its vertices at* $(0,0)$, $(1,0)$ *and* $(1,1)$.

(ii) *Over R, bounded by* $xy = 1, y = 0, y = x, x = 2$.

(iii) *Over the region in the positive quadrant for which* $x + y \le 1$.

(iv) *Over R, where R is the region bounded by* $x = 0, y = 0, x + y = 1$.

Solution: (i) Here the region of integration R is the $\triangle OAB$, where the vertices are $O(0,0), A(1,0)$ and $B(1,1)$. The equation of the straight line OB is $y = x$ [*vide.* **Figure 6.20**]. Therefore,

$$\iint_R (x^2 + y^2)\ dxdy = \int_0^1 dx \int_0^x (x^2 + y^2)\ dy = \int_0^1 dx \left[x^2 y + \frac{y^3}{3}\right]_0^x$$

$$= \int_0^1 \left[x^3 + \frac{x^3}{3}\right] dx = \left[\frac{x^4}{4} + \frac{x^4}{12}\right]_0^1 = \frac{1}{4} + \frac{1}{12} = \frac{1}{3}.$$

(ii) Here the region of integration is the region $R = OADCBO$ where $R_1 = $ region $OABO$ and $R_2 = $ region $ADCBA$ [*vide.* **Figure 6.21**]. So,

$$\iint_R (x^2 + y^2) dxdy = \iint_{R_1} (x^2 + y^2) dxdy + \iint_{R_2} (x^2 + y^2) dxdy$$

$$= \int_0^1 dx \int_0^x (x^2 + y^2) dy + \int_1^2 dx \int_0^{\frac{1}{x}} (x^2 + y^2) dy$$

$$= \frac{1}{3} + \int_1^2 \left[x^2 y + \frac{y^3}{3}\right]_0^{\frac{1}{x}} dx \text{ (Using the result of (i))}$$

$$= \frac{1}{3} + \int_1^2 \left(x + \frac{1}{3x^3}\right) dx = \frac{1}{3} + \left[\frac{x^2}{2} - \frac{1}{6x^2}\right]_1^2$$

$$= \frac{1}{3} + \left[\frac{1}{2} - \frac{1}{24} - \frac{1}{2} + \frac{1}{6}\right] = \frac{1}{3} + \frac{13}{8} = \frac{47}{24}.$$

(iii) Here the region of integration is $x + y = 1$ and the limits of integration can be expressed as $0 \le x \le 1, 0 < y < 1 - x$. Hence, the required integral

$$= \int_0^1 dx \int_0^{1-x} (x^2 + y^2)\ dy = \int_0^1 \left[x^2 y + \frac{y^3}{3}\right]_0^{1-x} dx$$

$$= \int_0^1 \left[x^2(1 - x) + \frac{1}{3}(1 - x)^3\right] dx = \int_0^1 (1 - x)\left[x^2 + \frac{1}{3}(1 - x)^2\right] dx$$

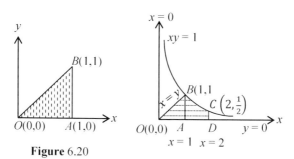

Figure 6.20

Figure 6.21

$$= \int_0^1 \frac{1}{3}[1 - 3x - 6x^2 - 4x^3]dx = \frac{1}{3}\left[x - \frac{3}{2}x^2 + 2x^3 - x^4\right]_0^1$$

$$= \frac{1}{3}\left[1 - \frac{3}{2} + 2 - 1\right] = \frac{1}{6}.$$

(iv) Same as (iii). □

Example 6.14.3 *Evaluate, by using suitable transformation, the **Example 6.14.2** (i)*

Solution: The region of integration R is bounded by the lines $y = 0, y = x$ and $x = 1$ and is in the positive quadrant.

We use the transformation $x = r\cos\theta$, $y = r\sin\theta$. Then the Jacobian $J = r$. From the **Figure** 6.20 it is clear that the limits of θ is from 0 to $\frac{\pi}{4}$ and r varies from 0 to $\sec\theta$. (The limits are obtained from the relations $r = x\sec\theta$ and $\theta = \tan^{-1}\frac{y}{x}$).

$$\therefore I = \iint_R (x^2 + y^2)dxdy = \iint_{R'} r^2 . rdrd\theta = \int_0^{\frac{\pi}{4}} d\theta \int_0^{\sec\theta} r^3 dr$$

$$= \int_0^{\frac{\pi}{4}} d\theta \left[\frac{r^4}{4}\right]_0^{\sec\theta} = \frac{1}{4}\int_0^{\frac{\pi}{4}} \sec^4\theta d\theta = \frac{1}{4}\left[\frac{1}{3}\tan^3\theta + \tan\theta\right]_0^{\frac{\pi}{4}} = \frac{1}{3}. \quad \square$$

Example 6.14.4 *Evaluate* $\iint_R \sqrt{\frac{a^2b^2 - b^2x^2 - a^2y^2}{a^2b^2 + b^2x^2 + a^2y^2}} \, dxdy$, *where R denotes the positive quadrant of the ellipse* $\frac{x^2}{a^2} + \frac{x^2}{a^2} = 1$.

Solution: We use the transformation $x = ar\cos\theta, y = br\sin\theta$ so that Jacobian $J = \frac{\partial(x,y)}{\partial(r,\theta)} = abr$.

$$\therefore I = \iint_{R'} \frac{ab\sqrt{1 - r^2}}{ab\sqrt{1 + r^2}} \, ab \, r \, drd\theta \text{ where } R' \text{ is the positive quadrant of the circle } r = 1.$$

$$\therefore I = ab\int_0^{\frac{\pi}{2}} d\theta \int_0^1 \sqrt{\frac{1 - r^2}{1 + r^2}} \, r \, dr$$

$$= ab \int_0^{\frac{\pi}{2}} d\theta \int_{\frac{\pi}{4}}^0 \sqrt{\frac{1 - \cos 2k}{1 + \cos 2k}} \, (-\sin 2k) \, dk \quad [\text{Putting } r^2 = \cos 2k]$$

$$= ab \int_0^{\frac{\pi}{2}} d\theta \int_0^{\frac{\pi}{4}} \tan k.2 \sin k \cos k \, dk = 2ab \int_0^{\frac{\pi}{2}} d\theta \int_0^{\frac{\pi}{4}} \sin^2 k \, dk$$

$$= 2ab \int_0^{\frac{\pi}{2}} d\theta \int_0^{\frac{\pi}{4}} \frac{1}{2}(1 - \cos 2k) \, dk = ab \int_0^{\frac{\pi}{2}} d\theta \left[k - \frac{\sin 2k}{2} \right]_0^{\frac{\pi}{4}}$$

$$= ab[\theta]_0^{\frac{\pi}{2}} \times \left[\frac{\pi}{4} - \frac{1}{2} \right] = \frac{\pi ab}{8}(\pi - 2). \qquad \square$$

Example 6.14.5 *Evaluate* $\displaystyle\iint_R \frac{dx\,dy}{(1 + x^2 + y^2)^2}$ *where R is the region taken over*

(i) the triangle with vertices $(0,0), (2,0)$ and $(1, \sqrt{3})$

(ii) a loop of the lemniscate $(x^2 + y^2)^2 - (x^2 - y^2) = 0$.

Solution: (i) Here the region of integration R is an equilateral $\triangle OAB$ bounded by the lines OA $(y = 0)$, OB $(y = \sqrt{3}x)$ and AB $(y = -\sqrt{3}x + 2\sqrt{3})$ [*vide.* **Figure 6.22**]. Let us use the polar coordinate transformation $x = r\cos\theta, y = r\sin\theta$ where the Jacobian $J = r$, i.e., we get

$$\iint_R \frac{dx\,dy}{(1 + x^2 + y^2)^2} = \iint_{R'} \frac{r\,dr\,d\theta}{(1 + r^2)^2}.$$

Here R is transformed to R' which is also an equilateral triangle with the lines in polar coordinate. The limits of θ is determined from the initial line OA to OB, i.e., $\theta = 0$ to $\theta = \dfrac{\pi}{3}$ (Since the equation of OB is $y = \sqrt{3}x \Rightarrow r\sin\theta = \sqrt{3}r\cos\theta \Rightarrow \tan\theta = \sqrt{3} \Rightarrow \theta = \dfrac{\pi}{3}$). The limits of r is determined from $r = 0$ to the value of r obtained from the equation of the line AB, i.e.,

$$y = -\sqrt{3}x + 2\sqrt{3} \quad \text{or,} \quad r\sin\theta = -\sqrt{3}r\cos\theta + 2\sqrt{3}$$

$$\text{or} \quad r(\sin\theta + \sqrt{3}\cos\theta) = 2\sqrt{3} \ \text{ or } r\left(\frac{1}{2}\sin\theta + \frac{\sqrt{3}}{2}\cos\theta \right) = \sqrt{3}$$

$$\text{or} \quad r\left(\sin\frac{\pi}{3}\cos\theta + \cos\frac{\pi}{3}\sin\theta \right) = \sqrt{3} \ \text{ or, } r\sin\left(\frac{\pi}{3} + \theta \right) = \sqrt{3}$$

$$\text{or} \quad r = \frac{\sqrt{3}}{\sin\left(\frac{\pi}{3} + \theta \right)}.$$

Hence the limits of r is from $r = 0$ to $r = \dfrac{\sqrt{3}}{\sin\left(\frac{\pi}{3} + \theta \right)}$.

$$\therefore \ I = \int_{\theta=0}^{\frac{\pi}{3}} d\theta \int_{r=0}^{\frac{\sqrt{3}}{\sin\left(\frac{\pi}{3}+\theta \right)}} \frac{r\,dr}{(1 + r^2)^2} = \int_{\theta=0}^{\frac{\pi}{3}} d\theta \int_{r=0}^{\frac{\sqrt{3}}{\sin\left(\frac{\pi}{3}+\theta \right)}} d\left[\frac{-1}{2(1 + r^2)} \right]$$

$$= \frac{1}{2} \int_0^{\frac{\pi}{3}} d\theta \left[\frac{-1}{(1 + r^2)} \right]_0^{\frac{\sqrt{3}}{\sin\left(\frac{\pi}{3}+\theta \right)}} = \frac{1}{2} \int_0^{\frac{\pi}{3}} d\theta \left[1 - \frac{1}{1 + \frac{3}{\sin^2\left(\frac{\pi}{3}+\theta \right)}} \right]$$

$$= \frac{1}{2} \int_0^{\frac{\pi}{3}} \left(1 - \frac{\sin^2\left(\frac{\pi}{3} + \theta \right)}{3 + \sin^2\left(\frac{\pi}{3} + \theta \right)} \right) d\theta$$

$$= \frac{1}{2} \int_0^{\frac{\pi}{3}} \frac{3 + \sin^2\left(\frac{\pi}{3} + \theta \right) - \sin^2\left(\frac{\pi}{3} + \theta \right)}{3 + \sin^2\left(\frac{\pi}{3} + \theta \right)} d\theta$$

$$= \frac{3}{2} \int_0^{\frac{\pi}{3}} \frac{1}{3 + \sin^2\left(\frac{\pi}{3} + \theta\right)} \, d\theta$$

$$= \frac{3}{2} \int_0^{\frac{\pi}{3}} \frac{d\theta}{3\sin^2\left(\frac{\pi}{3} + \theta\right) + 3\cos^2\left(\frac{\pi}{3} + \theta\right) + \sin^2\left(\frac{\pi}{3} + \theta\right)}$$

$$= \frac{3}{2} \int_0^{\frac{\pi}{3}} \frac{\sec^2\left(\frac{\pi}{3} + \theta\right) d\theta}{3 + 4\tan^2\left(\frac{\pi}{3} + \theta\right)} = \frac{3}{2} \int_{2\sqrt{3}}^{-2\sqrt{3}} \frac{1}{2} \frac{dz}{3 + z^2}$$

$$\left[\text{Putting } 2\tan\left(\frac{\pi}{3} + \theta\right) = z \ \therefore \ 2\sec^2\left(\frac{\pi}{3} + \theta\right) d\theta = dz\right]$$

$$= \frac{3}{4} \left[\frac{1}{\sqrt{3}} \tan^{-1} \frac{z}{\sqrt{3}}\right]_{2\sqrt{3}}^{-2\sqrt{3}} = \frac{\sqrt{3}}{4} \left[\tan^{-1}(-2) - \tan^{-1} 2\right]$$

$$= \frac{\sqrt{3}}{4} \left[\pi - \tan^{-1} 2 - \tan^{-1} 2\right] = \frac{\sqrt{3}}{4} \cdot 2 \left[\frac{\pi}{2} - \tan^{-1} 2\right] = \frac{\sqrt{3}}{2} \cot^{-1} 2$$

$$= \frac{\sqrt{3}}{2} \tan^{-1}\left(\frac{1}{2}\right).$$

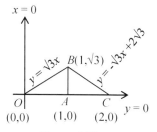

Figure 6.22   **Figure** 6.23

(ii) Changing to polar coordinates $x = r\cos\theta, y = r\sin\theta$, we get $J = r$ and R changes to R', where in R' the region in which r varies from 0 to $\sqrt{\cos 2\theta}$ and θ varies from $-\frac{\pi}{4}$ to $\frac{\pi}{4}$ [one loop of the lemniscate $r^2 = \cos 2\theta$] (See **Figure** 6.23, the upper half is being shaded).

$$\therefore \ I = \int_{-\frac{\pi}{4}}^{\frac{\pi}{4}} \int_0^{\sqrt{\cos 2\theta}} \frac{r\,dr}{(1 + r^2)^2} = 2 \int_0^{\frac{\pi}{4}} \left(-\frac{1}{2}\right) \left[\frac{1}{1 + r^2}\right]_0^{\sqrt{\cos 2\theta}} d\theta$$

$$= -\int_0^{\frac{\pi}{4}} \left(\frac{1}{1 + \cos 2\theta} - 1\right) d\theta = \int_0^{\frac{\pi}{4}} \left(1 - \frac{1}{2}\sec^2\theta\right) d\theta$$

$$= \left[\theta - \frac{1}{2}\tan\theta\right]_0^{\frac{\pi}{4}} = \frac{\pi}{4} - \frac{1}{2}. \qquad \square$$

Example 6.14.6 *Change the order of integration in* $\int_0^{2a} \int_{\frac{x^2}{4a}}^{3a-x} F(x,y) \, dx\,dy.$

Solution: Clearly the limits of integration is $x = 0$ to $x = 2a$ and $y = \frac{x^2}{4a}$ to $y = 3a - x$, i.e., the region of integration is bounded by the parabola $x^2 = 4ay$ and

the straight lines $x + y = 3a, x = 0$ and $x = 2a$. The region is the area $OABCO$
(See **Figure** 6.24). The point of intersection of the line BA and the parabola is
$(2a, a)$.

We have divided the region of integration in two parts - area OAC and
area CAB and the required integral is
the sum of the integrals in these two
regions in changed order.

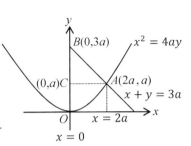

In region OAC, $\displaystyle\int_0^a dy \int_0^{\sqrt{4ay}} F(x,y)dx$

and

In region CAB, $\displaystyle\int_a^{3a} dy \int_0^{3a-y} F(x,y)dx.$

Hence by changing the order of integration we get

Figure 6.24

$$\int_0^{2a} \int_{\frac{x^2}{4a}}^{3a-x} F(x,y)\, dxdy = \int_0^a dy \int_0^{\sqrt{4ay}} F(x,y)dx + \int_a^{3a} dy \int_0^{3a-y} F(x,y)dx.$$

$\square$

Example 6.14.7 *Evaluate* $\displaystyle\int_0^1 dx \int_0^{\sqrt{1-x^2}} \frac{dy}{(1 + e^y)\sqrt{1 - x^2 - y^2}}$ *by changing the*
order of integration.

Solution: The limits of y are 0 to $\sqrt{1 - x^2}$ and those of the variable x are 0 to 1.
Hence, it is clear that the region of integration is the area bounded by the circle
$y = \sqrt{1 - x^2} \Rightarrow x^2 + y^2 = 1$ in the positive quadrant.

For changing the order of integration we consider a thin strip parallel to x
axis which will vary form $y = 0$ to $y = 1$ i.e., the limits of integration will be
$\left[0 \le x \le \sqrt{1 - y^2}; 0 \le y \le 1\right]$ in changed order. Hence the required integral takes
the form

$$\int_{y=0}^1 \left\{ \int_{x=0}^{\sqrt{1-y^2}} \frac{dx}{(1 + e^y)\sqrt{1 - x^2 - y^2}} \right\} dy$$

$$= \int_0^1 \frac{1}{1 + e^y} \left\{ \int_0^{\sqrt{1-y^2}} \frac{dx}{\sqrt{1 - x^2 - y^2}} \right\} dy$$

$$= \int_0^1 \frac{1}{1 + e^y} \left[\sin^{-1} \frac{x}{\sqrt{1 - y^2}} \right]_0^{\sqrt{1-y^2}} dy$$

$$= \int_0^1 \frac{1}{1 + e^y} \left[\sin^{-1}(1) - \sin^{-1}(0) \right] dy = \frac{\pi}{2} \int_0^1 \frac{1}{1 + e^y} dy$$

$$= \frac{\pi}{2} \int_0^1 \frac{1 + e^y - e^y}{1 + e^y} dy = \frac{\pi}{2} \left[\int_0^1 dy - \int_0^1 \frac{e^y}{1 + e^y} dy \right]$$

$$= \frac{\pi}{2} \left[[y]_0^1 - \int_0^1 \frac{d(1 + e^y)}{1 + e^y} \right] = \frac{\pi}{2} \left[1 - [\log(1 + e^y)]_0^1 \right]$$

$$= \frac{\pi}{2} \left[\log e - \log \frac{1 + e}{2} \right] = \frac{\pi}{2} \log \frac{2e}{1 + e}.$$

$\square$

Example 6.14.8 *By changing the order of integration, prove that*

$$\int_0^{2a} \int_0^{\sqrt{2ax-x^2}} \frac{(x^2+y^2)xf'(y)}{\sqrt{4a^2x^2-(x^2+y^2)^2}}\, dx\, dy = \pi a^2[f(a)-f(0)].$$

Solution: Clearly the region of integration is bounded by the straight lines $x=0$ and $x=2a$; and line $y=0$ and the curve $y=\sqrt{2ax-x^2}$ $\Rightarrow y^2=2ax-x^2$ $\Rightarrow (x-a)^2+y^2=a^2$ (which is a circle with radius a and centre at $(a,0)$), i.e., the region of integration is the upper half of the circle $(x-a)^2+y^2=a^2$. (See **Figure** 6.25)

Now we change the order of integration. Consider a thin strip parallel to x-axis which moves in the upper half and within the said circle as indicated in the given figure. Hence after changing the order of integration, the limits of y are 0 to a and the limit of x are from left circular part to right circular part of the said circle within the upper half, i.e.,

$$(x-a)^2+y^2=a^2$$
$$\Rightarrow \quad x-a=\pm\sqrt{a^2-y^2}$$
$$\Rightarrow \quad x=a\pm\sqrt{a^2-y^2}.$$

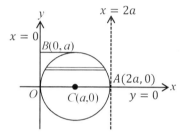

Figure 6.25

Hence the required integral is

$$\int_{y=0}^a dy \int_{a-\sqrt{a^2-y^2}}^{a+\sqrt{a^2-y^2}} \frac{(x^2+y^2)xf'(y)}{\sqrt{4a^2x^2-(x^2+y^2)^2}}\, dx$$

$$= \int_0^a f'(y)dy \int_{a-\sqrt{a^2-y^2}}^{a+\sqrt{a^2-y^2}} \frac{(x^2+y^2)x}{\sqrt{4a^2x^2-(x^2+y^2)^2}}\, dx$$

Now we calculate the inside integral first.

$$\int_{x=a-\sqrt{a^2-y^2}}^{a+\sqrt{a^2-y^2}} \frac{(x^2+y^2)x}{\sqrt{4a^2x^2-(x^2+y^2)^2}}\, dx$$

$$= \frac{1}{2}\int_{z=2a(a-\sqrt{a^2-y^2})}^{2a(a+\sqrt{a^2-y^2})} \frac{z.dz}{\sqrt{4a^2(z-y^2)-z^2}} \quad [\text{Putting } x^2+y^2=z]$$

$$= \frac{1}{2}\int_{2a(a-\sqrt{a^2-y^2})}^{2a(a+\sqrt{a^2-y^2})} \frac{z.dz}{\sqrt{-z^2+4a^2z-4a^2y^2}}$$

$$= \frac{1}{2}\times\left(-\frac{1}{2}\right)\left[\int_{2a(a-\sqrt{a^2-y^2})}^{2a(a+\sqrt{a^2-y^2})} \frac{(-2z+4a^2)dz}{\sqrt{-z^2+4a^2z-4a^2y^2}}\right.$$

$$\left. +(-4a^2)\int_{2a(a-\sqrt{a^2-y^2})}^{2a(a+\sqrt{a^2-y^2})} \frac{dz}{\sqrt{-z^2+4a^2z-4a^2y^2}}\right]$$

$$= \frac{1}{4}\left[-\int_{2a(a-\sqrt{a^2-y^2})}^{2a(a+\sqrt{a^2-y^2})} \frac{d(-z^2+4a^2z-4a^2y^2)}{\sqrt{-z^2+4a^2z-4a^2y^2}}\right.$$

$$\left. +4a^2 \int_{2a(a-\sqrt{a^2-y^2})}^{2a(a+\sqrt{a^2-y^2})} \frac{dz}{\sqrt{\left(2a\sqrt{a^2-y^2}\right)^2-(z-2a^2)^2}}\right]$$

$$= \frac{1}{4}\left[4a^2 \sin^{-1}\frac{z-2a^2}{2a\sqrt{a^2-y^2}} - 2\sqrt{-z^2+4a^2z-4a^2y^2}\right]_{2a(a-\sqrt{a^2-y^2})}^{2a(a+\sqrt{a^2-y^2})}$$

$$= \frac{1}{4}\left[4a^2 \sin^{-1}\frac{z-2a^2}{2a\sqrt{a^2-y^2}} - 2\sqrt{4a^2(a^2-y^2)-(z-2a^2)^2}\right]_{2a^2-2a\sqrt{a^2-y^2}}^{2a^2+2a\sqrt{a^2-y^2}}$$

$$= \frac{1}{4}\left[4a^2(\sin^{-1}(1)-\sin^{-1}(-1))\right] - 2.0 + 2.0$$

$$= \frac{1}{4}\left[4a^2\left\{\frac{\pi}{2}-\left(-\frac{\pi}{2}\right)\right\}\right] = \frac{1}{4} \times 4a^2\pi = a^2\pi.$$

$\therefore$ Required integral $= \int_0^a f'(y)dy \times a^2\pi = a^2\pi\,[f(y)]_0^a = \pi a^2[f(a)-f(0)].$ $\qquad\square$

Example 6.14.9 *Evaluate* $\iint_R [x+y]\, dxdy$ *over the rectangle* $R = [0,1;0,2]$ *where* $[x+y]$ *denotes the largest integer not larger than* $x+y$.

Solution: According to the given definition of the integrand function $[x+y]$ we can rewrite it in the given rectangular region of integration as

$$[x+y] \begin{cases} = 0 & \text{if } 0 \le x+y < 1 \\ = 1 & \text{if } 1 \le x+y < 2 \\ = 2 & \text{if } 2 \le x+y < 3. \end{cases}$$

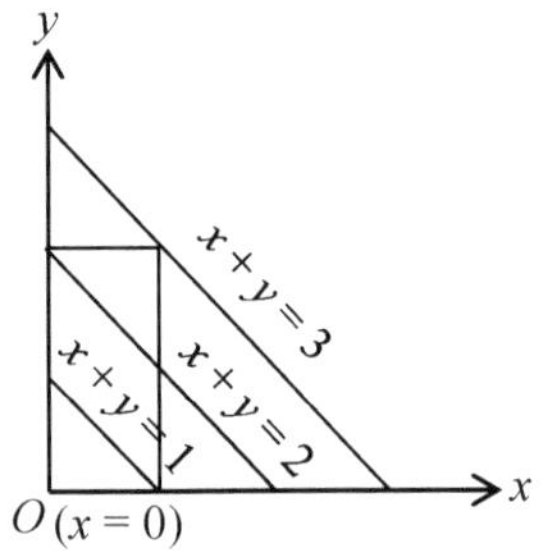

Hence the required integration is

Figure 6.26

$$\iint_R [x+y]\, dxdy$$

$$= \iint_{x+y<1} [x+y]\, dxdy$$

$$+ \iint_{1\le x+y<2} [x+y]\, dxdy + \iint_{2\le x+y<3} [x+y]\, dxdy$$

$$= \iint_{x+y<1} 0\, dxdy + \int_0^1 dx \int_{y=1-x}^{2-x} 1\, dy + \int_0^1 dx \int_{y=2-x}^2 2\, dy$$

$$= 0 + \int_0^1 [y]_{1-x}^{2-x}\, dx + \int_0^1 [y]_{1-x}^{2-x}\, dx$$

$$= \int_0^1 (2-x-1+x)\, dx + 2\int_0^1 (2-2+x)\, dx$$

$$= \int_0^1 dx + 2\int_0^1 x\, dx = 1 + 2\left[\frac{x^2}{2}\right]_0^1 = 1 + 2.\frac{1}{2} = 2. \qquad\square$$

Example 6.14.10 *Evaluate* $\iint_R |\cos(x+y)|\ dxdy$ *where* $R = \{(x,y)|\ 0 \le x \le \pi, 0 \le x \le \pi\}$.

Solution: We first rewrite the integrand function $f(x,y) = |\cos(x+y)|$ in the given square region of integration $[0,\pi;0,\pi]$.

$$|\cos(x+y)| \begin{cases} = & \cos(x+y) & \text{if } 0 \le x+y < \dfrac{\pi}{2} \\[2mm] = & -\cos(x+y) & \text{if } \dfrac{\pi}{2} \le x+y < \dfrac{3\pi}{2} \\[2mm] = & \cos(x+y) & \text{if } \dfrac{3\pi}{2} \le x+y < 2\pi. \end{cases}$$

$$\therefore \iint_R |\cos(x+y)|\ dxdy$$

$$= \iint_{0 \le x+y < \frac{\pi}{2}} |\cos(x+y)|\ dxdy$$

$$+ \iint_{\frac{\pi}{2} \le x+y < \frac{3\pi}{2}} |\cos(x+y)|\ dxdy$$

$$+ \iint_{\frac{3\pi}{2} \le x+y < 2\pi} |\cos(x+y)|\ dxdy$$

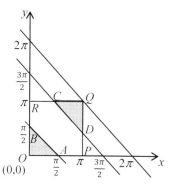

Figure 6.27

Clearly from the **Figure** 6.27, the region of the first and third integrals are $\triangle OAB$ and $\triangle CDQ$ respectively. The same for second integral is the region $APDCRBA$. Hence the required integral becomes

$$= \iint_{0 \le x+y < \frac{\pi}{2}} \cos(x+y)\ dxdy + \iint_{\frac{\pi}{2} \le x+y < \frac{3\pi}{2}} \{-\cos(x+y)\}\ dxdy$$

$$+ \iint_{\frac{3\pi}{2} \le x+y < 2\pi} \cos(x+y)\ dxdy$$

$$= \int_0^{\frac{\pi}{2}} dx \int_0^{\frac{\pi}{2}-x} \cos(x+y)\ dy - \int_0^{\frac{\pi}{2}} dx \int_{\frac{\pi}{2}-x}^{\pi} \cos(x+y)\ dy$$

$$- \int_{\frac{\pi}{2}}^{\pi} dx \int_0^{\frac{3\pi}{2}-x} \cos(x+y)\ dy + \int_{\frac{\pi}{2}}^{\pi} dx \int_{\frac{\pi}{2}-x}^{\pi} \cos(x+y)\ dy$$

$$= \int_0^{\frac{\pi}{2}} dx\ [\sin(x+y)]_0^{\frac{\pi}{2}-x} - \int_0^{\frac{\pi}{2}} dx\ [\sin(x+y)]_{\frac{\pi}{2}-x}^{\pi}$$

$$- \int_{\frac{\pi}{2}}^{\pi} dx\ [\sin(x+y)]_0^{\frac{3\pi}{2}-x} - \int_{\frac{\pi}{2}}^{\pi} dx\ [\sin(x+y)]_{\frac{3\pi}{2}-x}^{\pi}$$

$$= \int_0^{\frac{\pi}{2}} (1 - \sin x)\ dx - \int_0^{\frac{\pi}{2}} (-\sin x - 1)\ dx - \int_{\frac{\pi}{2}}^{\pi} (-1 - \sin x)\ dx$$

$$+ \int_{\frac{\pi}{2}}^{\pi} (-\sin x + 1)\ dx$$

$$= \int_0^{\frac{\pi}{2}} \{1 - \sin x - (-\sin x - 1)\} dx + \int_{\frac{\pi}{2}}^{\pi} \{-\sin x + 1 - (-\sin x - 1)\} dx$$

$$= \int_0^{\frac{\pi}{2}} 2dx + \int_{\frac{\pi}{2}}^{\pi} 2dx = 2\left[\frac{\pi}{2} + \left(\pi - \frac{\pi}{2}\right)\right] = 2\pi. \qquad \square$$

Example 6.14.11 *Evaluate* $\iiint_V (x^2 + y^2 + z^2)xyz\ dxdydz$ *taken through the sphere* $x^2 + y^2 + z^2 = 1$.

Solution: The region of integration R is the sphere $x^2 + y^2 + z^2 \le 1$. Transferring to spherical polar coordinates, i.e., by putting $x = r\sin\theta\cos\phi$, $y = r\sin\theta\sin\phi$ and $z = r\cos\theta$, we get the Jacobian $|J| = r^2\sin\theta$. Since the region of integration is the whole sphere so ϕ ranges form 0 to 2π, θ ranges from 0 to π and r ranges from 0 to 1 we get

$$\begin{aligned}
I &= \int_0^{2\pi} d\phi \int_0^{\pi} d\theta \int_0^1 r^3 \sin^2\theta\cos\theta\cos\phi\sin\phi . r^2 . r^2 \sin\theta\ dr \\
&= \int_0^{2\pi} \sin\phi\cos\phi\ d\phi \int_0^{\pi} \sin^3\theta\cos\theta\ d\theta \int_0^1 r^7 dr \\
&= \left[\frac{1}{2}\sin^2\phi\right]_0^{2\pi} \left[\frac{1}{4}\sin^4\theta\right]_0^{\pi} \left[\frac{1}{8}r^8\right]_0^1 = 0.0.\frac{1}{8} = 0. \qquad \square
\end{aligned}$$

Example 6.14.12 *Evaluate* $\iiint z^2\ dxdydz$, *where integration is taken over the hemisphere* $x \ge 0$, $x^2 + y^2 + z^2 \le a^2$.

Solution: Given integral

$$= \iint \left[\int_0^{\sqrt{a^2-x^2-y^2}} z^2\ dz\right] dxdy = \frac{1}{3} \iint (a^2 - x^2 - y^2)^{\frac{3}{2}}\ dxdy.$$

Now, we transform in polar coordinates $x = r\cos\theta, y = r\sin\theta$, we get $J = r$.

On the upper half of the hemisphere, r varies from 0 to a, θ from 0 to 2π. Therefore the required integral is

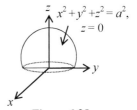

Figure 6.28

$$\begin{aligned}
&= \frac{1}{3} \int_{r=0}^{a} (a^2 - r^2)^{\frac{3}{2}} r\,dr \int_0^{2\pi} d\theta \\
&= \frac{2\pi}{3} \cdot \left(-\frac{1}{2}\right) \int_0^{a} (a^2 - r^2)^{\frac{3}{2}} (-2r)dr \\
&= -\frac{\pi}{3} \left[\frac{2}{5}(a^2-r^2)^{\frac{5}{2}}\right]_0^{a} = -\frac{2\pi}{15}(0 - a^5) = \frac{2\pi a^5}{15}.
\end{aligned}$$

Alternative Method (by transforming into spherical polar coordinates)

Here the region of integration R is the upper half of the hemisphere $z = \sqrt{a^2 - x^2 - y^2}$.

Let us transform the Cartesian coordinates to the spherical polar coordinates by the transformation $x = r\sin\theta\cos\phi$, $y = r\sin\theta\sin\phi$ and $z = r\cos\theta$, where the Jacobian $|J| = r^2\sin\theta$.

$$I = \iiint_R z^2 dxdydz = \iiint_{R'} (r^2\cos^2\theta)r^2\sin\theta\ drd\theta d\phi$$

where the region R transform to R' and in R', r varies from 0 to a, θ varies from 0 to π and ϕ varies from 0 to π. Hence we get

$$
\begin{aligned}
I &= \int_{r=0}^{a}\int_{\theta=0}^{\pi}\int_{\phi=0}^{\pi} r^4 \cos^2\theta \sin\theta \; dr d\theta d\phi \\
&= \int_0^\pi d\phi \int_0^\pi (\cos^2\theta\sin\theta)\, d\theta \int_0^a r^4 dr = \pi\int_0^\pi \cos^2\theta\, d(\cos\theta)\left[\frac{r^5}{5}\right]_0^a \\
&= \pi\frac{a^5}{5}\left[\frac{\cos^3\theta}{3}\right]_0^\pi = \frac{2\pi a^5}{15}.
\end{aligned}
$$

Alternative Method (By Direct Approach)

The region of integration R is the hemisphere $x^2+y^2+z^2 \le a^2$, $z \ge 0$, the limits of of x will be $-\sqrt{a^2-y^2-z^2}$ to $\sqrt{a^2-y^2-z^2}$, those of y will be $-\sqrt{a^2-z^2}$ to $\sqrt{a^2-z^2}$ and those of z will be 0 to a, i.e.,

$$
\begin{aligned}
I &= \int_0^a z^2 dz \int_{-\sqrt{a^2-z^2}}^{\sqrt{a^2-z^2}} dy \int_{-\sqrt{a^2-y^2-z^2}}^{\sqrt{a^2-y^2-z^2}} dx \\
&= \int_0^a z^2 dz \int_{-\sqrt{a^2-z^2}}^{\sqrt{a^2-z^2}} dy\, [x]_{-\sqrt{a^2-y^2-z^2}}^{\sqrt{a^2-y^2-z^2}} \\
&= 2\int_0^a z^2 dz \int_{-\sqrt{a^2-z^2}}^{\sqrt{a^2-z^2}} \sqrt{a^2-y^2-z^2}\, dy \\
&= 2\int_0^a z^2 dz \left[\frac{1}{2}\left\{ y\sqrt{a^2-y^2-z^2} + (a^2-z^2)\right.\right. \\
&\qquad\qquad\qquad\qquad \left.\left.\times \sin^{-1}\left(\frac{y}{\sqrt{a^2-x^2}}\right)\right\}\right]_{-\sqrt{a^2-z^2}}^{\sqrt{a^2-z^2}} \\
&= \int_0^a z^2 dz \left[\sqrt{a^2-z^2}.0 - \left(-\sqrt{a^2-z^2}\right).0 + (a^2-z^2)\sin^{-1}1\right. \\
&\qquad\qquad\qquad\qquad \left. -(a^2-z^2)\sin^{-1}(-1)\right] \\
&= \int_0^a z^2\left[(a^2-z^2)\frac{\pi}{2} - (a^2-z^2)\left(-\frac{\pi}{2}\right)\right] dz = \pi\int_0^a z^2(a^2-z^2)dz \\
&= \pi\left[\frac{a^2 z^3}{3} - \frac{z^5}{5}\right]_0^a = \frac{\pi}{15}(5a^5 - 3a^5) = \frac{2\pi a^5}{15}. \qquad \square
\end{aligned}
$$

Example 6.14.13 *Find the volume of the region bounded by the cylinder $x^2+y^2 = 16$ and points $z = 0$ to $z = 3$.*

Solution: Here the limits of z are given from 0 to 3. Also the limits of y are from $-\sqrt{16-x^2}$ to $\sqrt{16-x^2}$ and those of x are from -4 to 4. Therefore, required

$$
\text{volume} = V = \int_{x=-4}^{4}\int_{y=-\sqrt{16-x^2}}^{\sqrt{16-x^2}}\int_{z=0}^{3} dx dy dz
$$

[Putting $f(x,y,z) = 1$ for volume, *vide.* **Remark 2** of § 6.10]

$$
= 4\int_0^4\int_0^{\sqrt{16-x^2}} [z]_0^3\, dxdy = 12\int_0^4 dx \int_0^{\sqrt{16-x^2}} dy = 12\int_0^4 [y]_0^{\sqrt{16-x^2}} dx
$$

$$= 12 \int_0^4 \sqrt{16 - x^2} \, dx = 12 \left[\frac{1}{2} x \sqrt{4^2 - x^2} + \frac{1}{2} . 4^2 \sin^{-1} \left(\frac{x}{4} \right) \right]_0^4$$

$$= 12 . \frac{1}{2} . 16 . \sin^{-1}(1) = 96 . \frac{\pi}{2} = 48\pi. \qquad \square$$

Example 6.14.14 *Stating the region of validity of differentiation under the sign of integration, prove that*

$$\int_0^{\frac{\pi}{2}} \log \left(1 - x^2 \sin^2 \theta \right) d\theta = \pi \log \left(1 + \sqrt{1 - x^2} \right) - \pi \log 2 \quad \text{where} \quad x^2 < 1.$$

Solution: The given function is defined in the rectangle $-1 < x < 1$ and $0 < \theta < \dfrac{\pi}{2}$ and it satisfies the conditions of Leibnitz's rule.

$$\text{Let} \quad F(x) = \int_0^{\frac{\pi}{2}} \log \left(1 - x^2 \sin^2 \theta \right) d\theta \qquad (6.16)$$

Differentiating under the integral sign with respect to x, we get

$$
\begin{aligned}
F'(x) &= \int_0^{\frac{\pi}{2}} \frac{1}{1 - x^2 \sin^2 \theta} \left(-2x \sin^2 \theta \right) d\theta \\
&= \frac{2}{x} \int_0^{\frac{\pi}{2}} \frac{1 - x^2 \sin^2 \theta - 1}{1 - x^2 \sin^2 \theta} \, d\theta, \ (x \neq 0) \\
&= \frac{\pi}{x} - \frac{2}{x} \int_0^{\frac{\pi}{2}} \frac{d\theta}{1 - x^2 \sin^2 \theta} = \frac{\pi}{x} - \frac{2}{x} \int_0^{\frac{\pi}{2}} \frac{\operatorname{cosec}^2 \theta}{\operatorname{cosec}^2 \theta - x^2} \, d\theta \\
&= \frac{\pi}{x} - \frac{2}{x} \int_0^{\infty} \frac{dz}{1 + z^2 - x^2} \quad \text{[by substituting } \cot \theta = z] \\
&= \frac{\pi}{x} - \left[\frac{2}{x\sqrt{1 - x^2}} \tan^{-1} \frac{z}{\frac{2}{x\sqrt{1-x^2}}} \right]_0^{\infty} = \frac{\pi}{x} - \frac{2}{x\sqrt{1 - x^2}} . \frac{\pi}{2} \\
&= \frac{\pi}{x} - \frac{\pi}{x\sqrt{1 - x^2}}
\end{aligned}
$$

Integrating with respect to x, we obtain,

$$
\begin{aligned}
F(x) &= \int \frac{\pi}{x} dx - \int \frac{\pi}{x} \frac{dx}{\sqrt{1 - x^2}} = \pi \log x - \pi \log \frac{1 - \sqrt{1 - x^2}}{x} + C \\
&\qquad\qquad \text{(where } C \text{ is the constant of integration)} \\
&= \pi \log \frac{x^2}{1 - \sqrt{1 - x^2}} + C = \pi \log \frac{x^2 \left(1 + \sqrt{1 - x^2} \right)}{1 - (1 - x^2)} + C \\
&= \pi \log \left(1 + \sqrt{1 - x^2} \right) + C \qquad (6.17)
\end{aligned}
$$

But $F(0) = 0$ [by (6.16)] $\therefore$ $\pi \log 2 + C = 0$ [by (6.17)] $\Rightarrow$ $C = -\pi \log 2$. Hence

$$F(x) = \int_0^{\frac{\pi}{2}} \log \left(1 - x^2 \sin^2 \theta \right) d\theta = \pi \log \left(1 + \sqrt{1 - x^2} \right) - \pi \log 2. \qquad \square$$

6.15 Exercises

1. Evaluate the followings:

(i) $\iint_R xy(x^2 + y^2)\, dxdy$ over $R: \{0 \le x \le a; 0 \le y \le b\}$.

(ii) $\iint_R x^3 y\, dxdy$ over $R: \{0 \le x \le 1; 0 \le y \le 1\}$.

(iii) $\int_0^{\frac{\pi}{2}} \int_0^{\frac{\pi}{2}} \sin(x + y)\, dxdy$

(iv) $\int_1^2 \int_3^5 (y - 2x)\, dxdy$

(v) $\int_0^1 \int_1^2 (x^2 + y^2)\, dxdy$

(vi) $\int_0^{\log_e 2} \int_{-1}^1 y e^{xy}\, dxdy$

(vii) $\int_0^4 \int_0^1 xy(x - y)\, dydx$

(viii) $\int_0^{\pi} \int_0^{a\cos\theta} r\sin\theta\, drd\theta$

2. Evaluate: (i) $\int_0^1 \int_0^{1-y^2} \left[(x - 1)^2 + y^2\right]\, dxdy$,

(ii) $\int_0^a \int_0^{\sqrt{a^2-y^2}} \sqrt{a^2 - x^2 - y^2}\, dxdy$.

3. Evaluate:

(i) $\iint xy\, dxdy$ over the positive quadrant of the circle $x^2 + y^2 = a^2$.

(ii) $\iint y\, dxdy$ over the region bounded by the ellipse $\dfrac{x^2}{a^2} + \dfrac{y^2}{b^2} = 1$.

(iii) $\iint xy(x + y)\, dxdy$ over the area bounded by $y = x^2$ and $y = x$.

(iv) $\iint x\, dxdy$ over the ellipse $b^2x^2 + a^2y^2 = 1$.

(v) $\iint (x^2 + y^2)\, dxdy$ over the region R bounded by $y = x^2, x = 2$ and $y = 1$.

4. Evaluate, by using suitable transformation:

(i) $\int_0^a \int_0^{\sqrt{a^2-y^2}} y^2 \sqrt{x^2 + y^2}\, dydx$

[*Hints:* Use the transformation $x = r\cos\theta, y = r\sin\theta$.]

(ii) $\iint \sqrt{x^2 + y^2}\, dxdy$, the field of integration being R, the region in xy-plane bounded by the circles $x^2 + y^2 = 1$ and $x^2 + y^2 = 4$.

(iii) $\iint \{2a^2 - 2a(x + y) - (x^2 + y^2)\}\, dxdy$, the region of integration being the circle $x^2 + y^2 + 2a(x + y) = 2a^2$.

[*Hints:* Put $x + a = r\cos\theta, y + a = r\sin\theta$.]

(iv) $\iint xy(x^2 + y^2)^{\frac{n}{2}}\, dxdy$, over the positive quadrant of the circle $x^2 + y^2 = a^2, (n + 3 > 0)$.

(v) $\iint (x+y)^2 \, dxdy$ over the ellipse $\dfrac{x^2}{a^2} + \dfrac{y^2}{b^2} = 1$.

(vi) $\iint_E \sqrt{x(2a-x) + y(2b-y)} \, dxdy$ where E is the region bounded by the circle $x^2 + y^2 - 2ax - 2by = 0$.

(vii) $\iint xy \, dxdy$ over the region bounded by the parabolas $y^2 = 4x, y^2 = 8x, x^2 = 4y$ and $x^2 = 8y$.

[*Hints:* Using the transformation $\dfrac{y^2}{x} = u, \dfrac{x^2}{y} = v$, we get $x = v^{\frac{2}{3}} u^{\frac{1}{3}}, y = u^{\frac{2}{3}} v^{\frac{1}{3}}$ where Jacobian is

$$\frac{\partial(x,y)}{\partial(u,v)} = \begin{vmatrix} \dfrac{2}{3} v^{-\frac{1}{3}} u^{\frac{1}{3}} & \dfrac{1}{3} v^{\frac{2}{3}} u^{-\frac{2}{3}} \\[2mm] \dfrac{1}{3} u^{\frac{2}{3}} v^{-\frac{2}{3}} & \dfrac{2}{3} u^{-\frac{1}{3}} v^{\frac{1}{3}} \end{vmatrix} = \frac{4}{9} - \frac{1}{9} = \frac{1}{3}.$$

Now the region R transforms to R' which is a square bounded by the lines $u = 4, u = 8, v = 4$ and $v = 8$. Therefore

$$I = \int_4^8 \int_4^8 \frac{1}{3} uv \, dvdu = \frac{1}{3} \int_4^8 udu \int_4^8 vdv = \frac{1}{3} \left[\frac{u^2}{2}\right]_4^8 \left[\frac{v^2}{2}\right]_4^8$$

$$= \frac{1}{12} \times 48 \times 48 = 192.]$$

5. (i) Using the transformation $x + y = u, y = uv$, show that

$$\int_0^1 dx \int_0^{1-x} e^{\frac{y}{x+y}} \, dy \text{ becomes } \iint_E e^{\frac{y}{x+y}} \, dxdy \text{ where } E \text{ is the triangle bounded by}$$

$x = 0, y = 0, x + y = 1$ and whose value is $\dfrac{1}{2}(e-1)$.

(ii) Evaluate $\displaystyle\int_0^\infty \int_0^\infty e^{-(x^2 + 2xy \cos \alpha + y^2)} \, dxdy, \quad (0 \leq \alpha \leq \pi)$.

(iii) By using the transformation $y - x = 2u, y + x = 2v$, show that

$$\iint_\triangle e^{\frac{y-x}{y+x}} \, dxdy = \frac{1}{4}(e - e^{-1}),$$

where $\triangle$ denotes the triangle with vertices $(0,0), (1,0)$ and $(0,1)$.

6. Change the order of integration in the following cases:

(i) $\displaystyle\int_0^a dx \int_0^x f(x,y) \, dy$ (ii) $\displaystyle\int_0^1 dx \int_x^{\sqrt{x}} f(x,y) \, dy$

(iii) $\displaystyle\int_0^1 dy \int_0^{\sqrt{y}} f(x,y) \, dx$ (iv) $\displaystyle\int_0^{2a} dx \int_0^{\sqrt{2ax-x^2}} (a - \sqrt{a^2 - y^2}) \, dy$

7. Evaluate the following integrals by changing the order of integrals.

(i) $\displaystyle\int_0^a dx \int_{\frac{x^2}{2}}^{2a-x} xy \, dy$ (ii) $\displaystyle\int_0^1 dy \int_y^1 e^{x^2} dx$ (iii) $\displaystyle\int_0^3 dy \int_1^{\sqrt{4-y}} (x+y) \, dy$.

8. Prove, by evaluating the repeated integrals, that

(i) $\displaystyle\int_0^1 dx \int_0^1 \frac{x^2 - y^2}{(x^2 + y^2)^2} \, dy \neq \int_0^1 dy \int_0^1 \frac{x^2 - y^2}{(x^2 + y^2)^2} \, dx$

(ii) $\displaystyle\int_0^1 dx \int_0^1 \frac{x^2 - y^2}{x^2 + y^2} \, dy = \int_0^1 dy \int_0^1 \frac{x^2 - y^2}{x^2 + y^2} \, dx$

(iii) $\displaystyle\int_0^1 dx \int_0^1 \frac{x - y}{(x + y)^3} \, dy \neq \int_0^1 dy \int_0^1 \frac{x - y}{(x + y)^3} \, dx$

9. Show that $\displaystyle\int_\lambda^1 dx \int_\lambda^1 \frac{y^2 - x^2}{(y^2 + x^2)^2} \, dy = \int_\lambda^1 dy \int_\lambda^1 \frac{y^2 - x^2}{(y^2 + x^2)^2} \, dx$, but

$\displaystyle\int_0^1 dx \int_0^1 \frac{y^2 - x^2}{(y^2 + x^2)} \, dy = \int_0^1 dy \int_0^1 \frac{y^2 - x^2}{y^2 + x^2)} \, dx.$

10. Find, by double integration, the area of the region bounded by the curves given below:

(i) the ellipse $\dfrac{x^2}{a^2} + \dfrac{y^2}{b^2} = 1$.

(ii) $y^2 = 4x, y^2 = 16x, x = 1$ and $x = 16$ in the positive quadrant.

(iii) $y^2 = 8x, y^2 = 16x, xy = 25$ and $xy = 16$ in the positive quadrant.

(iv) $y = \dfrac{3x}{x^2 + 2}$, $4y = x^2$.

(v) the ellipse $\dfrac{x^2}{a^2} + \dfrac{y^2}{b^2} = 1$ and its auxiliary circle.

(vi) $x^2 + y^2 = 100, x^2 + y^2 = 64, y = \sqrt{3}x$ and $\sqrt{3}y = x$.

(vii) $r = a(1 + \cos\theta)$.

[*Hints:* (i) From the **Figure** 6.29, required area $= 4\times$(area of the quadrant $OABO$ of the ellipse)

$\displaystyle = 4 \int_{x=0}^a \int_{y=0}^{f(x)} dx dy = 4 \int_0^a dx \int_0^{f(x)} dy = 4 \int_0^a [y]_0^{f(x)}$

$\displaystyle = 4 \int_0^a \left[\frac{b}{a}\sqrt{a^2 - x^2} \right] dx \quad \left(\because y = f(x) = \frac{b}{a}\sqrt{a^2 - x^2} \right)$

$\displaystyle = \frac{4b}{a} \left[\frac{1}{2}x\sqrt{a^2 - x^2} + \frac{1}{2}a^2 \sin^{-1}\frac{x}{a} \right]_0^a = \frac{2b}{a}[0 + a^2 \sin^{-1}(1)]$

$\displaystyle = \frac{2b}{a}.a^2.\frac{\pi}{2} = ab\pi.$

(iii) Required area $\displaystyle\iint_R dx dy$.

To evaluate the double integration, we use the transformation $\dfrac{y^2}{x} = u, xy = v$ whence $x = v^{\frac{2}{3}} u^{-\frac{1}{3}}$

and $y = u^{\frac{1}{3}} v^{\frac{1}{3}}$ giving $J = \dfrac{1}{3u}$.

Now R transforms to R', which is a rectangle bounded by the lines $u = 8, u = 16, v = 16$ and $v = 25$.

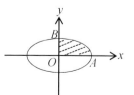

Figure 6.29

Therefore required area $= \dfrac{1}{3} \displaystyle\int_8^{16} \dfrac{1}{u} du \int_{16}^{25} dv = 3 \log_e 2$ (after calculation).

(iv) The region is given by $4y = x^2$ and $y = \dfrac{3x}{x^2+2} \;\Rightarrow\; 4y = \dfrac{12x}{x^2+2}$

$\Rightarrow x^2 = \dfrac{12x}{x^2+2} \;\Rightarrow\; x^4 + 2x^2 - 12x = 0 \;\Rightarrow\; x(x^3 + 2x - 12) = 0 \;\Rightarrow\; x = 0, 2.$

Required area $= \displaystyle\int_{x=0}^{2} \int_{y=\frac{x^2}{4}}^{\frac{3x}{x^2+2}} dx\,dy = \dfrac{3}{2} \log 3 - \dfrac{2}{3}$ (after calculation).

(v) Required area $= \displaystyle\iint_R dx\,dy = \iint_{R_1} dx\,dy - \iint_{R_2} dx\,dy$, where R_1 is the re-

gion of the auxiliary circle $x^2 + y^2 = a^2$ and R_2 is the region of the ellipse

$\dfrac{x^2}{a^2} + \dfrac{y^2}{b^2} = 1$. For R_1 we use the transformation $x = r\cos\theta, y = r\sin\theta$, i.e., $J = r$

and for R_2 we use $x = ar\cos\theta, y = br\sin\theta$ i.e., $J = abr$ and therefore required

area $= \displaystyle\int_{\theta=0}^{2\pi} \int_{r=0}^{a} r\,dr\,d\theta - \int_{\theta=0}^{2\pi} \int_{r=0}^{1} abr\,dr\,d\theta = a\pi(a-b).$

(vi) Required area $= \displaystyle\iint_R dx\,dy.$

Use the transformation $x = r\cos\theta, y = r\sin\theta$, i.e., $J = r$. Region R transforms

R' which is bounded by $r = 10, r = 8; \tan\theta = \sqrt{3}, \tan\theta = \dfrac{1}{\sqrt{3}}$, i.e., r varies from

8 to 10 and θ varies from $\dfrac{\pi}{6}$ to $\dfrac{\pi}{3}$.

$\therefore$ Required Area $\displaystyle\int_{\frac{\pi}{6}}^{\frac{\pi}{3}} d\theta \int_8^{10} r\,dr = [\theta]_{\frac{\pi}{6}}^{\frac{\pi}{3}} \left[\dfrac{r^2}{2}\right]_8^{10} = 3\pi.$

(vii) From the **Figure** 6.30, required

area $= 2\times$area $OABO = 2\displaystyle\iint r\,dr\,d\theta =$

$2\displaystyle\int_0^{\pi} d\theta \int_0^{f(\theta)} r\,dr$, where $r = f(\theta)$ and $r =$

$a(1 + \cos\theta)$ is the equation of the curve.

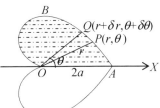

Figure 6.30

Therefore, required area

$= 2 \displaystyle\int_0^{\pi} d\theta \int_0^{a(1+\cos\theta)} r\,dr$

$= 2 \displaystyle\int_0^{\pi} d\theta \left[\dfrac{r^2}{2}\right]_0^{a(1+\cos\theta)}$

$= 2 \displaystyle\int_0^{\pi} \dfrac{a^2}{2}(1+\cos\theta)^2 d\theta = a^2 \int_0^{\pi} \left(2\cos^2\dfrac{\theta}{2}\right)^2 d\theta = 4a^2 \int_0^{\pi} \cos^4\dfrac{\theta}{2} d\theta$

$= 8a^2 \displaystyle\int_0^{\frac{\pi}{2}} \cos^4\phi\,d\phi$ [Putting $\theta = 2\phi$] $= 8a^2 \cdot \dfrac{3}{4} \cdot \dfrac{1}{2} \cdot \dfrac{1}{2} \cdot \pi = \dfrac{3}{2} a^2 \pi.$]

11. Evaluate $\displaystyle\iint_{\triangle} x^{\frac{1}{2}} y^{\frac{1}{3}} (1 - x - y)^{\frac{2}{3}} dx\,dy$, where $\triangle$ is the triangle bounded by

the lines $x = 0, \; y = 0, \; x + y = 1.$

12. If E is the region bounded by the circle $x^2 + y^2 + 2a(x + y) = 2a^2$, prove that $\iint_E \{2a^2 - 2a(x + y) - (x^2 + y^2)\}\, dx dy = 8\pi a^4$.

13. Evaluate $\displaystyle\int_0^1 dx \int_0^{\sqrt{1-x^2}} \frac{dy}{(1 + e^y)\sqrt{1 - x^2 - y^2}}$, by changing the order of integration.

14. Evaluate $\displaystyle\iint_R ye^{xy}\, dx dy$, where $R = \{(x, y) : 0 \le x \le a, 0 \le y \le b\}$.

15. Evaluate the followings:

(i) $\displaystyle\iint x\, dx dy$ over the region $r = 2a(1 + \cos\theta)$.

(ii) $\displaystyle\int_0^\infty \int_0^\infty e^{-(x^2 + y^2)}\, dx dy$ \qquad (iii) $\displaystyle\int_0^\infty \int_0^\infty e^{-(x^2 + 2xy + 2y^2)}\, dx dy$.

16. Evaluate the following triple integrals:

(i) $\displaystyle\int_{x=0}^1 \int_{y=0}^2 \int_{z=1}^2 x^2 yz\, dx dy dz$ \qquad (ii) $\displaystyle\int_0^a \int_0^a \int_0^a (x^2 + y^2 + z^2)\, dx dy dz$

(iii) $\displaystyle\int_0^1 \int_{y^2}^1 \int_0^{1-x} x\, dz dy dx$ \qquad (iv) $\displaystyle\int_0^{\log a} \int_0^x \int_0^{x+y} e^{x+y+z}\, dx dy dz$

(v) $\displaystyle\int_0^2 \int_0^x \int_0^{x\sqrt{3}} \frac{x}{x^2 + y^2}\, dy dx dz$ \qquad (ii) $\displaystyle\int_0^1 \int_0^{1-x} \int_0^{1-x-y} \frac{dx dy dz}{(1 + x + y + z)^3}$.

17. Evaluate, by suitable transformation:

(i) $\displaystyle\iiint (x^2 + y^2 + z^2)\, dx dy dz$ through the sphere $x^2 + y^2 + z^2 \le 1$.

(ii) $\displaystyle\iiint \frac{dx dy dz}{x^2 + y^2 + \left(z - \frac{1}{2}\right)^2}$ extended over the sphere $x^2 + y^2 + z^2 \le 1$.

(iii) $\displaystyle\iiint \sqrt{\frac{x^2}{a^2} + \frac{y^2}{b^2} + \frac{z^2}{c^2}}\, dx dy dz$ over the ellipsoid $\frac{x^2}{a^2} + \frac{y^2}{b^2} + \frac{z^2}{c^2} \le 1$.

(vi) $\displaystyle\iiint (ax^2 + by^2 + cz^2)\, dx dy dz$ over the sphere $x^2 + y^2 + z^2 \le R^2$.

18. Evaluate the followings:

(i) $\displaystyle\iiint x^2\, dx dy dz$ extended over the volume of the ball $x^2 + y^2 + z^2 \le a^2$.

(ii) $\displaystyle\iiint z^2\, dx dy dz$ over the region common to the surfaces $x^2 + y^2 + z^2 = a^2$ and $x^2 + y^2 = ax$.

[*Hints:* The region is bounded below and above by the surfaces $z = -\sqrt{a^2 - x^2 - y^2}$ and $z = \sqrt{a^2 - x^2 - y^2}$.

Its projection on the xy-plane is the circular domain $R' = x^2 + y^2 = ax$, therefore

$$\iiint z^2\, dx dy dz = \iint_{R'} dx dy \int_{-\sqrt{a^2 - x^2 - y^2}}^{\sqrt{a^2 - x^2 - y^2}} z^2\, dz = \frac{2}{3} \iint_{R'} (a^2 - x^2 - y^2)^{\frac{3}{2}}\, dx dy$$

Transforming to polar coordinates, i.e., by putting $x = r\cos\theta, y = r\sin\theta$ whence $J = r$, we get transformed region R' as the circle $r = a\cos\theta$ where $0 \leq \theta \leq \pi, 0 \leq r \leq a\cos\theta$. Therefore, required integral

$$= \frac{2}{3}\int_0^\pi d\theta \int_0^{a\cos\theta} (a^2 - r^2)^{\frac{3}{2}} r\, dr = \frac{2}{15}a^5 \int_0^{\frac{\pi}{2}}(1-\sin^5\theta)d\theta = \frac{2a^5(15\pi - 16)}{225}.\Big]$$

(iii) $\iiint z^2 \, dxdydz$ over the hemisphere $x \geq 0$, $x^2 + y^2 + z^2 \leq a^2$.

(iv) $\iiint_V e^{\sqrt{\frac{x^2}{a^2} + \frac{y^2}{b^2} + \frac{z^2}{c^2}}} \, dxdydz$ where V is the region $\frac{x^2}{a^2} + \frac{y^2}{b^2} + \frac{z^2}{c^2} \leq 1$.

(v) $\iiint \frac{dxdydz}{\sqrt{1 - x^2 - y^2 - z^2}}$, integral being taken to all the positive values of

the variables for which the expression is real.

19. Show that

(i) if R be the ellipsoid $\frac{x^2}{a^2} + \frac{y^2}{b^2} + \frac{z^2}{c^2} \leq 1$,

$$\iiint_R \sqrt{a^2b^2c^2 - b^2c^2x^2 - c^2a^2y^2 - a^2b^2z^2} \, dxdydz = \frac{1}{4}\pi^2 a^2 b^2 c^2.$$

(ii) $\iiint (x + y + z)x^2y^2z^2 \, dxdydz = \frac{1}{50400}$, taken throughout the tetrahe-

dron bounded by the planes $x = 0, y = 0, z = 0$ and $x + y + z = 1$.

(iii) $\iiint_R u^2v^2w \, dudvdw = \frac{\pi}{48}$ where R is the region $u^2 + v^2 \leq 1, 0 \leq w \leq 1$.

(iv) $\iiint (z^2 + z) \, dxdydz = 0$ over the sphere $x^2 + y^2 + z^2 = 1$.

20. Evaluate the triple integrals for the following regions:

(i) bounded by the cylinder $x^2 + y^2 = 16$ and the points $z = 0$ to $z = 3$.

(ii) a tetrahedron, bounded by the coordinate planes and the plane $x+y+z = 1$.

(iii) bounded by the sphere $x^2 + y^2 + z^2 = 4$ and the surface of the paraboloid $x^2 + y^2 = 3z$.

(iv) bounded by a cone $x^2 + y^2 = z^2$ and the plane $z = 1$.

(v) bounded by an ellipsoid $\frac{x^2}{a^2} + \frac{y^2}{b^2} + \frac{z^2}{c^2} = 1$.

21. If $f(x,y)$ be a continuous function of x and y on the rectangle $R = \{(x,y)|a \leq x \leq b, c \leq y \leq d\}$ and $f_y(x,y)$ exists and is continuous on R, show that

$$\frac{d}{dy}\left\{\int_a^b f(x,y) \, dx\right\} = \int_a^b \left\{\frac{\partial}{\partial y}f(x,y)\right\} dx.$$

[*Hints*: Differential under the integral sign.]

22. Obtain a sufficient condition under which

$$\int_a^b \left\{\int_c^d f(x,y) \, dy\right\} dx = \int_c^d \left\{\int_a^b f(x,y)dx\right\} dy.$$

[*Hints*: § 6.13: Corollary.]

23. Use differentiation under the sign of integral to prove that

(i) $\displaystyle\int_0^\pi \frac{\log(1 + \sin\alpha\cos x)}{\cos x}\,dx = \pi\alpha.$

(ii) $\displaystyle\int_0^\theta \log(1 + \tan\theta\tan x)dx = \log\sec\theta, \quad \left(-\frac{\pi}{2} < \theta < \frac{\pi}{2}\right).$

(iii) $\displaystyle\int_0^{\frac{\pi}{2}} \log(1 - x^2\cos^2\theta)\,d\theta = \pi\log(1 + \sqrt{1 - x^2}) - \pi\log 2,$ when $x^2 < 1.$

(iv) $\displaystyle\int_0^\pi \log(1 + a\cos x)\,dx = \pi\log\frac{1 + \sqrt{1 - a^2}}{2}, \ |a| < 1.$

(v) $\displaystyle\int_{\frac{1}{2}\pi - \alpha}^{\frac{1}{2}\pi} \sin\theta\cos^{-1}(\cos\alpha\ \mathrm{cosec}\ \theta)\,d\theta = \frac{\pi}{2}(1 - \cos\alpha).$

[*Hints:* (i) Take $a = \sin\alpha$ in **Example 6.13.1**. Also noticing that $|a| < 1 \Rightarrow |\sin\alpha| < 1$ is true, we get

$$\int_0^\pi \frac{\log(1 + \sin\alpha\cos x)}{\cos x}\,dx = \pi\sin^{-1}(\sin\alpha) = \pi\alpha.$$

(iii) Using the important property $\displaystyle\int_0^a f(x)dx = \int_0^a f(a - x)dx$ of definite integral, we get

$$\int_0^{\frac{\pi}{2}} \log(1 - x^2\cos^2\theta)\,d\theta = \int_0^{\frac{\pi}{2}} \log\left(1 - x^2\cos^2\overline{\frac{\pi}{2} - \theta}\right)\,d\theta$$

$$= \int_0^{\frac{\pi}{2}} \log(1 - x^2\sin^2\theta)\,d\theta = \pi\log(1 + \sqrt{1 - x^2}) - \pi\log 2$$

(following the **Example 6.14.14**).]

ANSWERS

1. (i) $\frac{1}{8}a^2b^2(a^2 + b^2)$; (ii) $\frac{1}{2}$; (iii) 2; (iv) 2; (v) $\frac{8}{3}$; (vi) $\frac{1}{2}$; (vii) 8; (viii) $\frac{a^2}{3}$; **2.** (i) $\frac{44}{105}$; (ii) $\frac{\pi a^3}{6}$; **3.** (i) $\frac{a^4}{8}$; (ii) $\frac{4}{3}ab^2$; (iii) $\frac{3}{56}$; (iv) 0; (v) 1006; **4.** (i) $\frac{\pi a^5}{20}$; (ii) $\frac{14}{3}\pi$; (iii) $8\pi a^4$; (iv) $\frac{1}{2}\frac{a^{n+4}}{n+4}$; (v) $\frac{\pi ab}{4}(a^2 + b^2)$; **5.** (ii) $\frac{\alpha}{2\sin\alpha}$; **6.** (i) $\int_0^a dy \int_y^a f(x, y)dx$; (ii) $\int_0^1 dy \int_{y^2}^y f(x, y)dx$; (iii) $\int_0^1 dx \int_{x^2}^1 f(x, y)dy$; (iv) $\int_0^a dy \int_{a - \sqrt{a^2 - y^2}}^{a + \sqrt{a^2 - y^2}}(a - \sqrt{a^2 - y^2})dx$; **7.** (i) $\frac{a^3}{24}(5a + 8)$; (ii) $\frac{1}{2}(e - 1)$; (iii) $\frac{8}{15}(16 - 7\sqrt{2})$; **10.** (i) πab; (ii) 84; (iii) $3\log_e 2$; (iv) $\frac{3}{2}\log 3 - \frac{2}{3}$; (v) $a(a - b)\pi$; (vi) 3π; **11.** $\frac{\Gamma(\frac{5}{3})\Gamma(\frac{4}{3})\Gamma(\frac{3}{2})}{\Gamma(\frac{9}{2})}$; **13.** $\frac{\pi}{2}\log\frac{2e}{1+e}$; **14.** $\frac{1}{a}(e^{ab} - ab - 1)$; **15.** (i) $10\pi a^3$; (ii) $\frac{\pi}{4}$; (iii) $\frac{\pi}{8}$; **16.** (i) 1; (ii) a^5; (iii) $\frac{4}{35}$; (iv) $\frac{1}{4}(a^4 - 6a^2 + 8a - 3)$; (v) $\frac{2\pi}{3}$; (vi) $\frac{1}{2}\left(\log 2 - \frac{5}{8}\right)$; **17.** (i) $\frac{4}{5}\pi$; (ii) $\left(2 + \frac{3}{2}\log 3\right)\pi$; (iii) πabc; (iv) $\frac{4}{15}(a + b + c)R^5$; **18.** $\frac{4}{15}\pi a^5$; (ii) $\frac{2(15\pi - 16)}{225}a^5$; (iii) $\frac{2\pi}{15}a^5$; (iv) $4\pi abc(e - 1)$; (v) $\frac{\pi^2}{8}$; **20.** (i) 48π; (ii) $\frac{1}{6}$; (iii) $\frac{19}{6}\pi$; (iv) $\frac{\pi}{3}$; (v) $\frac{4}{3}\pi abc.$

Chapter 7

Line, Surface and Volume Integrals

7.1   Introduction

In **Chapter 6**, the integrals of functions of several variables involving multiple integrals, i.e., double and triple integrals have been discussed in details. In this chapter, we shall study few new concepts of integrals - *Line, Surface* and *Volume integrals* and their interrelationship. We shall also discuss the applications of double and triple integrals over region, with line integrals over its boundary, referred to as *Green's Theorem* (or *Green's Theorem in a plane*), and can be generalized to cover *Stoke's Theorem* and *Gauss Divergence Theorem*.

Roughly speaking, *Line integrals* are integrals of functions, defined over curves. In *Riemann integral*, it is known to us that the region of integration is an interval $a \leq x \leq b$ and an ordinary single integral $\int_a^b f(x)\, dx$ is an integral which is defined along a line segment, i.e., interval of a coordinate axis. If however, the integral is defined along an arc of a curve in two or three dimensions, the integral can be still defined over that region, which is called a *line integral* or a *curvilinear integral over the arc*. Similarly, the integrals of functions defined on *surfaces* and *volumes* are respectively called *Surface integrals* and *Volume integrals*. Volume integrals or Triple integrals are a straight and simple extension of the ideal of double integrals and are in many respects almost completely analogous to them.

7.2   The Curve

In space geometry, according to their position the curves are of two types: (i) Plane curves and (ii) Space curves.

A plane curve

A plane curve in the xy plane is a set of points (x, y) for which

$$x = \phi(t),\ y = \psi(t);\ a \leq t \leq b$$

where $\phi(t)$ and $\psi(t)$ are two real-valued continuous functions within the interval $a \leq t \leq b$.

A space curve

A curve in space is a set of points (x, y, z) for which

$$x = \phi(t), \ y = \psi(t), \ z = \theta(t); \ a \leq t \leq b$$

where $\phi(t)$, $\psi(t)$ and $\theta(t)$ are real valued continuous functions in three dimensional space having x, y and z axes as the axes of coordinates within the interval $a \leq t \leq b$.

7.3 Line Integral

The corresponding kind of integral for a function which is defined along a curve is usually called a *line integral*. Here line means a curved line. This curve may be a *plane curve* or a *space curve*.

As for example, determination of the mass of a material line from its density, computation of the work of a field of force along a path etc. require the introduction of the so called *line integral*.

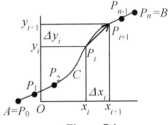

Figure 7.1

Let us consider a continuous vector function $\vec{F}(P)$ which is defined at each point of a curve C. To be more specific, let a point $P(x, y)$ be in motion along a plane curve C from the point A to the point B. A force $\vec{F}$ which varies in magnitude and direction is applied to P with its motion. It is thus same function of the coordinates of P, i.e., $\vec{F} = \vec{F}(P)$.

Let us now find the work done W by the applied force $\vec{F}$ in transferring the particle P from A to B. For this, we divide the portion AB of the curve C into n arbitrary points given by $A = P_0, P_1, P_2, ..., P_i, P_{i+1}, ..., P_{n-1}, P_n = B$ in the direction from A to B (*vide.* **Figure** 7.1).

Let the vector $\overrightarrow{P_i P_{i+1}}$ be denoted by $\overrightarrow{\triangle s_i}$. Let F_i be the magnitude of the force $\vec{F}$ at P_i. Then the scalar product $F_i \triangle s_i$ is regarded as an elementary approximate work done by the force $\vec{F}$ along the line $P_i P_{i+1}$.

Let $\vec{F} = X(x, y) \ \hat{i} + Y(x, y) \ \hat{j}$, where $X(x, y)$ and $Y(x, y)$ are the projections of the vector $\vec{F}$ on the coordinate axes of x and y respectively.

Let $\triangle x_i$ and $\triangle y_i$ be the increments of the coordinates x_i and y_i respectively when changing from the point P_i to the point P_{i-1}. So we get

$$\triangle s_i \ = \ \triangle x_i \ \hat{i} + \triangle y_i \ \hat{j}$$
$$\therefore \ F_i \ \triangle s_i \ = \ X(x_i, y_i) \triangle x_i + Y(x_i, y_i) \triangle y_i.$$

We now consider the sum

$$\sum_{i=1}^{n} F_i \ \triangle s_i = \sum_{i=1}^{n} \left[X(x_i, y_i) \triangle x_i + Y(x_i, y_i) \triangle y_i \right].$$

The limit of this sum, if it exits, as the number of sub-divisions increases indefinitely so that length of each subdivisional arc decreases to zero is called the *line integral* of $X(x,y)$ and $Y(x,y)$ along the curve C and this limit expresses the work of the force $\vec{F}$ over the curve C from the point A to the point B. Thus

$$W = \lim_{\substack{\triangle x_i \to 0 \\ \triangle y_i \to 0}} \sum_{i=1}^{n} [X(x_i, y_i)\triangle x_i + Y(x_i, y_i)\triangle y_i].$$

We denote the above *line integral* of $X(x,y)$ and $Y(x,y)$ along the curve C as

$$W = \int_C [X(x,y)dx + Y(x,y)dy] = \int_A^B [X(x,y)dx + Y(x,y)dy].$$

Note 7.3.1 *If the curve C be a space curve, then the line integral of three functions $X(x,y,z), Y(x,y,z)$ and $Z(x,y,z)$ may be defined similarly as*

$$\int_C [X(x,y,z)dx + Y(x,y,z)dy + Z(x,y,z)dz]$$

$$= \int_A^B [X(x,y,z)dx + Y(x,y,z)dy + Z(x,y,z)dz]$$

$$= \lim_{\substack{\triangle x_i \to 0 \\ \triangle y_i \to 0 \\ \triangle z_i \to 0}} \sum_{i=1}^{n} [X(x_i, y_i, z_i)\triangle x_i + Y(x_i, y_i, z_i)\triangle y_i + Z(x_i, y_i, z_i)\triangle z_i],$$

if this limit exists.

7.4   Existence of a Line Integral

Let the curve C be represented by equation in parametric form: $x = \phi(t)$, $y = \psi(t)$. Below we state only the existence theorem without proof.

Theorem 7.4.1 *Let the functions $\phi(t)$ and $\psi(t)$ are continuous having continuous derivatives $\phi'(t)$ and $\psi'(t)$. Also $X[\phi(t), \psi(t)]$ and $Y[\phi(t), \psi(t)]$ are continuous functions of t on the interval $[a,b]$, then the following limits*

$$\left. \begin{aligned} L_1 &= \lim_{\triangle x_i \to 0} \sum_{i=1}^{n} X(\bar{x}_i, \bar{y}_i)\triangle x_i \\ L_2 &= \lim_{\triangle y_i \to 0} \sum_{i=1}^{n} Y(\bar{x}_i, \bar{y}_i)\triangle y_i \end{aligned} \right\}$$

where $\bar{x}_i$ and $\bar{y}_i$ are the coordinates of some point lying on the arc $\triangle s_i$, exists.

Note 7.4.1 *The above two line integrals are denoted as $L_1 = \int_C X(x,y) \, dx$ and $L_2 = \int_C Y(x,y) \, dy$.*

Note 7.4.2 *The above theorem makes it possible to develop a method for computing a line integral.*

Note 7.4.3 *In a similar manner we compute the line integrals $\int X dx$, $\int Y dy$, $\int Z dz$ and $\int (X dx + Y dy + Z dz)$ etc. over the space curve defined by the equations $x = \phi(t)$, $y = \psi(t)$ and $z = \chi(t)$.*

Remark: In the above discussions, the limit of the integral sum is to be understood in the same sense as in the case of usual definite integral. *A line integral changes sign when the sense of integration is reversal*, since in that case the vector $\triangle s$ and hence, its projections $\triangle x$ and $\triangle y$ changes sign. Further we are to note that the definition of a line integral holds true also for the case when the curve C is closed. In this case, since the initial and terminal points coincide we cannot write the line integral as $\int_A^B (X\,dx + Y\,dy)$, but only $\int_C (X\,dx + Y\,dy)$, indicating the *direction of circulation*. The line integral over a closed curve C is frequently denoted by $\oint_C (X\,dx + Y\,dy)$ when the line integral of a vector function $\vec{F}$ is taken along a closed curve C. This line integral is also called a *circulation* of the vector function $\vec{F}$ over the closed contour C.

7.5 Properties of a Line Integral

Since a line integral can be treated as an ordinary definite integral, the basic properties of the line integral are almost completely analogous to those of the definite integral. Here the some of those:

Let f, g be integrable functions on C and k is any constant then

(i) $\displaystyle\int_C k f(x, y)\ dx = k \int_C f(x, y)\ dx$

(ii) $\displaystyle\int_C (f \pm g)\ dx = \int_C f\ dx \pm \int_C g\ dx$

(iii) Let the arc $C \equiv \widehat{AB}$ be composed of two arcs $\widehat{AD}$ and $\widehat{DB}$, then

$$\int_C f(x, y)\ dx = \int_{\widehat{AB}} f(x, y)\ dx = \int_{\widehat{AD}} f(x, y)\ dx + \int_{\widehat{DB}} f(x, y)\ dx.$$

This may be extended to a finite number of arc.

7.6 Illustrative Examples

Example 7.6.1 *Evaluate the line integral of a pair of functions $6x^2y$ and $10xy^2$ along a plane curve $y = x^3$ from the point $A(1,1)$ to the point $B(2,8)$.*

Solution: The required integral $= \displaystyle\int_{A(1,1)}^{B(2,8)} (6x^2y\ dx + 10xy^2\ dy).$

Here the explicitly defined equation of the curve is $y = x^3$, so the parametric equations of the curves are $x = x$, $y = x^3$ and the parameter x varies from 1 to 2. Hence the required integral

$$= \int_1^2 (6x^2.x^3 + 10x.x^6.3x^2)dx = \int_1^2 (6x^5 + 30x^9)dx = \left[x^6 + 3x^{10}\right]_1^2 = 3132.$$

$\square$

Example 7.6.2 *Compute the line integral of three functions $x^2, 3x^2z, -xy^2$ along the line segment from the point $A(1,2,3)$ to the point $B(0,0,0)$.*

Solution: The equation of the line along which the integration is to be performed is obtained as $\dfrac{x}{1} = \dfrac{y}{2} = \dfrac{z}{3}$. Hence, its parametric equation may be given by $x = t, y = 2t, z = 3t$, t being the parameter.

Obviously, to the origin of the segment AB corresponds the value of the parameter $t = 1$ and to the terminus, the value $t = 0$.

Therefore, required integral

$$= \int_A^B (x^2 dx + 3x^2 z dy - xy^2 dz) = \int_{t=1}^0 (t^2.dt + 3t^2.3t.2dt - t.4t^2.3dt)$$

$$= \int_1^0 (t^2 + 18t^3 - 12t^3)dt = \int_1^0 (t^2 + 6t^3)dt = -\left[\frac{t^3}{3} + 6\frac{t^4}{4}\right]_0^1$$

$$= -\left(\frac{1}{3} + \frac{3}{2}\right) = -\frac{11}{6}. \qquad \square$$

Note 7.6.1 *The above three functions can be written in its equivalent vector function as $x^2 \hat{i} + 3x^2 z \hat{j} - xy^2 \hat{k}$. For this, consult our vector calculus book entitled "Elements of Vector Calculus".*

Example 7.6.3 *Evaluate the line integrals (i)* $\displaystyle\int_C (3x^2 - y^2)dx$ *and (ii)* $\displaystyle\int_C (3x^2 - y^2)dy$, *where C is the arc of the parabola $y^2 = 4ax$ form $(0,0)$ to $(a, 2a)$.*

Solution: Here C is the arc $\overset{\frown}{OP}$ (See **Figure** 7.2). The parametric equation of the parabola $y^2 = 4ax$ are $x = \phi(t) = at^2$, $y = \psi(t) = 2at$, then clearly t varies from 0 to 1.

Now $f(x, y) = 3x^2 - y^2$ and from the parametric equation we get $dx = 2at\, dt$, $dy = 2a\, dt$.

∴ (i) $\displaystyle\int_C (3x^2 - y^2)dx = \int_0^1 (3a^2 t^4 - 4a^2 t^2)2at\, dt$

$= 2a^3 \displaystyle\int_0^1 (3t^5 - 4t^3)dt = 2a^3 \left[3.\frac{t^6}{6} - 4.\frac{t^4}{4}\right]_0^1$

$= 2a^3 \left(\dfrac{1}{2} - 1\right) = -a^3$

and (ii) $\displaystyle\int_C (3x^2 - y^2)dy = \int_0^1 (3a^2 t^4 - 4a^2 t^2)2a\, dt$

$= 2a^3 \displaystyle\int_0^1 (3t^4 - 4t^2)dt = 2a^3 \left[3.\frac{t^5}{5} - 4.\frac{t^3}{3}\right]_0^1 = 2a^3 \left(\dfrac{3}{5} - \dfrac{4}{3}\right) = -\dfrac{22a^3}{15}. \quad \square$

P(a,2a) $y^2 = 4ax$, $(a,0)$, **Figure 7.2**

Example 7.6.4 *Evaluate the line integral* $\displaystyle\int_C [3x^2 dx + (2xz - y)dy + z dz]$ *along (a) the straight line from $(0,0,0)$ to $(2,1,3)$ and (b) the curve defined by $x^2 = 4y$, $3x^3 = 8z$ from $x = 0$ to $x = 2$.*

Solution: (a) The equations of the straight line form $(0,0,0)$ to $(2,1,3)$ are

$$\frac{x}{2} = \frac{y}{1} = \frac{z}{3} = t \text{ (say)}.$$

Therefore $x = 2t$, $y = t$, $z = 3t$ are its parametric equations. The points $(0,0,0)$ and $(2,1,3)$ corresponds to $t = 0$ and $t = 1$ respectively. Therefore the required integral is

$$\int_0^1 [3(2t)^2.2dt + \{(4t)(3t) - t\}dt + (3t)3dt] = \int_0^1 (36t^2 + 8t)dt = 16.$$

(b) Let $x = t$ in $x^2 = 4y$, $3x^3 = 8z$.

Then the parametric equations of C are $x = t, y = \dfrac{t^2}{4}, z = \dfrac{3t^3}{8}$ and t varies from 0 to 2. Therefore the required integral is

$$\int_0^2 \left[3t^2 dt + \left(2t.\frac{3t^3}{8} - \frac{t^2}{4} \right) d\left(\frac{t^2}{4} \right) + \frac{3t^3}{8} \, d\left(\frac{3t^3}{8} \right) \right]$$

$$= \int_0^2 \left(3t^2 - \frac{t^3}{8} + \frac{51t^5}{64} \right) dt = \left[t^3 - \frac{t^4}{32} + \frac{17}{128}t^6 \right]_0^2 = 16. \qquad \square$$

Note 7.6.2 *In vector analysis, above example may be given in the following alternative way:*

Example 7.6.5 *Find the work done in moving a particle in the force field $\vec{F} = 3x^2 \, \hat{i} + (2xz - y) \, \hat{j} + z \, \hat{k}$ along*

(a) *the straight line from $(0,0,0)$ to $(2,1,3)$ and*

(b) *the curve defined by $x^2 = 4y$, $3x^3 = 8z$ from $x = 0$ to $x = 2$.*

Solution: Here we observe that

$$\int_C \vec{F}.d\vec{r} = \int_C \{3x^2 \, \hat{i} + (2xz - y) \, \hat{j} + z \, \hat{k}\}.(dx \, \hat{i} + dy \, \hat{j} + dz \, \hat{k})$$

$$= \int_C [3x^2 dx + (2xz - y)dy + zdz].$$

Now the problem goes to foregoing **Example 7.5.4**. $\qquad \square$

Example 7.6.6 *Evaluate $\displaystyle\int_C [(3x^2 + 6y)dx - 14yzdy + 20xz^2 dz]$ from the point $(0,0,0)$ to the point $(1,1,1)$ along the path $C : x = t, y = t^2, z = t^3$.*

Solution: Here the curve $C : (t, t^2, t^3)$. So $t = 0$ and $t = 1$ correspond to the points $(0,0,0)$ and $(1,1,1)$ respectively. Hence the required integral is

$$\int_C [(3x^2 + 6y)dx - 14yzdy + 20xz^2 dz]$$

$$= \int_0^1 [(3t^2 + 6t^2)dt - 14t^2.t^3.2tdt + 20.t.(t^3)^2.3t^2 dt]$$

$$= \int_0^1 (9t^2 - 28t^6 + 60t^9)dt = \left[9\frac{t^3}{3} - 28\frac{t^7}{7} + 60\frac{t^{10}}{10} \right]_0^1 = 5. \qquad \square$$

Note 7.6.3 *For more examples of line integrals involving vector treatment one can consult our book '**Elements of Vector Calculus**'.*

7.7 Concept of Surfaces

In mathematics, a *surface* is a two-dimensional manifold. Some surfaces arise as boundaries of three-dimensional solids, for example, the sphere is the boundary of a solid ball. Other surfaces arise as graphs of functions of two variables.

The simplest mathematical surfaces are planes and spheres in the Euclidean 3-space. A surface is a two dimensional space which means that a moving point on a surface may move in two directions.

7.8 Equation of a Surface

Any surface which is given by an equation $F(x, y, z) = 0$ and which does not contain any critical point must have two sides, the function F is positive on one side and negative on the other side.

7.9 Surface Integrals

The theory of *surface integrals* is in many aspects analogous to the theory of line integrals as stated above. To study integrals of functions defined on *surfaces* are called *Surface integrals*. In many physical problems we encounter functions defined on various surfaces. For example, velocity of a particle of a fluid passing through a surface density of a charge distribution over the surface of a conductor may be cited.

Surface Integral is a generalization of multiple integrals to integration over a surface. Given a surface, one may integrate a *scalar field* or a *vector field*. A surface integral of a point function is definite double integral taken over a surface within a scalar or vector field determined by the point function.

If the vector field $\vec{F}$ represents the flow of a fluid then the surface integral of $\vec{F}$ will represent the amount of fluid flowing through the surface. The amount of fluid flowing through the surface (per unit time) is also called the *flux* of the fluid through the surface.

Let $\hat{n}$ be a outward drawn unit normal to a regular surface S at the point $P(x, y, z)$ on S. Then the surface integral of a vector function $\vec{F}$ over the surface S is defined by the double integral $\iint_S (\vec{F}.\hat{n}) \, dS$ which is same as $\iint_S \vec{F}.d\vec{S}$. It represents the *flux* of the function through S, dS denoting the magnitude of the area of the sub-surface $\triangle S$.

Note 7.9.1 $\vec{F}.\hat{n} =$ *the normal component of* $\vec{F}$.

Note 7.9.2 *The actual calculation of a surface integral is frequently accomplished by reducing it to a double integral. Let $z = f(x, y)$ represents the surface S, where (x, y) lies on a region R in the xy-plane. We suppose that $z = f(x, y)$ has continuous derivatives in R. Then we can write*

$$\iint_S \phi(x, y, z) \, ds = \iint \phi(x, y, z) \sqrt{1 + \left(\frac{\partial z}{\partial x}\right)^2 + \left(\frac{\partial z}{\partial y}\right)^2} \, dxdy$$

This result holds when S is projected on to a region R of the xy-plane. Sometimes, it may, however be convenient to project S on the yz-plane or zx-plane and modifications are to be made accordingly.

7.10 Relationships among Line, Surface and Volume Integrals with Simple Integrals and other Multiple Integrals

From the results obtained so far for different integral types (including the integral types of **Chapter 6**), the following similarities or dissimilarities are seen among these

(i) A line integral is the generalization of simple integral.

(ii) A surface integral is the generalization of double integral.

(iii) A volume integral is the generalization of triple integral.

(iv) A multiple integral is any type of integral.

Let us go a little deeper. For simplicity, we will restrict our discussion to only Cartesian coordinates; but, the same holds for other coordinates systems as well.

7.10.1 Simple integral vs Line integral

A *simple integral* is evaluated along a particular axis of the coordinates, i.e., along x-axis or along y-axis. Thus $\int_a^b f(x)\,dx$ denotes an integration of $f(x)$ along x-axis. The points which go into evaluation of this integral is the *region of integration* comes only from the x-axis. But a *line integrals* generalizes the idea of a simple integral. In a line integral, the curve along which the integral is evaluated, is not necessarily a coordinate axis or not even a straight line. It can be any curve lying in higher dimensional space, though the curve itself is a two-dimensional identity by definition.

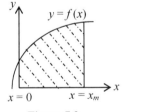

Figure 7.3

Figure 7.4

Thus $\int_0^{x_m} f(x)\,dx$ is along the x-axis only, which necessitates that the y-coordinate is zero along the path of integration (**Figure** 7.3). On the other hand, for a line integration $\int_C \vec{F}.d\vec{s}$, the path of integration is along the curve C, which is not the x-axis (or y-axis) in general. **Figure** 7.4 shows a closed curve, but the concept of the line integrals is valid for open curves as well.

The idea, that a line integral is a generalization of simple integral, is critical to our evaluation of a line integral.

We evaluate the line integral by expressing the curve C in parametric forms, in terms of the parameter $'t'$, say.

Geometrically, this is equivalent to stretching the curve in straight line on the t-axis. This reduces the line integral into a simple integral in terms of the variate t as below

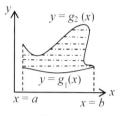

Figure 7.5

$$\int_C \vec{F}\, dS = \int_a^b \vec{F}\,(c(t))\, c'(t)\, dt,$$

The line integral of a vector valued function $\vec{F}$ has been expressed as simple integral in terms of t, $c(t)$ is a parametric representation of the curve C, $c'(t)$ denoting its derivative. A similar expression for a scalar function f is shown below:

$$\int_C f\, ds = \int_a^b f(c(t))|c'(t)|dt.$$

7.10.2   Surface integral vs Double integral

Just as a *line integral* extends the idea of *simple integral to general curves*, a *surface integral* extends the idea of *double integral to general surface*. In a double integral, the points which go into the evaluation of the integration come from a two-dimensional planar surface $\iint_D f(x,y)\, dA$. The domain of integration, region D in the expression above is a two-dimensional planar region over which the integration is calculated. The said region is shown by the shaded portion of **Figure** 7.5, which lies on the xy-plane and is bounded by $a < x < b$; $y = g_1(x)$ and $y = g_2(x)$.

However a plane is only one kind of surface, there are other surfaces as well, like a paraboloid, ellipsoid, sphere etc. What should we do if we have to calculate integrations over these non-planar surfaces? For example the *flux* of an electric field through a sphere.

This is what leads to the concept of a surface integral $\iint \vec{F}.d\vec{S}$, the surface S can be any surface in space [See **Figure** 7.6]. Even in multi-dimensional space, a surface is a two-dimensional entity and can be expressed completely and uniquely by two parameters u and v (say). This helps us to convert a surface integral into a double integral. Geometrically, it is akin to stretching and flattening out a surface into a plane having coordinate axes u and v.

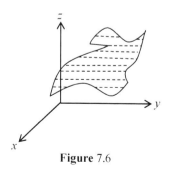

Figure 7.6

7.11   Calculating Area using Double Integrals

We know that one use of the single variable integral is to calculate the area under a
curve $f(x)$ over some interval $[a, b]$ by integrating it over the interval. The definite integral $\int_a^b f(x)\ dx$, therefore, represents the area
bounded the curve $y = f(x)$, the x-axis and two
fixed ordinates $x = a$ and $x = b$.

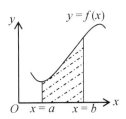

Figure 7.7

Sometimes we use the double integrals to
calculate area as well. But the *approach is quite
different*. It's fairly simple to see the trick ac-
complish once we can imagine how to use a single integral to *calculate the length of the interval.*

What happens if we integrate the function $f(x) = 1$ over the interval $[a, b]$?
We can calculate that

$$\int_a^b f(x)\ dx = \int_a^b 1\ dx = \int_a^b dx = [x]_a^b = b - a \tag{7.1}$$

Thus the integral of the function $f(x) = 1$ over the interval $[a, b]$ is just the length
of the interval $[a, b]$. It also happens to be the *area of the rectangle* of unit breadth
(or height) and length $(b - a)$, but we can interpret it as the *length of the interval*
(See **Figure** 7.7).

We can do the same trick for double integrals. The integral of a function $f(x, y)$ over the
region D can be interpreted as the *volume under
the surface* $z = f(x, y)$ over the region D (discussed in § 7.11). As we did above, we can try
to trick of integrating the function $f(x, y) = 1$
over the region. But the integral of $f(x, y) = 1$
is also the *area of the region* D. This can be a
nifty way of calculating the area of the region
D, we can write this as $A = \iint_D dA$. Thus in
the definition of a double integral, if we put
$f(x, y) = 1$, we have the area A of the region
bounded by the curves $y = f(x)$, $y = \phi(x)$, the straight lines $x = a$, $x = b$

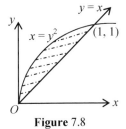

Figure 7.8

i.e., $\quad A = \int_a^b \int_{f(x)}^{\phi(x)} dy dx \tag{7.2}$

To illustrate the above, we take the following example:

Example 7.11.1 *Find the area of region bounded by the parabola* $x = y^2$ *and the
line* $y = x$.

Solution: The region is pictured (shaded) in **Figure** 7.8. We will let y go from 0
to 1 and then x goes from y^2 to y. The required area is

$$\int_0^1 \int_{y^2}^y dx dy = \int_0^1 [x]_{y^2}^y dy = \int_0^1 (y - y^2) dy = \left[\frac{y^2}{2} - \frac{y^3}{3} \right]_0^1 = \frac{1}{2} - \frac{1}{3} = \frac{1}{6}. \quad \square$$

Note 7.11.1 *(Volume using triple integral)*

*Triple integral can be used to find the **volume** just like the **double integral** is used to find the **area**. Thus in the triple integral*

$$\iiint_D f(x,y,z)\ dxdydz,$$

if we take $f(x,y,z) = 1$, the triple integral over the region D expresses the volume of D, i.e.,

$$V = \iiint_D dxdydz \tag{7.3}$$

7.12 Calculating Volume using Double Integral

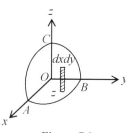

Figure 7.9

Let us consider an elementary area $dxdy$ on the plane $z = 0$ through each point on the boundary on this small area [**Figure** 7.9]. Lines are drawn parallel to z-axis and a small cylinder is constructed whose base is the area to z-axis. This cylinder cuts the given surface whose volume $= z.dxdy$. Therefore volume of the solid

$$= \iint z.dxdy.$$

Remarks:

(i) By considering area $dydz$ on the plane $x = 0$, the volume of the solid

$$= \iint x.dydz.$$

(ii) By considering area $dxdz$ on the plane $y = 0$, the volume of the solid

$$= \iint y.dxdz.$$

(iii) Let z-axis be taken perpendicular to the plane of a finite region R. If lines parallel to z-axis be drawn from points on the boundary of R up to the surface $z = f(x,y)$, we get a cylinder. Geometrically $\iint_R f(x,y)\ dxdy$ represents the volume of the given cylinder.

Note 7.12.1 *If we make $f(x,y) = 1$ over R, then $= \iint_R dxdy$ represents the area of the plane region bounded by R.*

Note 7.12.2 *In polar coordinates, the area bounded by curve $r = f(\theta)$, where $f(\theta)$ is a single valued function of θ in the domain (α, β) with the radii vectors $\theta = \alpha$ and $\theta = \beta$ is $\displaystyle\int_{\theta=\alpha}^{\beta} \int_{r=\theta}^{f(\theta)} r\ d\theta dr.$*

7.13   Illustrative Examples

Example 7.13.1 *Find the area of the surface $z^2 = 2xy$ included between planes $x = 0, x = a, y = 0, y = b$.*

Solution: The given surface is $z^2 = 2xy$.

$$\therefore \quad 2z\frac{\partial z}{\partial x} = 2y \quad \therefore \quad \frac{\partial z}{\partial x} = \frac{y}{z} \text{ and similarly } \frac{\partial z}{\partial y} = \frac{x}{z}.$$

$$\therefore \quad \text{Required area of the surface} = \iint \sqrt{1 + \left(\frac{\partial z}{\partial x}\right)^2 + \left(\frac{\partial z}{\partial y}\right)^2}\, dxdy$$

$$\text{(Taking } \phi(x, y, z) = 1 \text{ in } \textbf{Note 7.9.2})$$

$$= \int_{x=0}^{a}\int_{y=0}^{b}\sqrt{1 + \left(\frac{y}{z}\right)^2 + \left(\frac{x}{z}\right)^2}\, dxdy = \int_{x=0}^{a}\int_{y=0}^{b}\sqrt{\frac{x^2 + y^2 + z^2}{z^2}}\, dxdy$$

$$= \int_{x=0}^{a}\int_{y=0}^{b}\sqrt{\frac{x^2 + y^2 + 2xy}{2xy}}\, dxdy = \int_{x=0}^{a}\int_{y=0}^{b}\frac{x+y}{\sqrt{2xy}}\, dxdy$$

$$= \frac{1}{\sqrt{2}}\int_{x=0}^{a}\int_{y=0}^{b}\left(\sqrt{x}\frac{1}{\sqrt{y}} + \sqrt{y}\frac{1}{\sqrt{x}}\right) dxdy$$

$$= \frac{1}{\sqrt{2}}\int_{0}^{a}\sqrt{x}\,[2\sqrt{y}]_0^b\, dx + \frac{1}{\sqrt{2}}\int_{0}^{a}\frac{1}{\sqrt{x}}\left[\frac{2}{3}y^{\frac{3}{2}}\right]_0^b dx$$

$$= \sqrt{2b}\int_{0}^{a}\sqrt{x}dx + \frac{\sqrt{2}}{3}b^{\frac{3}{2}}\int_{0}^{b}\frac{1}{\sqrt{x}}dx = \sqrt{2b}\left[\frac{2}{3}x^{\frac{3}{2}}\right]_0^a + \frac{1}{3}\sqrt{2b^3}[2\sqrt{x}]_0^a$$

$$= \frac{2}{3}\sqrt{2b}a^{\frac{3}{2}} + \frac{2}{3}\sqrt{2b}b\sqrt{a} = \frac{2}{3}\sqrt{2ab}(a + b). \qquad \square$$

Example 7.13.2 *Find the area of the surface of the paraboloid $x^2 + y^2 = az$ which lies between the planes $z = 0$ and $z = a$.*

Solution: The projection of gives surface between the planes $z = 0$ and $z = a$ on the xy-plane is $x^2 + y^2 = a^2$, $z = 0$. Here the surface is the paraboloid $x^2 + y^2 = az$ (*vide.* **Figure** 7.10).

$$\therefore \quad \frac{\partial z}{\partial x} = \frac{2x}{a}, \frac{\partial z}{\partial y} = \frac{2y}{a}.$$

$$\therefore \quad S = \iint_R \sqrt{1 + \left(\frac{\partial z}{\partial x}\right)^2 + \left(\frac{\partial z}{\partial y}\right)^2}\, dxdy,$$

where R is the circle $x^2 + y^2 = a^2$.

$$= \iint_R \sqrt{1 + \frac{4x^2}{a^2} + \frac{4y^2}{a^2}}\, dxdy$$

$$= \frac{1}{a}\iint_R \sqrt{a^2 + 4x^2 + 4y^2}\, dxdy.$$

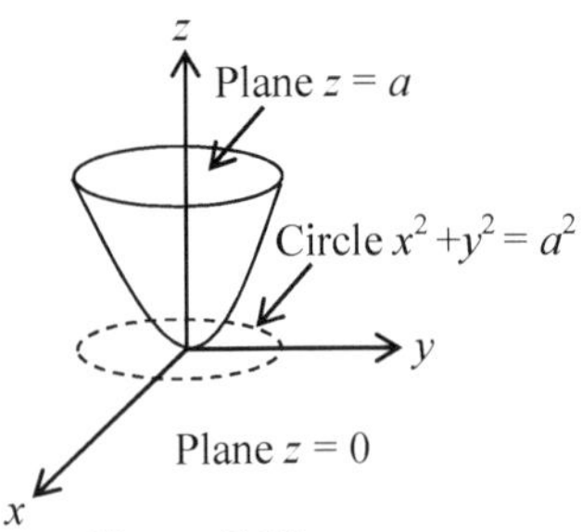

Figure 7.10

Now we transform it into polar coordinates by putting $x = r\cos\theta$, $y = r\sin\theta$ with Jacobian $J = r$, we get

$$S = \frac{1}{a}\int_{\theta=0}^{2\pi}\int_{r=0}^{a}\sqrt{a^2 + 4r^2}\, r.d\theta dr = \frac{1}{a}\int_{0}^{2\pi}d\theta\int_{0}^{a}\sqrt{a^2 + 4r^2}\, rdr.$$

Put $a^2 + 4r^2 = p^2$ $\therefore$ $8rdr = 2pdp$ i.e., $rdr = \frac{1}{4}p\,dp$. When $r = 0$, $p = a$ and $r = a \Rightarrow p = a\sqrt{5}$.

$$\therefore S = \frac{1}{a}\int_0^{2\pi} d\theta \int_a^{a\sqrt{5}} \frac{1}{4}p^2 dp = \frac{1}{4a}\int_0^{2\pi} \left[\frac{p^3}{3}\right]_a^{a\sqrt{5}} d\theta = \frac{1}{4a}\int_0^{2\pi} \frac{a^3}{3}(5\sqrt{5}-1)d\theta$$

$$= \frac{5\sqrt{5}-1}{12}a^2[\theta]_0^{2\pi} = \frac{5\sqrt{5}-1}{12}a^2 \cdot 2\pi = \frac{\pi}{6}(5\sqrt{5}-1)a^2. \qquad \square$$

Example 7.13.3 *Find the volume of a solid bounded by the surfaces* $x = 0, y = 0, z = 0, x + y + z = 1$.

Solution: Clearly $V = \iint_R z\,dxdy$ [using § 7.11] $= \iint_R (1 - x - y)\,dxdy$ where R is the triangular region in the xy-plane bounded by the straight lines $x = 0, y = 0, x + y = 1$ (*vide.* **Figure** 7.11). Then

$$V = \int_0^1 dx \int_0^{1-x} (1 - x - y)dy = \int_0^1 \frac{1}{2}(1-x)^2 dx = \frac{1}{2}\int_0^1 (1 - 2x + x^2)dx$$

$$= \frac{1}{2}\left[x - x^2 + \frac{x^3}{3}\right]_0^1 = \frac{1}{2}\left(1 - 1 + \frac{1}{3}\right) = \frac{1}{6}. \qquad \square$$

Example 7.13.4 *Find the volume bounded by the coordinate planes and the plane* $\frac{x}{a} + \frac{y}{b} + \frac{z}{c} = 1$.

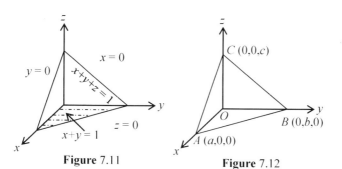

Figure 7.11   **Figure 7.12**

Solution: The planes cut the coordinate axes at the points $A(a,0,0)$, $B(0,b,0)$ and $C(0,0,c)$ (*vide.* **Figure** 7.12). Therefore, required volume is

$$\int_0^a \int_0^{b\left(1-\frac{x}{a}\right)} \int_0^{c\left(1-\frac{x}{a}-\frac{y}{b}\right)} dxdydz, \text{ [Using Note 7.10.1]}$$

$$= \int_0^a \int_0^{b\left(1-\frac{x}{a}\right)} c\left(1 - \frac{x}{a} - \frac{y}{b}\right) dydx$$

$$= c\int_0^a dx \int_0^{b\left(1-\frac{x}{a}\right)} \left(1 - \frac{x}{a} - \frac{y}{b}\right) dy = c\int_0^a dx \left[y - \frac{x}{a}y - \frac{y^2}{2b}\right]_0^{b\left(1-\frac{x}{a}\right)}$$

$$= c\int_0^a \left[b\left(1 - \frac{x}{a}\right) - \frac{x}{a}b\left(1 - \frac{x}{a}\right) - \frac{1}{2b}b^2\left(1 - \frac{x}{a}\right)^2\right] dx$$

$$= cb \int_0^a \left[1 - \frac{x}{a} - \frac{x}{a} + \frac{x^2}{a^2} - \frac{1}{2}\left(1 - 2\frac{x}{a} + \frac{x^2}{a^2} \right) \right] dx$$

$$= bc \int_0^a \left(\frac{1}{2} - \frac{x}{a} + \frac{1}{2}\frac{x^2}{a^2} \right) dx = bc \left[\frac{1}{2}x - \frac{1}{a}\frac{x^2}{2} + \frac{1}{2a^2}\frac{x^3}{3} \right]_0^a$$

$$= abc \left[\frac{1}{2} - \frac{1}{2} + \frac{1}{6} \right] = \frac{abc}{6}.$$

Alternatively, we can use § 7.12 to solve above **Example 7.13.4**, i.e., we can follow the procedure of **Example 7.13.3** as given below:

Arguing as in **Example 7.13.3** above, in this case, we get

$$V = \iint_R z \, dxdy = \iint_R c\left(1 - \frac{x}{a} - \frac{y}{b} \right) dxdy$$

$$= c \int_0^a dx \int_0^{b\left(1-\frac{x}{a}\right)} \left(1 - \frac{x}{a} - \frac{y}{b} \right) dy = c \int_0^a \left[y - \frac{x}{a}y - \frac{1}{b}\cdot\frac{y^2}{2} \right]_0^{b\left(1-\frac{x}{a}\right)} dx$$

$$= c \int_0^a \left[b\left(1 - \frac{x}{a} \right) - \frac{x}{a}.b\left(1 - \frac{x}{a} \right) - \frac{1}{2b}.b^2\left(1 - \frac{x}{a} \right)^2 \right] dx$$

$$= cb \int_0^a \left(1 - \frac{x}{a} \right) \left\{ \left(1 - \frac{x}{a} \right) - \frac{1}{2}\left(1 - \frac{x}{a} \right) \right\} dx$$

$$= bc \int_0^a \left(1 - \frac{x}{a} \right) \left(1 - \frac{x}{a} \right) \left(1 - \frac{1}{2} \right) dx$$

$$= \frac{bc}{2} \int_0^a \left(1 - \frac{2x}{a} + \frac{x^2}{a^2} \right) dx = \frac{bc}{2} \left(x - \frac{x^2}{a} + \frac{x^3}{3a^2} \right)_0^a$$

$$= \frac{bc}{2}\left(a - a + \frac{a}{3} \right) = \frac{abc}{6}. \qquad \square$$

7.14   Green's Formula or Green's Theorem in a Plane

Statement of Green's Theorem

Theorem 7.14.1 *In the xy-plane, let there be given a domain D, which is regular both in the direction of the x-axis and the y-axis, bounded by a closed curve Γ (also called a closed contour Γ). If X and Y are continuous functions of x and y having continuous derivatives in D, then*

$$\int_\Gamma (X\,dx + Y\,dy) = \iint_D \left(\frac{\partial Y}{\partial x} - \frac{\partial X}{\partial y} \right) dxdy.$$

Proof: Let the domain D be bounded below by curve $y = y_1(x)$ and above by the curve $y = y_2(x)$, $y_1(x) \le y_2(x)$, $(a \le x \le b)$ [See **Figure** 7.13]. Together both these curves represent the closed contour Γ.

Within the region D, continuous functions $X(x,y)$ and $Y(x,y)$ have continuous partial derivatives. We consider the integral $\iint_D \frac{\partial X}{\partial y} dxdy$. Representing it in the form of a two fold iterated integral we find

$$\iint_D \frac{\partial X}{\partial y} \, dxdy = \int_a^b \left[\int_{y(x)}^{y_2(x)} \frac{\partial X}{\partial y} \, dy \right] dx = \int_a^b [X(x,y)]_{y(x)}^{y_2(x)} \, dx$$

$$= \int_a^b [X(x, y_2(x)) - X(x, y_1(x))]\, dx$$

$$= \int_a^b X(x, y_2(x))\, dx - \int_a^b X(x, y_1(x))\, dx$$

$$= \int_{MPN} X(x, y)\, dx - \int_{MQN} X(x, y)\, dx$$

where, for the curve MPN, its parametric equations are taken as $x = x$, $y = y_2(x)$, x being a parameter and similarly, the parametric equations of the curve MQN are $x = x$, $y = y_1(x)$. So we get

$$\iint_D \frac{\partial X}{\partial y}\, dxdy = \int_{MPN} X(x, y)\, dx - \int_{MQN} X(x, y)\, dx$$

$$= \int_{MPN} X(x, y)\, dx + \int_{NQM} X(x, y)\, dx \qquad (7.4)$$

$$\left[\because \int_{MQN} X(x, y)\, dx = - \int_{NQM} X(x, y)\, dx, \text{ see Remark } \S\, 7.3. \right]$$

Now the sum of the line integrals on the R.H.S. of (7.4) is equal to the line integral taken along the entire closed curve Γ in the clockwise direction. Hence it can be reduced to the form

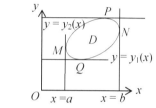

Figure 7.13

$$\iint_D \frac{\partial X}{\partial y}\, dxdy = \int_{MPNQM} X(x, y) dx$$

$$= \int_{\Gamma(\text{in clockwise sense})} X(x, y) dx \quad (7.5)$$

Analogously, we find

$$\iint_D \frac{\partial Y}{\partial x}\, dxdy = \int_{\Gamma \text{ (in clockwise sense)}} Y(x, y) dy \qquad (7.6)$$

Now subtracting (7.6) from (7.5) we get

$$\iint_D \left(\frac{\partial X}{\partial y} - \frac{\partial Y}{\partial x} \right) dxdy = \int_{\Gamma \text{ (in clockwise sense)}} (X dx + Y dy) \qquad (7.7)$$

If the sense of line integral in the R.H.S. of (7.7) is reversed, i.e., if the contour is traversed in the counter-clockwise sense[1] then we get

$$\int_\Gamma X dx + Y dy = \iint_D \left(\frac{\partial Y}{\partial x} - \frac{\partial X}{\partial y} \right) dxdy \qquad (7.8)$$

This equation (7.8) is known as Green's formula[2].   □

---

[1] If in a line integration along a closed contour, the direction of the circulation is not specified, it is assumed be counter-clockwise. That's why equation (7.8) is written so. If the direction of circulation is clockwise, it must be specied.

[2] The above formula (7.8) is known as Green's formula, named after the English physicist and mathematician George Green (1793 - 1841) who taught at Cambridge and known for his work on potential theory in connection with waves, vibrations, elasticity, electricity and magnetism. This formula is a special case of a more general formula discovered by the Russian mathematician M.V. Ostrogradsky.

Note 7.14.1 *If the part of the boundary line segment be parallel to the y-axis,*
then $\int X(x,y)\ dx = 0$ *and equation (7.5) holds true in this case as well.*

Note 7.14.2 *Green's formula establishes a connection between* **a line integral**
along a closed counter with a double integral over the domain bounded by
that contour.

Note 7.14.3 *In Green's theorem, we have assumed that the domain D is regular.*
But it may be shown that the same also holds true for any domain that may be
divided into regular domains.

Note 7.14.4 *In vector notation, Green's formula is written as*

$$\int_C \vec{F}.d\vec{r} = \iint_R (\vec{\nabla} \times \vec{F}).\hat{k}\ dR = \iint_R curl\vec{F}.\hat{k}\ dR$$

where $\vec{F} = \hat{i}X + \hat{j}Y + \hat{k}Z$, $\vec{r} = \hat{i}x + \hat{j}y + \hat{k}z$ *and C is traversed in counter-clockwise*
sense.

[For details consult Authors' Book, "Element of Vector Calculus"]

Example 7.14.1 *Verify Green's Theorem for*

$$\oint_C \{(x^2 - xy)\ dx + (y - x^2)\ dy\}$$

where C is the closed curve of the region bounded by $y = x^3$ *and* $y = x$.

Solution: Points of intersection of $y = x^3$ and $y = x$ are $(0,0)$ and $(1,1)$. Let R
be the region whose boundary is the closed curve $C : arc\ OABDO$ (*vide*. **Figure**
7.14).

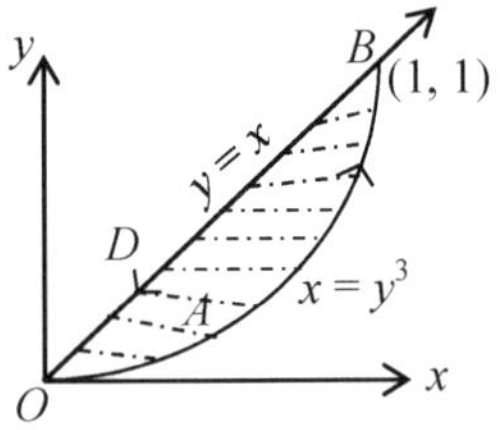

$$\therefore \oint_C \{(x^2 - xy)\ dx + (y - x^2)\ dy\}$$

$$= \int_{OAB} \{(x^2 - xy)\ dx + (y - x^2)\ dy\}$$

$$+ \int_{BDO} \{(x^2 - xy)\ dx + (y - x^2)\ dy\}$$

Figure 7.14

$$= \int_0^1 \{(x^2 - x^4) + (x^3 - x^2).3x^2\}dx +$$

$$\int_0^1 \{(x^2 - x^2) + (x - x^2)\}dx$$

$$= \int_0^1 (3x^5 - 4x^4 + x^2)dx + \int_1^0 (x - x^2)dx$$

$$= \left[3\frac{x^6}{6} - 4\frac{x^5}{5} + \frac{x^3}{3}\right]_0^1 + \left[\frac{x^2}{2} - \frac{x^3}{3}\right]_1^0$$

$$= \frac{1}{2} - \frac{4}{5} + \frac{1}{3} - \frac{1}{2} + \frac{1}{3} = \frac{2}{3} - \frac{4}{5} = -\frac{2}{15}.$$

Again $\iint_R \left\{\frac{\partial}{\partial x}(y - x^2) - \frac{\partial}{\partial y}(x^2 - xy)\right\} dxdy = \iint_R (-2x + x)\ dxdy$

$$= -\int_0^1 x\,dx \int_{y=x^3}^{y=x} dy = -\int_0^1 x(x - x^3)dx = -\int_0^1 (x^2 - x^4)dx$$

$$= \int_0^1 (x^4 - x^2)dx = \left[\frac{x^5}{5} - \frac{x^3}{3}\right]_0^1 = \frac{1}{5} - \frac{1}{3} = -\frac{2}{15}.$$

Hence Green's Theorem is verified. □

Example 7.14.2 *Use Green's theorem to evaluate*

$$\oint_\Gamma \{(y - \sin x)\ dx + \cos x\ dy\},$$

where Γ *is the triangle enclosed by the lines* $y = 0$, $x = \pi$, $y = \dfrac{2}{\pi}x$.

Solution: Here the region D in the xy-plane is bounded by the lines $y = 0$, $x = \pi$, $y = \dfrac{2}{\pi}x$.

Green's theorem in the xy-plane states that

$$\int_\Gamma (X\,dx + Y\,dy) = \iint_D \left(\frac{\partial Y}{\partial x} - \frac{\partial X}{\partial y}\right) dx\,dy.$$

So using this result, here we get

$$\int_\Gamma \{(y - \sin x)\ dx + \cos x\ dy\} = \iint_D \left\{\frac{\partial}{\partial x}(\cos x) - \frac{\partial}{\partial y}(y - \sin x)\right\} dx\,dy.$$

$$= \iint_D (-\sin x - 1)\ dx\,dy = -\int_0^\pi (1 + \sin x)\ dx \int_0^{\frac{2}{\pi}x} dy$$

$$= -\frac{2}{\pi}\int_0^\pi x(1 + \sin x)\ dx$$

$$= -\frac{2}{\pi}\int_0^\pi x\ dx - \frac{2}{\pi}\int_0^\pi x\sin x\ dx$$

$$= -\frac{\pi^2}{\pi} - \frac{2}{\pi}[-x\cos x]_0^\pi + \frac{2}{\pi}\int_0^\pi -\cos x\ dx$$

$$= -\pi - \frac{2}{\pi}\pi = -(\pi + 2).$$

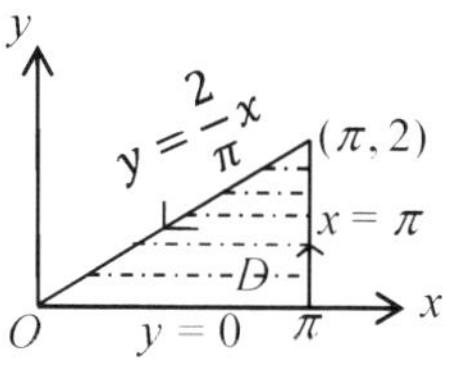

Figure 7.15 □

Note 7.14.5 *We can verify the above by direct integration as below.*

$$\int_\Gamma \{(y - \sin x)\ dx + \cos x\ dy\}$$

$$= -\int_0^\pi \sin x\ dx - \int_0^2 dy + \int_\pi^0 \left(\frac{2}{\pi}x - \sin x + \frac{2}{\pi}\cos x\right) dx$$

$$= -4 - \left[\frac{x^2}{\pi} + \cos x + \frac{2}{\pi}\sin x\right]_0^\pi = -4 - (\pi - 1 - 1) = -(\pi + 2).$$

7.15 Gauss's Theorem or Gauss's Divergence Theorem (Second Generalization of Green's Theorem)

Green's theorem in a plane establishes a *connection between a line integral and a double integral.* Now it is possible to generalize the theorem in which we can get a *connection between a double integral and a triple integral* over a three dimensional region. This is called **Gauss's theorem**. We state below the same without proof.

Statement of Gauss's Theorem

Theorem 7.15.1 *Let S be any closed surface and D be a three dimensional region such that S encloses D and let f, g, h be three functions with continuous partial derivatives $\frac{\partial f}{\partial x}, \frac{\partial g}{\partial y}, \frac{\partial h}{\partial z}$ at each point of D and S. Then*

$$\iiint_D \left(\frac{\partial f}{\partial x} + \frac{\partial g}{\partial y} + \frac{\partial h}{\partial z} \right) dx dy dz = \iint_S (f \, dy dz + g \, dz dx + h \, dx dy)$$

where the surface integral is taken over the exterior of S.

7.15.1 Application of Gauss's Theorem

We know that a surface integral can generally be evaluated by *reducing it to the corresponding double integral.* Using Gauss's theorem, we can replace a *double integral by a triple integral* over a three dimensional region in terms of a surface integral taken over the bounding surface of the region. Using Gauss's theorem, the *complicated evaluation of surface integrals* over a closed surface, sometimes, can thus be avoided by evaluating *comparatively easier volume integrals.*

7.15.2 Alternative Statement of Gauss's Theorem in Vector form

Theorem 7.15.2 *The normal surface integral of a vector function $\vec{F}$ over the boundary of a closed region is equal to the volume integral of the divergence of $\vec{F}$ taken throughout the enclosed surface.*

In vector notation, if $\vec{F}(x, y, z)$ and $\vec{\nabla}.\vec{F}$ are continuous over the closed regular surface S and its interior V, then the divergence theorem asserts that

$$\iint_S \vec{F}.\hat{n} \, dS = \iiint_V div \, \vec{F} \, dV$$

where $\hat{n}$ is the unit normal to the surface in the outward direction.

Note 7.15.1 *For proof consult authers' book "**Elements of Vector Calculus**".*

Example 7.15.1 *Verify Gauss's divergence theorem for the surface integral*

$$\iint_S \left\{ (2x - z) \, dy dz + x^2 y \, dz dx - z^2 x \, dx dy \right\}$$

taken over the region of the surface S bounded by a unit cube.

Solution: Gauss's divergence theorem is

$$\iiint_D \left(\frac{\partial f}{\partial x} + \frac{\partial g}{\partial y} + \frac{\partial h}{\partial z} \right) dxdydz = \iint_S (f\ dydz + g\ dzdx + h\ dxdy).$$

In this case the L.H.S. volume integral is

$$\iiint_D \left\{ \frac{\partial}{\partial x}(2x - z) + \frac{\partial}{\partial y}(x^2 y) - \frac{\partial}{\partial z}(z^2 x) \right\} dxdydz$$

$$= \iiint_D (2 + x^2 - 2zx)\ dxdydz = \int_0^1 dx \int_0^1 dy \int_0^1 (2 + x^2 - 2zx)dz$$

$$= \int_0^1 dx \int_0^1 dy\ [2z + x^2 z - z^2 x]_0^1 = \int_0^1 dx \int_0^1 (2 + x^2 - x)\ dy$$

$$= \int_0^1 [2y + x^2 y - xy]_0^1\ dx = \int_0^1 (2 + x^2 - x)\ dx$$

$$= \left[2x + \frac{x^3}{3} - \frac{x^2}{2} \right]_0^1 = 2 + \frac{1}{3} - \frac{1}{2} = \frac{11}{6}$$

and the R.H.S. surface integral is

$$\iint_S \{(2x - z)\ dydz + x^2 y\ dzdx - z^2 x\ dxdy\}.$$

We are to calculate this surface integral over the six faces of the cube as shown in **Figure** 7.16, the coordinates of whose vertices are given by $O(0,0,0)$, $O'(1,1,1)$, $A(1,0,0)$, $A'(0,1,1)$, $B(0,1,0)$, $B'(1,0,1)$, $C(0,0,1)$, $C'(1,1,0)$. Now

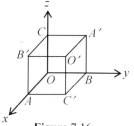

Figure 7.16

i) For the face $OAC'B$, $z = 0$ $\therefore$ Integration
$$I_1 = -\int_0^1 \int_0^1 0.dxdy = 0$$

ii) For the face $CB'O'A'$, $z = 1$ $\therefore$ $I_2 =$
$$\int_0^1 \int_0^1 -x.dxdy = -\frac{1}{2}$$

iii) For the face $OAB'C$, $y = 0$ $\therefore$ $I_3 = -\int_0^1 \int_0^1 0.dzdx = 0$

iv) For the face $O'A'BC'$, $y = 1$ $\therefore$ $I_4 - \int_0^1 \int_0^1 x^2.dxdz = \frac{1}{3}$

v) For the face $OBA'C$, $x = 0$ $\therefore$ $I_5 = -\int_0^1 \int_0^1 -z.dydz = \frac{1}{2}$

vi) For the face $O'C'AB'$, $x = 1$ $\therefore$ $I_6 = \int_0^1 \int_0^1 (2 - z).dydz = \frac{3}{2}.$

$\therefore$ R.H.S. $= -\frac{1}{2} + \frac{1}{3} + \frac{1}{2} + \frac{3}{2} = \frac{11}{6} = $ L.H.S.

Hence Gauss's theorem is verified. □

Note 7.15.2 *It should be carefully noticed that whenever we are performing the surface integration over the faces $x = 0, y = 0$ and $z = 0$ respectively, in each* *case a negative sign is taken because of the fact that outward drawn normals are negative here and opposite to that of the faces $x = 1, y = 1$ and $z = 1$ respectively.*

Example 7.15.2 *Use Gauss's theorem to evaluate*

$$\iint_S \{x \; dydz + y \; dzdx + z^2 \; dxdy\}$$

where S denotes the closed surface bounded by the cone $x^2 + y^2 = 1$ and the plane $z = 1$.

Solution: Gauss's divergence theorem is

$$\iint_S (f \; dydz + g \; dzdx + h \; dxdy) = \iiint_V \left(\frac{\partial f}{\partial x} + \frac{\partial g}{\partial y} + \frac{\partial h}{\partial z} \right) dxdydz.$$

Here $f = x, \; g = y, \; h = z^2 \; \therefore \frac{\partial f}{\partial x} = \frac{\partial g}{\partial y} = 1$ and $\frac{\partial h}{\partial z} - 2z.$

So we get the given integral $= \iiint_V (2 + 2z) \; dxdydz,$

where V is bounded by the domain $x^2 + y^2 = 1$ and $z = 1$, i.e., $z : \sqrt{x^2 + y^2} = 1.$

$\therefore$ Given integral

$$= \iint_{D:x^2+y^2=1} dxdy \int_{\sqrt{x^2+y^2}}^1 (2 + 2z)dz$$

$$= \iint_{D:x^2+y^2=1} \left[3 - 2\sqrt{x^2 + y^2} - (x^2 + y^2) \right] dxdy$$

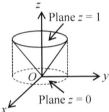

Figure 7.17

Now we put $x = r\cos\theta, \; y = r\sin\theta, \; dxdy = |J|drd\theta = rdrd\theta.$

Required integral $= \int_0^{2\pi} d\theta \int_{r=0}^1 (3 - 2r - r^2)rdr$

$$= \int_0^{2\pi} d\theta \int_0^1 (3r - 2r^2 - r^3)dr = \int_0^{2\pi} \left[3\frac{r^2}{2} - 2\frac{r^2}{3} - \frac{r^4}{4} \right]_0^1 d\theta$$

$$= \int_0^{2\pi} \left(\frac{3}{2} - \frac{2}{3} - \frac{1}{4} \right) d\theta = \frac{7}{12}.2\pi = \frac{7}{6}\pi. \qquad \square$$

7.16   Stoke's Theorem

This theorem is so named after the name of the English mathematician physicist **G. G. Stokes (1819 - 1903)**. This is an extension of *Green's theorem to surface integrals* and is analogue in two dimensions of the divergence theorem, which establishes a connection between a line integral over a closed space curve to a surface integral over a surface spanning the curve. We shall state the theorem here without proof.

Theorem 7.16.1 *Let S be a smooth oriented surface[3] bounded by a curve C oriented in the same sense and let $f(x,y,z), g(x,y,z), h(x,y,z)$ be three continuous functions having continuous first order partial derivatives at each point of domain D containing S. Then*

$$\int_C (f\ dx + g\ dy + h\ dz) = \iint_S \left[\left(\frac{\partial h}{\partial y} - \frac{\partial g}{\partial z} \right) dydz + \left(\frac{\partial f}{\partial z} - \frac{\partial h}{\partial x} \right) dzdx \right.$$
$$\left. + \left(\frac{\partial g}{\partial x} - \frac{\partial f}{\partial y} \right) dxdy \right] \quad (7.9)$$

An Alternative Statement in Vector Form:

Theorem 7.16.2 *The line integral of the tangential component of a vector $\vec{F}$ along a simple closed curve C is equal to the surface integrals of the normal component of curl $\vec{F}$, taken over any surface S having C as its boundary.*

In vector notation above can be written as

$$\int_C \vec{F}.d\vec{r} = \iint_S (curl\ \vec{F}).\hat{n}\ dS \quad (7.10)$$

where C is traversed in the positive direction (i.e., counter-clockwise direction) and $\vec{F}$ has continuous first order partial derivatives.

Note 7.16.1 *In mechanics, equation (7.10) interprets that circulation of a vector $\vec{F}$ around a curve C is equal to the outflow of curl $\vec{F}$ (flux) through an arbitrary surface S bounded by C.*

Note 7.16.2 *If the surface S is a piece of plane parallel to the xy-plane then $dz = 0$ and we get Green's formula as a special case of Stoke's formula.*

Important observations:

1. From formula (7.9), it follows that, if

$$\frac{\partial h}{\partial y} - \frac{\partial g}{\partial z} = \frac{\partial f}{\partial z} - \frac{\partial h}{\partial x} = \frac{\partial g}{\partial x} - \frac{\partial f}{\partial y} = 0 \quad (7.11)$$

then the line integral, along any closed curve C is zero, i.e.,

$$\int (f\ dx + g\ dy + h\ dz) = 0 \quad (7.12)$$

whence it follows that the line integral is independent of the shape of the curve of integration.

2. The above conditions (7.11) and (7.12) are not only sufficient but are necessary also.

3. In the fulfilment of the above conditions, the expression under the integral sign is an exact differential of some function $u(x,y,z)$, i.e., $f\ dx + g\ dy + h\ dz = du(x,y,z)$ and consequently,

$$\int_A^B (f\ dx + g\ dy + h\ dz) = \int_A^B du = u(B) - u(A)$$

ensuring that the integration does not depend on the shape, and depends only on the initial and final positions.

---

[3]A **Oriented surface** is a smooth surface together with a particular choice of orienting unit normal vector field. For more details consult Authors' *"Complex Analysis"* book.

Note 7.16.3 *Conditions (7.11) and (7.12) play a very important role in mechanics, viz., hydrodynamics, fluid mechanics etc.*

Example 7.16.1 *Verify Stoke's theorem in computing the integration*

$$\int_C \{(2x - y)\,dx - yz^2\,dy - y^2z\,dz\}$$

where the surface S is the upper half surface of the sphere $x^2 + y^2 + z^2 = 1$ and C is its boundary.

Solution: Stoke's theorem is

$$\int_C (f\,dx + g\,dy + h\,dz) = \iint_S \left[\left(\frac{\partial h}{\partial y} - \frac{\partial g}{\partial z} \right) dydz + \left(\frac{\partial f}{\partial z} - \frac{\partial h}{\partial x} \right) dzdx \right.$$
$$\left. + \left(\frac{\partial g}{\partial x} - \frac{\partial f}{\partial y} \right) dxdy \right] \quad (7.13)$$

Here C is the boundary of the upper half of the sphere $x^2 + y^2 + z^2 = 1$, which is clearly the circle $x^2 + y^2 = 1$ in the xy-plane [**Figure** 7.18].

Its parametric equations are $x = \cos\theta$, $y = \sin\theta$, $z = 0$, $0 \le \theta \le 2\pi$.

$\therefore$ L.H.S. of (7.13) $= \displaystyle\int_C (2x - y)\,dx$ [Taking $f = 2x - y$, $g = -yz^2$, $h = -y^2z$ and observing that $z = 0$ implies $dz = 0$.]

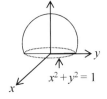

Figure 7.18

$$= \int_0^{2\pi} (2\cos\theta - \sin\theta).(-\sin\theta\,d\theta)$$

$$= \int_0^{2\pi} (-2\sin\theta\cos\theta + \sin^2\theta)d\theta$$

$$= \int_0^{2\pi} \left\{ -\sin 2\theta + \frac{1}{2}(1 - \cos 2\theta) \right\} d\theta = \left[\frac{\cos 2\theta}{2} + \frac{1}{2}\theta - \frac{\sin 2\theta}{4} \right]_0^{2\pi} = \pi.$$

For the R.H.S. of (7.13) we get $\dfrac{\partial f}{\partial y} = -1$, $\dfrac{\partial f}{\partial z} = 0$, $\dfrac{\partial g}{\partial x} = 0$, $\dfrac{\partial g}{\partial z} = -2yz$, $\dfrac{\partial h}{\partial x} = 0$, $\dfrac{\partial h}{\partial y} = -2yz$.

$\therefore$ we get R.H.S. $= \displaystyle\iint_S \{(-2yz + 2yz)dydz + 0.dzdx + (0 + 1)dxdy\}$

$$= \iint_S dxdy = \int_{x=-1}^{1} \int_{y=-\sqrt{1-x^2}}^{\sqrt{1-x^2}} dxdy = 4\int_0^1 dx \int_0^{\sqrt{1-x^2}} dy$$

$$= 4\int_0^1 \left[\frac{x}{2}\sqrt{1-x^2} + \frac{1}{2}\sin^{-1}x \right]_0^1 = 4 \times \frac{1}{2} \times \frac{\pi}{2} = \pi.$$

i.e., L.H.S. = R.H.S. of the relation (7.13) and hence Stoke's theorem is verified. $\qquad\square$

Example 7.16.2 *Using Stoke's theorem, show that*

$$\iint_S (y - z)dydz + (z - x)dzdx + (x - y)dxdy = a^3\pi$$

where S is the portion of the surface $x^2 + y^2 - 2ax + az = 0, z \ge 0$.

Solution: Applying Stoke's theorem, the given integral is reduced to

$$\frac{1}{2} \int_C \{(y^2 + z^2)\, dx + (z^2 + x^2)\, dy + (x^2 + y^2)\, dz\}$$

$$\left[\text{Taking } \frac{\partial h}{\partial y} - \frac{\partial h}{\partial y} = 2(y - z) \text{ etc.} \right]$$

where C is the curve $x^2 + y^2 - 2ax + az = 0, z = 0$

$$\text{or, } (x - a)^2 + y^2 = a^2, \ z = 0.$$

Let us put $x = a + a\cos\theta, \ y = a\sin\theta, \ z = 0$

$\therefore$ above integral

$$= \frac{1}{2} \int_{C:(x-a)^2+y^2=a^2, z=0} \{(y^2 + z^2)\, dx + (z^2 + x^2)\, dy + (x^2 + y^2)\, dz\}$$

$$= \frac{1}{2} \int_{C:(x-a)^2+y^2=a^2} (y^2\, dx + x^2\, dy) \quad [\because z = 0 \Rightarrow dz = 0 \text{ also}]$$

$$= \frac{1}{2} \int_{-\pi}^{\pi} a^2 \sin^2\theta(-a\sin\theta)\, d\theta + a^2(1 + \cos\theta)^2 a\cos\theta\, d\theta$$

$$= \frac{1}{2} \int_{-\pi}^{\pi} a^3(-\sin^3\theta + \cos^3\theta + 2\cos^2\theta + \cos\theta)\, d\theta$$

$$= \frac{a^3}{2} \left[0 + 2\int_0^\pi \frac{1}{4}(\cos 3\theta + 3\cos\theta)\, d\theta + 4\int_0^\pi \frac{(1 + \cos 2\theta)}{2}\, d\theta + 2\int_0^\pi \cos\theta\, d\theta \right]$$

$$= \frac{a^3}{2} \left[0 + 4.\frac{\pi}{2} + 0 \right] = a^3\pi. \qquad \square$$

7.17 Relationship between the Integral Theorems (Green's, Stoke's and Gauss's Divergence Theorems)

In § 7.9, we have discussed the relationship between the line integral, simple integral and the multiple integrals. There we have observed that a *line integral* extends the idea of *simple integrals to general curves* whereas a *surface integral* extends the idea of *double integral to general surfaces*. Now we shall discuss the same for the three integral theorems, viz., Green's Theorem, Stoke's Theorem and Gauss's Theorems.

In simplest terms, the integral theorems can be described as follows:

i) *Green's theorem* equates the *single integral* of a function f around a simple closed curve C in a plane to a *double integral* over the plane region R, bounded by C. It is a special two dimensional case of the more general Stoke's theorem which is as follows:

ii) *Stoke's theorem* equates the *single integral* of a function f along the boundary of a *surface itself*. To summarize the relationship we have the following pictorial diagram:

Explanation: *Green's theorem* gives the relationship between a *line integral* around a simple closed curve C *in a plane* and a *double integral* over the *plane region R*, bounded by C. *Stoke's theorem* gives a relationship between a *line integral* around a simple closed curve C, *in space* and a *surface integral* over

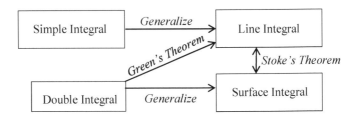

piecewise, smooth *surface*. Green's theorem, in fact, the special case of Stoke's theorem in which the surface lies entirely in the plane. Thus when we are applying Green's theorem, we technically applying Stoke's theorem as well. Especially, when we have vector field in the plane, the *curl* of the vector field is always vertical vector, so it makes sense to identify this with a scalar quantity and this scalar quantity is precisely the derivative appearing the double integral in Green's theorem.

iii) *Gauss's theorem* (or Gauss's divergence theorem) equates the *double integral* of a function f along a closed surface which is the boundary of a three dimensional region with a *triple integral* of some kind of derivative of f along the region itself. Thus the situation in Gauss's theorem is "*One dimensional up*" form the situation in *Stoke's theorem*. So it should be easy to figure out which of these results applies. If we see a three dimensional region bounded by a closed surface or if we see a triple integral, it must be Gauss's theorem that we want. On the other hand, if we see a two dimensional region bounded by a closed curve or if we a single integral (in deed a line integral) then it must be Stoke's theorem that we want.

From above discussion, we can think of the fact that *all the three theorems are of the same thing*. Or, rather, Green's theorem and the Divergence theorem are both *special cases* of Stoke's theorem in two and three dimensions respectively.

7.18   Miscellaneous Illustrative Examples

Example 7.18.1 *Evaluate* $\displaystyle\int_{(1,-2,1)}^{(3,1,4)} \left[(2xy + z^3)\, dx + x^2\, dy + 3xz^2\, dz\right]$.

Solution: $\displaystyle I = \int_{(1,-2,1)}^{(3,1,4)} d(x^2 y + xz^3) = \left[x^2 y + xz^3\right]_{(1,-2,1)}^{(3,1,4)} = 202.$ □

Example 7.18.2 *Compute* $\displaystyle\int_C (x^2 y\, dx + xy^2\, dy)$, *where the integration is taken in the clockwise sense along the hexagon whose vertices are* $(\pm 3a, 0)$, $(\pm 2a, \pm\sqrt{3}a)$.

Solution: $\displaystyle I = \int_C f = \int_{AB} f + \int_{BC} f + \int_{CD} f + \int_{DE} f + \int_{EF} f + \int_{FA} f,$

where $f = x^2 y\, dx + xy^2\, dy$.

Now (i) Equation of AB is $y = \sqrt{3}(x - 3a)$; x varies from $3a$ to $2a$; and $dy = \sqrt{3}x\, dx$.

$$\therefore \int_{AB} f = \int_{3a}^{2a} \{x^2.\sqrt{3}(x - 3a) + 3(x - 3a)^2.x\sqrt{3}\}dx$$

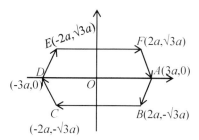

Figure 7.19

$$= \sqrt{3} \int_{3a}^{2a} (x-3a)(x^2+3x^2-9ax)\, dx = \sqrt{3} \int_{3a}^{2a} (x-3a)(4x^2-9ax)\, dx$$

$$= \sqrt{3} \int_{3a}^{2a} (4x^3 - 21ax^2 + 27a^2 x)\, dx = \sqrt{3}\left[x^4 - 7ax^3 + \frac{27a^2 x^2}{2} \right]_{3a}^{2a}$$

$$= \sqrt{3}\left[16a^4 - 56a^4 + 54a^4 - 81a^4 + 189a^4 - \frac{243}{2}a^4 \right]$$

$$= \sqrt{3}a^4\left[122 - \frac{243}{2} \right] = \frac{\sqrt{3}}{2}a^4.$$

(ii) Equation of BC is $y = -\sqrt{3}a$; x varies from $2a$ to $-2a$; $dy = 0$.

$$\therefore \int_{BC} f = \int_{2a}^{-2a} x^2(-\sqrt{3}a)dx = \sqrt{3}a \int_{-2a}^{2a} x^2 dx = \sqrt{3}a\left[\frac{x^3}{3} \right]_{-2a}^{2a} = \frac{16}{3}\sqrt{3}a^4.$$

(iii) Equation of CD is $y = -\sqrt{3}(x+3a)$; x varies from $-2a$ to $-3a$; $dy = -\sqrt{3}\, dx$.

$$\therefore \int_{CD} f = \int_{-2a}^{-3a} \left\{ -x^2\sqrt{3}(x+3a) + 3(x+3a)^2.x(-\sqrt{3}) \right\} dx$$

$$= \sqrt{3}\int_{-2a}^{-3a} -(x+3a)(4x^2+9ax)\, dx$$

$$= \sqrt{3}\int_{-3a}^{-2a} (4x^3 + 21x^2 a + 27a^2 x)\, dx = \frac{\sqrt{3}}{2}a^4.$$

(iv) Equation of DE is $y = \sqrt{3}(x+3a)$; x varies from $-3a$ to $-2a$; $dy = \sqrt{3}\, dx$.

$$\therefore \int_{DE} f = \int_{-3a}^{-2a} \left\{ x^2\sqrt{3}(x+3a) + 3(x+3a)^2.x(-\sqrt{3}) \right\} dx$$

$$= \sqrt{3}\int_{-3a}^{-2a} (x+3a)(4x^2+9ax)\, dx = \frac{\sqrt{3}}{2}a^4. \text{ [See (iii)]}$$

(v) Equation of EF is $y = \sqrt{3}a$; x varies from $-2a$ to $2a$; $dy = 0$.

$$\therefore \int_{EF} f = \int_{-2a}^{2a} x^2\sqrt{3}a\, dx = \frac{16}{3}\sqrt{3}a^4.$$

(vi) Equation of FA is $y = -\sqrt{3}(x - 3a)$; x varies from $2a$ to $3a$; $dy = -\sqrt{3}\, dx$.

$$\therefore \int_{FA} f = \int_{2a}^{3a} \left\{ -x^2\sqrt{3}(x - 3a) + 3(x - 3a)^2.x(-\sqrt{3}) \right\} dx$$

$$= \sqrt{3} \int_{3a}^{2a} (x - 3a)(x^2 + 3x^2 - 9ax)\, dx$$

$$= \sqrt{3} \int_{3a}^{2a} (4x^3 - 21x^2a + 27a^2x)\, dx = \frac{\sqrt{3}}{2}a^4. \text{ [See (i)]}$$

Therefore $\int_C (x^2y\, dx + xy^2\, dy) = $ the sum of the results of (i), (ii), ..., (vi)

$$= \frac{\sqrt{3}}{2}a^4 + \frac{16}{3}\sqrt{3}a^4 + \frac{\sqrt{3}}{2}a^4 + \frac{\sqrt{3}}{2}a^4 + \frac{16}{3}\sqrt{3}a^4 + \frac{\sqrt{3}}{2}a^4$$

$$= 2\sqrt{3}a^4 + \frac{32}{\sqrt{3}}a^4 = \sqrt{3}a^4\left(2 + \frac{32}{3}\right) = \frac{38a^4}{\sqrt{3}}. \qquad \square$$

Example 7.18.3 *Find the area of an ellipse* $\dfrac{x^2}{a^2} + \dfrac{y^2}{b^2} = 1 \ (a, b > 0).$

Solution: **First Method:** (Using the definite integral $\displaystyle\int_a^b f(x)\, dx$ of a single variable discussed in § 7.11.)

Clearly the area OAB is bounded by the curve

$y = \dfrac{b}{a}\sqrt{a^2 - x^2}$, the x-axis and the ordinates at

$x = 0$ and $x = a$.

$$\therefore \triangle OAB = \int_0^a y\, dx = \int_0^a \frac{b}{a}\sqrt{a^2 - x^2}\, dx$$

[See **Figure 7.20**]

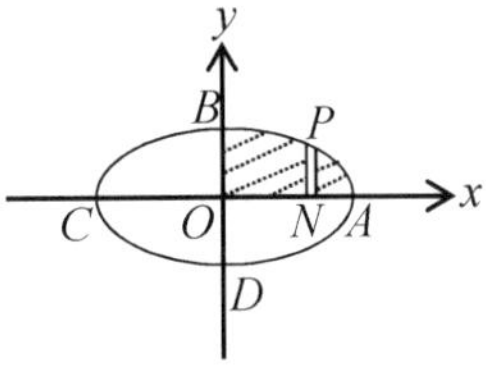

$$= \frac{b}{a} \int_0^{\frac{\pi}{2}} a\cos\theta.a\cos\theta\, d\theta$$

[Putting $x = a\cos\theta$]

$$= \frac{ab}{2} \int_0^{\frac{\pi}{2}} (1 + \cos 2\theta)\, d\theta = \frac{ab}{2}\left[\theta + \frac{\sin 2\theta}{2}\right]_0^{\frac{\pi}{2}} = \frac{ab}{2}.\frac{\pi}{2} = \frac{\pi ab}{4}.$$

Figure 7.20

Therefore the area of the whole ellipse is equal to four times of the area $OAB =$

$4.\dfrac{1}{4}\pi ab = \pi ab.$

Second Method: (Using the definition of double integral by putting $f(x, y) = 1$)

By putting $f(x, y) = 1$, the entire area $ABCDA$ of the ellipse is

$$= 4 \int_{x=0}^{x=a} \int_{y=0}^{y=f(x)} dx\, dy, \quad \text{where } y = f(x) = \frac{b}{a}\sqrt{a^2 - x^2}$$

$$= 4 \int_0^a [y]_0^{f(x)}\, dx = 4 \int_0^a f(x)\, dx = 4 \int_0^a \frac{b}{a}\sqrt{a^2 - x^2}\, dx$$

$$= \frac{4b}{a}\left[\frac{x\sqrt{a^2 - x^2}}{2} + \frac{a^2}{2}\sin^{-1}\left(\frac{x}{a}\right)\right]_0^a = \frac{2b}{a}\left[0 + a^2\sin^{-1}1\right] = \frac{2b}{a}.a^2.\frac{\pi}{2}$$

$$= \pi ab. \qquad \square$$

Example 7.18.4 *Find the surface area of the sphere $x^2 + y^2 + z^2 = 9$ lying inside the cylinder $x^2 + y^2 = 3y$.*

Solution: Given the sphere $x^2 + y^2 + z^2 = 9$.

$$\therefore \quad 2x + 2z\frac{\partial z}{\partial x} = 0 \quad \Rightarrow \quad \frac{\partial z}{\partial x} = -\frac{x}{z}$$

$$2y + 2z\frac{\partial z}{\partial y} = 0 \quad \Rightarrow \quad \frac{\partial z}{\partial y} = -\frac{y}{z}$$

$$\therefore \quad \sqrt{1 + \left(\frac{\partial z}{\partial x}\right)^2 + \left(\frac{\partial z}{\partial x}\right)^2} = \sqrt{1 + \frac{x^2}{z^2} + \frac{y^2}{z^2}} = \frac{\sqrt{x^2 + y^2 + z^2}}{z}$$

$$= \frac{3}{z} = \frac{3}{\sqrt{9 - (x^2 + y^2)}}.$$

Bounding surface is $x^2 + y^2 = 3y$.

Putting $x = r\cos\theta$, $y = r\sin\theta$ $\Rightarrow$ $r = 3\sin\theta$, $J = r$.

So the required surface area $= \iint_R \sqrt{1 + \left(\frac{\partial z}{\partial x}\right)^2 + \left(\frac{\partial z}{\partial x}\right)^2}\, dxdy.$

[Taking $\phi(x, y, z) = 1$ in **Note 7.9.2**]

$$= \iint_R \frac{3}{\sqrt{9 - (x^2 + y^2)}}\, dxdy = 4\int_{\theta=0}^{\frac{\pi}{2}} d\theta \int_{r=0}^{3\sin\theta} \frac{3}{\sqrt{9 - r^2}} r.dr$$

$$= 12\int_0^{\frac{\pi}{2}} d\theta \int_0^{3\sin\theta} \frac{rdr}{\sqrt{9 - r^2}}.$$

Now put $9 - r^2 = p^2$ $\therefore$ $rdr = -pdp$. When $r = 0$, $p = 3$ and when $r = 3\sin\theta$, $p = 3\cos\theta$.

Therefore required area $= -12\int_0^{\frac{\pi}{2}} d\theta \int_3^{3\cos\theta} dp = -12\int_0^{\frac{\pi}{2}} d\theta\, [p]_3^{3\cos\theta}$

$$= -12\int_0^{\frac{\pi}{2}} (3\cos\theta - 3)\, d\theta = -36[\sin\theta - \theta]_0^{\frac{\pi}{2}} = -36\left(1 - \frac{\pi}{2}\right) = 18(\pi - 2). \quad \square$$

Example 7.18.5 *Evaluate the surface integral*

$$\iint_S (x^3\, dydz + y^3\, dzdx + z^3\, dxdy),$$

with the help of Gauss's theorem over the sphere $x^2 + y^2 + z^2 = a^2$.

Solution: Gauss's divergence theorem is

$$\iiint_D \left(\frac{\partial f}{\partial x} + \frac{\partial g}{\partial y} + \frac{\partial h}{\partial z}\right) dxdydz = \iint_S (f\, dydz + g\, dzdx + h\, dxdy) \quad (7.14)$$

Here $f = x^3$, $g = y^3$, $h = z^3$ $\therefore$ $\frac{\partial f}{\partial x} = 3x^2$, $\frac{\partial g}{\partial y} = 3y^2$, $\frac{\partial h}{\partial z} = 3z^2$ and so by (7.14), we get

$$I = \iiint_D 3(x^2 + y^2 + z^2)\, dxdydz, \quad \text{where } D : x^2 + y^2 + z^2 = a^2.$$

Changing spherical polar coordinates $x = r \sin\theta \cos\phi$, $y = r \sin\theta \sin\phi$, $z = r \cos\theta$, whereby $J = r^2 \sin\theta$, we get

$$I = 3 \int_0^{2\pi} d\phi \int_0^\pi \sin\theta \, d\theta \int_0^a r^2.r^2 dr = 3 \int_0^{2\pi} d\phi \int_0^\pi \sin\theta \, d\theta \left[\frac{r^5}{5}\right]_0^a$$

$$= \frac{3a^5}{5} \int_0^{2\pi} d\phi \, [-\cos\theta]_0^\pi = \frac{3a^5}{5}.2[\phi]_0^{2\pi} = \frac{12}{5}\pi a^5. \qquad \square$$

Example 7.18.6 *Find the volume bounded by the cylinder $x^2 + y^2 = 4$ and the hyperboloid $x^2 + y^2 - z^2 = -1$.*

Solution: Here, surface $x^2 + y^2 = 4$ and $x^2 + y^2 - z^2 = -1$ are symmetrical about all the axes. Therefore

$$\text{Volume} = V = \iint z \, dxdy \quad [\text{See } \S\ 7.11] = 8 \int_0^2 dx \int_0^{\sqrt{4-x^2}} \sqrt{x^2 + y^2 + 1} \, dy.$$

Put $x = r \cos\theta$, $y = r \sin\theta$, $J = r$ and changing the limits of integrations for the region of quadrant $r = 2$ and $\theta = 0$ to $\frac{\pi}{2}$ we get

$$V = \int_0^{\frac{\pi}{2}} d\theta \int_0^2 r\sqrt{r^2 + 1} \, dr = 8 \int_0^{\frac{\pi}{2}} \frac{1}{3} \left[(r^2 + 1)^{\frac{3}{2}}\right]_0^2 d\theta$$

$$= \frac{8}{3}(5\sqrt{5} - 1) \int_0^{\frac{\pi}{2}} d\theta = \frac{8}{3}(5\sqrt{5} - 1)\frac{\pi}{2} = \frac{4\pi}{3}(5\sqrt{5} - 1). \qquad \square$$

Example 7.18.7 *Find the volume within the cylinder $x^2 + y^2 = a^2$ between the planes $y + z = b^2$ and $z = 0$.*

Solution: The cylindrical solid is bounded above by the surface $z = b^2 - y \equiv f(x, y)$ and below the disc $R \equiv x^2 + y^2 - a^2$. Therefore the required volume

$$= \iint_R f(x,y) \, dxdy \quad [\text{See } \S\ 7.12 \text{ **Remark (iii)**}]$$

$$= \int_{-a}^a dx \int_{-\sqrt{a^2-x^2}}^{\sqrt{a^2-x^2}} (b^2 - y) \, dy = \int_{-a}^a dx \left[b^2 y - \frac{y^2}{2}\right]_{-\sqrt{a^2-x^2}}^{\sqrt{a^2-x^2}}$$

$$= \int_{-a}^a dx \left[2b^2 \sqrt{a^2 - x^2}\right] = 2b^2 \int_{-a}^a \sqrt{a^2 - x^2} \, dx$$

$$= 2b^2 \left[\frac{x}{2}\sqrt{a^2 - x^2} + \frac{a^2}{2}\sin^{-1}\frac{x}{a}\right]_{-a}^a = 2b^2 \left[\frac{a^2}{2}.\frac{\pi}{2} + \frac{a^2}{2}.\frac{\pi}{2}\right] = \pi a^2 b^2.$$

Hence the required volume is $\pi a^2 b^2$. $\qquad \square$

Example 7.18.8 *Compute the volume of the ellipsoid $\frac{x^2}{a^2} + \frac{y^2}{b^2} + \frac{z^2}{c^2} = 1$.*

Solution: See **Example 6.11.7**. $\qquad \square$

Example 7.18.9 *Apply Green's theorem to evaluate*

$$\int_C \{(1 - x^2)y \, dx + (1 + y^2)x \, dy\}$$

where C is the circle $x^2 + y^2 = a^2$ in the xy-plane described in the positive direction.

Solution: Green's theorem in the plane of xy is given by

$$\iint_D \left(\frac{\partial Y}{\partial x} - \frac{\partial X}{\partial y} \right) dxdy = \int_C (X dx + Y dy) \tag{7.15}$$

Here $X = (1 - x^2)y$ $\therefore \dfrac{\partial X}{\partial y} = 1 - x^2$

$Y = (1 + y^2)x$ $\therefore \dfrac{\partial Y}{\partial x} = 1 + y^2.$

$\therefore$ Applying (7.15), we get

$$\int_C \left\{ (1 - x^2)y \ dx + (1 + y^2)x \ dy \right\} = \iint_D \left\{ (1 + y^2) - (1 - x^2) \right\} dxdy$$

where D is bounded by $C : x^2 + y^2 = a^2$. Therefore the required integral

$$= \iint_{x^2+y^2=a^2} (1 + y^2 - 1 + x^2) \ dxdy = \iint_{x^2+y^2=a^2} (x^2 + y^2) \ dxdy$$

Put $x = r \cos\theta$, $y = \sin\theta$, $J = r$, r varies from 0 to a and θ varies form 0 to 2π. Therefore the required integral is

$$= \int_{r=0}^{a} \int_{\theta=0}^{2\pi} r^2 . r \ drd\theta = \int_0^a r^3 dr . 2\pi = 2\pi \left[\frac{r^4}{4} \right]_0^a = 2\pi \frac{a^4}{4} = \frac{\pi a^4}{2}. \qquad \square$$

Example 7.18.10 *State Green's theorem in $\mathbb{R}^2$ and prove, by using Green's theorem, that* $\displaystyle\int_\Gamma \frac{x dy - y dx}{x^2 + y^2} = 2\pi$, *if Γ be a positively described closed contour round the origin.*

Solution: **1st Part:** See § 7.14.

2nd Part: Here the origin is inside the closed contour Γ. We consider a circle of contour C with centre at the origin O and of radius, say 'a' so that C does not intersect Γ.

Let D be the region bounded by Γ and C.

Here $X = -\dfrac{y}{x^2 + y^2}$ $\therefore \dfrac{\partial X}{\partial y} = -\left[\dfrac{1.(x^2 + y^2) - 2y.y}{(x^2 + y^2)^2} \right] = \dfrac{y^2 - x^2}{(x^2 + y^2)^2}$

and $Y = \dfrac{x}{x^2 + y^2}$ $\therefore \dfrac{\partial Y}{\partial x} = \dfrac{1.(x^2 + y^2) - 2x.x}{(x^2 + y^2)^2} = -\dfrac{x^2 - y^2}{(x^2 + y^2)^2}.$

So, using Green's theorem, we get

$$\int_\Gamma \frac{x dy - y dx}{x^2 + y^2}$$

$$= \iint_D \left\{ \frac{y^2 - x^2}{(x^2 + y^2)^2} - \frac{x^2 - y^2}{(x^2 + y^2)^2} \right\} dxdy$$

$$= \iint_{\Gamma-C} 0.dxdy = 0$$

$$\Rightarrow \int_{\Gamma-C} \frac{x dy - y dx}{x^2 + y^2} = 0$$

$$\Rightarrow \int_\Gamma \frac{x dy - y dx}{x^2 + y^2} = \int_C \frac{x dy - y dx}{x^2 + y^2}.$$

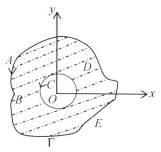

Figure 7.21

Now $C : x^2 + y^2 = a^2$. Putting $x = r \cos \theta$, $y = \sin \theta$ we get

$$\int_C \frac{x dy - y dx}{x^2 + y^2} = \int_0^{2\pi} \frac{a^2 \cos^2 \theta + a^2 \sin^2 \theta}{a^2} d\theta = \int_0^{2\pi} d\theta = 2\pi.$$

Therefore $\int_\Gamma \frac{x dy - y dx}{x^2 + y^2} = 2\pi.$ □

Example 7.18.11 *Show using Gauss's theorem that the surface integral*

$$\iint_S (ax \, dydz + by \, dzdx + cz \, dxdy) = \frac{4}{3}\pi(a + b + c)$$

where S is the surface of the sphere $x^2 + y^2 + z^2 = 1$.

Solution: Gauss's divergence theorem is

$$\iiint_D \left(\frac{\partial f}{\partial x} + \frac{\partial g}{\partial y} + \frac{\partial h}{\partial z} \right) dxdydz = \iint_S (f \, dydz + g \, dzdx + h \, dxdy) \quad (7.16)$$

Here $f = ax$, $g = by$, $h = cz$ $\therefore \frac{\partial f}{\partial x} = a$, $\frac{\partial g}{\partial y} = b$, $\frac{\partial f}{\partial x} = c$.

$\therefore$ Using (7.16) given integral $= \iiint_{x^2+y^2+z^2=1} \left(\frac{\partial f}{\partial x} + \frac{\partial g}{\partial y} + \frac{\partial h}{\partial z} \right) dxdydz$

$$= \iiint_{x^2+y^2+z^2=1} (a + b + c) \, dxdydz = (a + b + c) \iiint_{x^2+y^2+z^2=1} dxdydz.$$

Now we put $x = r \sin \theta \cos \phi$, $y = r \sin \theta \sin \phi$, $z = r \cos \theta$, whereby $J = r^2 \sin \theta$, r varies form 0 to 1, θ from 0 to π and ϕ form 0 to 2π.

$$\therefore \quad \text{required integral} = (a + b + c) \int_{r=0}^1 \int_{\theta=0}^\pi \int_{\phi=0}^{2\pi} r^2 \sin \theta \, drd\theta d\phi$$

$$= (a + b + c) \int_0^1 r^2 dr \int_0^\pi \sin \theta d\theta \int_0^{2\pi} d\phi$$

$$= 2\pi(a + b + c) \int_0^1 r^2 dr [-\cos \theta]_0^\pi = -2\pi(a + b + c) \int_0^1 -2r^2 dr$$

$$= 4\pi(a + b + c) \left[\frac{r^3}{3} \right]_0^1 = \frac{4\pi}{3}(a + b + c).$$ □

Example 7.18.12 *Use Stoke's theorem to find the line integral*

$$\int_C (x^2 y^3 \, dx + dy + z \, dz),$$

where C is the circle $x^2 + y^2 = a^2$, $z = 0$.

Solution: Stoke's theorem is

$$\int_C (f \, dx + g \, dy + h \, dz) = \iint_S \left[\left(\frac{\partial h}{\partial y} - \frac{\partial g}{\partial z} \right) dydz + \left(\frac{\partial f}{\partial z} - \frac{\partial h}{\partial x} \right) dzdx \right.$$

$$\left. + \left(\frac{\partial g}{\partial x} - \frac{\partial f}{\partial y} \right) dxdy \right] \quad (7.17)$$

Here $f = x^2 y^3$, $g = 1$, $h = z$.

$$\therefore \frac{\partial f}{\partial y} = 3x^2 y^2, \ \frac{\partial f}{\partial z} = 0, \ \frac{\partial g}{\partial x} = 0 = \frac{\partial g}{\partial z}, \ \frac{\partial h}{\partial y} = 0, \ \frac{\partial h}{\partial x} = 0.$$

$\therefore$ By Stoke's theorem, we get

$$\int_C (x^2 y^3 \ dx + dy + z \ dz) = \iint_S -3x^2 y^2 \ dxdy = -3\iint_S x^2 y^2 \ dxdy.$$

where S is the circle $x^2 + y^2 = a^2$ in the xy plane.

Put $x = r\cos\theta, \ y = \sin\theta, \ J = r$ and we get

$$I = -3 \int_{-\pi}^{\pi} \cos^2 \theta \sin^2 \theta \ d\theta \int_0^a r^5 dr = \frac{-3a^6}{6} \int_{-\pi}^{\pi} \frac{1}{4} \sin^2 2\theta \ d\theta$$

$$= -\frac{a^6}{8} \int_{-\pi}^{\pi} \sin^2 2\theta \ d\theta = -\frac{a^6}{8}.2.\frac{1}{2} \int_0^{\pi} (1 - \cos 4\theta) \ d\theta = -\frac{a^6}{8}\pi. \qquad \square$$

Example 7.18.13 *Find the volume of the portion of the sphere $x^2 + y^2 + z^2 = a^2$ lying inside the cylinder $x^2 + y^2 = ax$.*

Solution: See **Example 6.12.1**. $\qquad \square$

Example 7.18.14 *Find the volume of the region bounded by the cylinder $x^2 + y^2 = 16$ and points $z = 0$ to $z = 3$.*

Solution: See **Example 6.14.13**. $\qquad \square$

7.19   Exercises

1. Evaluate the integrals (i) $\int_C (x^2 + y^2) \, dx$, (ii) $\int_C (x^2 + y^2) \, dy$, where C is the arc of the parabola $y^2 = 4ax$ form $(0,0)$ to $(a, 2a)$.

2. Evaluate the integral $\int_C (x + y) \, dx$, where (i) C is arc of the unit circle $x^2 + y^2 = 1$ in the first quadrant, (ii) C is the two line segments $y = 0$, $0 \le x \le 1$ and $x = 0$, $0 \le y \le 1$.

3. Compute $\int_C \{xy \, dx + (x^2 + y^2) \, dy\}$ where C consists of a part of the x-axis form $x = 2$ to $x = 4$ and then a portion of the line $x = 4$ from $y = 0$ to $y = 12$.

4. Compute $\int_{A(1,1)}^{B(2,8)} (6x^2y \, dx + 10xy^2 \, dy)$ along a plane curve $y = x^3$.

5. Compute the line integral $\int_A^B (x^2 dx + 3x^2 z \, dy - xy^2 dz)$ along the line segment from the point $A(1, 2, 3)$ to the point $B(0, 0, 0)$.

6. Compute (i) $\int_C \{xy \, dx + (x + y) \, dy\}$ and (ii) $\int_C \{(x + y) \, dx + (x - y) \, dy\}$, where C is the arc AB in the first quadrant of the unit circle $x^2 + y^2 = 1$ from $A(1, 0)$ to $B(0, 1)$.

7. Compute $\int_C \{xy \, dx + (x^2 + y^2) \, dy\}$ where C is the arc of the parabola $y = x^2 - 4$ from $(2, 0)$ to $(4, 12)$ in the xy plane.

8. Evaluate $\int_C \left[(3x^2 + 6y) \, dx - 14yz \, dy + 20xz^2 dz\right]$ (i) from the point $(0, 0, 0)$ to the point $(1, 1, 1)$ along the path $C : x = t, \, y = t^2, \, z = t^3$, (ii) along the straight line form $(0, 0, 0)$ to $(1, 0, 0)$, then to $(1, 1, 0)$ and then to $(1, 1, 1)$, and (iii) along the straight line form $(0, 0, 0)$ to $(1, 1, 1)$.

9. Show that $\int_C (y \, dx + x \, dy) = 0$ where C is given by $x = \cos\theta, \, y = \sin\theta$, $0 \le \theta \le \dfrac{\pi}{2}$.

10. Compute $\int_C \{(x^2 + y) \, dx + (2x + y^2) \, dy\}$ over the boundary of the square with vertices $(1, 1), (1, 2), (2, 2)$ and $(2, 1)$ in the clockwise sense.

11. Prove that $\int_{(1,2)}^{(3,4)} \{(6xy^2 - y^3) \, dx + (6x^2y - 3xy^2) \, dy\}$ is independent of the path joining $(1, 2)$ and $(3, 4)$. Also evaluate the integral.

[*Hints:* First Part. Take $X = 6xy^2 - y^3$ and $Y = 6x^2y - 3xy^2$. Then clearly

$$\frac{\partial X}{\partial y} = 12xy - 3y^2 = \frac{\partial Y}{\partial x}$$ which implies $X dx + Y dy$ is an exact differential. Hence the line integral is independent of the path.

Second Part. Since the integral is independent of the path, we take the path AB form $(1, 2)$ to $(3, 2)$ and then BC form $(3, 2)$ to $(3, 4)$ [See **Figure 7.22**].

Now along AB, $y = 2$, $dy = 0$ and along BC, $x = 3$, $dx = 0$. Therefore required

integral$= \int_1^3 (24x - 8) \, dx + \int_2^4 (54y - 9y^2) \, dy = 256.$]

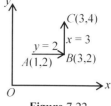

Figure 7.22

12. Show that $\iint_S f(x, y, z) \, dS$, where S is the surface of the paraboloid $x^2 + y^2 = 2 - z$ above the xy-plane is equivalent to

$$\iint_R f(x, y, z)\sqrt{1 + 4x^2 + 4y^2} \, dxdy$$

where R is the projection of S on the xy-plane.

Hence evaluate the surface integral when (i) $f = 1$, (ii) $f = x^2 + y^2$, (iii) $f = 3z$.

[*Hints:* R is the projection of S on the xy-plane. The required surface integral is equal to

$$\iint_R f(x, y, z)\sqrt{1 + \left(\frac{\partial z}{\partial x}\right)^2 + \left(\frac{\partial z}{\partial y}\right)^2} \, dxdy \quad \text{[See \textbf{Note} 7.9.2]}$$

and R is given by $z = 0$, $x^2 + y^2 = 2$.

From $x^2 + y^2 = 2 - z$, we get $\dfrac{\partial z}{\partial x} = -2x$, $\dfrac{\partial z}{\partial y} = -2y$. So the surface integral

becomes $\iint_R f(x, y, z)\sqrt{1 + 4x^2 + 4y^2} \, dxdy.$

Now (i) when $f(x, y, z) = 1$ we get the surface integral

$$\iint_R \sqrt{1 + 4x^2 + 4y^2} \, dxdy; \quad R: x^2 + y^2 = 2$$

To evaluate the integral, we put $x = r\cos\theta$, $y = r\sin\theta$, $\therefore J = r$ and R changes to R', r form 0 to $\sqrt{2}$, θ changes form 0 to 2π.

$\therefore$ Given integral is $\displaystyle\int_0^{2\pi} d\theta \int_0^{\sqrt{2}} r\sqrt{1 + 4r^2} \, dr = \frac{1}{4}\int_0^{2\pi} d\theta \int_1^3 p.p \, dp$

[Putting $1 + 4r^2 = p^2 \Rightarrow 8rdr = 2pdp \Rightarrow rdr = \dfrac{1}{4}pdp.$

$r = 0 \Rightarrow p = 1$, $r = \sqrt{2} \Rightarrow p = 3.$]

$$= \frac{1}{4}\int_0^{2\pi} d\theta \left[\frac{p^3}{3}\right]_1^3 = \frac{1}{12}\int_0^{2\pi} 26 \, d\theta = \frac{1}{12} \times 26 \times 2\pi = \frac{13\pi}{3}.$$

(ii) When $f(x, y, z) = x^2 + y^2$, we get

$$\iint_R f(x, y, z)\sqrt{1 + 4x^2 + 4y^2} \, dxdy = \iint_R (x^2 + y^2)\sqrt{1 + 4x^2 + 4y^2} \, dxdy$$

$$= \int_0^{2\pi} d\theta \int_0^{\sqrt{2}} r\sqrt{1 + 4r^2} \, dr = \frac{149\pi}{30}.$$

[Proceeding as (i) by the same substitution $1 + 4r^2 = p^2$.]

(iii) When $f(x,y,z) = 3z$, we get $\iint_R f(x,y,z)\sqrt{1+4x^2+4y^2}\,dxdy$

$= \iint_R 3\{2-(x^2+y^2)\}\sqrt{1+4x^2+4y^2}\,dxdy = \dfrac{111\pi}{10}$. (Proceeding as (i))]

13. Find the area of the surface $z^2 = 2xy$ included between the planes $x = 0, x = a, y = 0, y = b$.

14. Evaluate the volume of the tetrahedron bounded by the coordinate planes and the plane $x+y+z = 1$.

15. Find the volume within the cylinder $x^2+y^2 = a^2$ between the planes $y+z = b^2$ and $z = 0$.

16. State Green's theorem in the plane of xy. Verify Green's theorem in the plane $\int_C \{(x^2-xy^3)\,dx + (y^2-2xy)\,dy\}$ where C is the square with vertices at $(0,0),(2,0),(2,2),(0,2)$.

[*Hints:* Common value = 8.]

17. Verify Green's theorem for $\oint_C \{(x^2-xy)\,dx + (y-x^2)\,dy\}$, where C is the closed curve of the region bounded by $y = x^3$ and $y = x$.

[*Hints:* Common value = $-\dfrac{2}{15}$.]

18. Prove that $\int_{(1,0)}^{(2,1)} \{(2xy-y^4+3)\,dx + (x^2-4xy^3)\,dy\}$, is independent of the path joining $(1,0)$ and $(2,1)$. Find also the value of the integral.

19. Using line integral find the area of the ellipse $x = a\cos\theta,\ y = b\sin\theta$.

20. Find the area of the region bounded by

(i) the parabola $y^2 = 2x$ and the line $y = x$.

(ii) the curves $y = \sin x,\ y = \cos x, x = 0, 0 \le x \le \dfrac{\pi}{2}$.

(iii) the curves $\sqrt{x}+\sqrt{y} = \sqrt{a},\ x+y = a$.

21. Show that the volume of the solid bounded by the following surfaces (i) $x^2+y^2 = 1$, $z = 0$, $x+y+z = 3$ is 3π and (ii) the cylinder $x^2+y^2 = 16$ and the point $z = 0$ to $z = 3$ is 48π.

22. State Gauss's theorem and apply it to evaluate the following surface integral $\iint_S (x^3\,dydz + y^3\,dzdx + z^3\,dxdy)$ over the sphere $x^2+y^2+z^2 = a^2$.

23. Evaluate $\iint_S (x\,dydz + y\,dzdx + z\,dxdy)$ taken over the outer surface of the cube of side 'a' unit.

24. Prove that the volume of the ellipsoid $\dfrac{x^2}{a^2}+\dfrac{y^2}{b^2}+\dfrac{z^2}{c^2} = 1$ is $\dfrac{4}{3}\pi abc$.

25. Compute the area of a loop of the curve $\dfrac{x^2}{a^2}+\dfrac{y^2}{b^2} = \dfrac{2xy}{c^2}$.

[*Hints:* Put $x = ar\cos\theta,\ y = br\sin\theta$.]

26. State Gauss's theorem and apply it to evaluate the following surface integral $\iint_S \{(x^3 - yz)\, dydz - 2x^2y\, dzdx + 2\, dxdy\}$ taken over the outer surface of the cube bounded by the planes $x = 0$, $x = a$, $y = 0$, $y = a$, $z = 0$, $z = a$.

27. Using Gauss's theorem, show that the volume V of a region R bounded by a surface S is given by $V = \dfrac{1}{3} \iint (x\, dydz + y\, dzdx + z\, dxdy)$.

28. Verify Gauss's theorem for the surface integral

$$\iint_S \{(2xy + z)\, dydz + y^2 dzdx - (x + 3y)\, dxdy\}$$

taken over the region bounded by $2x + 2y + z = 6$, $x = 0$, $y = 0$, $z = 0$.

[*Hints:* Common value = 27.]

29. Verify Gauss's theorem for the surface integral

$$\iint_S \{(2x - z)\, dydz + x^2y\, dzdx - z^2x\, dxdy\}$$

taken over the region bounded by $x = 0$, $x = a$, $y = 0$, $y = a$, $z = 0$, $z = a$.

$\left[\text{\textit{Hints:} Common value} = \dfrac{11}{6}.\right]$

30. Verify Stoke's theorem

(i) in evaluating $\displaystyle\int_C (3y\, dx - zx\, dy + yz^2\, dz)$, where S is the surface of the paraboloid $x^2 + y^2 = 2z$ bounded by $z = 2$ of which C is the boundary.

(ii) for the line integral $\displaystyle\int_C (x^2\, dx + xy\, dy)$, where C denotes the square in the plane $z = 0$ with sides along the lines $x = 0$, $x = a$, $y = 0$, $y = a$.

$\left[\text{\textit{Hints:} (i) Common value} = -20\pi, \text{ (ii) Common value} = \dfrac{a^3}{2}.\right]$

31. State Stoke's theorem and verify it in computing $\displaystyle\int_C (x^2y^3 dx + dy + z\, dz)$ where C is the circle $x^2 + y^2 = a^2$, $z = 0$.

$\left[\text{\textit{Hints:} Common value} = -\dfrac{\pi}{8}a^6.\right]$

32. Using Stoke's theorem, show that

$$\iint_S \{(y - z)\, dydz + (z - x)\, dzdx + (x - y)\, dxdy\} = a^3\pi$$

where S is the portion of the surface $x^2 + y^2 - 2ax + az = 0, z \geq 0$.

33. Apply Stoke's theorem to evaluate $\displaystyle\int_C (y\, dx + z\, dy + x\, dz)$, where C is the curve of intersection of $x^2 + y^2 + z^2 = a^2$ and $x + z = a$.

34. Evaluate $\displaystyle\int_\Gamma (\sin z\, dx - \cos x\, dy + \sin y\, dz)$ where Γ is the boundary of the triangle: $0 \leq x \leq \pi$, $0 \leq y \leq 1$, $z = 3$.

ANSWERS

1. (i) $\frac{7a^3}{3}$; (ii) $\frac{46a^3}{15}$; **2.** (i) $-\frac{1}{2}\left(1+\frac{\pi}{2}\right)$; (ii) $-\frac{1}{2}$; **3.** 768; **4.** 3132; **5.** $-\frac{11}{6}$; **6.** (i) $\frac{\pi}{4}+\frac{1}{6}$; (ii) -1; **7.** 732; **8.** (i) 5; (ii) $7\frac{2}{3}$; (iii) $4\frac{1}{3}$; **10.** -1; **12.** (i) $\frac{13\pi}{3}$; (ii) $\frac{149\pi}{30}$; (iii) $\frac{111\pi}{10}$; **13.** $\frac{2\sqrt{2}}{3}(a+b)\sqrt{ab}$; **14.** $\frac{1}{6}$; **15.** $\pi a^2 b^2$; **18.** 5; **19.** πab; **20.** (i) $\frac{2}{3}$; (ii) $\sqrt{2}-1$; (iii) $\frac{a^2}{3}$; **22.** $\frac{12}{5}\pi a^5$; **23.** $3a^3$; **25.** $\frac{a^2 b^2}{c^2}$; **26.** $\frac{1}{3}a^5$; **33.** $\frac{-\pi a^2}{\sqrt{2}}$; **34.** 2.

Chapter 8

Dirichlet's Theorem and Liouville's Extension

8.1 Introduction

In this chapter, we will discuss two important theorems involving multiple integrals, viz., Dirichlet's theorem and Liouville's extension of Dirichlet's theorem. We know that integrations are basically used for finding the areas of the two-dimensional regions and computing volumes of three-dimensional objects. Several physical applications of the definite integral are common in Engineering and Physics. The values of the integrals can be derived using various ways. In **Chapter 6** and **Chapter 7**, we have considered and solved many such problems some of which can also be solved by using these two theorems, i.e., by Dirichlet's theorem and Liouville's theorem in comparatively easy manner. However, the knowledge of Beta and Gamma functions along with their properties and important results are necessary to have the concepts of these two theorems.

8.2 Dirichlet's Theorem

Theorem 8.2.1 *If the integral is extended to all positive values of the variables* $x_1, x_2, ..., x_n$ *subject to the condition* $x_1 + x_2 + ... + x_n \leq 1$, *then*

$$\iint \cdots \int x_1^{l_1-1} x_2^{l_2-1} ... x_n^{l_n-1} dx_1 dx_2 ... dx_n = \frac{\Gamma(l_1)\Gamma(l_2)...\Gamma(l_n)}{\Gamma(l_1 + l_2 + ... + l_n + 1)}.$$

Proof: We shall prove this theorem by using the principle of mathematical induction. For this, we first prove it for two variables, i.e., for $n = 2$, by considering the integral

$$I_2 = \iint x_1^{l_1-1} x_2^{l_2-1} dx_1 dx_2 \quad \text{where} \quad x_1 + x_2 \leq 1.$$

Here, the region of integration of I_2 is bounded by the straight lines $x_1 = 0, x_2 = 0$ and $x_1 + x_2 = 1$ and the region is expressed as $0 \leq x_1 \leq 1, 0 \leq x_2 \leq 1 - x_1$. So,

$$
\begin{aligned}
I_2 &= \int_0^1 \int_0^{1-x_1} x_1^{l_1-1} x_2^{l_2-1} dx_1 dx_2 = \int_0^1 x_1^{l_1-1} \left[\frac{x_2^{l_2}}{l_2} \right]_0^{1-x_1} dx_1 \\
&= \frac{1}{l_2} \int_0^1 x_1^{l_1-1} (1-x_1)^{l_2} dx_1 = \frac{1}{l_2} \int_0^1 x_1^{l_1-1} (1-x_1)^{l_2+1-1} dx_1
\end{aligned}
$$

$$= \frac{1}{l_2}B(l_1, l_2 + 1) = \frac{1}{l_2}\frac{\Gamma(l_1)\Gamma(l_2 + 1)}{\Gamma(l_1 + l_2 + 1)} = \frac{1}{l_2}\frac{\Gamma(l_1)l_2\Gamma(l_2)}{\Gamma(l_1 + l_2 + 1)}$$

$$= \frac{\Gamma(l_1)\Gamma(l_2)}{\Gamma(l_1 + l_2 + 1)}. \qquad\qquad [\because \ \Gamma(n + 1) = n\Gamma(n)]$$

Thus, we see that the theorem holds good for $n = 2$, i.e., we have

$$I_2 = \iint x_1^{l_1-1}x_2^{l_2-1}dx_1dx_2 = \frac{\Gamma(l_1)\Gamma(l_2)}{\Gamma(l_1 + l_2 + 1)} \qquad (8.1)$$

As a particular case, we consider the double integral as follows

$$J_2 = \iint x_1^{l_1-1}x_2^{l_2-1}dx_1dx_2 \ \text{ where } x_1 + x_2 \leq h \text{ (instead of } x_1 + x_2 \leq 1).$$

We put $x_1 = hu_1, x_2 = hu_2$, i.e., $u_1 = \dfrac{x_1}{h}, u_2 = \dfrac{x_2}{h}$ so that $u_1 + u_2 \leq 1$. Also, $dx_1 = hdx_1$ and $dx_2 = hdx_2$. Now, we get

$$J_2 = \iint (hu_1)^{l_1-1}(hu_2)^{l_2-1}(hdu_1)(hdu_2)$$

$$= h^{l_1+l_2}\iint u_1^{l_1-1}u_2^{l_2-1}du_1du_2, \ \text{ where } \ u_1 + u_2 \leq 1$$

$$\therefore I_2 = h^{l_1+l_2}\frac{\Gamma(l_1)\Gamma(l_2)}{\Gamma(l_1 + l_2 + 1)}, \ \text{ where } \ x_1 + x_2 \leq h \qquad (8.2)$$

i.e., the change of condition from $x_1 + x_2 \leq 1$ to $x_1 + x_2 \leq h$ gives rise to the factor $h^{l_1+l_2}$ in the result.

Now, we shall consider the theorem for three variables, given by

$$I_3 = \iiint x_1^{l_1-1}x_2^{l_2-1}x_3^{l_3-1}dx_1dx_2dx_3,$$

where $x_1 + x_2 + x_3 \leq 1$, i.e., $x_2 + x_3 \leq 1 - x_1$.

$$\therefore \ I_3 = \int_0^1 x_1^{l_1-1}dx_1 \iint x_2^{l_2-1}x_3^{l_3-1}dx_2dx_3 \ \text{where } x_2 + x_3 \leq 1 - x_1$$

$$= \int_0^1 x_1^{l_1-1}dx_1(1 - x_1)^{l_1+l_2}\frac{\Gamma(l_2)\Gamma(l_3)}{\Gamma(l_2 + l_3 + 1)} \ \text{[by (8.2) as } h = 1 - x_1]$$

$$= \frac{\Gamma(l_2)\Gamma(l_3)}{\Gamma(l_2 + l_3 + 1)}\int_0^1 x_1^{l_1-1}(1 - x_1)^{(l_2+l_3+1)-1}dx_1$$

$$= \frac{\Gamma(l_2)\Gamma(l_3)}{\Gamma(l_2 + l_3 + 1)}.B(l_1, l_2 + l_3 + 1)$$

$$= \frac{\Gamma(l_2)\Gamma(l_3)}{\Gamma(l_2 + l_3 + 1)}.\frac{\Gamma(l_1)\Gamma(l_2 + l_3 + 1)}{\Gamma(l_1 + l_2 + l_3 + 1)} = \frac{\Gamma(l_1)\Gamma(l_2)\Gamma(l_3)}{\Gamma(l_1 + l_2 + l_3 + 1)}.$$

Thus we get $\quad I_3 = \dfrac{\Gamma(l_1)\Gamma(l_2)\Gamma(l_3)}{\Gamma(l_1 + l_2 + l_3 + 1)} \qquad (8.3)$

Therefore the theorem holds good for three variables.

Let us assume that the theorem holds for n variables

$$I_n = \iint \cdots \int x_1^{l_1-1}x_2^{l_2-1}\cdots x_n^{l_n-1}dx_1dx_2\dots dx_n = \frac{\Gamma(l_1)\Gamma(l_2)\dots\Gamma(l_n)}{\Gamma(l_1 + l_2 + \dots + l_n + 1)} \qquad (8.4)$$

where $x_1 + x_2 + \dots + x_n \leq 1$.

We shall now prove it for $(n + 1)$ variables, where

$$x_1 + x_2 + ... + x_n + x_{n+1} \leq 1 \quad \text{or,} \quad x_2 + x_3 + ... + x_{n+1} \leq (1 - x_1).$$

Now,

$$
\begin{aligned}
I_{n+1} &= \iint ... \int x_1^{l_1-1} x_2^{l_2-1} ... x_n^{l_n-1} x_{n+1}^{l_{n+1}-1} dx_1 dx_2 ... dx_n dx_{n+1} \\
&= \int x_1^{l_1-1} dx_1 \iint ... \int x_2^{l_2-1} x_3^{l_3-1} ... x_{n+1}^{l_{n+1}-1} dx_2 dx_2 ... dx_{n+1} \\
&= \int_0^1 x_1^{l_1-1} dx_1 (1-x_1)^{l_2+l_3+...+l_{n+1}} \times \frac{\Gamma(l_2)\Gamma(l_3)...\Gamma(l_{n+1})}{\Gamma(l_2+l_3+...+l_{n+1}+1)} \\
&= \frac{\Gamma(l_2)\Gamma(l_3)...\Gamma(l_{n+1})}{\Gamma(l_2+l_3+...+l_{n+1}+1)} \int_0^1 x_1^{l_1-1}(1-x_1)^{(l_2+l_3+...+l_{n+1}+1)-1} dx_1 \\
&= \frac{\Gamma(l_2)\Gamma(l_3)...\Gamma(l_{n+1})}{\Gamma(l_2+l_3+...+l_{n+1}+1)} \times \frac{\Gamma(l_1)\Gamma(l_2+l_3+...+l_{n+1}+1)}{\Gamma(l_1+l_2+l_3+...+l_{n+1}+1)} \\
&= \frac{\Gamma(l_1)\Gamma(l_2)...\Gamma(l_{n+1})}{\Gamma(l_1+l_2+...+l_{n+1}+1)}.
\end{aligned}
$$

Above shows that if the theorem holds for n variables, then it holds also for $(n + 1)$ variables. Also, we have shown that the theorem holds for two and three variables. Therefore, by the principle of mathematical induction, we conclude that the theorem holds for any number of variables. $\qquad\square$

8.3 Liouville's Theorem (Extension of Dirichlet's Theorem)

Theorem 8.3.1 *If x, y, z are positive variables where $h_1 < x + y + z < h_2$, then*

$$\iiint f(x+y+z)x^{l-1}y^{m-1}z^{n-1}dxdydz = \frac{\Gamma(l)\Gamma(m)\Gamma(n)}{\Gamma(l+m+n)} \int_{h_1}^{h_2} f(u)u^{l+m+n-1}du.$$

Proof: Let $x+y+z \leq u$. Then under the condition, we have by Dirichlet's theorem

$$I = \iiint x^{l-1}y^{m-1}z^{n-1}dxdydz = \frac{\Gamma(l)\Gamma(m)\Gamma(n)}{\Gamma(l+m+n+1)}u^{l+m+n} \qquad (8.5)$$

Again, if $x + y + z \leq u + \delta u$, then

$$I = \frac{\Gamma(l)\Gamma(m)\Gamma(n)}{\Gamma(l+m+n+1)}(u+\delta u)^{l+m+n} \qquad (8.6)$$

So for $u < x + y + z < u + \delta u$,

$$
\begin{aligned}
&\iiint x^{l-1}y^{m-1}z^{n-1}dxdydz \\
&= \frac{\Gamma(l)\Gamma(m)\Gamma(n)}{\Gamma(l+m+n+1)}\left[(u+\delta u)^{l+m+n} - u^{l+m+n}\right] \\
&= \frac{\Gamma(l)\Gamma(m)\Gamma(n)}{\Gamma(l+m+n+1)}u^{l+m+n}\left[\left(1+\frac{\delta u}{u}\right)^{l+m+n} - 1\right] \\
&= \frac{\Gamma(l)\Gamma(m)\Gamma(n)}{\Gamma(l+m+n+1)}u^{l+m+n}\left[1 + (l+m+n)\frac{\delta u}{u} + ... - 1\right] \\
&= \frac{\Gamma(l)\Gamma(m)\Gamma(n)}{\Gamma(l+m+n+1)}u^{l+m+n}(l+m+n)\frac{\delta u}{u}
\end{aligned}
$$

(considering up to a first order of approximation)

$$= \frac{\Gamma(l)\Gamma(m)\Gamma(n)}{(l+m+n)\Gamma(l+m+n)} u^{l+m+n-1}(l+m+n)\delta u \; [\because \Gamma(n+1) = n\Gamma(n)]$$

$$= \frac{\Gamma(l)\Gamma(m)\Gamma(n)}{\Gamma(l+m+n)} u^{l+m+n}\delta u.$$

Therefore, $\iiint f(x+y+z)x^{l-1}y^{m-1}z^{n-1}dxdydz$ takes between the same limits

for its value $\dfrac{\Gamma(l)\Gamma(m)\Gamma(n)}{\Gamma(l+m+n)} f(u)u^{l+m+n-1}\delta u.$

When $h_1 < x+y+z < h_2$, the value of the integral

$$= \frac{\Gamma(l)\Gamma(m)\Gamma(n)}{\Gamma(l+m+n)} \int_{h_1}^{h_2} f(u)u^{l+m+n-1}du$$

($\because$ when $x+y+z$ lies between u and $u+\delta u$, the value of $f(x+y+z)$ differs from $f(u)$ by a small quantity as higher powers of δu is neglected.) $\square$

8.4 Miscellaneous Illustrative Examples

Example 8.4.1 *Evaluate* $\iiint x^{l-1}y^{m-1}z^{n-1}dxdydz$, *where* x, y, z *are all positive, subject to the following given condition* $\left(\dfrac{x}{a}\right)^p + \left(\dfrac{y}{b}\right)^q + \left(\dfrac{z}{c}\right)^r \leq 1.$

Solution: We are to first reduce the above integral to Dirichlet's form. To do this we put $\left(\dfrac{x}{a}\right)^p = u \Rightarrow x = au^{\frac{1}{p}} \; \therefore \; dx = \dfrac{a}{p}u^{\frac{1}{p}-1}.$

$$\therefore \; x^{l-1}dx = a^{l-1}.u^{\frac{l-1}{p}} \times \frac{a}{p} \times u^{\frac{1-p}{p}} du = \frac{a^l}{p}u^{\frac{l}{p}-1}du.$$

Similarly, if we put $\left(\dfrac{y}{b}\right)^q = v$ and $\left(\dfrac{z}{c}\right)^r = w$ then we get

$$y^{m-1}dy = \frac{b^m}{q}v^{\frac{m}{q}-1}dv \text{ and } z^{n-1}dz = \frac{c^n}{r}w^{\frac{n}{r}-1}dw.$$

Hence subject to the condition $u+v+w \leq 1$, the given integral becomes

$$I = \frac{a^l b^m c^n}{pqr} \iiint u^{\frac{l}{p}-1} v^{\frac{m}{q}-1} w^{\frac{n}{r}-1} dudvdw$$

$$= \frac{a^l b^m c^n}{pqr} . \frac{\Gamma\left(\frac{l}{p}\right)\Gamma\left(\frac{m}{q}\right)\Gamma\left(\frac{n}{r}\right)}{\Gamma\left(\frac{l}{p}+\frac{m}{q}+\frac{n}{r}+1\right)} \quad \text{[by Dirichlet's theorem.]} \qquad \square$$

Example 8.4.2 *Calculate the Dirichlet's integral*

$$\iiint x^\alpha y^\beta z^\gamma (1-x-y-z)^\lambda dxdydz$$

over the tetrahedron formed by the plane $x+y+z = 1$ *and the coordinate planes.*

Solution: Let us put $x+y+z = u$, $x+y = uv$, $x = uvw$, i.e., $x = uvw$, $y = uv - uvw = uv(1-w)$, $z = u - uv = u(1-v)$ and $u = x+y+z$, $v = \dfrac{x+y}{u} = \dfrac{x+y}{x+y+z}$, $w = \dfrac{x}{uv} = \dfrac{x}{x+y}.$

The region of integration over the tetrahedron, given by $x = 0, y = 0, z = 0$ and $x + y + z = 1$ is transformed into $u = 0, v = 0, w = 0, u = 1, v = 1, w = 1$, i.e., $u \in [0, 1]; v \in [0, 1]; w \in [0, 1]$, i.e., a unit cube in (u, v, w)-space. The Jacobian is given by

$$J = \frac{\partial(x, y, z)}{\partial(u, v, w)} = \begin{vmatrix} \dfrac{\partial x}{\partial u} & \dfrac{\partial x}{\partial v} & \dfrac{\partial x}{\partial w} \\ \dfrac{\partial y}{\partial u} & \dfrac{\partial y}{\partial v} & \dfrac{\partial y}{\partial w} \\ \dfrac{\partial z}{\partial u} & \dfrac{\partial z}{\partial v} & \dfrac{\partial z}{\partial w} \end{vmatrix} = \begin{vmatrix} vw & uw & uv \\ v(1-w) & u(1-w) & -uv \\ 1-v & -u & 0 \end{vmatrix}$$

$$= uv \begin{vmatrix} v(1-w) & u(1-w) \\ 1-v & -u \end{vmatrix} + uv \begin{vmatrix} vw & uw \\ 1-v & -u \end{vmatrix}$$

$$= uv(1-w) \begin{vmatrix} v & u \\ 1-v & -u \end{vmatrix} + uvw \begin{vmatrix} v & u \\ 1-v & -u \end{vmatrix}$$

$$= uv(1-w)(-uv - u + uv) + uvw(-uv - u + uv)$$

$$= uv(1-w)(-u) + uvw(-u) = -u^2 v + u^2 vw - u^2 vw = -u^2 v.$$

$$\therefore \quad I = \int_0^1 u^{\alpha+\beta+\gamma+2}(1-u)^\lambda du . \int_0^1 v^{\alpha+\beta+1}(1-v)^\gamma dv . \int_0^1 w^\alpha(1-w)^\beta dw$$

$$[\because |J| = |-u^2 v| = u^2 v]$$

$$= B(\alpha+\beta+\gamma+3, \lambda+1).B(\alpha+\beta+2, \gamma+1).B(\alpha+1, \beta+1)$$

$$= \frac{\Gamma(\alpha+1)\Gamma(\beta+1)\Gamma(\gamma+1)\Gamma(\lambda+1)}{\Gamma(\alpha+\beta+\gamma+\lambda+4)}. \qquad \square$$

Note 8.4.1 *For an alternative method follow **Example 8.4.10**.*

Example 8.4.3 *Evaluate* $\iiint xyz\, dxdydz$ *where the integral taken throughout the ellipsoid* $\dfrac{x^2}{a^2} + \dfrac{y^2}{b^2} + \dfrac{z^2}{c^2} \le 1.$

Solution: Putting $\dfrac{x^2}{a^2} = u$, i.e., $x = au^{\frac{1}{2}}$, we get $dx = \dfrac{a}{2} u^{-\frac{1}{2}} du.$

$$\therefore \quad xdx = \frac{a^2}{2} du = \frac{a^2}{2} u^{1-1} du.$$

Similarly, putting $\dfrac{y^2}{b^2} = v$ and $\dfrac{z^2}{c^2} = w$, we get respectively $ydy = \dfrac{b^2}{2} v^{1-1} dv$ and $zdz = \dfrac{c^2}{2} w^{1-1} dw.$

Let us now evaluate the given integral for positive octant (i.e., first octant) first, which gives

$$I = \iiint \frac{a^2}{2} . \frac{b^2}{2} . \frac{c^2}{2} u^{1-1} v^{1-1} w^{1-1} dudvdw$$

subject to the condition $u + v + w \le 1.$

i.e., $\quad I = \dfrac{a^2 b^2 c^2}{8} . \dfrac{\Gamma(1)\Gamma(1)\Gamma(1)}{\Gamma(1+1+1+1)}$ [by Dirichlet's theorem]

$$= \frac{a^2 b^2 c^2}{8} . \frac{1}{\Gamma(4)} = \frac{a^2 b^2 c^2}{8} \frac{1}{\lfloor 3} = \frac{a^2 b^2 c^2}{48}.$$

Hence for the whole ellipsoid, the given integral $= 8 \times \dfrac{a^2b^2c^2}{48} = \dfrac{a^2b^2c^2}{6}$, as there are eight octants. $\qquad\square$

Example 8.4.4 *Find the volume enclosed by the surface*

$$\left(\frac{x}{a}\right)^{2n} + \left(\frac{y}{b}\right)^{2n} + \left(\frac{z}{c}\right)^{2n} = 1.$$

Solution: The given surface is symmetrical in all the eight octants. We first want to obtain its volume in the positive octant and we find

$$V = \iiint dxdydz, \text{ subject to the condition } \left(\frac{x}{a}\right)^{2n} + \left(\frac{y}{b}\right)^{2n} + \left(\frac{z}{c}\right)^{2n} = 1.$$

Let us put $\left(\dfrac{x}{a}\right)^{2n} = u$, i.e., $x = au^{\frac{1}{2n}}$ $\therefore$ $dx = \dfrac{a}{2n}u^{\frac{1}{2n}-1}du$ and similarly if

we put $\left(\dfrac{y}{b}\right)^{2n} = v$ i.e., $y = bv^{\frac{1}{2n}}$ and $\left(\dfrac{z}{c}\right)^{2n} = w$, i.e., $z = cw^{\frac{1}{2n}}$ we shall get

$dy = \dfrac{b}{2n}v^{\frac{1}{2n}-1}dv$ and $dz = \dfrac{c}{2n}w^{\frac{1}{2n}-1}dw$ respectively.

$\therefore$ $V = \dfrac{abc}{8n^3}\iiint u^{\frac{1}{2n}-1}v^{\frac{1}{2n}-1}w^{\frac{1}{2n}-1}dudvdw$, where $u+v+w \le 1$ for positive octant.

$$= \frac{abc}{8n^3}\frac{\Gamma\left(\frac{1}{2n}\right)\Gamma\left(\frac{1}{2n}\right)\Gamma\left(\frac{1}{2n}\right)}{\Gamma\left(\frac{1}{2n}+\frac{1}{2n}+\frac{1}{2n}+1\right)} = \frac{abc}{8n^3}\frac{\left\{\Gamma\left(\frac{1}{2n}\right)\right\}^3}{\frac{3}{2n}\Gamma\left(\frac{3}{2n}\right)}.$$

Hence the total volume enclosed by the given surface

$$= 8V = \frac{2abc}{3n^3}\frac{\left\{\Gamma\left(\frac{1}{2n}\right)\right\}^3}{\Gamma\left(\frac{3}{2n}\right)}. \qquad\square$$

Example 8.4.5 *Evaluate* $\displaystyle\iiint dxdydz$, *where* $\dfrac{x^2}{a^2}+\dfrac{y^2}{b^2}+\dfrac{z^2}{c^2} \le 1.$

Solution: Proceeding as above **Example 8.4.4**, we get

$$I = \iiint \frac{a}{2}\cdot\frac{b}{2}\cdot\frac{c}{2}.u^{\frac{1}{2}-1}v^{\frac{1}{2}-1}w^{\frac{1}{2}-1}dudvdw \text{ where } u+v+w \le 1$$

$$= \frac{abc}{8}\frac{\Gamma\left(\frac{1}{2}\right)\Gamma\left(\frac{1}{2}\right)\Gamma\left(\frac{1}{2}\right)}{\Gamma\left(\frac{1}{2}+\frac{1}{2}+\frac{1}{2}+1\right)} = \frac{abc}{8}\frac{(\sqrt{\pi})^3}{\Gamma\left(\frac{3}{2}+1\right)} = \frac{abc}{8}\frac{\pi\sqrt{\pi}}{\frac{3}{2}\cdot\frac{1}{2}.\Gamma\left(\frac{1}{2}\right)}$$

$$= 4.\frac{abc}{8}\frac{\pi\sqrt{\pi}}{3.1.\sqrt{\pi}} = \frac{abc}{6}\pi. \qquad\square$$

Note 8.4.2 *Above integration gives the portion of the volume of the first octant enclosed by the ellipsoid* $\dfrac{x^2}{a^2}+\dfrac{y^2}{b^2}+\dfrac{z^2}{c^2}=1.$

Hence the total volume of the ellipse $= 8.\dfrac{abc}{6}.\pi = \dfrac{4}{3}\pi abc.$ *[See **Example 7.17.8** to compute the volume of the ellipse as an alternative method.]*

Example 8.4.6 *Find the volume of the tetrahedron bounded by the coordinate planes and the plane* $\dfrac{x}{a}+\dfrac{y}{b}+\dfrac{z}{c}=1.$

Solution: We know that the volume of a small amount at a point $(x, y, z) = dxdydz$. Therefore, the volume of the tetrahedron $\iiint dxdydz$, where the integral is extended to all positive values of the variables x, y, z.

Putting $\dfrac{x}{a} = u, \dfrac{y}{b} = v, \dfrac{z}{c} = w$ and proceeding as above, we get the required

volume $= \iiint abc \, dudvdw$, where $u + v + w \leq 1$

$$= abc \iiint u^{1-1} v^{1-1} w^{1-1} dudvdw = abc.\frac{\Gamma(1)\Gamma(1)\Gamma(1)}{\Gamma(1+1+1+1)} = \frac{abc}{\Gamma(4)} = \frac{abc}{\underline{3}} = \frac{abc}{6}.$$

$\square$

Note 8.4.3 *This problem has already been solved in* **Example 7.12.4** *using two different methods.*

Example 8.4.7 *Using the definitions of Beta and Gamma functions evaluate*

$$\iint_E \sqrt{xy(1 - x - y)} \, dxdy,$$

taking over the interior of the triangle bounded by the lines $x = 0, \ y = 0, \ x + y = 1.$

Solution: Take $x + y = u$ and $x = uv$, then $x = uv \Rightarrow y = u - uv = u(1 - v)$. The Jacobian J is given by

$$J = \frac{\partial(x, y)}{\partial(u, v)} = \begin{vmatrix} \dfrac{\partial x}{\partial u} & \dfrac{\partial x}{\partial v} \\ \dfrac{\partial y}{\partial u} & \dfrac{\partial y}{\partial v} \end{vmatrix} = \begin{vmatrix} v & u \\ 1 - v & -u \end{vmatrix} = -u \therefore |J| = u.$$

The integral over the region bounded by $x + y = 1, x = 0$ and $y = h$ exists if $h \to 0$. The transformed region will now be bounded by $u = 1, v = 0, u(1 - v) = h$ and when $h \to 0$, the region in the uv-plane is given by $u = 0, v = 0, u = 1, v = 1$ and we get

$$\begin{aligned}
I &= \iint_E \sqrt{xy(1 - x - y)} \, dxdy = \int_0^1 \int_0^1 \sqrt{uv.u(1 - v)(1 - u)}.|J|.dudv \\
&= \int_0^1 \int_0^1 u\sqrt{v(1 - v)(1 - u)}.u.dudv \\
&= \int_0^1 u^2 (1 - u)^{\frac{1}{2}} du \times \int_0^1 v^{\frac{1}{2}} (1 - v)^{\frac{1}{2}} dv \\
&= \int_0^1 u^{3-1} (1 - u)^{\frac{3}{2}-1} du \times \int_0^1 v^{\frac{3}{2}-1} (1 - v)^{\frac{3}{2}-1} dv
\end{aligned}$$

[Writing in Beta function form]

$$\begin{aligned}
&= B\left(3, \frac{3}{2}\right).B\left(\frac{3}{2}, \frac{3}{2}\right) = \frac{\Gamma(3)\Gamma\left(\frac{3}{2}\right)}{\Gamma\left(3 + \frac{3}{2}\right)}.\frac{\Gamma\left(\frac{3}{2}\right)\Gamma\left(\frac{3}{2}\right)}{\Gamma\left(\frac{3}{2} + \frac{3}{2}\right)} \\
&= \frac{\Gamma(3)\Gamma\left(\frac{3}{2}\right)}{\Gamma\left(\frac{9}{2}\right)}.\frac{\Gamma\left(\frac{3}{2}\right)\Gamma\left(\frac{3}{2}\right)}{\Gamma(3)} = \frac{\left\{\Gamma\left(\frac{3}{2}\right)\right\}^3}{\frac{7}{2}.\frac{5}{2}.\frac{3}{2}.\Gamma\left(\frac{3}{2}\right)} = \frac{\left\{\Gamma\left(\frac{3}{2}\right)\right\}^2}{\frac{105}{8}} \\
&= \left\{\frac{1}{2}\Gamma\left(\frac{1}{2}\right)\right\}^2 \times \frac{8}{105} = \frac{2\pi}{105} \qquad \left[\because \Gamma\left(\frac{1}{2}\right) = \sqrt{\pi}.\right]
\end{aligned}$$

$\square$

Example 8.4.8 *Find the value of* $\displaystyle\iint \ldots \int dx_1 dx_2 \ldots dx_n$, *extended to positive values of the variables subject to* $x_1^2 + x_2^2 + \ldots + x_n^2 < R^2$.

Solution: We put $\dfrac{x_1^2}{R^2} = u_1 \;\Rightarrow\; x_1 = Ru_1^{\frac{1}{2}} \;\Rightarrow\; dx_1 = \dfrac{1}{2}Ru_1^{-\frac{1}{2}}du_1$.

Similarly, put $\dfrac{x_2^2}{R^2} = u_2 \;\Rightarrow\; x_2 = Ru_2^{\frac{1}{2}} \;\Rightarrow\; dx_2 = \dfrac{1}{2}Ru_2^{-\frac{1}{2}}du_2$ and so on.

Then the given integral becomes,

$$
\begin{aligned}
I &= \iint \ldots \int dx_1 dx_2 \ldots dx_n \\
&= \iint \ldots \int \frac{1}{2}Ru_1^{-\frac{1}{2}}du_1 \frac{1}{2}Ru_2^{-\frac{1}{2}}du_2 \ldots \frac{1}{2}Ru_n^{-\frac{1}{2}}du_n \\
&= \left(\frac{1}{2}\right)^n R^n \iint \ldots \int u_1^{-\frac{1}{2}} u_2^{-\frac{1}{2}} \ldots u_n^{-\frac{1}{2}}du_1 du_2 \ldots du_n \\
&= \left(\frac{R}{2}\right)^n \iint \ldots \int u_1^{\frac{1}{2}-1} u_2^{\frac{1}{2}-1} \ldots u_n^{\frac{1}{2}-1}du_1 du_2 \ldots du_n
\end{aligned}
$$

with respect to the condition $u_1 + u_2 + \ldots + u_n < 1$.

$$
\begin{aligned}
&= \left(\frac{R}{2}\right)^n \frac{\Gamma\left(\frac{1}{2}\right).\Gamma\left(\frac{1}{2}\right)\ldots\Gamma\left(\frac{1}{2}\right)}{\Gamma\left(\frac{1}{2}+\frac{1}{2}+\ldots+\frac{1}{2}+1\right)} \quad \text{[By Dirichlet's theorem]} \\
&= \left(\frac{R}{2}\right)^n \frac{\left\{\Gamma\left(\frac{1}{2}\right)\right\}^n}{\Gamma\left(\frac{n}{2}+1\right)} = \left(\frac{R}{2}\right)^n \frac{(\sqrt{\pi})^n}{\Gamma\left(\frac{n}{2}+1\right)} = \left(\frac{R}{2}\right)^n \frac{\pi^{\frac{n}{2}}}{\Gamma\left(\frac{n}{2}+1\right)}. \qquad \square
\end{aligned}
$$

Example 8.4.9 *Evaluate* $\displaystyle\iiint e^{x+y+z} dx\,dy\,dz$, *taken over the positive octant.*

Solution: In the positive octant variables x, y, z are all positive the condition for which is $x + y + z \leq 1$, which also can be written as $0 < x + y + z \leq 1$.

So, using Liouville's extension theorem, we have

$$
\begin{aligned}
I &= \iiint e^{x+y+z} dx\,dy\,dz = \frac{\Gamma(1)\Gamma(1)\Gamma(1)}{\Gamma(1+1+1)} \int_0^1 e^u u^{1+1+1-1}du \quad \text{[See § 8.10]} \\
&= \frac{1}{\Gamma(3)} \int_0^1 u^2 e^u du = \frac{1}{\lfloor 2} \left[(u^2 e^u)_0^1 - \int_0^1 2ue^u du \right] \\
&= \frac{1}{2}\left[e - 2\left\{(ue^u)_0^1 - \int_0^1 e^u du\right\}\right] = \frac{1}{2}\left[e - 2\left\{e - (e^u)_0^1\right\}\right] \\
&= \frac{1}{2}[e - 2(e - e + 1)] = \frac{1}{2}(e - 2). \qquad \square
\end{aligned}
$$

Example 8.4.10 *Use Liouville's extension of Dirichlet's theorem to prove that*

$$
\iiint x^\alpha y^\beta z^\gamma (1 - x - y - z)^\lambda dx\,dy\,dz = \frac{\Gamma(\alpha+1)\Gamma(\beta+1)\Gamma(\gamma+1)\Gamma(\lambda+1)}{\Gamma(\alpha+\beta+\gamma+\lambda+4)}
$$

where the region of integration is bounded by the coordinate planes and the plane $x + y + z = 1$.

Solution: The region of integration is bounded by the planes $x = 0, y = 0, z = 0$ and $x + y + z = 1$. The variables all have positive values subject to the condition $0 < x + y + z < 1$. Therefore, the given integral is

$$\iiint x^{(\alpha+1)-1} y^{(\beta+1)-1} z^{(\gamma+1)-1} [1 - (x + y + z)]^{\lambda} \, dxdydz$$

$$= \frac{\Gamma(\alpha + 1)\Gamma(\beta + 1)\Gamma(\gamma + 1)}{\Gamma(\alpha + \beta + \gamma + 3)} \int_0^1 u^{\alpha+1+\beta+1+\gamma+1-1}(1 - u)^{\lambda} du$$

[By Liouville's extension theorem of § 8.10]

$$= \frac{\Gamma(\alpha + 1)\Gamma(\beta + 1)\Gamma(\gamma + 1)}{\Gamma(\alpha + \beta + \gamma + 3)} \int_0^1 u^{(\alpha+\beta+\gamma+3)-1}(1 - u)^{(\lambda+1)-1} du$$

$$= \frac{\Gamma(\alpha + 1)\Gamma(\beta + 1)\Gamma(\gamma + 1)}{\Gamma(\alpha + \beta + \gamma + 3)} \cdot B(\alpha + \beta + \gamma + 3, \lambda + 1)$$

$$= \frac{\Gamma(\alpha + 1)\Gamma(\beta + 1)\Gamma(\gamma + 1)}{\Gamma(\alpha + \beta + \gamma + 3)} \cdot \frac{\Gamma(\alpha + \beta + \gamma + 3)\Gamma(\lambda + 1)}{\Gamma(\alpha + \beta + \gamma + \lambda + 4)}$$

$$= \frac{\Gamma(\alpha + 1)\Gamma(\beta + 1)\Gamma(\gamma + 1)\Gamma(\lambda + 1)}{\Gamma(\alpha + \beta + \gamma + \lambda + 4)}. \qquad \square$$

Note 8.4.4 *This problem has already been solved in **Example 8.4.2** using Dirichlet's integral.*

Example 8.4.11 *Evaluate* $\iiint \log(x + y + z)dxdydz$, *the integral being extend over all positive values of x, y, z, subject to the condition $x + y + z < 1$.*

Solution: Applying Liouville's theorem in $0 < x + y + z < 1$, we get

$$\iiint \log(x + y + z) \, dxdydz$$

$$= \iiint \log(x + y + z)x^{1-1}y^{1-1}z^{1-1} \, dxdydz$$

$$= \frac{\Gamma(1)\Gamma(1)\Gamma(1)}{\Gamma(3)} \int_0^1 \log u.u^{1+1+1-1} \, du = \frac{1}{2\Gamma(1)} \int_0^1 u^2 \log u \, du$$

$$= \frac{1}{2}\left[\log u.\frac{u^3}{3} - \frac{1}{3}.\frac{u^3}{3}\right]_0^1 = \frac{1}{2}\left(-\frac{1}{9}\right) = -\frac{1}{18}. \qquad \square$$

Example 8.4.12 *Prove that*

$$\iint f(x + y).x^{m-1}y^{n-1}dxdy = \frac{\Gamma(m)\Gamma(n)}{\Gamma(m + n)} \int_0^1 f(u)u^{m+n-1}du,$$

where the domain of integration is given by

$$D \equiv \{(x, y) : x \geq 0, \ y \geq 0, \ x + y \leq 1\}.$$

Solution: By Liouville's theorem, we get

$$\iint f(x + y).x^{m-1}y^{n-1}dxdy = \frac{\Gamma(m)\Gamma(n)}{\Gamma(m + n)} \int_0^1 f(u)u^{m+n-1}du,$$

where $0 < x + y < 1$. $\qquad \square$

Example 8.4.13 *Evaluate* $\iiint xyz \sin(x+y+z)\,dxdydz$, *subject to the condition* $x+y+z \le \dfrac{\pi}{2}$ *for all positive values of the variables.*

Solution: In $x+y+z \le \dfrac{\pi}{2}$, applying Liouville's theorem, we get

$$\iiint xyz \sin(x+y+z)\,dxdydz$$

$$= \iiint \sin(x+y+z)\,x^{2-1}y^{2-1}z^{2-1}\,dxdydz$$

$$= \frac{\Gamma(2)\Gamma(2)\Gamma(2)}{\Gamma(2+2+2)} \int_0^{\frac{\pi}{2}} \sin u.u^{2+2+2-1}\,du = \frac{1}{\Gamma(6)} \int_0^{\frac{\pi}{2}} u^5 . \sin u\, du.$$

Integrating successively by parts, we get form the reduced form of the given integral

$$= \frac{1}{\underline{|5}}[u^5(-\cos u) - 5u^4(-\sin u) + 20u^3(\cos u) - 60u^2(\sin u)$$

$$+120u(-\cos u) - 120(-\sin u)]_0^{\frac{\pi}{2}}$$

$$= \frac{1}{120}\left[-5\left(\frac{\pi}{2}\right)^4(-1) - 60\left(\frac{\pi}{2}\right)^2.1 - 120(-1)\right]$$

$$= \frac{1}{120}\left[\frac{5\pi^4}{16} - 15\pi^2 + 120\right] = \frac{1}{24 \times 16}(\pi^4 - 48\pi^2 + 384). \qquad \square$$

Example 8.4.14 *Evaluate (a)* $\iint x^{\frac{1}{2}}y^{\frac{1}{2}}(1-x-y)^{\frac{2}{3}}\,dxdy$, *over the domain bounded by the lines* $x=0, y=0$ *and* $x+y=1$.

(b) $\iiint x^{-\frac{1}{2}}y^{-\frac{1}{2}}z^{-\frac{1}{2}}(1-x-y-z)^{\frac{1}{2}}\,dxdydz$, *extended to all positive values of the variables subject to the condition* $x+y+z < 1$.

Solution: *(a)* Applying Liouville's theorem when $0 < x+y < 1$

$$I = \iint x^{\frac{3}{2}-1}y^{\frac{3}{2}-1}(1-x-y)^{\frac{2}{3}}\,dxdy$$

$$= \frac{\Gamma\left(\frac{3}{2}\right)\Gamma\left(\frac{3}{2}\right)}{\Gamma\left(\frac{3}{2}+\frac{3}{2}\right)} \int_0^1 u^{\frac{3}{2}+\frac{3}{2}-1}(1-u)^{\frac{2}{3}}\,du$$

$$= \frac{\{\Gamma\left(\frac{3}{2}\right)\}^2}{\Gamma(3)} \int_0^1 u^{3-1}(1-u)^{\frac{5}{3}-1}\,du = \frac{\{\frac{1}{2}\Gamma\left(\frac{1}{2}\right)\}^2}{\Gamma(3)} \times B\left(3, \frac{5}{3}\right)$$

$$= \frac{\{\frac{1}{2}\sqrt{\pi}\}^2}{\Gamma(3)}.\frac{\Gamma(3)\Gamma\left(\frac{5}{3}\right)}{\Gamma\left(3+\frac{5}{3}\right)} = \frac{\pi}{4}\frac{\frac{2}{3}\Gamma\left(\frac{2}{3}\right)}{\Gamma\left(\frac{14}{3}\right)} = \frac{\pi}{4}.\frac{\frac{2}{3}\Gamma\left(\frac{2}{3}\right)}{\frac{11}{3}.\frac{8}{3}.\frac{5}{3}.\frac{2}{3}\Gamma\left(\frac{2}{3}\right)} = \frac{27\pi}{1760}.$$

(b) Proceeding exactly in a similar manner as above, when $0 < x+y+z < 1$, we get

$$I = \iiint x^{\frac{1}{2}-1}y^{\frac{1}{2}-1}z^{\frac{1}{2}-1}(1-x-y-z)^{\frac{1}{2}}\,dxdydz$$

$$= \frac{\Gamma\left(\frac{1}{2}\right)\Gamma\left(\frac{1}{2}\right)\Gamma\left(\frac{1}{2}\right)}{\Gamma\left(\frac{1}{2}+\frac{1}{2}+\frac{1}{2}\right)} \int_0^1 u^{\frac{1}{2}+\frac{1}{2}+\frac{1}{2}-1}(1-u)^{\frac{1}{2}}\,du$$

$$= \frac{\pi\sqrt{\pi}}{\Gamma\left(\frac{3}{2}\right)} \int_0^1 u^{\frac{3}{2}-1}(1-u)^{\frac{3}{2}-1} du = \frac{\pi\sqrt{\pi}}{\Gamma\left(\frac{3}{2}\right)} \times B\left(\frac{3}{2}, \frac{3}{2}\right)$$

$$= \frac{\pi\sqrt{\pi}}{\Gamma\left(\frac{3}{2}\right)} \cdot \frac{\Gamma\left(\frac{3}{2}\right)\Gamma\left(\frac{3}{2}\right)}{\Gamma\left(\frac{3}{2}+\frac{3}{2}\right)} = \frac{\pi\sqrt{\pi}\Gamma\left(\frac{3}{2}\right)}{\Gamma(3)} = \frac{\pi\sqrt{\pi}\cdot\frac{1}{2}\Gamma\left(\frac{1}{2}\right)}{\lfloor 2} = \frac{\pi^2}{4}.$$

$$\left[\because \ \Gamma\left(\frac{1}{2}\right) = \sqrt{\pi} \ and \ \Gamma(3) = \lfloor 2\right]. \qquad \square$$

Example 8.4.15 *Show that* $\iiint \dfrac{dxdydz}{(1+x+y+z)^3} = \dfrac{1}{2}\log 2 - \dfrac{5}{16}$*, where the integral is extended over the volume of the tetrahedron bounded by the coordinate planes and the plane* $x + y + z = 1$.

Solution: Just as in previous examples, when $0 < x + y + z < 1$, by Liouville's theorem, we have

$$I = \iiint \frac{x^{1-1}y^{1-1}z^{1-1}\,dxdydz}{(1+x+y+z)^3} = \frac{\Gamma(1)\Gamma(1)\Gamma(1)}{\Gamma(3)} \int_0^1 \frac{u^{1+1+1-1}}{(1+u)^3}\,du$$

$$= \frac{1}{\lfloor 2} \int_0^1 \frac{u^2}{(1+u)^3}\,du = \frac{1}{\lfloor 2} \int_1^2 \frac{(v-1)^2}{v^3}\,dv = \frac{1}{2} \int_1^2 \left(\frac{1}{v} - \frac{2}{v^2} + \frac{1}{v^3}\right)dv$$

$$\text{[Put } u + 1 = v \ \Rightarrow \ du = dv]$$

$$= \frac{1}{2}\left[\log v + \frac{2}{v} - \frac{1}{2v^2}\right]_1^2 = \frac{1}{2}\left[\log 2 + 2\left(\frac{1}{2} - 1\right) - \frac{1}{2}\left(\frac{1}{4} - 1\right)\right]$$

$$= \frac{1}{2}\log 2 - \frac{5}{16}. \qquad \square$$

Note 8.4.5 *The problem has already been solved in **Example 6.11.5**, and obviously the method adopted here is a different one.*

Example 8.4.16 *Prove that* $\iiiint dxdydzdt$*, for all values of the variables for which* $a^2 < x^2 + y^2 + z^2 + t^2 \le b^2$ *is* $\dfrac{\pi^2\left(b^4 - a^4\right)}{32}$.

Solution: Put $x^2 = u_1$, i.e., $x = \sqrt{u_1} \ \Rightarrow \ dx = \frac{1}{2}u_1^{-\frac{1}{2}}\,du_1 = \frac{1}{2}u_1^{\frac{1}{2}-1}\,du_1$ and similarly put $y^2 = u_2, z^2 = u_3$ and $t^2 = u_4$, correspondingly we get by applying Liouville's theorem

$$\iiiint dxdydzdt = \frac{1}{16} \iiiint u_1^{\frac{1}{2}-1}u_2^{\frac{1}{2}-1}u_3^{\frac{1}{2}-1}u_4^{\frac{1}{2}-1}\,du_1du_2du_3du_4$$

$$= \frac{1}{16}\cdot\frac{\Gamma\left(\frac{1}{2}\right)\Gamma\left(\frac{1}{2}\right)\Gamma\left(\frac{1}{2}\right)\Gamma\left(\frac{1}{2}\right)}{\Gamma\left(\frac{1}{2}+\frac{1}{2}+\frac{1}{2}+\frac{1}{2}\right)} \int_{a^2}^{b^2} u^{\frac{1}{2}+\frac{1}{2}+\frac{1}{2}+\frac{1}{2}-1}\,du = \frac{1}{16}\cdot\frac{(\sqrt{\pi})^4}{\Gamma(2)} \int_{a^2}^{b^2} u\,du$$

$$= \frac{1}{16}\cdot\frac{\pi^2}{1}\left[\frac{u^2}{2}\right]_{a^2}^{b^2} = \frac{\pi^2\left(b^4 - a^4\right)}{32}. \qquad \square$$

8.5 Exercises

1. Apply Dirichlet's theorem to find the entire volume of the solid bounded by the surface $\left(\dfrac{x}{a}\right)^{\frac{2}{3}} + \left(\dfrac{y}{b}\right)^{\frac{2}{3}} + \left(\dfrac{z}{c}\right)^{\frac{2}{3}} = 1$.

2. Find the whole volume of the solid bounded by the surface whose equation is $\dfrac{x^2}{a^2} + \dfrac{y^2}{b^2} + \dfrac{z^4}{c^4} = 1$.

[*Hints:* Volume $V = 8 \displaystyle\iiint dxdydz$. Put $\dfrac{x^2}{a^2} = u$; $\dfrac{y^2}{b^2} = v$; $\dfrac{z^4}{c^4} = w$ where $u + v + w \leq 1$.

$$\therefore V = 8 \iiint \frac{abc}{4.2.2} \cdot u^{\frac{1}{2}-1} v^{\frac{1}{2}-1} w^{\frac{1}{4}-1} \, dudvdw$$

$$= \frac{abc}{2} \cdot \frac{\Gamma\left(\frac{1}{2}\right)\Gamma\left(\frac{1}{2}\right)\Gamma\left(\frac{1}{4}\right)}{\Gamma\left(\frac{1}{2}+\frac{1}{2}+\frac{1}{4}+1\right)} \quad \text{[By Dirichlet's theorem]}$$

$$= \frac{abc}{2} \cdot \frac{\pi\Gamma\left(\frac{1}{4}\right)}{\frac{5}{4}\cdot\frac{1}{4}\cdot\Gamma\left(\frac{1}{4}\right)} = \frac{8\pi abc}{5}.]$$

3. Show that if l, m, n are all positive,

$$\iiint x^{l-1} y^{m-1} z^{n-1} \, dxdydz = \frac{a^l b^m c^n}{8} \cdot \frac{\Gamma\left(\frac{l}{2}\right)\Gamma\left(\frac{m}{2}\right)\Gamma\left(\frac{n}{2}\right)}{\Gamma\left(\frac{l+m+n+2}{2}\right)}$$

where the region of the multiple integral is taken throughout that part of the ellipsoid $\dfrac{x^2}{a^2} + \dfrac{y^2}{b^2} + \dfrac{z^2}{c^2} = 1$ which lies in the positive octant.

[*Hints:* For points inside the positive octant of the ellipsoid, we have $\dfrac{x^2}{a^2} + \dfrac{y^2}{b^2} + \dfrac{z^2}{c^2} \leq 1$. Hence proceeding as illustrative **Example 8.4.1** or putting $p = q = r = 2$ in its result we get the required proof.]

4. Evaluate $\displaystyle\iint dxdy$ over the region of integration in the positive quadrant for which $x + y \leq 1$.

5. Find the volume of the tetrahedron bounded by the plane $\dfrac{x}{a} + \dfrac{y}{b} + \dfrac{z}{c} = 1$ and the coordinate planes.

6. Prove that the area in the positive quadrant between the curve $x^n + y^n = a^n$ and the axis is $\dfrac{a^2}{2n} \dfrac{\left\{\Gamma\left(\frac{1}{n}\right)\right\}^2}{\Gamma\left(\frac{2}{n}\right)}$.

7. Use Liouville's extension of Dirichlet's theorem to prove that

$$\iiint e^{x+y+z} \, dxdydz = \frac{1}{2}e - 1$$

where the integral is taken over the positive octant such that $x + y + z \leq 1$.

8. Apply Liouville's theorem to prove that (*i*) $\iiint_V xyz\,dxdydz = \dfrac{1}{720}$ and (*ii*) $\iiint_V \dfrac{dxdydz}{(1+x+y+z)^3} = \dfrac{1}{16}\log\dfrac{256}{e^5}$ where V is the tetrahedron bounded by the planes $x=0, y=0, z=0$ and $x+y+z=1$.

9. Prove that

$$\int_0^\infty \int_0^\infty \phi(x+y)x^\alpha y^\beta\,dxdy = \frac{\Gamma(\alpha+1)\Gamma(\beta+1)}{\Gamma(\alpha+\beta+2)}\int_0^\infty \phi(u)u^{\alpha+\beta+1}\,du.$$

10. Reduce the integral $\displaystyle\iint\ldots\int \frac{dx_1dx_2\ldots dx_n}{\sqrt{a^2-x_1^2-x_2^2-\ldots-x_n^2}}$ to the standard form by the substitutions $x_1^2=u_1, x_2^2=u_2, \ldots, x_n^2=u_n$ and then evaluate it, the integral being extended to all positive values of the variables for which the expression is real.

[*Hints:* The expression will be real if $a^2-x_1^2-x_2^2-\ldots-x_n^2 > 0$ or $0 < x_1^2+x_2^2+\ldots+x_n < a^2$.

Now by the substitution given in the problem, the given integral will reduce to the standard form

$$I = \left(\frac{1}{2}\right)^n \iint\ldots\int \frac{u_1^{\frac{1}{2}-1}u_2^{\frac{1}{2}-1}\ldots u_n^{\frac{1}{2}-1}}{\sqrt{a^2-(u_1+u_2+\ldots+u_n)}}\,du_1du_2\ldots du_n,$$

where $0 < u_1+u_2+\ldots+u_n < a^2$

$$= \left(\frac{1}{2}\right)^n \frac{\Gamma\left(\frac{1}{2}\right)\Gamma\left(\frac{1}{2}\right)\ldots\Gamma\left(\frac{1}{2}\right)}{\Gamma\left(\frac{1}{2}+\frac{1}{2}+\ldots+\frac{1}{2}\right)}\int_0^{a^2}\frac{u^{\frac{n}{2}-1}}{\sqrt{a^2-u}}\,du \quad \text{[By Liouville's theorem]}$$

To evaluate the integral portion, we put $u = a^2\sin^2\theta$ and get

$$\int_0^{a^2}\frac{u^{\frac{n}{2}-1}}{\sqrt{a^2-u}}\,du = \int_0^{\frac{\pi}{2}}\frac{(a^2\sin^2\theta)^{\frac{n}{2}-1}}{\sqrt{a^2-a^2\sin^2\theta}}\,2a^2\sin\theta\cos\theta\,d\theta$$

$$= 2a^{n-1}\int_0^{\frac{\pi}{2}}\sin^{n-1}\theta d\theta = 2a^{n-1}\frac{\Gamma\left(\frac{n}{2}\right)\Gamma\left(\frac{1}{2}\right)}{2\Gamma\left(\frac{n+1}{2}\right)} \quad \text{[by \textbf{Note 0.3.2}]}$$

Hence we get $I = \left(\dfrac{1}{2}\right)^n \dfrac{(\sqrt{\pi})^n}{\Gamma\left(\frac{n}{2}\right)}.2a^{n-1}\dfrac{\Gamma\left(\frac{n}{2}\right)\Gamma\left(\frac{1}{2}\right)}{2\Gamma\left(\frac{n+1}{2}\right)} = \dfrac{\pi^{\frac{n+1}{2}}a^{n-1}}{2^n\Gamma\left(\frac{n+1}{2}\right)}.$]

11. Following method of the forgoing **Exercise 10** prove that

(*a*) $\displaystyle\iiint \frac{dxdydz}{\sqrt{a^2-x^2-y^2-z^2}} = \frac{\pi^2 a^2}{8}$, (*b*) $\displaystyle\iiint \frac{dxdydz}{\sqrt{1-x^2-y^2-z^2}} = \frac{\pi^2}{8}$.

12. Evaluate $\displaystyle\iint_E \sqrt{\frac{a^2b^2-b^2x^2-a^2y^2}{a^2b^2+b^2x^2+a^2y^2}}\,dxdy$, the field of integration E being taken as the positive quadrant of the ellipse $\dfrac{x^2}{a^2}+\dfrac{y^2}{b^2}=1$.

[*Hints:* In order to reduce the standard form we put $\dfrac{x^2}{a^2}=u$ etc. and re-write the integral as $I = \displaystyle\iint_E \sqrt{\dfrac{1-\frac{x^2}{a^2}-\frac{y^2}{b^2}}{1+\frac{x^2}{a^2}+\frac{y^2}{b^2}}}\,dxdy$. Now proceed as **Exercise 10**.]

13. Evaluate $\iiint \sqrt{\dfrac{1 - x^2 - y^2 - z^2}{1 + x^2 + y^2 + z^2}} \, dx\,dy\,dz$, integration being taken over all positive values of x, y, z, such that $x^2 + y^2 + z^2 \leq 1$.

14. Evaluate $\iiint_V (x+y+z+1)^2 dx\,dy\,dz$, where V is the tetrahedron bounded by the planes $x = 0, y = 0, z = 0$ and the plane $x + y + z = 1$.

15. Show that $\iint x^{\frac{1}{2}} y^{\frac{1}{3}} (1 - x - y)^{\frac{1}{2}} \, dx\,dy$ over the triangle bounded by $x = 0, y = 0$ and $x + y = 1$ is $\dfrac{\Gamma\left(\frac{5}{3}\right) \Gamma\left(\frac{4}{3}\right) \Gamma\left(\frac{3}{2}\right)}{\Gamma\left(\frac{9}{2}\right)}$.

16. Evaluate $\iiint \sqrt{a^2 b^2 c^2 - b^2 c^2 x^2 - c^2 a^2 y^2 - a^2 b^2 z^2} \, dx\,dy\,dz$, taken throughout the ellipsoid $\dfrac{x^2}{a^2} + \dfrac{y^2}{b^2} + \dfrac{z^2}{c^2} = 1$.

17. Evaluate $\iiint (x + y + z) \, dx\,dy\,dz$ over the tetrahedron $x = 0, y = 0, z = 0$ and $x + y + z \leq 1$.

18. Prove that $\iint_R e^{-x^2 - y^2} dx\,dy = \dfrac{\pi}{4}\left(1 - e^{-R^2}\right)$ where R is the region defined by $x \geq 0, y \geq 0, x^2 + y^2 \geq R^2$.

19. With certain restrictions on the values of a, b, m and n, prove that

$$\int_0^\infty \int_0^\infty e^{-(ax^2 + by^2)} x^{2m-1} y^{2n-1} dx\,dy = \frac{\Gamma(m)\Gamma(n)}{4a^m b^n}.$$

[*Hints:* Put $ax^2 = p,\ by^2 = q\ \therefore\ x = \sqrt{\dfrac{p}{a}} \Rightarrow dx = \dfrac{1}{2\sqrt{a}} p^{\frac{1}{2}-1} dp$ etc.

Also p and q are all to be positive and hence a and b should be positive, therefore $0 < p + q < \infty$.

$$
\begin{aligned}
\therefore\ I &= \iint \frac{e^{-(p+q)}}{4a^m b^n} p^{m-1} q^{n-1} dp\,dq \\
&= \frac{1}{4a^m b^n} \frac{\Gamma(m)\Gamma(n)}{\Gamma(m+n)} \int_0^\infty e^{-u} u^{m+n-1} du \quad \text{[by Liouville's theorem]} \\
&= \frac{1}{4a^m b^n} \frac{\Gamma(m)\Gamma(n)}{\Gamma(m+n)} \cdot \Gamma(m+n) \quad \left[\because \int_0^\infty e^{-u} u^{m+n-1} du = \Gamma(m+n)\right] \\
&= \frac{\Gamma(m)\Gamma(n)}{4a^m b^n}.
\end{aligned}
$$
]

20. Evaluate the double integral $\iint x^{\frac{1}{2}} y^{\frac{1}{2}} (1 - x - y)^{\frac{2}{3}} dx\,dy$ over the region R bounded by the lines $x = 0, y = 0, x + y = 1$.

ANSWERS

1. $\frac{4}{35}abc\pi$; **2.** $\frac{8\pi abc}{5}$; **4.** $\frac{1}{24}$; **5.** $\frac{\pi abc}{6}$; **10.** $\frac{\pi^{\frac{n+1}{2}}a^{n-1}}{2^n\Gamma\left(\frac{n+1}{2}\right)}$; **12.** $\frac{\pi}{4}\left(\frac{\pi}{2}-1\right)ab$; **13.** $\frac{\pi}{4}\left[B\left(\frac{3}{4},\frac{1}{2}\right)-B\left(\frac{5}{4},\frac{1}{2}\right)\right]$; **14.** $\frac{31}{60}$; **16.** $\frac{\pi^2 a^2 b^2 c^2}{4}$; **17.** $\frac{1}{8}$; **20.** $\frac{27\pi}{1760}$.

Bibliography

[1] Spivak M., *Calculus on Manifolds: A Modern Approach to Classical Theorems of Advanced Calculus*, Massachusetts, Addison-Wesley Publishing Company, 1965.

[2] Piskunov N., *Differential and Integral Calculus*, Moscow, Mir Publishers, 1969.

[3] Edwards C. H., *Advanced Calculus of Several Variables*, London, Academic Press, 1973.

[4] Khanna M. L., *Integral Calculus*, Meerut City, Jai Prakash Nath & Co., 1976.

[5] Craven B. D., *Functions of Several Variables*, London, Chapman and Hall, 1981.

[6] Lang S., *Calculus of Several Variables, 3rd ed.*, New York, Springer, 1987.

[7] Courant R. and John F., *Introduction to Calculus and Analysis, Vol. II*, New York, Springer-Verlag, 1989.

[8] Marsden J. E., Tromba A. and Weinstein A., *Basic Multivariable Calculus*, New York, Springer, 1993.

[9] Apostol T. M., *Calculus, Vol. II, 2nd ed.*, New Delhi, John Wiley & Sons, 2003.

[10] Chatterjee U., *Advanced Mathematical Analysis: Theory and Problems*, Kolkata, Academic Publishers, 2008.

[11] Stewart J., *Multivariable Calculus, 7th ed.*, Belmont, USA, Cengage Learning, 2012.

[12] Choudhury A. K. and Mondal P., *Mathematical Analysis*, Kolkata, New Central Book Agency, 2013.

[13] Dineen S., *Multivariate Calculus and Geometry, 3rd ed.*, London, Springer, 2014.

[14] Das A. N., *Mathematical Analysis, Vol. I*, Kolkata, Books & Allied Pvt. Ltd., 2014.

[15] Mukhopadhyay S. N. and Mitra S., *Mathematical Analysis, Vol. II*, Kolkata, U. N. Dhur & Sons Pvt. Ltd., 2014.

[16] Das B. C. and Mukherjee B. N., *Differential Calculus*, Kolkata, U. N. Dhur & Sons Pvt. Ltd., 2015.

[17] Das B. C. and Mukherjee B. N., *Integral Calculus*, Kolkata, U. N. Dhur & Sons Pvt. Ltd., 2015.

[18] Ghosh R. K. and Maity K. C., *An Introduction to Analysis: Integral Calculus*, Kolkata, New Central Book Agency, 2016.

[19] Mallik S. C. and Arora S., *Mathematical Analysis*, New Delhi, New Age International Publishers, 2017.

[20] Bandyopadhyay S., *Mathematical Analysis: Problems and Solutions*, Kolkata, Academic Publishers, 2017.

[21] Ghosh R. K. and Maity K. C., *An Introduction to Analysis: Differential Calculus, Part I*, Kolkata, New Central Book Agency, 2017.

[22] Ghosh R. K. and Maity K. C., *An Introduction to Analysis: Differential Calculus, Part II*, Kolkata, New Central Book Agency, 2018.

[23] Pundir S. K., *Mathematical Analysis*, New Delhi, CBS Publishers Pvt. Ltd., 2019.

[24] Kar B. K., *An Introduction to Modern Analysis, Vol. I, 2nd ed.*, Kolkata, Books & Allied Pvt. Ltd., 2021.

Index

Printed in the United States
by Baker & Taylor Publisher Services